Electronic Properties of Multilayers and Low-Dimensional Semiconductor Structures

NATO ASI Series

Advanced Science Institutes Series

A series presenting the results of activities sponsored by the NATO Science Committee, which aims at the dissemination of advanced scientific and technological knowledge, with a view to strengthening links between scientific communities.

The series is published by an international board of publishers in conjunction with the NATO Scientific Affairs Division

A	**Life Sciences**	Plenum Publishing Corporation
B	**Physics**	New York and London
C	**Mathematical**	Kluwer Academic Publishers
	and Physical Sciences	Dordrecht, Boston, and London
D	**Behavioral and Social Sciences**	
E	**Applied Sciences**	
F	**Computer and Systems Sciences**	Springer-Verlag
G	**Ecological Sciences**	Berlin, Heidelberg, New York, London,
H	**Cell Biology**	Paris, and Tokyo

Recent Volumes in this Series

Series B: Physics

Electronic Properties of Multilayers and Low-Dimensional Semiconductor Structures

Edited by

J. M. Chamberlain and L. Eaves

Nottingham University
Nottingham, United Kingdom

and

J.-C. Portal

CNRS–INSA
Toulouse, France
and CNRS–SNCI
Grenoble, France

Plenum Press
New York and London
Published in cooperation with NATO Scientific Affairs Division

Proceedings of a NATO Advanced Study Institute on
Electronic Properties of Multilayers and
Low-Dimensional Semiconductor Structures,
held September 11–22, 1989,
at Château de Bonas, Castéra-Verduzan, France

Library of Congress Cataloging-in-Publication Data

NATO Advanced Study Institute on Electronic Properties of Multilayers
  and Low-Dimensional Semiconductor Structures (1989 : Castéra
  -Verduzan, France)
   Electronic properties of multilayers and low-dimensional
semiconductor structures / edited by J.M. Chamberlain and L. Eaves
and J.C. Portal.
       p.   cm. -- (NATO ASI series. Series B, Physics ; vol. 231)
   Proceedings of a NATO Advanced Study Institute on Electronic
Properties of Multilayers and Low-Dimensional Semiconductor
Structures, held Sept. 11-22, 1989 in Castéra-Verduzan, France.
   "Published in cooperation with NATO Scientific Affairs Division."
   Includes bibliographical references and index.
   ISBN 0-306-43662-0
   1. Semiconductors--Congresses. 2. Layer structure (Solids)-
-Congresses. 3. Tunneling (Physics)--Congresses. 4. Superlattices
as materials--Congresses.   I. Chamberlain, J. M. II. Eaves, L.
III. Portal, J. C. IV. Title. V. Series: NATO ASI series. Series
B, Physics ; v. 231.
QC610.9.N3644 1989
537.6'226--dc20   90-44242
                                                      CIP

© 1990 Plenum Press, New York
A Division of Plenum Publishing Corporation
233 Spring Street, New York, N.Y. 10013

Printed in the United States of America

ORGANISING COMMITTEE

Honorary Chair

Leo Esaki
(IBM, Yorktown Heights, USA)

Co-Directors

Laurence Eaves
University of Nottingham
United Kingdom

Jean-Claude Portal
CNRS-INSA, Toulouse
& CNRS-SNCI, Grenoble, France

Secretary

Martyn Chamberlain
University of Nottingham
United Kingdom

Treasurer

Peter Main
University of Nottingham
United Kingdom

DANIEL PAQUET

DEDICATION

This volume is dedicated to the memory of

<u>Daniel Paquet</u>

Our colleague Daniel Paquet died suddenly on August 13 1989 after a heart attack, whilst on a mountain walk.

In the course of his scientific career, so abruptly interrupted, he made original contributions on difficult theoretical problems and had thus acquired a wide renown in many areas of solid state physics. Initially, his interest was centred on metal-insulator phase transitions in magnetic materials and on the effect of disorder in alloy semiconductors. More recently, he worked on different aspects of semiconductor superlattice properties: lattice dynamics, *ab initio* calculations of electronic band structures, Fibonacci stackings and the electronic properties in magnetic fields. The properties of electrons in superlattices in the presence of a magnetic field was the subject on which Daniel Paquet had been invited to speak at this Advanced Study Institute.

His broad understanding and his deep interest in the work of his colleagues rendered him an invaluable partner for discussions and collaborations. He would always contribute with his innovative insight and would invariably throw a new light on the subject. The courses he taught and the seminars he gave were a testimony to his love for communicating and sharing his knowledge as well as his passion for research. They were full of humour, as those who heard him lecture on the complex problem of Fibonacci stackings, can remember.

We share the sorrow of his family and of all people who had become his friends in the course of professional contacts. It is the wish of the organisers, lecturers and participants to dedicate the Proceedings of this Advanced Study Institute to the memory of Daniel Paquet.

PREFACE

This Advanced Study Institute on the Electronic Properties of Multilayers and Low Dimensional Semiconductor Structures focussed on several of the most active areas in modern semiconductor physics. These included resonant tunnelling and superlattice phenomena and the topics of ballistic transport, quantised conductance and anomalous magnetoresistance effects in laterally gated two-dimensional electron systems. Although the main emphasis was on fundamental physics, a series of supporting lectures described the underlying technology (Molecular Beam Epitaxy, Metallo-Organic Chemical Vapour Deposition, Electron Beam Lithography and other advanced processing technologies). Actual and potential applications of low dimensional structures in optoelectronic and high frequency devices were also discussed.

The ASI took the form of a series of lectures of about fifty minutes' duration which were given by senior researchers from a wide range of countries. Most of the lectures are recorded in these Proceedings. The younger members of the Institute made the predominant contribution to the discussion sessions following each lecture and, in addition, provided most of the fifty-five papers that were presented in two lively poster sessions.

The ASI emphasised the impressive way in which this research field has developed through the fruitful interaction of theory, experiment and semiconductor device technology. Many of the talks demonstrated both the effectiveness and limitations of semiclassical concepts in describing the quantum phenomena exhibited by electrons in low dimensional structures.

The meeting took place over a period of two weeks in the tranquil surroundings of the Chateau de Bonas, near the town of Castera-Verduzan in south-west France. Just over one hundred participants, from all of the NATO countries, attended. The organisers wish to acknowledge the strong support given to the ASI by Nobel Laureates Professor Leo Esaki and Professor Klaus von Klitzing, both of whom gave review lectures on the subjects that they have done so much to develop. Professor Esaki also acted as the Honorary Chairman of the ASI.

The Organisers wish to express their sincere thanks to the speakers for contributing the papers to this volume, and to the workbook which proved so valuable at the meeting itself. It is hoped that this Proceedings will provide a useful reference work for researchers in this field.

ACKNOWLEDGEMENTS

The Organisers wish to acknowledge the following Meeting Sponsors: their generous help enabled a large number of graduate students to attend and a wide programme of talks to be presented:

North Atlantic Treaty Organisation
Air-Inter - French Domestic Airways
Centre National d'Études des Télécommunications (France-Télécom)
Chambre de Commerce et d'Industrie de Toulouse
Conseil Régional, Région Midi-Pyrénées
Délegation Génerale pour l'Armement, Direction des Recherches Études et Techniques
ISA-RIBER S.A. (France)
Ministère de l'Education Nationale de la Jeunesse et des Sports
(Direction de la Recherche et des Études Doctorales)
National Science Foundation (USA)
SNCF - French Railways
Technopole de l'Agglomeration Toulousaine
Thomson S.A. (France)

In addition, the Organisers wish to thank the following who have assisted with the ASI organisation and the preparation of this Proceedings: Margaret Carter, Edmund and Ralph Chamberlain, Terry Davies, Laurence Kirk and Helen Smith. The great efficiency and co-operation of Colonel Stockmann and his staff at the Chateau de Bonas was also greatly appreciated.

CONTENTS

THE EVOLUTION OF SEMICONDUCTOR QUANTUM STRUCTURES IN REDUCED

DIMENSIONALITY - DO-IT-YOURSELF QUANTUM MECHANICS

L. Esaki

IBM Thomas J. Watson Research Center
Yorktown Heights, New York 10598, U.S.A.

ABSTRACT - Following the past twenty-year evolutionary path in the interdisciplinary research of semiconductor superlattices and other quantum structures, significant milestones are presented with emphasis on experimental achievements in the physics of reduced dimensionality associated with technological advances.

I. INTRODUCTION - INCEPTION

In 1969, research on quantum structures was initiated with a proposal of an ''engineered'' semiconductor superlattice by Esaki and Tsu (1) (2). In anticipation of advancement in epitaxy, we envisioned two types of superlattices with alternating ultrathin layers: doping and compositional, as shown at the top and bottom of Fig. 1, respectively.

This was, perhaps, the first proposal of ''designed semiconductor quantum structures'': namely, we asserted that confined or miniband states could be produced if potential barriers and wells were created by means of successive deposition of different semiconductor layers with thicknesses smaller than the phase-coherent length of electrons. Since the electronic properties characteristic of semiconductor structures are mainly governed by such quantum states, it was predicted that new electronic materials can be designed and engineered to obtain desired transport and optical properties through tailoring the band structure with the control of the potential profile during the layer deposition.

It was thought that semiconductors and technologies developed with them, might be called for in order to demonstrate the quantum wave nature of electrons associated with the interference phenomena, since their small Fermi energies due to low carrier densities help make the de Broglie wavelength relatively large. Namely, the Fermi wavelength $\lambda_f = 2\pi/k_f$, where k_f is the magnitude of the Fermi wavevector, is given as a function of the carrier density n, as follows: $\lambda_f = (8\pi/3n)^{1/3}$ for a three-dimensional (3D) system; $(2\pi/n)^{1/2}$ for a two-dimensional (2D) system; $(4/n)$ for a one-dimensional (1D) system.

In this context, we first examined the feasibility of structural formation by heteroepitaxy for barriers and wells, thin enough to exhibit resonant electron tunneling through them (3). Figure 2 shows the calculated transmission coefficient as a function of electron energy for a double barrier. When the well width L, the barrier width and height, and the electron effective mass m* are assumed to be 50Å, 20Å, 0.5eV and one tenth of the free electron mass m_o, respec-

Electronic Properties of Multilayers and Low-Dimensional Semiconductors Structures
Edited by J. M. Chamberlain *et al.*, Plenum Press, New York, 1990

1

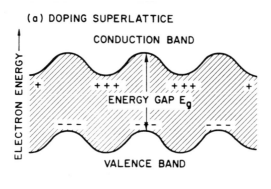

(a) DOPING SUPERLATTICE

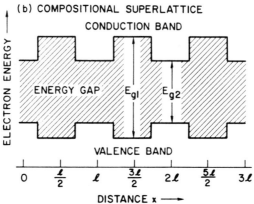

(b) COMPOSITIONAL SUPERLATTICE

Fig. 1. Spacial variations of the conduction and valence bandedges
in two types of superlattices: doping (top) and compositional.

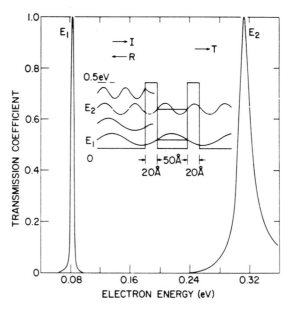

Fig. 2. Transmission coefficient versus election energy for a double
barrier shown in the insert. At the resonant condition that
the energy of incident elections coincides with one of those
of the bound states, E_1 and E_2, the electrons tunnel through
both barriers without attenuation.

2

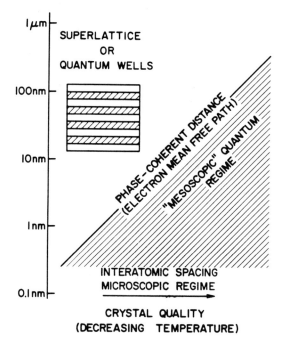

Fig. 3. Schematic illustration of a ''mesoscopic'' quantum regime
(hatched) with a superlattice or quantum wells in the insert.

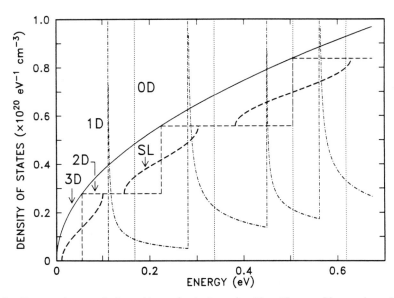

Fig. 4. Comparison of density of states in the three-dimensional (3D)
electron system with those of a superlattice, and the two-
dimensional (2D), one-dimensional (1D), and zero-dimensional
(0D) electron systems.

tively, the bound energies, E_1 and E_2, in the quantum well are derived to be 0.08 and 0.32eV. As shown in the insert of the figure, if the energy of incident electrons coincides with such bound energies, the electrons tunnel through both barriers without attenuation. Such unity transmissivity at the resonant condition arises from the destructive interference of reflected electron waves from inside (R) with incident waves (I), so that only transmitted waves (T) remain.

The idea of the superlattice (SL) occurred to us as a natural extension of double- and multi-barrier structures where quantum effects are expected to prevail. An important parameter relevant to the observation of such effects is the phase-coherent distance or, roughly, the electron inelastic mean free path, which depends heavily on bulk and interface quality of crystals and also on temperature and values of the effective mass. As schematically illustrated in Fig. 3, if characteristic dimensions such as SL periods and well widths are reduced to less than the phase-coherent distance, the entire electron system will enter a ''mesoscopic'' quantum regime of reduced dimensionality, being placed in the scale between the macroscopic and the microscopic.

In the early 1970s, we initiated the attempt of the seemingly formidable task of engineering nanostructures in the search for novel quantum phenomena (4). It was theoretically shown that the introduction of the SL potential perturbs the band structure of the host materials, yielding unusual electronic properties of quasi-two-dimensional character (1) (2). Figure 4 shows the density of states $\rho(E)$ for electrons with $m^* = 0.067m_0$ in an SL with a well width of 100Å and the same barrier width, where the first three subbands are indicated with dashed curves. The figure also includes, for comparison, a parabolic curve $E^{1/2}$ for 3D, a steplike density of states for 2D (quantum well), a curve $\Sigma(E - E_m - E_n)^{-1/2}$ for 1D (quantum line or wire), and a delta function $\sum_{mn}\delta(E - E_1 - E_m - E_n)$ for 0D system (quantum box or dot) where the quantum unit is taken to be 100Å for all cases and the barrier height is assumed to be infinite in obtaining the quantized energy levels, E_1, E_m and E_n. Notice that the ground state energy increases with decrease in dimensionality if the quantum unit is kept constant. Each quantized energy level in 2D, 1D and 0D is identified with the one, two and three quantum numbers, respectively. The unit for the density of states here is normalized to $eV^{-1}cm^{-3}$ for all the dimensions, although $eV^{-1}cm^{-2}$, $eV^{-1}cm^{-1}$ and eV^{-1} may be commonly used for 2D, 1D and 0D, respectively.

The analysis of the electron dynamics in the SL direction predicted an unusual current-voltage characteristic including a negative differential resistance, and even the occurrence of ''Bloch oscillations.'' The calculated Bloch frequency f is as high as 250 GHz from the equation $f = eFd/h$, for an applied field F and a superlattice period d of 10^3V/cm and 100Å, respectively.

Esaki, Chang and Tsu (5) reported an experimental result on a GaAs-GaAsP SL with a period of 200Å synthesized with CVD (chemical vapor deposition) by Blakeslee and Aliotta (6). Although transport measurements failed to show any predicted effect, this system probably constitutes the first strained-layer SL having a lattice mismatch 1.8% between GaAs and $GaAs_{0.5}P_{0.5}$.

Esaki et al. (7) found that an MBE (molecular beam epitaxy)-grown GaAs-GaAlAs SL exhibited a negative resistance in its transport properties, which was, for the first time, interpreted in terms of the above-mentioned SL effect. Although our early efforts focused on transport measurements, Tsu and Esaki (8) calculated optical nonlinear response of conduction electrons in an SL medium. Since the first proposal and early attempt, the field of semiconductor SLs and quantum wells (QWs) has proliferated extensively in a cross-disciplinary environment (9).

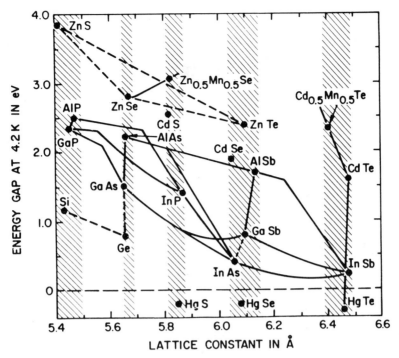

Fig. 5. Plot of energy gaps at 4.2K versus lattice constants.

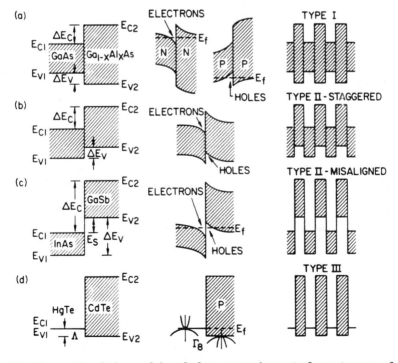

Fig. 6. Discontinuities of bandedge energies at four types of hetero-interfaces: band offsets (left), band bending and carrier confinement (middle), and superlattice (right).

II. EPITAXY AND SUPERLATTICE GROWTH

Heteroepitaxy is of fundamental interest for the SL and QW growth. Innovations and improvements in growth techniques such as MBE (10), MOCVD (metalorganic chemical vapor deposition) (11) and MOMBE or CBE during the last decade have made possible high-quality heterostructures. Such structures possess predesigned potential profiles and impurity distributions with dimensional control close to interatomic spacing and with virtually defect-free interfaces, particularly, in a lattice-matched case such as GaAs – $Ga_{1-x}Al_xAs$. This great precision has cleared access to a ''mesoscopic'' quantum regime.

The semiconductor SL structures have been grown with III-V, II-VI and IV-VI compounds, as well as elemental semiconductors. Figure 5 shows the plot of energy gaps at 4.2K versus lattice constants for zinc-blende semiconductors together with Si and Ge. Joining lines represent ternary alloys except for Si-Ge, GaAs-Ge and InAs-GaSb. The introduction of II-VI compounds apparently extended the available range of energy gaps in both the high and the low direction: that of ZnS is as high as 3.8eV and all the Hg compounds have a negative energy gap or can be called zero-gap semiconductors. The magnetic compounds, CdMnTe and ZnMnSe are relative newcomers in the SL and QW arena.

Semiconductor hetero-interfaces exhibit an abrupt discontinuity in the local band structure, usually associated with a gradual band-bending in its neighborhood which reflects space-charge effects. According to the character of such discontinuity, known hetero-interfaces can be classified into four kinds: type I, type II-staggered, type II-misaligned, and type III, as illustrated in Fig. 6(a)(b)(c)(d): band offsets (left), band bending and carrier confinement (middle), and SLs (right).

The bandedge discontinuities, ΔE_c for the conduction band and ΔE_v for the valence band, at the hetero-interfaces obviously command all properties of QWs and SLs, and thus constitute the most relevant parameters for device design. For such fundamental parameters, ΔE_c and ΔE_v , however, the predictive qualities of theoretical models are not very accurate and precise experimental determination cannot be done without great care. In this regard, the GaAs-GaAlAs system is most extensively investigated with both spectroscopic and electrical measurements. Recent experiments have reduced an early established value of $\Delta E_c/\Delta E_g$, 85 percent to somewhat smaller values in the range between 60 and 70 percent.

III. RESONANT TUNNELING VIA QUANTUM WELLS

Tsu and Esaki (12) computed the resonant transmission coefficient as a function of electron energy for multi-barrier structures from the tunneling point of view, leading to the derivation of the current-voltage characteristics. The SL band model previously presented, assumed an infinite periodic structure, whereas, in reality, not only a finite number of periods is prepared with alternating epitaxy, but also the phase-coherent distance is limited. Thus, this multibarrier tunneling model provided useful insight into the transport mechanism and laid the foundation for the following experiment.

Chang, Esaki and Tsu (13) observed resonant tunneling in GaAs-GaAlAs double-barriers, and subsequently, Esaki and Chang (14) measured quantum transport with local resonant tunneling for a GaAs-AlAs SL having a tight-binding potential. The current and conductance versus voltage curves for a double barrier are shown in Fig. 7. The energy diagram is shown in the inset where resonance is achieved at such applied voltages as to align the Fermi level of the electrode with the bound states, as shown in cases (a) and (c). The current and conductance versus voltage characteristics for an SL are shown in Fig. 8. The energy diagrams for band-type conduction, (a),

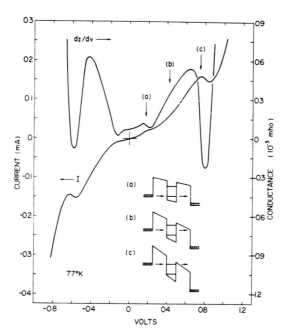

Fig. 7. Current- and conductance- voltage characteristics of a
GaAs-GaAlAs double-barrier structure. Conditions at resonance (a) and (c), and at off-resonance (b), are indicated
by arrows in the insert.

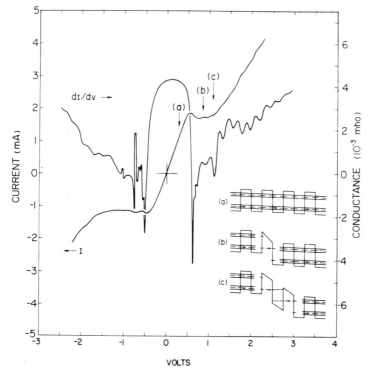

Fig. 8. Current- and conductance- voltage characteristics of a
GaAs-AlAs superlattice. Band-type conduction (a) and resonant tunneling at an expanding high-field domain (b) and (c),
are indicated in the insert.

and resonant tunneling at an expanding high-field domain, (b) and (c), are shown in the inset. Both of those quantum transport measurements, probably constitute the first observation of man-made bound states in both single and multiple QWs.

The technological advance in MBE for the last decade resulted in dramatically-improved characteristics in resonant tunneling, as shown in Fig. 9, as well as in SL transport, which spurred renewed interest in such structures (15). The observation of resonant tunneling in p-type double barrier structures (16) revealed fine structure corresponding to each bound state of both heavy and light holes. The existence of optical phonon-assisted resonant electron tunneling in the valley current region was found (17), (18), for which a theoretical approach was proposed (19). Discrete electronic states in a three-dimensionally confined quantum well (quantum dot) were observed in double-barrier heterostructures which had been shaped into slender columns of less than 2500 Å in diameter by advanced processing techniques (20).

A variety of tunneling heterostructures have been developed with a combination of materials other than GaAs/AlAs, such as InGaAs/InAlAs (21) and InAs/AlSb (22). Research in polytype heterostructures of InAs/AlSb/GaSb (23), (24), has been rejuvenated with the observation of a substantial negative differential resistance based on resonant interband tunneling in GaSb/AlSb/InAs/AlSb/GaSb (25) and InAs/AlSb/GaSb/AlSb/InAs (26). Those multi-heterojunctions, indeed, correspond to a portion of the proposed polytype superlattices (23). The emerging studies on one- and two- dimensional electron tunneling in lateral structures will be discussed in section IX.

A great deal of engineering interest in resonant tunneling, obviously, has arisen from its potential application to high-speed semiconductor devices. Picosecond bistable operation has been observed in a double-barrier resonant tunneling diode, where a rise time of 2 ps is said to be the fastest switching yet measured for an electronic device (27). The high-speed performance of this device is apparently related to the lifetime of the quasibound-state in the quantum well (28). The evidence of sequential tunneling due to intersubband scattering was also observed (29).

IV. OPTICAL ABSORPTION, PHOTOCURRENT SPECTROSCOPY, PHOTOLUMINESCENCE AND STARK EFFECT

Optical investigation on the quantum structures during the last two decades has revealed the salient features of quantum confinement. Dingle et al. (30) (31) observed pronounced structure in the optical absorption spectrum, representing bound states in isolated and double QWs. In low-temperature measurements for such structures, several exciton peaks, associated with different bound-electron and bound-hole states, were resolved. The spectra clearly indicate the evolution of resonantly split, discrete states into the lowest subband of an SL. van der Ziel et al. (32) observed optically pumped laser oscillation from GaAs-GaAlAs QW structures at 15K. Since then, the application of such structures to semiconductor laser diodes has received considerable attention because of its superior characteristics, such as low threshold current, low temperature dependence, tunability and directionality (33). Advanced graded-index separate-confinement single quantum well heterostructure lasers (GRIN-SCH) with high performance have been developed (34) (35).

Tsu et al. (36) made photocurrent measurements on GaAs-GaAlAs SLs subject to an electric field perpendicular to the well plane with the use of a semitransparent Schottky contact. The photocurrents as a function of incident photon energy are shown in Fig. 10 for samples of three different configurations, designated A (35Å GaAs - 35Å GaAlAs), B (50Å GaAs - 50Å GaAlAs), and C (110Å GaAs - 110Å GaAlAs). The energy diagram is shown schematically in the upper part of the

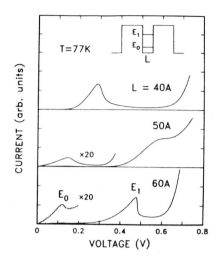

Fig. 9. Current-voltage curves for three double-barriers with well widths, L, of 40Å, 50Å, and 60Å. The bound energies, E_1 and E_2, are shown in the energy diagram of the upper part.

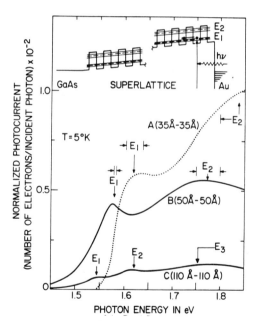

Fig. 10. Photocurrent versus photon energy for three superlattice samples A, B, and C. Calculated energies and bandwidths are indicated. The energy diagram of a Schottky-barrier structure is shown in the upper part.

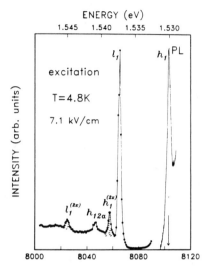

Fig. 11. Excitation spectrum with the photoluminescence of the
heavy-hole exciton for a 160-Å quantum well at 7.1 kV/cm,
where h_1, $h_1^{(2x)}$, l_1 and $l_1^{(2x)}$ denote the ground and excited
states of the heavy- and light-hole excitons, respectively,
and h_{12a} corresponds to an exciton related to the n=1 conduction and n=2 heavy-hole valence bands.

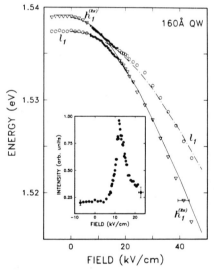

Fig. 12. Stark shifts of the light-hole exciton (open circles) and
excited states of the heavy-hole exciton (open triangles),
exhibiting strong coupling (solid circles) between them.
The integrated intensity of $h_1^{(2x)}$ normalized to l_1 is shown
as a function of electric field in the insert.

10

figure where quantum states created by the periodic potential are labeled E_1 and E_2. These states are essentially discrete if there is a relatively large separation between wells, as is the case in sample C, but broaden into subbands in samples A and B owing to an increase in overlapping of wave functions. Calculated energies and bandwidths for transitions shown in the figure are found to be in satisfactory agreement with the observation. Recently, Deveaud et al. (37) made transport measurements of 2D carriers in SL minibands by subpicosecond luminescence spectroscopy. Schneider et al. (38) studied the dynamics of electron transport along the growth direction of tight-binding GaAs-AlAs SLs by electrical and optical time-of-flight experiments.

In undoped high-quality GaAs-GaAlAs QWs, the main photoluminescence peak is attributed to the excitonic transition between 2D electrons and heavy holes. Mendez et al. (39) studied the field-induced effect on the photoluminescence in such wells: when the electric field, for the first time in luminescence measurements, was applied perpendicular to the well plane, pronounced field-effects, Stark shifts, were discovered. More recently, Vina et al. (40) studied such shifts on the excitonic coupling as indicated in Figures 11 and 12. Figure 11 shows excitation spectrum with the photoluminescence of the heavy-hole exciton, where h_1, $h_1^{(2x)}$, and $l^{(2x)}$ denote the ground and excited states of the heavy- and light-hole excitons, respectively, and h_{12a} corresponds to an exciton related to the n=1 conduction and n=2 heavy-hole valence bands. Figure 12 shows Stark shifts of the light-hole exciton (open circles) and excited states of the heavy-hole exciton (open triangles), exhibiting strong coupling (solid circles) between them. Chemla et al. (41) observed Stark shifts of the excitonic absorption peak, even at room temperature; Miller et al. (42) analyzed such electroabsorption spectra and demonstrated optical bistability, level shifting and modulation.

Although electric field-induced effects on a crystal such as Bloch oscillations and Stark ladder, have been studied since the early days of quantum mechanics, experimental verification was hardly possible. Semiconductor SLs, however, with their versatility in design, have provided a unique opportunity for this study. Recently it was theoretically shown (43) and experimentally observed (44) that the application of an electric field F along the growth direction of a semiconductor SL indeed induces the Stark localization of the eigenstates resulting in a blue shift of the optical-absorption edge and the presence of oscillations. The energy diagrams in Fig. 13 illustrate the localization process of the states with increase in applied electric fields.

V. RAMAN SCATTERING

Manuel et al. (45) reported the observation of enhancement in the Raman cross section for photon energies near electronic resonance in GaAs – Ga$_{1-x}$Al$_x$As SLs of a variety of configurations. Both the energy positions and the general shape of the resonant curves agree with theoretical values. Later, it was pointed out that resonant inelastic light scattering yields separate spectra of single particle and collective excitations which will lead to the determination of electronic energy levels in QWs as well as Coulomb interactions. Subsequently, this was experimentally confirmed and the technique has now widely been used as a spectroscopic tool to study electronic excitations (46).

Meanwhile, Colvard et al. (47) reported the observation of Raman scattering from folded acoustic longitudinal phonons in a GaAs(13.6Å)-AlAs(11.4Å) SL. The SL periodicity leads to Brillouin zone folding, resulting in the appearance of gaps in the phonon spectrum for wave vectors satisfying the Bragg condition, whereas Raman scattering for the optical modes revealed quasiconfined behavior (48).

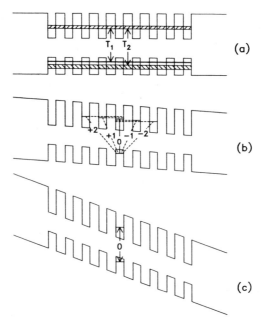

Fig. 13. Sketches of the conduction- and valence- band potential
profiles for GaAs – Ga$_{1-x}$Al$_x$As superlattice under a (a) small,
(b) moderate, and (c) high electric field, clad by thick
Ga$_{1-x}$Al$_x$As regions. The diagrams are approximately scaled
for a 30-35-Å superlattice with x=0.35, and fields of
2x10^3, 2x10^4, and 1x10^5 V/cm, respectively.

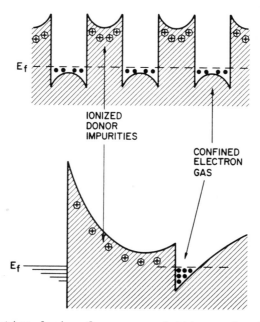

Fig. 14. Modulation doping for a superlattice and a heterojunction
with a Schottky junction.

Fasolino et al. (49) claimed that folded modes appear in the frequency range where the bulk dispersions of the two constituents overlap as in the longitudinal acoustical region, while modes confined in one layer and evanescent in the other, arise from the situation of no overlap as in the optical region of GaAs and AlAs.

VI. MODULATION DOPING

In SLs, in order to prevent usual impurity scattering, it is possible to spatially separate free carriers from their parent impurity atoms by doping impurities in the region of the potential hills. Though this concept was expressed in the original article, Esaki and Tsu, (1), Dingle et al. (50) successfully implemented such a concept in modulation-doped GaAs-GaAlAs SLs, as illustrated at the top of Fig. 14, achieving unprecedented electron mobilities far exceeding the Brooks-Herring predictions. Subsequently, the similar technique was used to form the high- mobility 2D hole gas. Recently, low-temperature electron mobilities as high as 5×10^6 cm^2/Vs at a carrier density of 1.6×10^{11}cm^{-2} (51), and subsequently, 10^7 cm^2/Vs at a density of 2.4×10^{11}cm^{-2} (52) have been reported at modulation-doped GaAs heterostructures: the latter, obviously, constitutes the highest ever measured in a solid.

The modulation-doping technique has exerted profound impact on the course of the evolution of the quantum structures in low dimensionality. The achievement of the high-mobility 2D carriers with long phase-coherent lengths has opened up the new genre of lateral quantum structures in dimensionality less than two, leading to further ramification in this research. The availability of such high-mobility 2D carriers has prompted the development of a high-speed field-effect transistor called HEMT (high electron mobility transistor); its band energy diagram is shown at the bottom of Fig. 14 (53).

VII. MAGNETOTRANSPORT, QUANTIZED HALL EFFECT AND FRACTIONAL FILLING

Chang et al. (54) observed Shubnikov-de Haas oscillations associated with the subbands of the nearly 2D character in GaAs-GaAlAs SLs. The oscillatory behaviors agreed well with those predicted from the Fermi surfaces derived from the SL configurations and the electron concentrations.

v. Klitzing et al. (55) demonstrated the interesting proposition that quantized Hall resistance could be used for precision determination of the fine structure constant α, using 2D electrons in the inversion layer of a Si MOSFET. Subsequently, Tsui and Gossard (56) found modulation-doped GaAs-GaAlAs heterostructures desirable for this purpose, primarily because of their high electron mobilities, which led to the determination of α with a great accuracy.

The quantized Hall effect in a 2D electron or hole system is observable at such high magnetic fields and low temperatures as to locate the Fermi level in the localized states between the extended states. Under these conditions, the magnetoresistance ρ_{xx} vanishes and the Hall resistance ρ_{xy} goes through plateaus. This surprising result can be understood by the argument that the localized states do not take part in quantum transport. At the plateaus, the Hall resistance is given by $\rho_{xy} = h/e^2\nu$ where ν is the number of filled Landau levels; h, Planck's constant; and e, the electronic charge.

Tsui, Stormer and Gossard (57) discovered a striking phenomenon, an anomalous quantized Hall effect, exhibiting a plateau in ρ_{xy}, and a dip in ρ_{xx}, at a fractional filling factor of 1/3 in the extreme quantum limit at temperatures lower than 4.2K. This discovery has spurred a large number of experimental and theoretical studies.

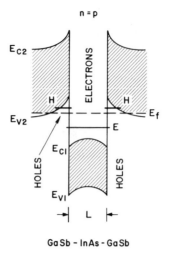

Fig. 15. Energy-band diagram of ideal GaSb-InAs-GaSb quantum wells for electrons and holes.

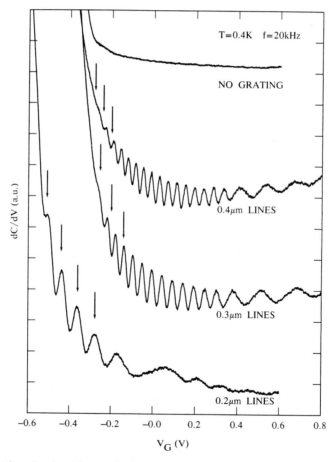

Fig. 16. The derivative of the capacitance versus gate voltage for three quantum wire samples and a control sample.

VIII. InAs-GaSb and GaSb-AlSb SYSTEMS

Since 1977, the InAs-GaSb system was investigated because of its extraordinary bandedge relationship at the interface, called type II-''misaligned'' in Fig. 6(c). The band calculation for InAs-GaSb SLs (58) revealed a strong dependence of the subband structure on the period: The semiconducting energy gap decreases when increasing the period, becoming zero at 170Å, corresponding to a semiconductor-to-semimetal transition. The electron concentration in SLs was measured as a function of InAs layer thickness, (59); it exhibited a sudden increase of an order-of-magnitude in the neighborhood of 100Å. Such increase indicates the onset of electron transfer from GaSb to InAs which is in good agreement with theoretical prediction. Far-infrared magneto-absorption experiments (60) were performed at 1.6K for semi-metallic SLs which confirmed their negative energy-gap.

MBE-grown GaSb-InAs-GaSb have been investigated, where the unique bandedge relationship allows the coexistence of electrons and holes across the two interfaces, as shown in Fig. 14. Prior to experimental studies, Bastard et al. (61) performed self-consistent calculations for the electronic properties of such QWs, predicting the existence of a semiconductor-to-semimetal transition as a result of electron transfer from GaSb at the threshold thickness of InAs; this is some-what similar to the mechanism in the InAs-GaSb SLs. Such a transition was confirmed by experiments carried out by Munekata et al. (62), although the electron and hole densities are not the same, probably because of the existence of some extrinsic electronic states. Anal-ysis (63) for such a 2D electron-hole gas elucidated, for the first time, the fact that the quantum Hall effect is determined by the de-gree of uncompensation of the system.

The GaSb-AlSb system, which belongs to type I in the relative position of the band gaps at the interface, has been studied. Optical absorption measurements in GaSb-AlSb SLs revealed the effect of strain induced by a lattice mismatch of 0.65%, resulting in the re-versal of light and heavy valence subbands (64): Raman scattering confirmed the magnitude of the strain (65). More recently, photoreflectance and photoluminescence studies under hydrostatic pressure at cryogenic temperature have been performed for GaSb-AlSb multiple QWs (66).

The polytype heterostructures and SLs involving InAs, GaSb and AlSb, were proposed on the consideration that the differences in lattice constants (6.058Å, 6.095Å and 6.136Å) are not excessive, and yet their band parameters are significantly differing among them (23).

IX. LATERAL QUANTUM STRUCTURES

It is of significance that the extraordinary high mobility, μ, for a 2D carrier density, n, at modulation doped GaAs heterostructures has made the elastic mean-free-path length, ℓ, greater than the pat-tern dimensions of small devices accessible with the present-day microfabrication technique (67), where $\ell = h/e\mu(n/2\pi)^{1/2} = 1.65 \times 10^{-15}\mu n^{1/2}$. Namely, if $n = 2.5 \times 10^{11} cm^{-2}$ ($\lambda_f = 50nm$), the magnitude, ℓ, reaches 0.8, 8, and 80 μm for $\mu = 10^5$, 10^6 and 10^7 cm^2/Vs, respectively. Although these large ℓ values are derived, assuming no introduction of damages or scattering centers during the processing, it is now feasible to fabricate such devices that phase coherence of electrons is main-tained on the length scale of whole structures. On the background of the remarkable advances in MBE, dry-etching and electron-beam lithography, the physics of ''coherent small devices'' has gained a great deal of attention in the past several years, including studies on quantum interference (68), the Aharonov-Bohm effect (69), and quenching of the Hall effect (70).

In most cases, quasi ID and OD electronic systems were achieved with lateral confinement by depleting a high-mobility 2D gas in the surrounding area through a means, or simply by forming a small mesa through dry etching. A clear manifestation for spatial quantization

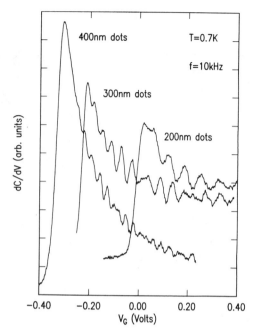

Fig. 17. The derivative of the capacitance versus gate voltage for three quantum dot samples. The main peak in the derivative is due to turn-on of the capacitors.

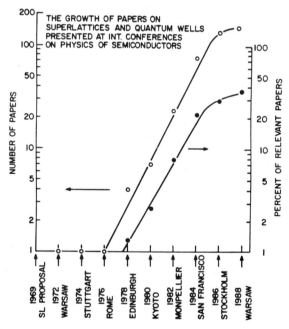

Fig. 18. Growth in the number of papers on SLs and QWs and the percentage of those to the total presented at the International Conference on the Physics of Semiconductors being held every other years.

was the capacitance measurements of the density of states of such ID and OD systems (71) (72). Figures 16 and 17 show capacitance spectra for samples with three different wire widths and three dot diameters, respectively, where the period of the oscillations duly decreases with increase in the size. With capacitance spectroscopy for the dot samples, Hansen et al (73) observed a pronounced magnetic-field-induced split of quantum levels into surface states and bulklike Landau states, indicating that spacial quantization effects become negligible with the magnetic length much smaller than the size of the confining potential.

With the ''coherent small devices'', the study of ballistic electron transport or 2D electron optics has been a rapidly growing field of research, which is in contrast with the traditional transport study in the diffusive regime where length scales are large in comparison with the electron mean-free-path. Recently, van Wees et al. (74), and subsequently Wharam et al. (75), clearly demonstrated a quantized conductance of $2e^2/h$ per occupied subband or wave guide mode in observing electron transport through a short pathway with a variable width of the order of the Fermi wavelength. The studies of magnetic focusing in such electron system have been reported (76) (77). ID electron resonant tunneling has been attempted through the OD state defined in a 0.42 μm square box (78). With reduction to ID, it is speculated that higher mobilities would be achieved because of the suppression of scattering (79), and also exciton binding energies and oscillator strengths could be enhanced (80) (81).

For the purpose of the development of novel field-effect transistors, a variety of lateral structures with grating-gate and grid-gate (82), and dual-gate (83) configuration have been explored, including the observation of 2D electron resonant tunneling through the ID state defined by the dual-gate. In principle, with reduction in dimensionality, the current-voltage characteristic should be sharpened in the consideration of the energy and parallel momentum conservation rule in resonant tunneling.

X. OTHER SUPERLATTICES

Since the inception of the superlattice, the major part of research has been centered on compositional structures. Nevertheless, GaAs doping SLs have exhibited an interesting optical property, i.e., tunability in emission wavelength as well as absorption coefficient, arising from long recombination lifetimes of photoexcited electrons and holes which are separated in the superlattice potential (84). The advent of the δ-doping technique (85) has contributed to the growth of high quality doping SLs (86). Recently, Schubert et al. (87) achieved tunable, stimulated emission of radiation from such a doping SL confined between GaAlAs barriers.

It is certainly desirable to select a pair of materials closely lattice-matched in order to obtain stress- and dislocation- free interfaces. Lattice-mismatched heterostructures, however, can be grown with essentially no misfit dislocations, if the layers are sufficiently thin, because the mismatch is accommodated by uniform lattice strain (88). On the basis of such premise, Osbourn and his co-workers (89) prepared strained-layer SLs from lattice-mismatch pairs, claiming the versatility in the choice of layer materials. A number of strained-layer SLs have been explored in the past several years.

Gossard et al. (90) achieved GaAs-AlAs epitaxial structures with alternate-atomic-layer composition modulation by MBE. Recently, there is a great deal of interest in optical properties of ultra-short-period SLs associated with their band structure (91): the direct-indirect (type I - type II) transition occurs with the varying of layer thickness (92) (93).

Kasper et al. (94) pioneered the MBE growth of Si-SiGe strained-layer SLs, and Manasevit et al. (95) observed unusual mobility enhancement in such structures. The growth of high-quality Si-SiGe SLs (96) attracted much interest in view of possible applications as well as scientific investigations. In short period Si-Ge SLs, the band structure is strongly influenced by induced strain due to a lattice mismatch of more than 4% and also predicted zone folding (97) resulting in the observation of new optical transitions (98). Resonant tunneling of holes with a negative differential resistance has been observed in Si-SiGe structures with unstrained (99) and strained (100) Si double-barriers.

Dilute-magnetic SLs such as CdTe-CdMnTe (101) (102) and ZnSe-ZnMnSe are contemporary additions to the superlattice family, and have already exhibited unique magneto-optical properties.

XI. CONCLUSION

We have witnessed remarkable progress of an interdisciplinary nature on this subject. A variety of ''engineered'' structures exhibited extraordinary transport and optical properties; some of them, such as ultrahigh carrier mobilities, semimetallic coexistence of electrons and holes, resonant tunneling and large Stark shifts on the optical properties, may not even exist in any ''natural'' crystal. Thus, this new degree of freedom offered in semiconductor research through advanced material engineering has inspired many ingenious experiments, resulting in observations of not only predicted effects but also totally unknown phenomenon. The growth of papers on the subject for the last decade is indeed phenomenal. For instance, the number of papers and the percentage of those to the total presented at the International Conference on the Physics of Semiconductors (ICPS) being held every other year, increased as shown in Fig. 18. After the incubation period of several years, it really took off, with a rapid annual growth rate, around ten years ago. It says that, currently, more than one third of semiconductor physicists in the world are working in this field.

Activities in this new frontier of semiconductor physics, in turn, give immeasurable stimulus to device physics, leading to unprecedented transport and optoelectronic devices or provoking new ideas for applications. Figure 19 illustrates a correlation diagram in such interdisciplinary research where beneficial cross-fertilizations are prevalent.

I hope this article, which cannot possibly cover every landmark, provides some flavor of the excitement in this field. Finally, I would like to thank the many participants in and out of superlattice research for their contributions, as well as the ARO's partial sponsorship from the initial day of our investigation.

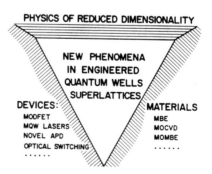

Fig. 19. Correlation diagram in a cross-disciplinary research environment.

REFERENCES

1. L. Esaki and R. Tsu, ''Superlattice and negative conductivity in semiconductors,'' IBM Research Note RC-2418 (1969).

2. L. Esaki and R.Tsu, ''Superlattice and negative differential conductivity in semiconductors,'' IBM J. Res. Develop. 14: 61 (1970).

3. D. Bohm, ''Quantum Theory:'' (Prentice Hall, Englewood Cliffs, N.J. 1951), p.283.

4. L. Esaki, ''Long journey into tunneling,'' Les Prix Nobel en 1973, Imprimerie Royale, P.A. Norstedt & Soner, Stockholm 1974, p. 66.

5. L. Esaki, L.L. Chang, and R. Tsu, ''A one-dimensional 'superlattice' in semiconductors,'' Proc. 12th Int. Conf. on Low Temperature Physics, Kyoto, Japan, 1970 (Keigaku Publishing Co.,Tokyo, Japan), p. 551.

6. A.E. Blakeslee and C.F. Aliotta, ''Man-made superlattice crystals,'' IBM J. Res. Develop. 14: 686 (1970).

7. L. Esaki, L.L. Chang, W.E. Howard, and V.L. Rideout, ''Transport properties of a GaAs-GaAlAs superlattice,'' Proc. 11th Int. Conf. on the Physics of Semiconductors, Warsaw, Poland, 1972, (PWN-Polish Scientific Publishers, Warsaw, Poland), p. 431.

8. R. Tsu and L. Esaki, ''Nonlinear optical response of conduction electrons in a superlattice,'' Appl. Phys. Lett. 19: 246 (1971).

9. L. Esaki, ''Semiconductor Superlattices and Quantum Wells,'' Proc. 17th Int. Conf. on the Physics of Semiconductors, San Francisco, 1984, (Springer-Verlag, New York, 1985), p.473; IEEE J. Quantum Electron., QE-22: 1611 (1986).

10. A. C. Gossard, ''Growth of Microstructures by Molecular Beam Epitaxy,'' IEEE J. Quantum Electron., QE-22: 1649 (1986).

11. M. Razeghi, ''The MOCVD challenge'', (Adam Hilger, Bristol and Philadelphia, 1989).

12. R. Tsu and L. Esaki, ''Tunneling in a finite superlattice,'' Appl. Phys. Lett. 22: 562 (1973).

13. L.L. Chang, L. Esaki, and R. Tsu, ''Resonant tunneling in semiconductor double barriers,'' Appl. Phys. Lett. 24: 593 (1974).

14. L. Esaki and L.L. Chang, ''New transport phenomenon in a semiconductor 'superlattice','' Phys. Rev. Lett. 33: 495 (1974).

15. F. Capasso, K. Mohammed and A. Y. Cho, ''Resonant tunneling through double barriers,'' IEEE J. Quantum Electron., QE-22: 1853 (1986).

16. E.E. Mendez, W.I. Wang, B. Ricco, and L. Esaki, ''Resonant Tunneling of holes in AlAS-GaAs-AlAS heterostructures,'' Appl. Phys. Lett. 47: 415 (1985).

17. V.J. Goldman, D.C. Tsui and J.E. Cunningham, ''Evidence for LO-phonon-emission-assisted tunneling in double-barrier heterostructures,'' Phys. Rev. B36: 7635 (1987).

18. M.L. Leadbeater, E.S. Alves, L. Eaves, M. Henini, O.H. Hughes, A. Celeste, J.C. Portal, G. Hill and M.A. Pate, ''Magnetic field studies of elastic scattering and optic-phonon emission in resonant-tunneling devies'', Phys. Rev. B 39: 3438 (1989).

19. W. Cai, T.F. Zheng, P. Hu, B. Yudanin and M. Lax, ''Model of phonon-associated electron tunneling through a semiconductor double barrier,'' Phys. Rev. Lett. 63: 418 (1989).

20. M.A. Reed, J.N. Randall, R.J. Aggarwal, R.J. Matyi, T.M. Moore and A.E. Wetsel, ''Observation of discrete electronic states in a zero-dimensional semiconductor nanostructure,'' Phys. Rev. Lett. 60: 535 (1988).

21. A. Tackeuchi, T. Inata, S. Muto and E. Miyauchi, ''Picosecond characterization of InGaAs/InAlAs resonant tunneling barrier di-

ode by electro-optic sampling,'' <u>Jpn. J. Appl. Phys.</u> 28: L750 (1989).

22. L.F. Luo, R. Beresford and W.I. Wang, ''Resonant tunneling in AlSb/InAs/AlSb double-barrier heterostructures,'' <u>Appl. Phys. Lett.</u> 53: 2320 (1988).

23. L. Esaki, L.L. Chang and E.E. Mendez, ''Polytype superlattices and multi-heterojunctions,'' <u>Jpn. J. Appl. Phys.</u> 20: L529 (1981).

24. H. Takaoka, Chin-An Chang, E.E. Mendez, L.L. Chang and L. Esaki, ''GaSb-AlSb-InAs multi-heterojunctions,'' <u>Physica</u> 117B &118B: 741 (1983).

25. L.F. Luo, R. Beresford and W.I. Wang, ''Interband tunneling in polytype heterostructures'', (to be published).

26. J.R. Soderstrom, D.H. Chow and T.C. McGill, ''A new negative differential resistance device based on resonant interband tunneling'' (to be published).

27. J.F. Whitaker, G.A. Mourou, T.C.L.G. Sollner and W.D. Goodhue, ''Picosecond switching time measurement of a resonant tunnel diode,'' <u>Appl. Phys. Lett.</u> 53: 385 (1988).

28. E.R. Brown, C.D. Parker and T.C.L.G. Sollner, ''Effect of quasibound-state lifetime on the oscillation power of resonant tunneling diodes,'' <u>Appl. Phys. Lett</u> 54: 934 (1989).

29. L. Eaves, G.A. Toombs, F.W. Sheard, C.A. Payling, M.L. Leadbeater, E.S. Alves, T.J. Foster, P.E. Simmonds, M. Henini, O.H. Hughes, J.C. Portal, G. Hill and M.A. Pate, ''Sequential tunneling due to intersubband scattering in double-barrier resonant tunneling devices'', <u>App. Phys. Lett.</u> 52: 212 (1988).

30. R. Dingle, W. Wiegmann and C.H. Henry, ''Quantum states of confined carriers in very thin $Al_xGa_{1-x}As$ – GaAs – $Al_xGa_{1-x}As$ heterostructures,'' <u>Phys. Rev. Lett.</u> 33: 827 (1974).

31. R. Dingle, A.C. Gossard and W. Wiegmann, ''Direct observation of superlattice formation in a semiconductor heterostructure,'' <u>Phys. Rev. Lett.</u> 34: 1327 (1975).

32. J.P. van der Ziel, R. Dingle, R.C. Miller, W. Wiegmann and W.A. Nordland Jr., ''Laser oscillation from quantum states in very thin GaAs – $Al_{0.2}Ga_{0.8}As$ multilayer structures,'' <u>Appl. Phys. Lett.</u> 26: 463 (1975).

33. Y. Arakawa and A. Yariv, ''Quantum well lasers,'' <u>IEEE J. Quantum Electron.</u>, QE-22: 1887 (1986).

34. W.T. Tsang, ''A graded-index waveguide separate- confinement laser with very low threshold and a narrow Gaussian beam,'' <u>Appl. Phys. Lett.</u> 39: 134 (1981).

35. P.L. Derry, A. Yariv, K.Y. Lau, N. Bar-Chaim, K. Lee and J. Rosenberg, ''Ultralow-threshold graded-index separate-confinement single quantum well buried heterostructure (Al,Ga)As lasers with high reflectivity coatings,'' <u>Appl. Phys. Lett.</u> 50: 1325 (1986).

36. R. Tsu, L.L. Chang, G.A. Sai-Halasz, and L. Esaki, ''Effects of quantum states on the photocurrent in a superlattice,'' <u>Phys. Rev. Lett.</u> 34: 1509 (1975).

37. B. Deveaud, J. Shah, T. C. Damen, B. Lambert and A. Regreny, ''Bloch transport of electrons and holes in superlattice minibands: direct measurement by subpicosecond luminescence spectroscopy,'' <u>Phys. Rev. Lett.</u> 58: 2582 (1987).

38. H. Schneider, W.W. Rühle, K. v.Klitzing and K. Ploog, ''Electrical and optical time-of-flight experiments in GaAs/AlAs superlattices,'' <u>Appl. Phys. Lett.</u> 54: 2656 (1989).

39. E.E. Mendez, G Bastard, L.L. Chang, and L. Esaki, ''Effect of an electric field on the luminescence of GaAs quantum wells,'' <u>Phys. Rev. B</u> 26: 7101 (1982).

40. L. Vina, R. T. Collins, E. E. Mendez and W. I. Wang, ''Excitonic coupling in GaAs/GaAlAs quantum wells in an electric field,'' <u>Phys. Rev. Lett.</u> 58: 832 (1987).

41. D.S. Chemla, T.C. Damen, D.A.B. Miller, A.C. Gossard, and W. Wiegmann, ''Electroabsorption by Stark effect on room-temperature excitons in GaAs/GaAlAs multiple quantum well structures,'' Appl. Phys. Lett. 42: 864 (1983).

42. D.A.B. Miller, J.S. Weiner, and D.S. Chemla, ''Electric-field dependence of linear optical properties in quantum well structures,'' IEEE J. Quantum Electron., QE-22: 1816 (1987).

43. P. Manuel, G.A. Sai-Halasz, L.L. Chang, Chin-An Chang, and L. Esaki, ''Resonant Raman scattering in a semiconductor superlattice,'' Phys. Rev. Lett. 37: 1701 (1976).

44. J. Bleuse, G. Bastard and P. Voison, ''Electric-field-induced localization and oscillatory electro-optical properties of semiconductor superlattices,'' Phys. Rev. Lett. 60: 220 (1988).

45. E.E. Mendez, F. Agullo-Rueda and J.M. Hong, ''Stark localization in GaAs-GaAlAs superlattices under an electric field,'', Phys. Rev. Lett. 60: 2426 (1988). L. Esaki, ''Resonant Raman scattering in a semiconductor superlattice,'' Phys. Rev. Lett. 37: 1701 (1976).

46. G. Abstreiter, R. Merlin, and A. Pinczuk, ''Inelastic light scattering by electronic excitations in semiconductor heterostructures,'' IEEE J. Quantum Electron., QE-22: 1771 (1987).

47. C. Colvard, R. Merlin, and M.V. Klein, and A.C. Gossard, ''Observation of folded acoustic phonons in a semiconductor superlattice,'' Phys. Rev. Lett. 45: 298 (1980).

48. M.V. Klein, ''Phonons in semiconductor superlattices,'' IEEE J. Quantum Electron., QE-22: 1760 (1987).

49. A. Fasolino, E. Marinari and J.C. Maan, ''Resonant quasiconfined optical phonons in semiconductor superlattices,'' Phys. Rev. B 39: 3923 (1989).

50. R. Dingle, H.L. Stormer, A.C. Gossard, W. Wiegmann, ''Electron mobilities in modulation-doped semiconductor heterojunction superlattices,'' Appl. Phys. Lett. 33: 665 (1978).

51. J.H. English, A.C. Gossard, H.L. Stormer, and K.W. Baldwin, ''GaAs structures with electron mobility of 5 x 10^6cm^2/Vs,'' Appl. Phys. Lett. 50 1826 (1987).

52. L. Pfeiffer, K.W. West, H.L. Stormer, and K.W. Baldwin, ''Electron mobility of 10^7 cm^2/Vs in modulation doped GaAs,'' Bull. Am. Phys. Soc. 34: 549 (1989).

53. M. Abe, T. Mimura, K. Nishiuchi, A. Shibatomi and M. Kobayashi, ''Recent advances in ultra-high-speed HEMT technology,'' IEEE J. Quantum Electron., QE-22: 1870 (1986).

54. L. L. Chang, H. Sakaki, C. A. Chang, and L. Esaki, ''Shubnikov-de Haas oscillations in a semiconductor superlattice,'' Phys. Rev. Lett. 38: 1489 (1977).

55. K.von Klitzing, G. Doreda, and M. Pepper, ''New method for high-accuracy determination of the fine-structure constant based on quantized hall resistance,'' Phys. Rev. Lett. 45: 494 (1980).

56. D.C. Tsui and A.C. Gossard, ''Resistance standard using quantization of the Hall resistance of GaAs – Al$_x$Ga$_{1-x}$As heterostructures,'' Appl. Phys. Lett. 38: 550 (1981).

57. D.C. Tsui, H.L. Stormer, and A.C. Gossard, ''Determination of the fine-structure constant using GaAs – Al$_x$Ga$_{1-x}$As heterostructures,'' Phys. Rev. Lett. 48: 1559 (1982).

58. G.A. Sai-Halasz, R. Tsu,and L.Esaki, ''A new semiconductor superlattice,'' App. Phys. Lett. 30: 651 (1977); G.A. Sai-Halasz, L. Esaki, and W.A. Harrison, ''InAs-GaSb superlattice energy structure and its semiconductor-semimetal transition,'' Phys. Rev. B 18: 2812 (1978).

59. L.L. Chang, N.J. Kawai, G.A. Sai-Halasz, R. Ludeke, and L. Esaki, ''Observation of semiconductor-semimetal transition in InAs-GaSb superlattices,'' Appl. Phys. Lett. 35: 939, (1979).

60. Y. Guldner, J.P. Vieren, P. Voisin, M. Voos, L.L. Chang, and L. Esaki, ''Cyclotron resonance and far-infrared magneto-absorption experiments on semimetallic InAs-GaSb superlattices,'' <u>Phys. Rev. Lett.</u> 45: 1719, (1980).

61. G. Bastard, E.E. Mendez, L.L. Chang, L. Esaki, ''Self-consistent calculations in InAs-GaSb heterojunctions,'' <u>J. Vac. Sci. Technol.</u> 21: 531 (1982).

62. H. Munekata, E.E. Mendez, Y. Iye, and L. Esaki, ''Densities and mobilities of coexisting electrons and holes in MBE grown GaSb-InAs-GaSb quantum well,'' <u>Surf. Sci. 174</u>: 449 (1986).

63. E.E. Mendez, L. Esaki, and L.L. Chang, ''Quantum Hall effect in a two-dimensional electron hole gas,'' <u>Phys. Rev. Lett.</u> 55: 2216 (1985).

64. P. Voisin, C. Delalande, M. Voos, L.L. Chang, A. Segmuller, C.A. Chang and L. Esaki, ''Light and heavy valence subband reversal in GaSb-AlSb superlattices,'' <u>Phys. Rev. B</u> 30: 2276 (1984).

65. B. Jusserand, P. Voisin, M. Voos, L.L. Chang, E.E. Mendez and L. Esaki, ''Raman scattering in GaSb-AlSb strained layer superlattices,'' <u>Appl. Phys. Lett.</u> 46: 678 (1985).

66. B. Rockwell, H.R. Chandrasekhar, M. Chandrasekhar, F.H. Pollak, H. Shen, L.L. Chang, W.I. Wang and L. Esaki, ''High pressure optical studies of GaSb-AlSb multiple quantum wells,'' (to be published).

67. J.M. Hong, T.P. Smith III, K.Y. Lee, C.M. Knoedler, S.E. Laux, D.P. Kern and L. Esaki, ''One- and zero- dimensional systems: fabrication and characterization,'' <u>J. Cryst. Growth</u>, 95 266 (1989); K.Y. Lee, T.P. Smith, III, H. Arnot, C.M. Knoedler, J.M. Hong, D.P. Kern and S.E. Laux, ''Fabrication and characterization of one- and zero- dimensional electron systems'' <u>J. Vac. Sci. Technol.</u> B 6: 1856 (1988).

68. T.J. Thornton, M. Pepper, H. Ahmed, D. Andrews and G.J. Davies, ''One-dimensional conduction in the 2D electron gas of a GaAs-AlGaAs heterojunction,'' <u>Phys. Rev. Lett.</u> 56: 1198 (1986).

69. G. Timp, A.M. Chang, J.E. Cunningham, T.Y. Chang, P. Mankiewich, R. Behringer and R.E. Howard, ''Observation of the Aharonov-Bohm effect for $\omega c_\tau > 1$,'' <u>Phys. Rev. Lett.</u> 58: 2814 (1987).

70. M.L. Roukes, A. Scherer, S.J. Allen, Jr., H.G. Craighead, R.M. Ruthen, E.D. Beebe and J.P. Harbison, ''Quenching of the Hall effect in a one-dimensional wire,'' <u>Phys. Rev. Lett.</u> 59: 3011 (1987).

71. T.P. Smith III, H. Arnot, J.M. Hong, C.M. Knoedler, S.E. Laux and H. Schmid, ''Capacitance oscillations in one-dimensional electron system,'' <u>Phys. Rev. Lett.</u> 59: 2802 (1987).

72. T.P. Smith III, K.Y. Lee, C.M. Knoedler, J.M. Hong and D.P. Kern, ''Electronic spectroscopy of zero-dimensional systems,'' <u>Phys. Rev B</u> 38: 2172 (1988).

73. W. Hansen, T.P. Smith III, K.Y. Lee, J.A. Brum, C.M. Knoedler, J.M. Hong and D.P. Kern, ''Zeeman Bifurcation of Quantum-Dot Spectra,'' <u>Phys. Rev. Lett.</u> 62: 2168 (1989).

74. B.J. van Wees, H. van Houten, C.W.J. Beenakker, J.G. Williamson, L.P. Kouwenhoven and D. van der Marel, ''Quantized conductance of point contacts in a two-dimensional electron gas,'' <u>Phys. Rev. Lett.</u> 60: 848 (1988).

75. D.A. Wharam, T.J. Thornton, R. Newbury, M. Pepper, H. Ahmed, J.E.F. Frost, D.G. Hasko, D.C. Peacook, D.A. Ritchie and G.A.C. Jones, ''One-dimensional transport and the quantization of the ballistic resistance,'' <u>J. Phys. C.: Solid State Phys.</u> 21: L 209 (1988).

76. H. van Houten, B.J. van Wees, J.E. Mooij, C.W.J. Beenakker, J.G. Williamson and C.T. Foxon, ''Coherent electron focusing in a two-dimensional electron gas,'' <u>Europhys. Lett.</u> 5: 721 (1988).

77. C.W.J. Beenakker, H. van Houton and B.J. van Wees, ''Mode inter-ference effect in coherent electron focusing,'' _Europhys. Lett._ 7: 359 (1988).

78. C.G. Smith, M. Pepper, H. Ahmed, J.E.F. Frost, D.G. Hasko, R. Newbury, D.C. Peacook, D. A. Ritchie and G.A.C. Jones, ''One di-mensional electron tunneling and related phenomena,'' (to be published).

79. H. Sakaki, ''Scattering suppression and high-mobility effect of size- quantized electrons in ultrafine semiconductor wire structures,'' _Jpn. J. Appl. Phys._ 19: L735 (1980).

80. Y-C Chang, L. L. Chang and L. Esaki, ''A new one-dimensional quantum well structure,'' _Appl. Phys. Lett._ 47: 1324 (1985).

81. D.A.B. Miller, D.S. Chemla and S. Schmitt-Rink, ''Electroabsorption of highly confined systems: Theory of the quantum-confined Franz-Keldysh effect in semiconductor quantum wires and dots'', _Appl. Phys. Lett._ 52: 2154 (1988).

82. K. Ismail, W. Chu, D.A. Antoniadis and H.I. Smith, ''Negative transconductance and negative differential resistance in a grid-gate modulation-doped field-effect transistor,'' _Appl. Phys. Lett._ 54: 460 (1989); K. Ismail, D.A. Antoniadis and H.I. Smith, ''One-dimensional subbands and mobility modulation in GaAs/AlGaAs quantum wires,'' _Appl. Phys. Lett._ 54: 1130 (1989).

83. S.Y. Chou, D.R. Allee, R.F.W. Pease and J.S. Harris, Jr., ''Ob-servation of electron resonant tunneling in a lateral dual-gate resonant tunneling field-effect transistor,'' _Appl. Phys. Lett._ 55: 176 (1989); K. Ismail, D.A. Antoniadis and H.I. Smith, ''Lateral resonant tunneling in a double-barrier field-effect transistor,'' _Appl. Phys. Lett._ 55: 589 (1989).

84. G.H. Dohler, H. Kunzel, D. Olego, K. Ploog, P. Ruden, H.J. Stolz, and G. Abstreiter, ''Observation of tunable band gap and two-dimensional subbands in a novel GaAs superlattice,'' _Phys. Rev. Lett._ 47: 864 (1981).

85. A. Zrenner, H. Reisinger, F. Koch and K. Ploog, ''Electron subband structure of a $\delta(z)$-doping layer in n-GaAs'', _Proc. 17th Int. Conf. on the Physics of Semiconductors,_ San Francisco, 1984, (Springer-Verlag, New York, 1985) p. 325.

86. E.F. Schubert, J.E. Cunningham and W.T. Tsang, ''Realization of the Esaki-Tsu-type doping superlattice'', _Phys. Rev. B_ 36: 1348 (1987).

87. E.F. Schubert, J.P. van der Ziel, J.E. Cunningham and T.D. Harris, ''Tunable stimulated emission of radiation in GaAs doping superlattices'', (to be published).

88. J.H. van der Merwe, ''Crystal interfaces,'' _J. Appl. Phys._ 34: 117 (1963).

89. G.C. Osbourn, R.M. Biefeld and P.L. Gourley, ''A $GaAs_xP_{1-x}$/GaP strained-layer superlattice,'' _Appl. Phys. Lett._ 41: 172 (1982); G.C. Osbourn, ''Strained-layer superlattices: a brief review'', _IEEE J. Quantum Electron_, QE-22: 1677 (1986).

90. A.C. Gossard, P.M. Petroff, W. Weigmann, R. Dingle, and S. Sav-age, ''Epitaxial structures with alternate-atomic-layer composi-tion modulation,'' _Appl. Phys. Lett._ 29: 323 (1976).

91. E. Finkman, M. D. Sturge, M-H. Meynadier, R.E. Nahary, M.C. Tamargo, D.M. Hwang and C.C. Chang, ''Optical properties and band structure of short-period GaAs/AlAs superlattices'', _J. Lum._ 39: 57 (1987).

92. T. Nakazawa, H. Fujimoto, K. Imanishi, K. Taniguchi, C. Hamaguchi, S. Hiyamizu and S. Sasa, ''Photoreflectance and photoluminescence study of $(GaAs)_m/(AlAs)_5(m = 3 - 11)$ superlattices: direct and indi-rect transition'', _J. Phys. Soc. Jap._, 58: 2192 (1989).

93. R. Cingolani, M. Holtz, R. Muralidharan, K. Ploog, K. Reiman and K. Syassen, ''Type I-type II transition in ultra short period

GaAs/AlAs superlattices revealed by luminescence under high-excitation intensity and high-pressure'', (to be published).

94. E. Kasper, H. J. Herzog and H. Kibbel, ''A one-dimensional SiGe superlattice grown by UHV epitaxy,'' <u>Appl. Phys.</u> 8: 199 (1975).

95. H. M. Manasevit, I. S. Gergis, and A. B. Jones, ''Electron mobility enhancement in epitaxial multilayer $Si - Si_{1-x} GE_x$ alloy films on (100) Si,'' <u>Appl. Phys. Lett.</u> 41: 464 (1982).

96. J.C. Bean, L.C. Feldman, A.T. Fiory, S. Nakahara, and J.D. Robinson, '' Ge_xSi_{1-x}/Si strained-layer superlattice grown by molecular beam epitaxy,'' <u>J. Vac. Sci. Technol.</u> A2, 436 (1984).

97. R. Zachai, E. Friess, G. Abstreiter, E. Kasper and H. Kibbel, ''Band structure and optical properties of strain symmetrized short period Si/Ge superlattices on Si (100) substrates'', <u>Proc. of the 19th Int. Conf. on the Physics of Semiconductors</u>, Warsaw, Poland, 1988, (Institute of Physics, Polish Academy of Sciences) p. 487;
G. Abstreiter, K. Eberl, E. Friess, W. Wegscheider and R. Zachai, ''Silicon/Germanium strained layer superlattices'', <u>J. Cryst. Growth</u> 95: 431 (1989).

98. U. Gnutzmann and K. Clausecker, ''Theory of direct optical transitions in an optical indirect semiconductor with a superlattice structure'', <u>Appl. Phys</u> 3: 9 (1974).

99. H.C. Liu, D. Landheer, M. Buchanan and D.C. Houghton, ''Resonant tunneling in $Si/Si_{1-x}Ge_x$ double-barrier structures'' <u>Appl. Phys. Lett.</u> 52: 1809 (1988).

100. S.S. Rhee, J.S. Park, R.P.G. Karunasiri, Q. Ye and K.L. Wang, ''Resonant tunneling through a $Si/Ge_xSi_{1-x}/Si$ heterostructure on a GeSi buffer layer'' <u>Appl. Phys. Lett.</u> 53: 204 (1988).

101. A.V. Nurmikko, R.L. Gunshor and L.A. Kolodziejski, ''Optical Properties of CdTe/CdMnTe multiple quantum wells,'' <u>IEEE J. Quantum Electron.</u> QE-22: 1785 (1986).

102. L.L. Chang, ''CdTe-CdMnTe superlattices'', <u>Superlatt. and Microstruct.</u> 6: 39 (1989).

THE QUANTUM HALL EFFECT AND RELATED PROBLEMS

Klaus v. Klitzing

Max-Planck-Institut für Festkörperforschung
Heisenbergstr. 1, D-7000 Stuttgart 80, FRG

INTRODUCTION

10 years after the discovery of the Quantum Hall Effect (QHE) the number of publications related to this phenomenon still increases. At present about 250 papers per year are published in this field and it is expected that this number increases further due to the application of the QHE in metrology and the discovery of new phenomena like ballistic transport in one-dimensional systems[1,2] and new phenomena in the field of the Fractional Quantum Hall Effect.[3,4]

Review books are available for both the Fractional Quantum Hall Effect[5] and the (Integer) Quantum Hall Effect[6] and a tutorial introduction into the physics and application of the Quantum Hall Effect has been published recently.[7]

Some new ideas in the interpretation of the QHE are based on a theory which relates the resistance between two reservoirs to transmission probabilities like in electron waveguides. This will be discussed in more detail by Büttiker in this book. In the following, I will focus on recent developments with respect to t he application of the QHE in metrology but will start with a short summary of the published ideas used for an interpretation of the Quantum Hall Effect.

QUANTUM HALL EFFECT

Starting from the classical expression for the Hall effect which relates the Hall voltage U_H of a metallic conductor with the threedimensional carrierdensity n_{3d}, the strength of the magnetic field B, and the current I through the conductor one ob tains (e=elementary charge)

$$U_H = \frac{B \cdot I}{n_{3d} \cdot d \cdot e} \tag{1}$$

which reduced to

$$U_H = \frac{B \cdot I}{n_s \cdot e} \tag{2}$$

if instead of the product of the threedimensional carrier density with the thickness d of the sample a twodimensional carrierdensity n_s is introduced. Two-dimensional electro-

Electronic Properties of Multilayers and Low-Dimensional Semiconductors Structures
Edited by J. M. Chamberlain *et al.*, Plenum Press, New York, 1990

nic systems are well known in semiconductor physics.[8] The simplest explanation of the QHE uses an ideal two-dimensional electronic system in strong magnetic fields at zero temperature where the energy spectrum consists of discrete energy levels (Landau levels). The energy spectrum itself, depends in a complicated way on different parameter (effective mass, g-factor, valley splitting) but the degeneracy N_L of each level is independent of the material and is given as

$$N_L = \frac{e \cdot B}{h} \tag{3}$$

which corresponds to the number of flux quanta per unit area.

If an integer number i of energy levels is fully occupied, i.e.

$$n_s = i \cdot N_L = i \cdot \frac{e \cdot B}{h} \tag{4}$$

then the Hall resistance $R_H = \frac{U_H}{I}$ becomes quantized:

$$R_H = \frac{h}{ie^2} \tag{5}$$

This result seems to be correct if one compares the mean value of the measured quantized Hall resistance of $R_H = (25812, 807 \pm 0, 005)\Omega$ with the recommended value for $\frac{h}{e^2} = (25812, 8056 \pm 0, 0012)\Omega$.

However, the simple explanation of the QHE on the basis of Egs (1)-(5) is not realistic and not useful for a microscopic interpretation of the quantum phenomenon. Especially the experimentally observed appearance of Hall plateaus can be explained only if one assumes that the carrier density n_s changes linearly with the magnetic field. Some theories use a reservoir outside the electronic system[9] (interface - or depletion charge, contacts, donors in the barrier of heterostructures) in order to stabilize the ratio $\frac{B}{n_s}$ at the value $\frac{h}{e}$. In principle, these reservoirs for electrons are able to produce Hall plateaus but for a quantitative explanation of the experiments localized states within the two- dimensional system seems to be necessary. Localization of electrons in strong magnetic fields is one of the topics of theories related to the QHE.[10,11,12,13] This localization is quite different from the localization without magnetic field since in the limit of very strong magnetic fields electrons move on equipotential lines, so that any closed equipotential line in a disordered system corresponds to a localized electronic state. Within the percolation picture[14,15,16] an infinitely large two-dimensional system has extended states (open equipotential lines) at exactly half-filled Landau levels so that only under this conditions energy dissipation is possible at zero temperature. This explains the experimental observation that the Shubnikov-de Haas oscillations show δ-like peaks in the resistivityρ_{xx} as a function of the magnetic field (Fig. 1). In the magnetic field regions where ρ_{xx} vanished, the Fermi energy is pinned in the tails of the Landau levels and quantized values for the Hall r esistance are observed. Measuremens[17,18,19] and calculations[20] of the density of states show that for the devices used in the experiments, no real energy gaps exist in the energy spectrum so that variable range hopping and at higher temperatures thermal excitation into extended states[21] leads always to a finite conductivity σ_{xx} (and therefore finite resistivity ρ_{xx}) at nonzero temperature. Laughlin[22] discussed the existence of the quantized Hall resistance in a more general form based on gauge invariance arguments, however this theory is not useful if energy dissipation is present. This means that no theory is available which predicts quantitatively the deviation of the Hallplateaus from the

26

quantized value for measurements on sample with finite size, a currentflow through the device of some μA and a nonvanishing energy dissipation due to the current injection and the finite resistive ρ_{xx}. However, for the application of the QHE as a resistance standard with an accuracy of better than 10^8 some guidelines have been published[23] which allow high precisions measurements of the quantized Hall resistance even if the experimental conditions are not ideal. These guidelines will be discussed in the next chapter in connection with the application of the QHE in metrology.

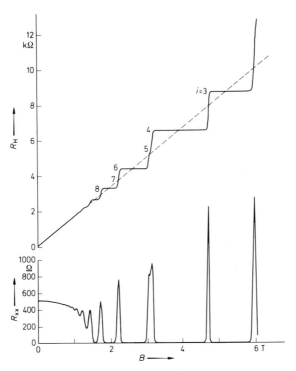

Fig.1. Experimental result for the Hall-resistance R_H and the longitudinance $R_{xx} \sim \rho_{xx}$ of a GaAs-AlGaAs heterostructure as a function of the magnetic field at $T=8mK$.

All experimentalists working in the field of the quantized Hall resistance know, that the topology of the sample is very important. In the quantum Hall regime the two-terminal resistance of a Corbino disc is infinite whereas the same measurement on a Hall device gives the quantized resistance value. It is extremely important whether the electrical contacts in a two-terminal measurement are located at a common edge or not. This result can be used as an indication, that the current flow is exclusively along the edges of the device corresponding to skipping orbits of the cyclotron motion of electrons in a magnetic field. All experiments can be explained if such edge currents are assumed but an accurate determination of the current distribution is not available up to now, both from the experimental and from the theoretical point of view. Recent experiments show that due to edge currents a description of the magnetotransport properties by tensor components is inadequate and also measurements of the quantum transport across a barrier can be nicely explained if edge channels are assumed.[24,25,26]

Despite the fact that a microscopic picture of the QHE is not available, the application of the quantized Hall resistance as a resistance standard has been accepted internationally. This decision is a result of the experimental fact that quantized Hall resistance measurements at different laboratories with different devices from different materials yield within the experimental uncertainty of about $2 \cdot 10^{-7}$ the same resistance value. Fig. 2 shows a summary of these results obtained at national laboratories in Australia (CSIRO), Japan (G.U. and ETL), France (LCIE and BIPM), USA (NBS), England (NPL), UdSSR (IMS), and China (NIM).

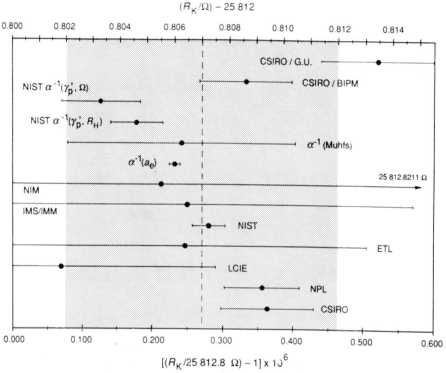

Fig.2. Comparison of the recommended value or R_K (vertical dashed line) and its standard deviation uncertainty (delimited by the shading) with the values of R_K and their standard deviation measured by different laboratories.

APPLICATION OF THE QHE IN METROLOGY

Starting from 1.1.1990 all calibrations of resistance will be based on the quantum Hall effect. A fixed value for the quantized Hall resistance (vertical dashed line in Fig. 2) with the value $R_{K-90} = 25812,807\Omega$ (von Klitzing constant) will be used. This decision is a result of the recommendations adopted by the Comité International des Poids et Mesures at its 77th Meeting (October 4-6, 1988). These recommendations are summarized in Appendix 1.

The technical guidelines for reliable measurements of the quantized Hall resistance include mainly the following points:

1. Silicon MOSFETs and III-V-heterostructures (mainly GaAs-AlGaAs) are the most suitable two-dimensional systems for thigh precision measurements of the QHE. Uncertainties smaller than 10^{-8} should be possible for measurements below T=1.5K, magnetic fields in the range of 10 Tesla and source-drain currents of about 20μA. In addition the device should fulfil the following conditions

2. The series resistance of the electrical contacts should be so small that all two-terminal resistances agree with the quantized value within $1°/_{oo}$. A special design of the source and drain contacts can be used to reduce the intrinsic energy dissipation at the contacts[27]

3. The resistivity ρ_{xx} should be smaller than 0.5 mΩ under the condition of quantized Hall resistance measurements. Extrapolation to $\rho_{xx} = 0$ should be used if the Hall resistance changes with ρ_{xx}

The extrapolation to vanishing resistivity ρ_{xx} seems to eliminate all known corrections to the quantized Hall resistance since the expected corrections $\Delta\rho_{xy}$ due to the finite size of the sample, finite temptc. is automatically combin ed with a finite ρ_{xx} and a relation $\Delta\rho_{xy} = S \cdot \rho_{xx}^{min}$ is found experimentally.[28] The coefficient S is a very complicated function of microscopic details of the two-dimensional electronic system and reliable calc ulations are not available.

If one believes that the measured value for the quantized Hall resistance R_K agrees with the fundamental constant $\frac{h}{e^2}$, one has the unsatisfactory situation that two different mean values for $\frac{h}{e^2}$ are recommended. The CODATA task group, which is responsible for the adjustment of fundamental constants, published on the basis of high precision measurements of the anomalous magnetic moment of the electron and the corresponding QED theory a value[29] $\frac{h}{e^2} = 25812,8056\Omega$ as the best value, whereas a value $R_K = 25812,807\Omega$ is fo und from QHE experiments. If the difference between these values is real one has to find out whether the QED theory is not accurate enough or whether corrections are necessary if one identifies measured quantized Hall resistance values with the fundamental constant $\frac{h}{e^2}$. However, independent of the development of high precision measurements and new calculations in the field of the Quantum Hall Effect, the quantized resistance value $R_K = 25812,807\Omega$ will play a fundamental role in metrology - at least for the next 20 years.

APPENDIX

Recommendations adopted by the Comité International des Poids et Mesures at its 77th Meeting (October 4-6, 1988)

Representation of the ohm by means of the quantum Hall effect
Recommendation 2 (CI-1988)

The Comité International des Poids et Mesures,

acting in accordance with instructions given in Resolution 6 of the 18th Conference Generale des Poids et Mesures concerning the forthcoming adjustment of the representations of the volt and the ohm,

considering
- that most existing laboratory reference standards of resistance change significantly with time,
- that a laboratory reference standard of resistance based on the quantum Hall effect would be stable and reprodubile,
- that a detailed study of the results of the most recent determinations leads to a value of 25 812,807Ω for the von Klitzing constand, R_K, that is to say, for the quotient of the Hall potential difference divided by current corresponding to the plateau $i = 1$ in the quantum Hall effect,
- that the quantum Hall effect, together with this value of R_K, can be used to establish a reference standard of resistance having a one-standard-deviation uncertainty with respect to the ohm estimated to be 2 parts in 10^7, and a reproducibility which is significantly better,

recommends
- that 25 812,807Ω exactly be adopted as a conventional value, denoted by R_{K-90}, for the von Klitzing constant, R_K,
- that this value be used from 1st January 1990, and not before, by all laboratories which base their measurements of resistance on the quantum Hall effect,
- that from this same date all other laboratories adjust the value of their laboratory reference standards to agree with R_{K-90},
- that in the use of the quantum Hall effect to establish a laboratory reference standard of resistance, laboratories follow the most recent edition of the technical guidelines for reliable measurements of the quantized Hall resistance drawn up by the Comité Consultatif d'Electricité and published by the Bureau International des Poids et Mesures,

and is of the opinion
- that no change in this recommended value of the von Klitzing constant will be necessary in the foreseeable future.

REFERENCES

[1] B.J. van Wees, H. van Houten, C.W.J. Beenakker, J.G. Williamson, L.P, Kouwenhoven, D. van der Marel and C.T. Foxon, Phys. Rev. Lett. 60, 848 (1988)

[2] D.A. Wharam, T.J. Thornton, R. Newbury, M. Pepper, H. Ahmed, J.E.F. Frost, D.G. Hasko, D.C. Peacock, D.A. Ritchie and G.A.C. Jones, J.Phys. C: Solid State Phys. 21, L209 (1988)

[3] R.G. Clark, S.R. Haynes, A.M. Suckling, J.R. Mallett, J.J. Harris, and C.T. Foxon, Phys. Rev. Letters 62 (1989) 1536

[4] J.P. Eisenstein, H.L. Störmer, L. Pfeiffer, and K.W. West, Phys. Rev. Letters 62 (1989) 1540

[5] Tapash Chakraborty and P. Pietiläinen, The Fractional Quantum Hall Effect (Springer, Heidelberg, 1988)

[6] R.E. Prange, S.M. Girvin (eds.): The Quantum Hall Effect (Springer, New York, Berlin, Heidelberg 1987)

[7] K. v. Klitzing, NATO ASI Series, Series B, Physics, vol. 170, p. 229 (1987)

[8] T. Ando, A.B. Fowler, and F. Stern, Rev. Mod. Phys. 54, 437 (1982)

[9] G.A. Baraff, and D.C. Tsui, Phys. Rev. B24, 2274 (1981)

[10] R.E. Prange, Phys. Rev. B23, 4802 (1981)

[11] H. Aoki and T. Ando, Solid State Commun. 38, 1079 (1981)

[12] W. Brenig, Z. Phys. B50, 305 (1983)

[13] J.T. Chalker, J. Phys. C16, 4297 (1983)

[14] S.V. Iordansky, Solid State Commun. 43, 1 (1982)

[15] R.F. Kazarinov and S. Luryi, Phys. Rev. B25, 7626 (1982)

[16] S.A. Trugman, Phys. Rev. B27, 7529 (1983)

[17] D. Weiss and K. von Klitzing in: High Magnetic Fields in Semiconductor Physics, ed. G. Landwehr, Vol. 71, Springer Series in Solid State Sciences (Berlin 1987), p. 57

[18] E. Gornik, R. Lassnig, G. Strasser, H.L. Störmer, A.C. Gossard, W. Wiegmann: Phys. Rev. Lett. 54, 1820 (1985)

[19] J.P. Eisenstein, H.L. Störmer, V. Narayanamurti, A.Y. Cho, A.C. Gossard, Phys. Rev. Lett. 55, 875 (1985)

[20] R.R. Gerhardts, V. Gudmundsson and U. Wulf Proc. 19th Int. Conf. on the Physics of Semiconductors, Warsaw 1988, Institute of Physics, Polish Academy of Science p. 205

[21] K. von Klitzing, G. Ebert, N. Kleinmichel, H. Obloh, G. Dorda and G. Weimann, Proc. 17th Int. Conf. on the Physics of Semiconductors, San Francisco 1984, Springer-Verlag New York, Inc (1985), p. 271

[22] R.R. Laughlin, Phys. Rev. B23, 5632 (1981)

[23] Technical guidelines for reliable measurements of the quantized Hall resistance. Document of the working group of the Comité Consultatif d'Electricité on the Quantum Hall Effect, ed. F. Delahaye, Bureau International des Poids et Mesures, Pavillon de Bretenil, F-92312 Sévres, Cedex

[24] R.J. Haug, A.H. MacDonald, P. Streda, and K. von Klitzing, Phys. Rev. Lett. 61, 2797 (1988)

[25] S. Washburn, A.B. Fowler, H. Schmid, and D. Kern, Phys. Rev. Lett. 61,2801 (1988)

[26] H. Hirai, S. Komiyama, S. Hiyamizu and S. Sasa Proc. 19th Int. Conf. Physics of Semiconductors, Warsaw 1988, Institute of Physics, Polish Academy of Sciences, p. 55

[27] D. Dominguez, K. von Klitzing, and K. Ploog, Metrologia (1988)

[28] M.E. Cage, B.F. Field, R.F. Dzinba, S.M. Girvin, A.C. Gossard, and D.C. Tsui, Phys. Rev. B30, 2286 (1984)

[29] E.R. Cohen and B.N. Taylor, Codata Bulletin 63, 1 (1986)

HIGH MAGNETIC FIELDS AND LOW DIMENSIONAL STRUCTURES:

A SURVEY OF TRANSPORT AND OPTICAL EFFECTS

G. Landwehr

Physikalisches Institut der Universität Würzburg
Am Hubland, D-8700 Würzburg, F. R. Germany

INTRODUCTION

In 1988 an international conference with the title "The Application of High Magnetic Fields in Semiconductor Physics" was held in Würzburg. The first meeting of this series was organized in Würzburg in 1972. At that time, only a few lectures were concerned with two-dimensional systems. The situation has changed, however, in the meantime. At the last conference, the majority of the invited talks and the contributed papers concerned 2D systems. As a matter of fact, high magnetic fields have become an indispensable tool for the research on 2D systems. This is also reflected by the programs of the recent international conferences which were devoted towards the investigation of the electronic structure of 2D systems.

There are several reasons for these developments. Two-dimensional systems can be characterized by the confinement of charge carriers to narrow potential wells. This results in boundary quantization effects. The carriers cannot move freely perpendicular to the surface or the interface so that standing wave patterns arise which lead to electric subbands. The application of high magnetic fields perpendicular to the interface leads to Landau quantization of the system. Due to the angular momentum quantization of the motion in the conducting layer only discrete energies are possible and the density of states consists of a series of more or less sharp peaks which are separated by an energy gap amounting to the cyclotron energy $\hbar\omega$ where $\hbar$ = Planck's constant/2π and ω is the cyclotron frequency eB/m. If one compares the density of states in three and two dimensions, one finds that there is no continuous background density of states in two dimensions. This can result in rather large quantum effects in the transport properties.

In order to observe quantum oscillations in the magneto resistance, the electronic system has to be degenerate in the sense that the Fermi level is located well inside the conduc-

Electronic Properties of Multilayers and Low-Dimensional Semiconductors Structures
Edited by J. M. Chamberlain *et al.*, Plenum Press, New York, 1990

33

tion or valence band. To achieve the necessary carrier concentration in bulk material rather heavy doping is necessary. This, however, results in a comparatively low mobility. In order to observe pronounced quantum effects in high magnetic fields, it is necessary that the condition $\omega\tau > 1$ holds, where τ is the carrier relaxation time. This is equivalent to the expression $\mu B > 1$ where μ stands for the carrier mobility.

In 2D devices like silicon metal-oxide semiconductor field-effect transistors (MOSFETs) the free carriers in the channel are induced by a surface capacitor in rather pure silicon. This has the effect that the ionized impurity scattering of the carriers at helium temperatures can be kept small so that high mobilities arise. For n-channel MOSFETs electron mobilities as high as 40000 cm²/Vs have been achieved. Extremely high mobilities of more than $3 \cdot 10^6$ cm²/Vs have been realized in modulation doped GaAs-(GaAl)As heterostructures. This is possible because the donor impurities are located in a heavily doped (GaAl)As-layer adjacent to the GaAs sheet of interest again with the consequence that the impurity scattering is small. Such high mobilities lead to very sharp Landau levels in magnetic fields of the order of 10 Tesla (which can easily be produced with superconducting magnets) resulting in readily observable quantum effects.

An effect which has been frequently studied in 2D heterostructres is the Shubnikov-de Haas effect, that is quantum oscillations in the magneto resistance. Subsequently it will be explained what kind of bandstructure information may be obtained from the study of the Shubnikov-de Haas effect. In the course of studying the magneto transport properties of silicon inversion layers the Quantum Hall Effect was discovered (K. von Klitzing 1980). It turned out that in high quality samples plateaus in the Hall resistance showed up which were quantized in whole fractions of h/e^2 where h = Planck's constant and e = electron charge. Subsequently, the fractional Quantum Hall Effect was discovered by Tsui and coworkers (1982). It became immediately clear that the Quantum Hall Effect has important implications for basic physics. Although in the meantime much work was devoted to experimental and theoretical studies of the integer and fractional Quantum Hall Effect (about one third of the papers of the 1988 Würzburg conference were dealing with the subject) our present understanding is still incomplete.

The Shubnikov-de Haas effect has not only been used to obtain band structure information, it has also been employed to study hot electron effects and scattering effects due to magnetic impurities. However, the scope of this article does not allow to treat these subjects in any detail.

It is well known that a wealth of information on semiconductors has been gained by the application of magneto-optical methods. This also holds for structures with reduced dimensionality. Various experimental methods have been used, which can be classified as interband and intraband techniques. The absorption of light increases strongly at band edges and also at Landau levels due to the enhanced density of states. Consequently, detailed band structure information can be obtained from these methods. Also, the emission of light after laser excitation has revealed a wealth of information about

band structure and impurities. The confinement of carriers to
quantum wells and heterostructures significantly influences
the optical spectra. Because the application of high magnetic
fields produces additional quantization they have been frequently used to assist in the detailed analysis of experimental data.

It is obvious that due to the wide scope of the field of
magneto transport and magneto optics in 2D systems it is
impossible to give a comprehensive review even for these
particular areas. An excellent review of the work up to 1980
has been written by Ando, Fowler and Stern (1982). An article
about quantum magnetotransport has been published by Hajdu and
Landwehr (1985). An excellent review on magnetospectroscopy of
confirmed semiconductor systems by Petrou and McCombe will be
published by 1990. Many good review articles on 2D-systems can
be found in the proceedings of the last two Würzburg conferences: Landwehr, ed. 1987 and 1989. Here it will be tried to
give a few typical examples which indicate the potential of
magneto-transport and -optics. The majority of the published
data concern a few materials and systems which can be considered as model systems. The first detailed information on a 2D
electronic system was obtained in silicon MOSFETs. Subsequently, with the development of new crystal growth techniques like
molecular beam epitaxy (MBE) high quality heterostructures
made of adjacent films of GaAs and (GaAl)As became available.
Most of the examples given will be for these systems. Recently, other III-V heterostructures like (InGa)As on InP have
been studied in detail by magneto-optical and magneto transport experiments. Also, II-VI heterostructures like HgTe/CdTe
superlattices have been investigated by these methods and
turned out to be interesting.

MAGNETO TRANSPORT PROPERTIES

The Shubnikov-de Haas effect is an oscillatory magneto
resistance as a function of magnetic field, provided the
conditions which are mentioned in the introduction are fulfil-
led. If the relation $\omega\tau > 1$ is obeyed a carrier performs
several revolutions perpendicular to the magnetic field before
it is scattered. The quasi-periodic motion causes energy and
momentum quantization. The density of states has maxima at the
Landau level energies $E = (N + \frac{1}{2})\hbar\omega$ with $\omega = eB/m$, N = Landau
quantum number. In this expression m is the cyclotron mass
which is the average of the effective mass along the orbit. If
the semiconductor is degenerate and the Fermi energy is E_F,
appreciable scattering occurs whenever a Landau level crosses
the Fermi level. Consequently, the quantum oscillations are
periodic in 1/B. The information which may be extracted immediately from the period $\Delta(1/B)$ is the carrier concentration n
given by the formula:

$$\Delta(1/B) = 2e/h \ (3\pi^2 n)^{-2/3}$$

The theoretical expression for the resistivity oscillations can be obtained from the review articles cited. (see,
eg. Hajdu and Landwehr (1985)). The 2D-formula differs somewhat from that which is valid in the 3D case in the dependence
of the amplitude on the magnetic field, due to differences in
the density of states in 2D and 3D. The temperature dependence

of the oscillations arises partly from the shape of the Fermi function around E_F for finite temperatures. The amplitude is also influenced by the broadening of the Landau levels which leads to the so called Dingle factor. It is a well established practice to derive the cyclotron mass from the temperature dependence of the amplitude of the quantum oscillations. Provided that the carrier mobility is of the order 10^4 cm²/Vs or higher, the effective mass can be determined with an accuracy of a few percent under the condition that only a single period is present.

Whenever carriers are confined in a potential well of a 2D structure discrete subbands arise. For semiconductors with a small effective mass the density of states is relatively low so that it takes several subbands to accomodate carrier concentrations of the order 10^{12}/cm². The spacing of the subbands can be of the order of 100 meV or larger. For energies of this magnitude nonparabolicities become usually important. This holds for narrow-gap materials like n-type HgTe as well as for n-type GaAs.

Recently it has become possible to prepare special GaAs heterostructures in which by MBE a single layer of silicon donors is introduced (Zrenner et al., 1984). By this technique called δ-doping a V-shaped potential well is generated. With electron concentrations of the order 10^{13}/cm² it takes four electric subbands to accomodate the carriers. The qualitative shape of the potential well and calculated subband levels are shown in Fig. 1 (Koch et al.). It has been possible to get quantitative information about the electric subbands by Shubnikov-de Haas experiments. In Fig. 2 the derivative of the magneto conductivity has been plotted as a function of the inverse magnetic field. In order to extract information about the oscillatory periods a Fourier analysis was made. From the insert in the upper part of Fig. 2 one can recognize immediately that four electric subbands are occupied. From the period of the oscillations the electron concentration in the different subbands can be determined rather accurately. It is obvious that it would be difficult to obtain this information with other methods.

It turns out that especially for p-type inversion layers the Shubnikov-de Haas effect is a very useful tool to extract information about electric subbands. The boundary quantization produced by the high electric field present at the interface lifts the degeneracy of light and heavy holes which is present at the top of the valence band of the elemental semiconductors Ge and Si and the III-V compounds. The light and heavy hole states become strongly mixed so that the usual distinction becomes meaningless. Under such circumstances it is not possible to infer the subband structure by simple extrapolation of the bulk properties. However, Shubnikov-de Haas experiments can give detailed insight. It is obvious that for the interpretation of experimental data it will always be necessary to have detailed theoretical calculations of the subband structure available. Such calculations have been performed by solving self-consistently Schrödinger's- and Poisson's equations (Stern and Howard, 1967), (Bangert et al., 1974), (Okhawa and Uemura, 1975).

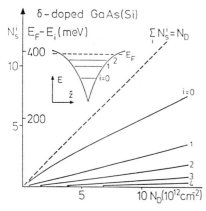

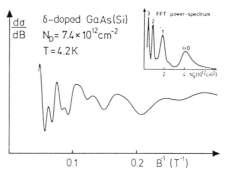

Fig. 1.
Calculated subband levels
and occupancies versus
donor density in a
δ-accumulation layer. A
parabolic conduction
band has been assumed.
After Koch 1987.

Fig. 2.
Shubnikov-de Haas oscillations
for a Si-doped layer with
$7.4 \cdot 10^{12}$ cm^{-2} donors. The
corresponding fast Fourier
transform (FFT) reveals four
individual subband contributions. After Koch 1987.

An example for a p-type subband structure with somewhat
unexpected properties can be found in a modulation doped p-
type GaAs-(GaAl)As heterostructure (Bangert and Landwehr,
1985). In this case the potential is roughly triangular as can
be seen from Fig. 3. The energy versus wave vector relation
has also been plotted together with the constant energy contours. It turns out that the Kramers degeneracy is lifted
for finite wave vectors and that two subbands are present
which can be attributed to two branches of the heavy-hole
band. The effective mass connected with the lighter branch is
rather low and leads to mobilities up to $2 \cdot 10^5$ cm^2/Vs in the
mK range. Consequently very pronounced quantum oscillations
can be observed at low temperatures. This can be seen in Fig.
5 in which both the transverse magneto resistance and the Hall
resistance have been plotted. Very well developed Quantum Hall
plateaus can be recognized. At higher magnetic fields it was
possible to observe the fractional Quantum Hall effect at mK
temperatures (Mendez 1986), (Reményi et al., 1986). In Fig. 5
one can see at magnetic fields below 1 T pronounced beating
effects. Detailed analysis shows that certain Shubnikov-de
Haas maxima are missing. The observations can be quantitatively explained by the theoretical calculations as will be outlined subsequently.

One point that should be mentioned here is that it is
necessary to take changes of the band structure into account
which are introduced by the application of a high magnetic
field. Usually it is anticipated in semiconductor problems
that application of a strong magnetic field results in a
splitting of a band into magnetic subbands due to Landau quantization which no change of the band parameters. However, this

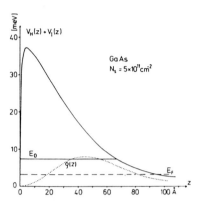

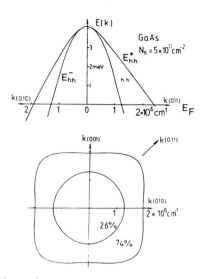

Fig. 3.
Hartree- and image poten-
tial of a (GaAl)As-GaAs
p-type heterojunction for a
hole concentration of
$5 \cdot 10^{11}/cm^2$. E_F = Fermi
energy. The dotted line gives
the charge distribution.
After Bangert and Landwehr
1985.

Fig. 4.
Top: Energy vs wave vector
relation for the heavy hole
subband in (100) GaAs for the
[110] and [010] directions.
Bottom: Constant energy contours with occupation percentage for the split heavy hole
subband. After Bangert and
Landwehr 1985.

approximation no longer holds for 2D structures for which the
subband splitting is comparable to the cyclotron energy and
especially for quantized p-type layers. In order to obtain
realistic predictions for electric subbands in high magnetic
fields one has to incorporate the magnetic field in the self-
consistent solution of Schrödinger's and Poisson's equation
making use of the Luttinger 4x4 matrix representation of the
valence band of GaAs (Bangert and Landwehr, 1986), (Broido and
Sham, 1985, Sham 1987). It is necessary to take the anisotropy of the valence bands into account otherwise no agreement
between theory and experiment can be obtained. The resulting
Landau level scheme is quite complicated as can be seen in
Fig. 6. The calculations were done for a hole concentration of
$2.3 \times 10^{11}/cm^2$ for T = 0 K. Two series of Landau levels can be
distinguished, the a-series can' be derived from the heavy
hole band with small curvature, and the b-series from the band
with the smaller effective mass at k = 0. The dashed lines
refer to calculations without inclusion of the anisotropy in
the interface plane. At low magnetic fields the Landau levels
belonging to the a- and b-series frequently cross. This can be
seen in the lower part of the figure, where the Landau levels
between 0.3 and 2.1 T have been plotted. Whenever the Fermi-
level passes such a crossing point, a Shubnikov-de Haas-mini-
mum should be missing. This gives rise to the beating effects
in the oscillatory magnetoresistance which can be recognized
in the lower field range of Fig. 5. In order to explain the
data quantitatively, it is necessary to calculate the density
of states and the Fermi-energy as a function of the magnetic
field. The results of such calculations (Fig. 7), which take
into account realistic level broadening, the Shubnikov-de Haas
data can be explained quantitatively.

38

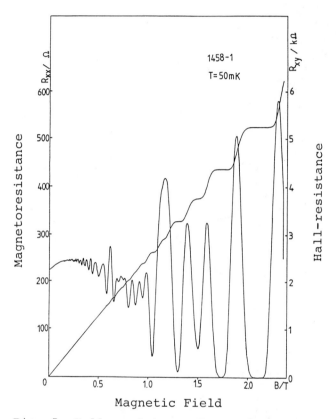

Fig. 5. Hall resistance R_{xy} and magnetoresistance R_{xx} of a p-type heterostructure at T = 50 mK.
After Reményi et al. (1986).

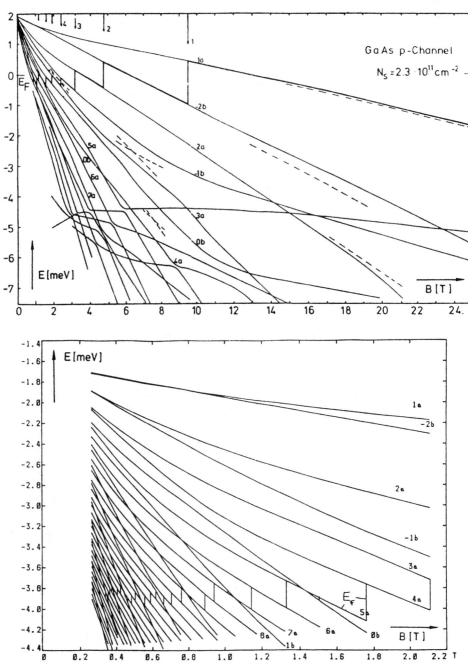

Fig. 6a (top) and 6b (bottom).
 a) Landau-levels and Fermi energy for a hetero-
    structure as a function of magnetic fields up
    to 25 T
 b) Details of the same calculations on an enlarged
    scale between 0.3 T and 2.1 T
    After Bangert and Landwehr 1986.

Recently the validity of the approach was confirmed by
Shubnikov-de Haas experiments on p-type GaAs-(GaAl)As heterojunctions in a wider hole concentration range between 1.5 x
10^{11} and 1.2 x 10^{12}/cm² (W. Heuring et al., 1989). Again, it
became clear that inclusion of the warping was necessary in
order to obtain agreement between theory and experiment. In
order to investigate the Landau levels in high magnetic fields
the transport experiments were performed in tilted magnetic
fields. It is well known from n-type heterostructures that a
magnetic field parallel to the interface plane enhances the
spin splitting. In the p-type case, however, there is a strong
mixing between the various levels and only those which have a
well-defined spin character are shifted by a parallel magnetic
field. This has been observed in experiments where only particular oscillations are affected by the longitudinal component
of the magnetic field. In order to make a comparison between
theory and experiment it was necessary to perform detailed
calculations. The obtained agreement was excellent. In Fig. 8
the relative change between Landau levels at a perpendicular
magnetic field of 5.4 T corresponding to a filling factor 2
has been plotted as a function of the parallel magnetic field
component. The good agreement between theory and experiment
indicates that the Landau level scheme is realistic.

MAGNETO-OPTICAL EFFECTS

Magneto-optical experiments have given very detailed
band structure information on many bulk semiconductors. All
the common techniques like magneto-interband absorption and -
reflection, magneto-photoconductivity, magneto-luminescence
and cyclotron resonance have been successfully applied to
superlattices and single heterostructures. The study of impurity effects with far-infrared techniques in magnetic fields
has given valuable information on the distribution of impurities in quantum wells. (See, e.g. Petrou and McCombe, 1990).
The confinement of carriers to narrow quantum wells leads to
new effects, which naturally are reflected in the optical
spectra. High magnetic fields have turned out to be a very
useful tool in modifying the structure of the electric subbands at will. It became clear, however, that through the
spatial confinement of carriers in quantum wells the valence
band properties get much more complicated than in bulk material. It should be remembered that excitons in high magnetic
fields pose already a serious problem in bulk material. The
calculation of the hydrogen-like bound states is by no means
trivial. When an exciton is confined in a quantum well with a
width which is smaller than the Bohr radius the computational
difficulties rise substantially. This is the reason why the
problem was addressed in its full complexity only recently
(Bauer and Ando, 1988). Studies of the excitonic luminescence
in high magnetic fields revealed that certain lines correspon-
and valence bands in quantum wells, it is no longer sufficient
to rely on the calculation of Landau levels. This is so because high magnetic fields change the mixing of the m = ±3/2 and
1/2 valence band states so much that the electric dipole
matrix elements which are a measure of the optical transition

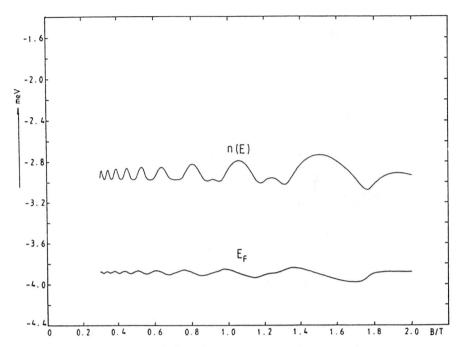

Fig. 7. Density of states at the Fermi
level for a p-type heterostructure
(arb. units) and Fermi energy as
function of B (E. Bangert).

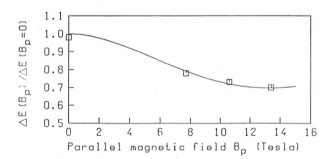

Fig. 8. Relative change of the energy difference between
Landau levels at a perpendicular magnetic field
corresponding to filling factor 2 due to an additional parallel magnetic field B_p. The hard line
gives results from calculations, the squares indicate experimental data.
W. Heuring et al. (1989).

probability are drastically influenced. Recent data on the excitonic luminescence in GaAs-(GaAl)As multiple quantum wells which demonstrate these points will be discussed subsequently. Also, cyclotron resonance experiments on p-type single hetero-structures will be treated. They demonstrate that in spite of the relatively complex Landau states considerable information of the system is available by now.

EXCITONS IN GaAs QUANTUM WELLS

In GaAs-(GaAl)As quantum wells at helium temperatures photo-excited electrons and holes form excitons which can in principle be described by a hydrogen-like system. Due to the removal of the valence band degeneracy at k = 0 there are two types of excitons which are usually called heavy and light hole excitons. This term refers to the effective masses in the direction of the quantum confinement. Excitons can move either freely in the plane of the 2D structure or they can be bound to impurities. Because transitions from the conduction band to impurity states and conduction- to valence band transitions are possible, photoluminescence spectra are usually very rich in structure. It turns out that high magnetic fields are very useful in distinguishing between Landau- and excitonic transitions. A typical magneto-luminescence spectrum (obtained at low excitation) of a multi-quantum well GaAs-(GaAl)As structure is shown in Fig. 9. The width of the quantum wells is 180 A and the magnetic field was oriented perpendicular to the growth direction of the sample. The experimental data were obtained by Ossau et al. (1986) and have been indicated by circles. The solid lines are theoretical curves as calculated by Bauer and Ando (1988). Not only the splitting of the light (l_1 (1s)) and heavy (h_1 (1s)) hole exciton states are correctly predicted, but also the excited states and their shift in energy with magnetic field. The rich structure observed in magneto-optical experiments on high quality quantum-well samples obviously requires detailed theoretical calculations for interpretation.

Two spectra obtained for GaAs-(GaAl)As samples with a quantum well width of 68 A is shown in Fig. 10 for two circu-lar polarizations of the exciting laser light. At first sight the data resemble a simple Landau fan. A closer inspection reveals, however, significant differences. For σ^+-polarization around a magnetic field of 4 T the transitions with the Landau quantum number n = 4 and n = 5 disappear and only a single transition is observed at higher magnetic fields in the corre-sponding energy range. No effect of this kind was observed for σ^- polarization.

These results clearly indicate that for the explanation of magneto-optical data in quantum wells it is necessary to calculate the dipole matrix elements for optical transitions because these change significantly with the quantum number. In Fig. 11 the Landau levels for the light and heavy hole states are shown for a quantum well sample with a well width of 80 A.

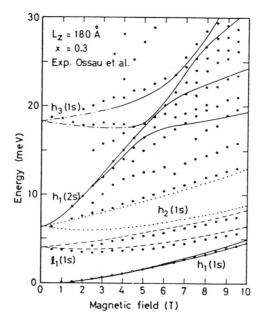

Fig. 9.
Photonenergies as a function of the magnetic
field for L_z = 18 nm. The solid lines represent
the calculations by Bauer and Ando (1988).
After Ossau 1986.

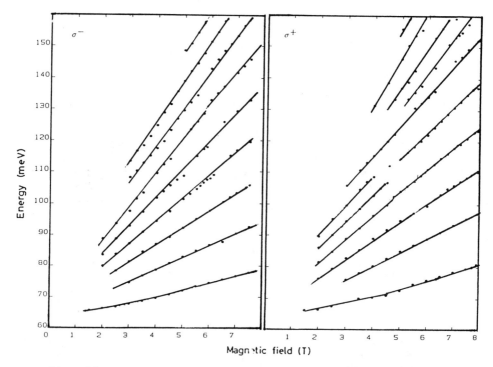

Fig. 10.
Luminescence Excitation Spectra for σ^+- and σ^--pola-
rization as a function of B for a 68 A quantum well
sample (Jäkel 1989).

One recognizes a quite complicated scheme for magnetic fields beyond 3 T with different high field behavior for the two polarization modes. The calculations alone can obviously not assist much in interpreting the data. It turns out however, that calculation of the matrix elements for optical dipole transitions is helpful. In Fig. 12 the square of matrix elements $|p_{if}|^2$ for allowed and forbidden transitions have been plotted as a function of the Landau quantum number N of the conduction band at B = 6 T. The quantum number of the conduction band has been chosen for the abscissa because the largest part of the total energy shift is caused by the conduction band due to the small electron effective mass. In the figure optical transitions have been plotted which lead to the same final state, for the initial states with the magnetic quantum numbers +3/2, +1/2, -1/2, -3/2.

It is obvious that the dipole matrix elements of all allowed transitions decrease with increasing Landau quantum number N. For the forbidden transitions $|p_{if}|^2$ increases continuously and can decrease at higher quantum numbers. The decrease can be explained by the growing interaction with Landau levels belonging to excited subbands. The transition from the +1/2 valence band states is very weak for N>3. The strong mixing of this state causes an intense forbidden transition, the intensity of which exceeds that of all allowed transitions in the σ^+-polarization.

Already at small Landau quantum numbers N the allowed transitions of the light hole systems decrease in intensity. Consequently, one cannot expect a pronounced Landau fan in the excitation spectra.

The decrease of the matrix elements of the allowed transitions around 4 T and the increase of the matrix elements (especially of the +1/2-state) in this magnetic field range explain the observed particular features in the excitation spectra show in Fig. 8 very well. It is obvious, that the distinction between allowed and forbidden transitions has to be considered with care in high magnetic fields and that special caution is necessary in magnetic fields of "medium" strength. It should be pointed out that the features discussed here are especially relevant for quantum wells with not too large width L. For quantum wells with L = 180 A the mixing of the different hole states is less pronounced.

CYCLOTRON RESONANCE IN N- and P-TYPE GaAs-(GaAl)As HETERO-STRUCTURES

The complex subband structure of the valence bands in a p-type GaAs-(GaAl)As heterostructure is reflected in a multi-line cyclotron resonance spectrum. It indicates immediately that the classification in light and heavy holes has lost its meaning. One has to discuss the results in terms of Landau transitions rather than effective masses attributed to the different lines. A typical spectrum obtained with a laser spectrometer at far-infrared frequencies is shown in Fig. 13 (Erhardt et al., 1986). Four lines are well resolved. The Landau level diagram shown in Fig. 5 corresponds to the carrier concentration of the sample on which the experiment was made. Again, it was necessary to calculate matrix elements for

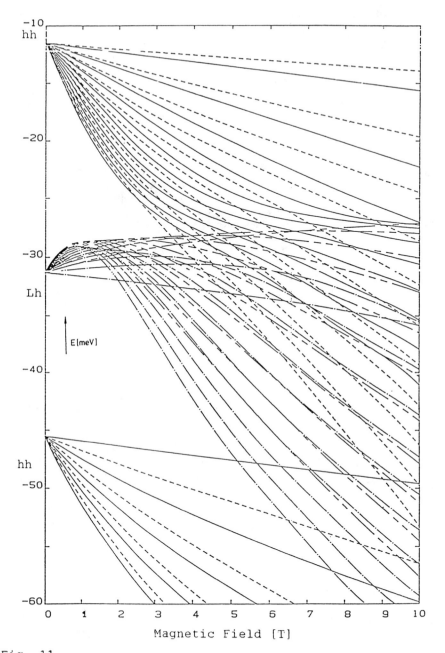

Fig. 11.
Landau levels for light and heavy holes for a 80 Å quantum
well sample (Jäkel 1989). Solid lines: σ^+-polarization;
broken lines: σ^--polarization.

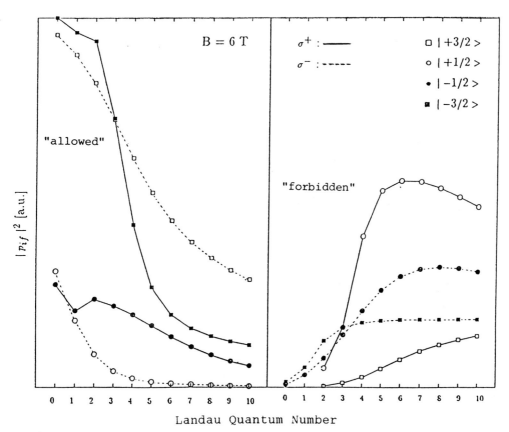

Fig. 12.
Matrix element $|p_{if}|^2$ as function of the Landau quantum number at B = 6 T for "allowed" and "forbidden" transitions (Jäkel 1989).

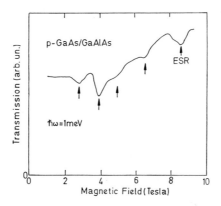

Fig. 13.
Cyclotron resonance at two fixed frequencies for a p-type heterostructure. After Erhardt et al.(1986)

47

the optical transitions in the Faraday configuration (Bangert and Landwehr, 1986). It turns out that the strongest experimentally observed transitions are correctly predicted by the theoretical calculations. However, several transitions which have relatively large matrix elements have not been observed so far. It could be that interference effects in the samples are responsable for this, because it is known that in spite of the small thickness of the conducting layers in heterostructures interference can play a role (von Ortenberg 1975).

Of interest is the dependence of the cyclotron mass on the magnetic field in n-type GaAs-(GaAl)As heterostructures. Due to the extremely high electron mobilities in high quality samples the cyclotron resonance lines are extremely sharp. Experimental data obtained by Thiele et al (1987) have been compared with theoretical calculations performed by Rössler and collaborators (Rössler, 1989). The calculations explain the experimental findings on the oscillatory behavior of the cyclotron mass in a satisfying manner, as well as the dependence of the effective mass on the electron concentration which is caused by non-parabolicity.

REFERENCES

Ando, T., Fowler, A.B., and Stern, F. (1982), Rev. of Modern Physics 54, 437.
Bangert, E., von Klitzing, K., Landwehr, G. (1974), Proc. 12th Int. Conf. on the Physics of Semiconductors, Stuttgart, M. Pilkuhn Ed., Teubner Verlag, p. 714.
Bangert, E., and Landwehr, G. (1985), Superlattices and Microstructures 1, 363.
Bangert, E., and Landwehr, G. (1986), Surface Science 170, 593.
Bauer, G.E.W., and Ando, T. (1988), Phys. Rev. B37, 3130.
Broido, D.A., and Sham, L.J. (1985), Phys. Rev. B31, 888.
Erhardt, W., Staguhn, P., Byszewski, P., von Ortenberg, M., Landwehr, G., Weimann, G., van Bockstal, L., Janssen, P., Herlach, G., and Witters, J., (1986), Surface Science, 170, 581.
Hajdu, J., and Landwehr, G. (1985), in: Strong and Ultrastrong Magnetic Fields, F. Herlach Ed. Topics in Appl. Physics 57, Springer Verlag, Berlin, Heidelberg, New York, Tokyo.
Heuring, W., Bangert, E., Grötsch, K., Landwehr, G., Weimann, G., Schlapp, W., Reetsma, J.-H., Heime, K. (1989), EP2DSVIII, Grenoble, to be published in Surface Science.
Jäkel, B. (1989), Dissertation Universität Würzburg.
von Klitzing, K., Dorda, G., and Pepper, M. (1980), Phys. Rev. Lett. 45, 494.
Koch, F., Zrenner, A., and Ploog, K. (1987), in: High Magnetic Fields in Semiconductor Physics, G. Landwehr Ed., Springer Verlag, p. 308.
Landwehr, G., Ed. (1987), High Magnetic Fields in Semiconductor Physics, Springer Series Solid State Sciences 71, Springer Verlag Berlin, Heidelberg, New York, Tokyo.
Landwehr, G., Ed. (1989), High Magnetic Fields in Semiconductor Physics II – Transport and Optics - Springer Series Solid State Sciences 87, Springer Verlag Berlin, Heidelberg, New York, Tokyo.

Mendez, G., (1986), Surface Science <u>170</u>, 593.

Ohkawa, F.J., and Uemura, Y. (1975), Progr. in Theoretical
    Physics, Suppl. 57, 164.

von Ortenberg, M. (1975), Solid State Comm. <u>17</u>, 1335.

Ossau, W., Jäkel, B., Bangert, E., Landwehr, G., and Weimann,
    G. (1986), Surface Science <u>174</u>, 188.

Reményi, G., Landwehr, G., Heuring, W., Bangert, E., Weimann,
    G., and Schlapp, W. (1986), Proc. 18th Int. Conf. on the
    Physics of Semiconductors, Stockholm, O. Engström Ed.,
    World Scientific, p. 417.

Rössler, U. (1989), <u>in</u>: High Magnetic Fields in Semiconductor
    Physics II – Transport and Optics – Springer Series
    Solid State Sciences 87, G. Landwehr Ed., Springer
    Verlag Berlin, Heidelberg, New York, Tokyo, p. 376.

Sham, L.J., (1987) <u>in</u>: High Magnetic Fields in Semiconductor
    Physics, G. Landwehr Ed., Springer Verlag, p. 288.

Stern, F., and Howard, W.E. (1987), Phys. Rev. <u>163</u>, 816.

Thiele, F., Merkt, U., Kotthaus, J.P., Lommer, G., Malcher,
    F., Rössler, U., and Weimann, G. (1987), Solid State
    Comm. <u>62</u>, 841.

Tsui, D.C., Störmer, H.L., and Gossard, A.C., (1982) Phys.
    Rev. Lett. <u>48</u>, 1559.

Zrenner, A., Reisinger, H., Koch, F., and Ploog, K., (1984),
    Proc. of the Int. Conf. on the Physics of Semiconduc-
    tors, San Francisco, J.P. Chadi and W.A. Harrison Eds.,
    Springer Verlag, New York, p. 325.

TRANSMISSION, REFLECTION AND THE RESISTANCE OF SMALL CONDUCTORS

M. Büttiker

IBM Research Division
Thomas J. Watson Research Center
Yorktown Heights, N.Y. 10598

I. INTRODUCTION

In this paper we present a simple discussion of the resistance of multi-probe conductors formulated in terms of transmission and reflection probabilities for carriers incident on the sample (Büttiker, 1986a; 1988a). It is transport in open conductors with current probes and voltage probes which is addressed. The formulation of resistances in terms of transmission of carriers through a conducting structure and reflection at the conductor stressing equilibrium electron reservoirs as carrier sources and sinks has been successfully applied to explain a number of transport phenomena in small conductors. These phenomena include the observed symmetries of the magneto-resistances at low fields in Aharonov-Bohm experiments (Büttiker, 1986a; Benoit et al. 1986), and in ballistic electron focusing experiments (van Houten et al. 1989) as well as in a number of other experiments both on macroscopic and microscopic conductors. This approach has also been applied to the phenomena of huge conductance fluctuations in multi-probe conductors (Büttiker, 1987; Baranger et al. 1988, Kane et al. 1989), and to low field magnetic anomalies in ballistic conductors (Roukes et al. 1987; Timp et al. 1988; Takagaki et al. 1988; Ford et al. 1989a; Avishai and Band, 1989; Kirczenow, 1989a; Ravenhall et al. 1989, Baranger and Stone, 1989a; Beenakker and van Houten, 1989a). Most important, the same formulae provide an especially clear discussion of the quantum Hall effect and lead to the prediction (Büttiker, 1988b) and experimental confirmation of quantized four-terminal resistances at values which differ from the bulk quantization (Washburn et al. 1988, Haug et al. 1988; van Wees et al. 1989a, Komiyama et al. 1989a). This approach and experiments for the first time shed light on the important role of contacts in the quantization at high magnetic fields, and thus have permitted a clearer understanding of the quantum Hall effect. The success of this approach is surprising since it is conceptually very simple: We deal with one-electron transport only and rather than having to treat the reservoirs in all their complexity they enter into the discussion only as a set of boundary conditions.

The approach discussed here is closely related to much of the literature on quantum tunneling which at least implicitly also uses equilibrium current sources and sinks (Duke, 1969). This literature, however, does not address the question which is at the root of our discussion: What determines the voltage difference measured at probes connected to the conductor? This question has been answered differently. One viewpoint is that a voltage difference, since it is an integral of the electrostatic field along a line connecting two points of measurement, should be discussed on the basis of the Poisson equation. This viewpoint holds that the voltage is determined by a charge neutrality condition, at least at points more than a screening length away from the obstacle. Such a viewpoint leads to the study of piled up charges

Electronic Properties of Multilayers and Low-Dimensional Semiconductors Structures
Edited by J. M. Chamberlain *et al.,* Plenum Press, New York, 1990

and the study of charge depletion near an obstacle obstructing free carrier motion. Such a view was advanced by Landauer in pioneering papers (see Landauer 1957; 1970; 1975) and provided a guideline for our initial work with Imry and Landauer in this field (see Büttiker et al. 1985). However, a number of experimental observations, such as the symmetry of measured resistances in the presence of a magnetic field, lead to an approach which treats all probes connected to a conductor (whether current sources and sinks or probes used to measure voltages) in an equivalent fashion using the same physical principles for all probes. It is such an invariant treatment of all the probes that leads to results which are in accordance with the basic reciprocity symmetries of electrical conductance (Casimir, 1945; van der Pauw 1958; Büttiker, 1986a; Benoit et al. 1986). This approach emphasizes that a measured resistance depends on the very detailed nature of all the probes. This observation is at the heart of many of the recently observed transport phenomena in small conductors and is the main subject of this paper. To begin, we start with the simpler case of a two-terminal conductor.

II. TWO-TERMINAL CONDUCTANCE

We view the sample as a target at which carriers are reflected or transmitted. Electric conduction is treated as a scattering problem. To formulate a scattering problem there must be "asymptotic" regions which permit the definition of incoming and outgoing states. We achieve that by imagining a perfect wire uniformly extended along the x-direction, to either side of a scattering region of interest, as shown in Fig. 1a. For simplicity, first we discuss the case of zero magnetic field and discuss the case, where the confining potential of the perfect wire is only a function of the transverse coordinate y (or y and z). Translational invariance along x immediately tells us that the solutions of the Schrödinger equation for the perfect wires are of the form

$$\psi_{n, \pm k} = e^{\pm ikx} f_{n, \pm k}(y) \tag{2.1}$$

The factor $e^{\pm ikx}$ is due to translational invariance. $f_{n, \pm k}$ is the transverse wave function. For fixed k the spectrum consists of a ladder of in general discrete states with energy $E_n(k) = \hbar^2 k^2/2m + E_n(0)$, called quantum channels. The energy $E_n \equiv E_n(0)$ is the threshold of the n-th quantum channel (see Fig. 1b).

The first physical question which we have to answer is as follows: How are the incident channels to the left and right of the scattering region populated? Here, we assume that we have a large source of electrons, an electron reservoir, on either side of the scattering region. The reservoir, a large conductor, if it is at equilibrium can be characterized by a chemical potential and an equilibrium Fermi function. The basic assumption (Büttiker et al. 1985) is now twofold; (a) The reservoir fills all the channels in the "asymptotic" region, i.e. in the perfect wires introduced above, equally and according to the Fermi function. (b) Every carrier incident on the reservoir can enter the reservoir. Let us, for simplicity, concentrate on the limit of a vanishingly small temperature. The reservoirs are then characterized by their chemical potentials μ_1 and μ_2 alone. We are also concerned with small differences in the chemical potentials, i. e. μ_1 and μ_2 never deviate much from the Fermi energy E_F in the absence of carrier flux. It is then sufficient to consider the solutions of the Schrödinger equation at the Fermi energy. The longitudinal momenta $k_n(E_F)$ at the Fermi energy are determined by $E_n(k) = E_F$. We have thus in our perfect wire n solutions with positive momenta and positive velocities (incident states)

$$v_n = (1/\hbar)(dE_n(k)/dk) \mid_{E_F}, \tag{2.2}$$

and similarly N states with negative momenta and negative velocities (outgoing states).

Assume now that the left hand side reservoir is at a chemical potential μ_1 and the right hand side reservoir at a chemical potential $\mu_2 < \mu_1$. Below μ_2 all states are fully occupied and the net current due to these states is zero. Thus we need to consider only the energy range between μ_1 and μ_2. The net current incident in channel j in the energy interval $\Delta\mu = \mu_1 - \mu_2$ is $I_j = ev_j dn_j$ where dn_j is the density of carriers in this energy interval, $dn_j = (dn_j/dE)\Delta\mu$. Here we have used the one-dimensional density of states for a quantum channel, $dn_j/dE = 1/2\pi\hbar v_j$. Thus the incident current is

$$I_j = (e/h)\Delta\mu \tag{2.3}$$

independent of the channel index j.

Elastic scattering at the sample can be described by transmission probabilities T_{mn} for carriers incident in channel n on the left hand side to emerge in channel m on the right hand side. The current incident in channel n gives rise to a current $I_{mn} = (e/h)T_{mn}\Delta\mu$ in channel m in the left hand side perfect conductor. To obtain the total current we have to sum over all channels to the left and right. We obtain,

$$I = (e/h) \sum_{m=1, n=1}^{m=N, n=N} T_{mn}\Delta\mu. \qquad (2.4)$$

The conductance due to the chemical potential drop $\Delta\mu = \mu_1 - \mu_2$ between the left hand side and the right hand side reservoir is thus

$$G = I/e(\mu_1 - \mu_2) = (e^2/h)T. \qquad (2.5)$$

Here we have introduced the total transmission probability

$$T = \sum_{m=1, n=1}^{m=N, n=N} T_{mn} \qquad (2.6)$$

We re-emphasize, that in this derivation we have assumed that the chemical potential difference is so small that the transmission probabilities can be evaluated at the Fermi energy. We have considered only the case of a linear relationship between the current and the applied voltage and have considered only the case of a temperature small enough to be neglected. It is easy to extend Eq. (2.5) to situations where these restrictions do not apply.

Eq. (2.5) was, it seems, first obtained by Anderson et al. (1980) in the limit of an opaque scatterer (small transmission). (For large transmission their result is incorrect. It predicts an infinite conductance even if transmission in only one channel is equal to 1). Subsequent papers by Economou and Soukoulis (1981) and Fisher and Lee (1981) derived Eq. (2.5) using a linear response approach. An important aspect of this work is the connection between transmission probabilities and Green's functions which facilitates numerical computations of transmission probabilities. Fisher and Lee are interested in the case where the total transmission probability T is small compared to the total number of channels N. They state in their paper that the reservoirs, taken to be an infinitely long perfect lead, have the same cross section as the disordered region. Clearly, as pointed out by Landauer (1981) under this circumstances Eq. (2.5) cannot be right in the limit of weak scattering (perfect transmission). Current is, after all, conserved, and in a conductor with a uniform cross section there is no region where equilibrium can be

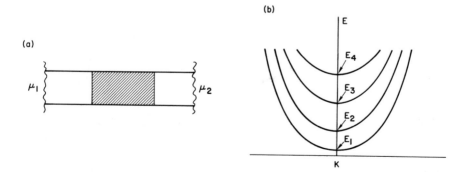

Figure 1. (a) Scattering region (shaded) connected to perfect conductors which in turn are connected to reservoirs at chemical potentials μ_1 and μ_2. (b) Energy dispersion of the quantum channels of a perfect wire. k is the wave vector along the wire and $E_n(0)$ is the threshold of the n-th quantum channel.

achieved. A paper by Thouless (1981) also tries to refute Eq. (2.5) but does so by attributing a resistance to a closed system without introducing a power dissipating mechanism. We mention the controversy surrounding this earlier appearances of $G \cong T$ mainly to demonstrate that at this time no clear understanding existed of the circumstances under which Eq. (2.5) is valid. In addition there appeared at this time a paper by Azbel (1981) who found a conductance formula which was later reestablished by Büttiker et al. (1985) and a paper by Langreth and Abrahams (1981) with results which differed from Eq. (2.5). For a number of years the question did not find further interest, perhaps under the wrong expectation that it is only the case of small transmission which is relevant. A key objection which was brought up against Eq. (2.5) is that a piece of perfect wire cannot have a resistance (Landauer, 1981). But Eq. (2.5), for perfect transmission, gives a resistance $h/e^2 1/N$, where N is the number of channels. Imry (1986) realized that the correct interpretation of this result is that of a contact resistance or Sharvin point contact resistance (Sharvin, 1965). In the reservoirs the carrier distribution is an equilibrium Fermi function. In the perfect conductor the carrier distribution must describe a current carrying state with positive velocity states occupied to a higher energy than negative velocity states. The transition from an equilibrium distribution to a current carrying distribution at the entrance to the perfect conductor and the transition at the exit back to an equilibrium distribution is dissipative.

According to Imry (1986), for Eq. (2.5) to be valid it is important that we have equilibrium reservoirs with well defined chemical potentials. Equilibrium requires that the current density in the reservoirs is negligibly small. Since the total current is conserved we can achieve equilibrium only by distributing the current among many channels. Therefore, we should not, as we have done above, just consider a perfect wire with a number of channels equal to that of the scattering region. To make sure that the wide regions to the left and right are reservoirs we must consider a large density of states, i. e. many channels in the asymptotic region. We are thus led to consider the conductor in Fig. 2a, with two wide regions to the left and right of a narrow conductor. To formulate a scattering problem we again assume that the wide region consists of perfect wires with a number of channels M that is much larger than than the number of channels N of the narrow conductor connecting the two wide regions. Obviously the result for the conductance is formally the same as Eq. (2.5) except that now the summation is over the number of channels in the wide region and not, as in the previous discussion, over the channels of the narrow conductor. The two wide regions need not be of the same shape, i.e. they can contain a different number of channels M_1 to the left and M_2 to the right. In this case the conductance is given by Eq. (2.5) with a total transmission probability,

$$T = \sum_{m=1, n=1}^{m=M_2, n=M_1} T_{mn}.$$ (2.7)

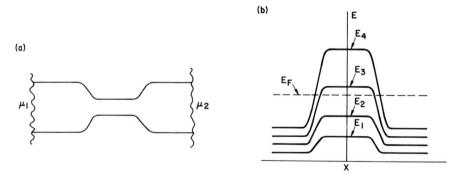

Figure 2.   (a) Model of a conductor which incorporates elastic scattering in the transition from a narrow conductor to wide reservoirs. (After Landauer, 1987).   (b) Position dependence of the threshold of the quantum channels for the conductor shown to the left. Two quantum channels are open. (After Glazman, 1988).

The wide regions to the left and right are now reservoirs with an accuracy (Büttiker, 1988a) of T/M, M = min(M_1, M_2). The situation depicted in Fig. 2a represents a step towards reality: the scattering at the transition from the wide region towards the narrow region is now included in the conductance. This is important since depending on the shape of this transition and the possibility of impurity scattering in this region the channels in the narrow conductor might now not all be populated equally (filled) despite the fact that the channels in the wide portions of the conductor are all fed from a common chemical potential. Landauer (1987) credits the author for pointing to the necessity of including the transition region from the reservoirs to the narrow conductor. Underlying this insight was the perception that what counts is a full volume of the size of the inelastic scattering length. A conductor of length L smaller than an inelastic scattering length doesn't really exist.

We might not be satisfied by a conductance formula which treats a wide conductor as a reservoir and, therefore, is accurate only to the order of T/M. That, of course, depends on the questions one is trying to answer. Finally, it is worthwhile mentioning, that since we deal with an electric problem, there must exist a self-consistent electrostatic potential which yields a vanishing electric field in the reservoirs and provides a smooth connection between the two, describing the potential drop across the constriction or barrier. This is guaranteed by the fact that we have equilibrium reservoirs with a large density of states (Büttiker, 1988a).

Recently, Eq. (2.5) has provided a model to discuss the step like rise of the conductance of a constriction as a function of gate voltage (van Wees et al. 1988; Wharam et al. 1988). We can cite here only a few of the recent papers which have investigated this phenomenon using Eq. (2.5) (van der Marel and Haanappel, 1989; Kirczenow, 1988; Szafer and Stone, 1989; Escapa and Garcia, 1989; Glazman et al. 1989; Tekman and Ciraci 1989; Streda, 1989, Landauer, 1989; Yacoby and Imry, 1989). In the simple model of Fig. 1a these steps simply arise due to the fact that more and more subbands become available as the width of the conductor d is increased. In the more realistic configuration of Fig. 2a the quantization is still observed, if the constriction is not too short to permit much tunneling, and if the transition from the wide region to the narrow region is not abrupt enough to permit multiple reflection due to scattering at the entrance and exit point of the constriction. If the width of the conductor changes slowly compared to a Fermi wave length an adiabatic approximation is appropriate (Glazman et al. 1989): In each segment of the conductor we can determine the spectrum of the quantum channels (see Fig. 1a) and obtain, due to the variation in width, channel thresholds which are a function of position (see Fig. 2b). Carriers entering the narrow conductor stay in the same quantum channel but increase their transverse energy at the expense of longitudinal energy.

It is surprising that it took a long time to understand Eq. (2.5) in view of the fact that related formulae have been used for a long time to discuss tunneling problems. Below we wish to briefly point to this connection. This work does not start from a situation with lateral confinement and often assumes special symmetries. Frenkel in his 1930 work discusses transmission through a potential barrier which has a height V(x) which depends only on x. Tsu and Esaki (1973) consider a superlattice with a potential profile which depends only on x but not on the transverse coordinates. Hence the Schrödinger equation is completely separable. The incident waves can be characterized by the momentum component parallel to the barrier k_y and the momentum component perpendicular to the barrier k_x. Due to the symmetry k_y is a good quantum number. It is only the transmission T_{k_y,k_y} which is non-zero and the total transmission probability is

$$T = \sum_{k_y} T_{k_y,k_y}.$$ (2.8)

The sum is over all k_y such that $k_F^2 = k_x^2 + k_y^2$. In these papers k_y is taken to be continuous and the sum in Eq. (2.8) is eventually written as an integral. In contrast, in the absence of symmetry, all transmission probabilities $T_{k',k}$ can be non-zero. Here $k = (k_x,k_y)$ characterizes the incident wave and k' characterizes the outgoing wave. Thus the difference between Eq. (2.5) and the widely used tunneling formulae is chiefly one of notation. Eq. (2.5) is more general since it does not require any special symmetry.

III. MULTI-TERMINAL CONDUCTORS

A conductance measurement or resistance measurement requires that both a current and a voltage difference are determined. In the two-terminal conductor discussed above the current is fed into the same reservoirs as are used to measure a voltage difference. It is this fact, that the same reservoir is used both to feed current and to measure a voltage, which leads to some of the difficulties addressed above. For the very same reason, to avoid dealing with contacts and their effect on what we actually want to study, it is desirable, to avoid, whenever possible, a two terminal measurement. Thus a wide variety of measurements are made on conductors with many probes. Fig. 3a shows a conductor with four probes and an Aharonov-Bohm flux through the hole of the loop. For a general discussion of Aharonov-Bohm effects in conductors we refer the reader to Imry (1986), Aronov and Sharvin (1987) and Webb and Washburn (1988). To discuss the general symmetry properties of the resistances we can either consider an Aharonov-Bohm flux or a uniform magnetic field: A magnetic field can be represented by a flux through each lattice cell, as long as the flux through the cell is small compared to a flux quantum hc/e, all the symmetry properties which apply to the Aharonov-Bohm flux through the hole of the conductor, also apply to a uniform magnetic field. A more explicit discussion of transport in high magnetic fields is given in Section V). In the conductor of Fig. 3a, two probes can be used as current source and sink and two probes are used to measure a voltage difference. It is, therefore, necessary to extend the discussion given above to deal with such many probe conductors. As in the two terminal case we imagine that each reservoir can be characterized by a perfect conductor with a number of channels M_i, $i=1,2,3,4$. Each of the reservoirs is characterized by a chemical potential μ_i, $i = 1,2,3,4...$ Consider now a carrier incident in channel n in probe j described by a plane wave of unit amplitude. Scattering in the conductor is again assumed to be elastic. The incident wave will typically give rise to transmitted waves into all channels in all leads and will give rise to reflected waves into all channels into the probe j from which the wave is incident. The elastic scattering properties are now described by transmission probabilities and reflection probabilities

$$T_{ij,mn}, \quad R_{jj,mn}, \tag{3.1}$$

which have four indices. A wave incident in channel n in probe j leads to transmission into channel m in probe i and leads to reflection into channel m in probe j. Fortunately, since all the incident channels in a given probe are fed by the same chemical potential, it is only the total transmission probabilities which we need. The total current from probe j to probe i due to carriers incident in probe i, if all the incident channels are full, is determined by

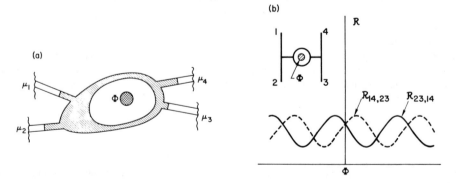

Figure 3. (a) Conductor with four terminals at chemical potentials μ_i and an Aharonov-Bohm flux. (After Büttiker, 1986a). (b) h/e-Aharonov-Bohm resistance oscillations for the conductor shown in the inset for two measurements which are related by reciprocity.

$$T_{ij} = \sum_{m=1}^{m=M_i} \sum_{n=1}^{n=M_j} T_{ij,mn}. \tag{3.2}$$

Similarly, the total reflected current in lead j due to carriers incident in lead j is determined by the total probability for reflection,

$$R_{jj} = \sum_{m=1}^{m=M_j} \sum_{n=1}^{n=M_j} R_{jj,mn}. \tag{3.3}$$

Using these probabilities we can calculate the net carrier flux in each probe of the conductor. As a convention we regard the velocities which give motion from the reservoir towards the conductor as positive. The incident flux in probe 1 is $M_1(\mu_1 - \mu_0)/h$, where μ_0 is a reference chemical potential which can be taken to be equal to the lowest of all the chemical potentials $\mu_0 = \min(\{\mu_i\})$. Of this flux a portion $R_{11}(\mu_1 - \mu_0)/h$ is reflected. The incident flux is further diminished by carriers which are incident in the other probes and are transmitted into probe 1. These fluxes are proportional to $T_{1j}(\mu_j - \mu_0)/h$. Thus the current in probe i is (Büttiker, 1986a),

$$I_i = \frac{e}{h} [(M_i - R_{ii})\mu_i - \sum_{j \neq i} T_{ij}\mu_j]. \tag{3.4}$$

The reference chemical potential does not appear: the coefficients which multiply the chemical potentials in Eq. (3.4) add to zero in each row of this matrix due to current conservation. (Due to time reversal invariance, simultaneous reversal of momentum and magnetic field, the columns of this matrix add also up to zero. See Appendix A).

Eq. (3.4) permits us to calculate the resistance in a four-terminal measurement. In a configuration where reservoirs m and n are used as a source and sink and contacts k and l are voltage probes, the resistance is $\mathscr{R}_{mn,kl} = (\mu_k - \mu_l)/eI$. Here $I = I_m = -I_n$ is the current impressed on the sample. At the voltage contacts, there is zero net current flow, $I_k = I_l = 0$. The voltmeter can be taken to have an infinite impedance. These conditions on the currents determine the resistance. A calculation gives (Büttiker, 1986a),

$$\mathscr{R}_{mn,kl} = (h/e^2)(T_{km}T_{ln} - T_{kn}T_{lm})/D. \tag{3.5}$$

Here D is a subdeterminant of rank three of the matrix formed by the coefficients in Eq. (3.4) which multiply the chemical potentials. It can be shown that all subdeterminants of rank three of this matrix are equal and independent of the indices m,n,k, and l.

By considering the scattering matrix S, one can show that microreversibility (current conservation and time reversal under simultaneous field inversion) implies (see Appendix A)

$$T_{ij}(B) = T_{ji}(-B), \quad R_{ii}(B) = R_{ii}(-B). \tag{3.6}$$

Using this in Eq. (3.4) gives rise to the reciprocity of four-terminal resistances,

$$\mathscr{R}_{kl,mn}(B) = \mathscr{R}_{mn,kl}(-B). \tag{3.7}$$

A very interesting test of reciprocity symmetry demonstrating the phase-sensitive nature of voltage measurements is provided in an experiment by Benoit et al. (1986). They investigated the symmetry of the Aharonov-Bohm oscillations of a metallic four-probe conductor at milli Kelvin temperatures. In metallic loops the Aharonov-Bohm oscillations manifest themselves in a small oscillatory contribution to the total resistance. According to Eq. (3.5) the h/e - oscillations need not be symmetric if the flux Φ through the loop is reversed but can appear with an arbitrary phase ϕ, (see Fig. 3b). But reciprocity requires that if the oscillatory contribution to $\mathscr{R}_{14,23}$ is $\Delta\mathscr{R} \cos(2\pi\Phi/\Phi_0 - \phi)$, where $\Phi_0 = h/e$ is the elementary flux quantum, then the oscillatory component in $\mathscr{R}_{23,14}$ is precisely $\Delta\mathscr{R} \cos(2\pi\Phi/\Phi_0 + \phi)$. Just such a behavior was observed in the experiments of Benoit et al. 1986. The significance of these papers was their demonstration of the phase-sensitivity of the voltage measurement in small conductors.

The observation of a phase different from zero or π needs quantum mechanical phase coherence (Büttiker, 1986a). Only if the probes are close to the loop or as in Fig. 3a directly connected to the loop can a non-trivial phase in the oscillations be observed. DiVincenzo and Kane (1989) have calculated the probability distribution of the phase ϕ as function of the phase breaking length for a conductor in the metallic diffusive limit.

Reciprocity theorems have a long history dating back to the last century. These symmetries are also closely connected to the work of Onsager and Casimir (Casimir, 1945). Casimir emphasizes the symmetries of resistivities and conductivities. Our derivation, limited to a conductor with elastic scattering only, emphasizes (global) resistances and is valid when the concepts of local resistivities and local conductivities fail.

Reciprocity symmetries have been tested in a number of experiments both in macroscopic conductors and submicron structures (see Büttiker, 1988a for references). Electron focusing experiments (van Houten et al. 1989, Beenakker et al. 1989) provide a demonstration which is striking and of immediate clarity.

A natural question is the relation of Eqs. (3.4 - 3.7) to the linear response formalism. We have already mentioned the connection between Eq. (2.5) and the linear response approach established by Fisher and Lee (1981). Proceeding in a similar way Stone and Szafer (1988) have given a linear response derivation of Eqs. (3.4-3.7) for conductors without a magnetic field. The inclusion of a magnetic field is more difficult and only recently has such derivation been found by Baranger and Stone (1989). It should be stressed, that linear response formalism does not provide a test of Eqs. (3.4-3.7). These results are a consequence of a number of physical assumptions, and to rederive them with a linear response approach the same assumptions need to be made. A major by-product of such a derivation is the expression of the transmission coefficients in terms of Green's functions. This is important for the computation of these coefficients starting from a given Hamiltonian.

IV. SIMPLE LIMITS

Eq. (3.4) is quite simple but contains a wealth of physics. To understand at least in part why this simple result is so powerful, we consider below a few simple limits and discuss briefly what can be learned from them.

A. Two-Terminal Conductor

It is obvious that Eq. (3.4) contains the two-terminal conductance discussed in Section II. If the conductor is only connected to two reservoirs current conservation requires (see Appendix A) $M_1 = R_{11} + T_{12}$ and $M_2 = R_{22} + T_{21}$. Here M_1 and M_2 are the number of channels with which reservoir 1 and 2 feed the conductor. Since the columns of Eq. (3.4) must also add up to zero, we find $M_1 = R_{11} + T_{21}$ and $M_2 = R_{22} + T_{12}$ and hence, $T_{21} = T_{12} \equiv T$. The total transmission probability is symmetric with regard to magnetic field reversal. Eq. (3.4) yields a two-terminal resistance,

$$\mathscr{R}_{12,12} = (\mu_1 - \mu_2)/eI = (h/e^2)(1/T). \tag{4.1}$$

In contrast to a four-probe measurement of the Aharonov-Bohm effect, the resistance of a loop connected to two electron baths can only have an oscillatory amplitude $\Delta\mathscr{R} \cos(2\pi\Phi/\Phi_0 - \phi)$ with ϕ either equal to π or zero. How the phase jumps between these two values as a function of Fermi energy is an interesting question which is discussed in Büttiker (1986c) and D'Amato et al. (1989).

B. Three Terminal Conductor

Consider a conductor connected to three terminals. An example is shown in Fig. 4a. Suppose that terminal 3 is a voltage probe. That is the current in probe 3 is zero. From Eq. (3.4) we immediately find that the chemical potential at probe 3 is (Büttiker, 1986b),

$$\mu_3 = \frac{T_{31}\mu_1 + T_{32}\mu_2}{T_{31} + T_{32}}, \tag{4.2}$$

The voltage measured at a terminal is not determined by piled up charges in the vicinity of the connection of the terminal to the conductor. Instead the voltage is determined by the probability with which carriers incident from terminal 1 and terminal 2 can reach terminal 3. Despite the fact that no net current flows into the probe used to make a voltage measurement, the probe is nevertheless dissipative. The dissipated energy at probe 3 (in the absence of a magnetic field) is (Büttiker, 1986b)

$$W = \frac{1}{h} \frac{T_{31}T_{32}}{T_{31} + T_{32}} (\mu_1 - \mu_2)^2. \tag{4.3}$$

If such a probe is dissipative, it can be used as a mechanism to bring dissipation into an otherwise coherent conduction process. Consider the two terminal conductance of this three terminal conductor. The net current from terminal 1 to terminal 2 can be calculated by eliminating μ_3 in the equations for I_1 and I_2. Since $I_3 = 0$, no carriers are lost in the voltage probe and hence $I = I_1 = -I_2$. A little algebra yields for the two terminal conductance

$$G_{12,12} = \left(\frac{e^2}{h} \right)(T_{21} + \frac{T_{23}T_{31}}{T_{31} + T_{32}}). \tag{4.4}$$

This equation is easy to understand. We have *two* mechanisms to bring carriers from terminal 1 to terminal 2. First, carriers can reach terminal 2 with probability T_{21} directly, without ever entering probe 3. This portion of the conductance process is coherent. Second, carriers from terminal 1 can enter probe 3 with probability T_{31}. In reservoir 3 they loose phase-memory. Since no net current flows into probe 3 each carrier entering reservoir 3 is replaced by a carrier reemerging with a phase and energy unrelated to that of the incident carriers. A fraction $T_{23}/T_{31} + T_{32}$ is eventually scattered into reservoir 2. This second process is phase-incoherent, or borrowing a term from the tunneling literature, sequential.

It is clear that if probe 3 is very narrow, then only a few carriers can reach probe 3, and the conduction process is predominantly coherent. If, on the other hand, probe 3 is very wide and free of obstacles, then almost all carriers will reach probe 3, and the conduction process is predominantly sequential.

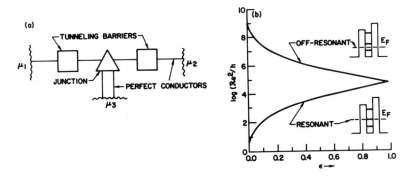

Figure 4. (a) Model of a three terminal double barrier structure. One of the probes can be used to model incoherent conduction. (After Büttiker et al. 1988c). (b) Resistance at resonance and off resonance (see inset) as a function of the probability ε for carriers approaching the junction to enter terminal 3. Increasing incoherent transmission (increasing ε) enhances the resistance at resonance and decreases the off resonant transmission. For $\varepsilon = 1$ conduction is completely incoherent and the total resistance is given by Eq. (4.5). (After Büttiker, 1988c).

Fig. 4a shows two barriers with a probe, acting as an inelastic scatterer, connected to the well of the double barrier. If the conduction process is completely coherent, the double well potential leads to resonances. In Fig. 4b the resistance is compared for the case that the Fermi energy is at a resonant level with the case that the Fermi level is between two resonant levels (off-resonant) and shown as a function of a parameter ε which characterizes the probability of carriers incident on the junction to enter reservoir 3. For $\varepsilon = 0$ the conduction process is completely coherent and for $\varepsilon = 1$ the conduction process is completely incoherent. In the completely incoherent case $\varepsilon = 1$ carriers enter probe 3 with probability 1. We have $T_{31} = T_1$ and $T_{32} = T_2$ and $T_{21} = 0$. Hence from Eq. (4.4) we obtain

$$\mathscr{R}_{12,12} = G_{12,12}^{-1} = \frac{h}{e^2} \left(\frac{1}{T_1} + \frac{1}{T_2} \right) \tag{4.5}$$

i.e. the classical law for addition of resistances.

For details of the calculation leading to Fig. 4b we refer the reader to Büttiker (1988c). The effect of incoherent events on the persistent current of a loop has been analyzed using this approach by Büttiker (1985) and Cheung et al. (1988). Recently, the same principles have been used in discussing the addition of point contacts in series by Beenakker and van Houten (1989). As a guide to the experimental work on the addition of point contacts, we refer the reader to Beton et al. (1989). To study distributed inelastic scattering many probes can be attached to a conductor. The contribution of incoherent conduction processes has been treated in this way by Datta (1989) and D'Amato and Patawskii (1989). It is clear that such an approach is more realistic if incoherence comes about through low energy excitations (acoustic phonons) and is not immediately applicable if we deal with high energy excitations (optical phonons) as pointed out by Sokolovski (1989).

There is another important consequence flowing from these results. Through elimination of the chemical potential μ_3 in Eq. (3.4) with the help of Eq. (4.2) we effectively reduced a three terminal conduction problem to a two-terminal conduction problem. The total transmission probability of this two-terminal conductor is just

$$\hat{T}_{21} = T_{21} + \frac{T_{23}T_{31}}{T_{31} + T_{32}} . \tag{4.6}$$

Similarly, we obtain a reflection probability

$$\hat{R}_{11} = R_{11} + \frac{T_{13}T_{31}}{T_{31} + T_{32}} . \tag{4.7}$$

for carriers incident on the left. The reflection probability for carriers incident from the right is

$$\hat{R}_{22} = R_{22} + \frac{T_{23}T_{32}}{T_{31} + T_{32}} . \tag{4.8}$$

Each of these probabilities is the sum of a coherent term and an incoherent or sequential term. A little exercise shows that these probabilities have the symmetry in the presence of a magnetic field and the current conservation laws appropriate for a two terminal conductor (see Section IVA). Thus by elimination of the chemical potential of a voltage probe in an n-terminal conductor we obtain a set of transmission and reflection probabilities that obey the current conservation and symmetry laws of Eq. (3.4) for an (n-1)-terminal conductor.

This mapping has two implications (Büttiker, 1988a): First, Eq. (3.5) and Eq. (3.6) which we derived for a four-terminal conductor are, in fact, valid for a conductor with an arbitrary number of terminals (also for transmission probabilities which are functions of the n-terminal conductor). Second, we can use terminals to make the conduction process incoherent. Thus the mapping discussed above shows that the reciprocity of resistances does not in any way depend on our assumption of elastic, coherent transmission, but is also valid for incoherent conduction as indeed it must be.

C. Four-Terminal Conductor with Weakly Coupled Probes

The third example which we discuss is a conductor with two weakly coupled probes (Fig. 5 and Fig. 6b). This can be achieved by narrowing two probes, by building a gate over two of the probes (application of a gate voltage creates a barrier), using split gates which leave a narrow channel which can be varied by application of a gate voltage or by using a tunnel microscope. Such a geometry can be used either to inject current through a weakly coupled probe or by using the probe to measure a voltage.

Consider the conductor in Fig. 5. Although a four-terminal problem, it is considerably simpler than the general case addressed in Section III. Transmission from probe 3 (and probe 4) to the other terminals is small. To take advantage of that, we introduce a coupling parameter, which we can take to be equal to the largest transmission probability in the set $T_{13,kl}, T_{23,kl}, T_{14,kl}, T_{42,kl}$ where k and l run over all channel indices in these leads. We denote the largest transmission probability in this set by ε and use ε as an expansion parameter. Thus the total probabilities which describe transmission into the leads 3 and 4 (or transmission from these leads) are of the order ε, except the transmission probabilities T_{43} and T_{34} which are of the order ε^2. The transmission from lead 1 to lead 2, on the other hand, contains a term which is independent of ε. Let us now consider the case where we use both weakly coupled probes as voltage probes (or by reciprocity as current injector and current sink). In this case transmission from probe 3 to probe 4 can be neglected, and the four-terminal problem decouples into two three-terminal problems. To obtain the voltage measured at probe 3, we can neglect the presence of probe 4, and obtain just,

$$\mu_3 = \frac{T_{31}\mu_1 + T_{32}\mu_2}{T_{31} + T_{32}}, \tag{4.9}$$

i.e. Eq. (4.2). Similarly, the chemical potential at probe 4 is determined by

$$\mu_4 = \frac{T_{41}\mu_1 + T_{42}\mu_2}{T_{41} + T_{42}}. \tag{4.10}$$

The current from terminal 1 to terminal 2 is to lowest order independent of ε, and as in the case of a two-terminal conductor, $I = (e/h)T(\mu_1 - \mu_2)$. Here, $T = T_{12} = T_{21}$ is the transmission probability between terminals 1 an 2 (to lowest order in ε). It is symmetric in the magnetic field. Using the expression for the current and using Eqs. (4.9, 4.10) to find the potential difference we obtain (Büttiker, 1988a)

$$\mathcal{R}_{12,34}(B) = \mathcal{R}_{34,12}(-B) = \left(\frac{h}{e^2}\right)\frac{1}{T}\frac{T_{31}T_{42} - T_{32}T_{41}}{(T_{31} + T_{32})(T_{41} + T_{42})}. \tag{4.11}$$

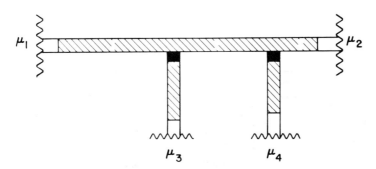

Figure 5.   Conductor with two weakly coupled probes (After Büttiker, 1988a).

Note that the denominator has precisely the form required by Eq. (3.5). The nominator is an explicit expression for this simple case of the determinant D. It is, as it must be, to order ε, a symmetric function in the magnetic field B. Taking into account that the denominator is as in Eq. (3.5) and the determinant D is given by the nominator of Eq. (4.11), it is easy to find all the possible four-terminal resistances of the conductor of Fig. 5. The relation of Eq. (4.11) to the Landauer formula which predicts a resistance proportional to 1-T/T (Landauer, 1970) is discussed in Büttiker (1988a) and Büttiker (1989d).

This example of a conductor with weakly coupled probes also teaches us an important lesson. Even if the probes are weakly coupled to the conductor the measured resistances depends on the precise way these probes couple to the conductor (Büttiker, 1988a). There is no unique answer to the question: What is the resistance of a small conductor? It is the conductor including the probes and the precise way in which they couple to the conductor which matters.

E. Conductors with Geometrical Symmetries

In disordered metallic conductors the geometrical symmetry of the conductor plays a role only if we are concerned with transport coefficients which are averaged over all possible impurity configurations. A typical impurity configuration breaks all symmetries. On the other hand, if we are concerned with ballistic conduction in the limit where the mean free path due to impurity scattering is large compared to the size of the conductor, it makes sense to study at least as a first approximation, conductors with special symmetries. This leads to very simple expressions for the four-terminal resistances, Eq. (3.5), as we now briefly discuss. Fig. 6a shows a perfect four-probe conductor with two narrow probes and two wide probes such that there are two symmetry axes. In the presence of a magnetic field, we can discuss the transmission behavior of such a conductor in terms of the following probabilities: The probability for carriers incident in a wide lead to be scattered into another probe in the direction favored by the Lorentz force is denoted by T_+, scattering into the probe against the Lorentz force by T_- and if the carrier traverses the conductor from wide probe to wide probe by T_d. (The index d stands for *direct* transmission). Similarly, we denote the transmission probabilities for carriers incident from one of the narrow probes by $T_\wedge, T_\wedge$, and $T_\wedge$. Clearly, the geometrical symmetry of the conductor implies $T_+(B) = T_-(\overset{+}{-}B)$ and microreversibility, Eq. (3.6), requires $T_-(-B) = T_\wedge(B)$. Hence $T_+ \equiv T_\wedge$, and a similar consideration also yields, $T_- \equiv T_\wedge$. Thus the transmission behavior of the conductor in Fig. 6a is determined by four different transmission probabilities only, T_+, T_-, T_d and $T_\wedge$. The determinant of rank three of the matrix in Eq. (3.4) is

$$D = (T_+ + T_-)(2T_d T_\wedge^d + (T_d + T_\wedge^d)(T_+ + T_-) + T_+^2 + T_-^2) \qquad (4.12)$$

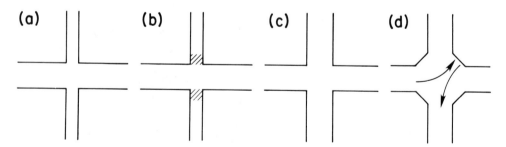

Figure 6.   Four probe conductors with symmetry: (a) Two narrow and two wide probes. (b) Two weakly coupled probes and two wide probes. (c) Four identical probes (nominally perfect junction). (d) Four probe junction with a widened central section (After Ford et al. 1989a).

The Hall resistance is anti-symmetric,

$$\mathscr{R}_{13,42}(B) = -\mathscr{R}_{42,13}(B) = (\frac{h}{e^2})(T_+^2 - T_-^2)/D. \tag{4.13}$$

Taking into account that D is proportional to $T_+ + T_-$, it follows that the Hall resistance is proportional to $T_+ - T_-$, i.e. the transmission asymmetry created by the magnetic field. The longitudinal resistances, also referred to as bend resistances, are

$$\mathscr{R}_{12,43} = \mathscr{R}_{14,23} = (\frac{h}{e^2})(T_- T_+ - T_d T_\wedge^d)/D. \tag{4.14}$$

In the limit where transmission into two leads is made small due to barriers in two of the Hall bars, i.e. if two leads are weakly coupled, we can proceed as in Section C above. We have $T_\pm \cong \epsilon$ and $T_\wedge^d \cong \epsilon^2$, giving rise to a Hall voltage determined by

$$\mathscr{R}_{13,42}(B) = -\mathscr{R}_{42,13}(B) = (\frac{h}{e^2})\frac{1}{T_d}\frac{(T_+ - T_-)}{(T_+ + T_-)}. \tag{4.15}$$

The discovery by Roukes et al. 1987 that in nominally four-fold symmetric crosses the Hall resistance is quenched (approximately zero) has drawn a lot of attention to the case of four-fold symmetric crosses (Fig. 6c). In the work of Kirczenow (1989a), Ravenhall et al. (1989), Baranger and Stone (1989a), and Beenakker and van Houten (1989b), Eq. (4.13) with $T_d = T_\wedge$ provides the starting point. The simpler limit, Eq. (4.15), was analyzed by Peeters (1988). In conjunction with the quenching of the Hall effect, it is observed (see Takagaki et al. 1988, 1989; Avishai and Band, 1989; Kirczenow, 1989b; Ford et al. 1989b) that the bend resistances, Eq. (4.14), are negative, i.e. at low fields the direct transmission T_d is much stronger than transmission into the Hall probes. Clearly, this is a manifestation of ballistic transport. These results are dominated by the geometry of the conductor. If the cross is widened as in Fig. 6d, the Hall resistance is not zero but negative (Ford et al. 1989a). In this case, for a small region of field, the transmission T_- is dominated by carriers which initially follow the Lorentz force but strike one of the reflecting walls and as a consequence end up in the wrong probe (Ford et al. 1989a). Possibly, even crosses which are nominally composed of straight wires, are in reality closer to the structure shown in Fig. 6d than to Fig. 6c. The self-consistent potential is a smooth function of the position coordinates. This suggests that scattering into the "wrong" probe is a relevant process also in the quenching of the Hall effect. Both Ford et al. (1989a, 1989b) and Chang et al. (1989) have investigated the Hall resistance in a cross with two narrow probes and two wide probes in addition to a number of other interesting

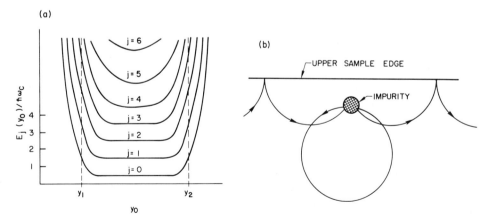

(a)

(b)

Figure 7. (a) Magneto-elastic subbands for a perfect strip with hard walls at y_1 and y_2 as a function of y_0, the minimum of the magnetic energy (Eq. (5.1)). (b) Suppression of backscattering: Carriers on a skipping orbit after (multiple) scattering of an impurity continue to skip along the edge. (After Büttiker, 1988b).

geometries. The picture which emerges from these experiments and the recent work by Baranger and Stone (1989a) and Beenakker and van Houten (1989b) is that the widening of the probes which occurs as the junction is approached leads to an enhanced injection of carriers into the forward direction. If due to this collimation effect the carriers cannot be scattered directly into the side probes but only via reflection at internal surfaces, the Hall effect can be quenched or even negative as found by Ford et al. (1989a; 1989b).

V. THE QUANTUM HALL EFFECT

The quantum Hall effect is one of the major discoveries in Solid State Physics in this decade (von Klitzing et al. 1980). If the previous discussion is indeed general, then it must in particular also apply to the quantum Hall effect. This is indeed the case. In the following Section, we present a simple discussion which appeals on intuition, but hopefully is free of some of the mysteries with which the quantum Hall effect is sometimes surrounded.

In Section II, we have discussed the concept of quantum channels in terms of a perfect wire at zero magnetic field. We now extend this discussion and include explicitly a magnetic field. Consider a two-dimensional strip in a magnetic field extended in the x-direction and a confining potential in the y-direction. Application of a magnetic field leads to a new term in the Schrödinger equation for the "transverse" wave functions. The Schrödinger equation contains an energy term which is governed by the cyclotron frequency $\omega_c = e \mid B \mid /mc$ where m is the effective mass of the carriers. This additional energy term takes the form of a harmonic potential centered at $y_0 = - l_B^2 k$, where $l_B = (\hbar c/e \mid B \mid)^{1/2}$ is the magnetic length and k is the "longitudinal" momentum in the x-direction,

$$E_M = (1/2)m\omega_c^2(y - y_0)^2. \tag{5.1}$$

Instead of k, the spectrum of such a wire is often shown as a function of y_0 as in Fig. 7a. The confining potential in Fig. 7a is assumed to be a hard wall potential. Electrons in a region where the potential is flat perform a cyclotron motion with frequency ω_c. The center of these orbits is stationary. On the other hand electrons closer to the potential wall of the strip are reflected at the wall and as a consequence describe a skipping orbit. These carriers propagate. Thus it is the quantum mechanical equivalent of the skipping orbits, the states near the edge of the sample (Halperin, 1982; MacDonald and Streda, 1984), which allow propagation. The velocity of carriers in these states is given by

$$v_j = (1/\hbar)(dE_j/dk) = (dE_j/dy_0)(dy_0/dk) \tag{5.2}$$

and the density of states is $dn_j/dE = 1/hv_j$. Thus the product of the velocity and the density of states is for each channel 1/h just as in the case of zero magnetic field. The current which can be injected by a reservoir into one quantum channel is again given by Eq. (2.3). Therefore, all arguments used to derive the results of Section II can be repeated for the case of magneto-electric subbands. For this reason, Eqs. (3.4-3.7) are valid also in high magnetic fields (Büttiker, 1988b). Thus the approach out-lined in Section II permits us to give a four-terminal formulation of the quantum Hall effect and to discuss the effect of probes on the quantization (Büttiker 1988b).

A. The Quantum Hall Effect In Open Conductors

It is now essential to understand that the edge states exist even in the presence of disorder (Halperin 1982, Büttiker, 1988b). This is because an impurity can at best scatter a carrier a distance of a cyclotron radius away from the impurity. Hence, the trajectory, after possibly impinging a number of times on the impurity, will eventually return to the potential wall and continue on a skipping path (see Fig. 7b). As long as the impurities are far apart compared to a cyclotron radius there is no effective backscattering of carriers. Alternatively, we can consider a potential which varies smoothly on the length scale of a cyclotron radius. The states at the Fermi energy E_F are then determined by the solution of the equation, $E_F = \hbar\omega_c(n + 1/2) + eU(x,y)$, where n is a positive integer and x,y is a path in two-dimensional space (an equipotential line). We emphasize, that it is the equilibrium potential eU which matters, i.e. the potential in the absence of a net current flow ($\mu_1 = \mu_2 = \mu_3 = \mu_4$). There are two types of solutions: There are open paths (the edge states) which necessarily originate and terminate at a contact. In addition, there are closed paths. In Fig. 8a the open paths are indicated by faint solid lines. At zero temperature and in the absence of tunneling between open states and the closed paths it is only the open paths (edge states) which contribute to electric conduction. It is the connection of the contacts via edge states (open paths) which determines the measured resistance. (Kazarinow and Luryi (1982), who also consider a weakly fluctuating potential, neglect the open states connecting contacts, and as a consequence obtain a quantum Hall effect only above a critical voltage). Since there is no backscattering, we find that if each of the incident edge states carries a unit current (is full), all outgoing edge states carry also a unit current (are full). Given N edge states, the total transmission probabilities in the conductor of Fig. 8a are $T_{41} = N$, $T_{34} = N$, $T_{23} = N$, and $T_{12} = N$. All other T_{ij} are zero. The total reflection probabilities in the absence of internal reflection are $R_{ii} = M_i - N$. With the transmission and reflection probabilities as specified above it is now easy to calculate the four-terminal resistances, Eq. (3.5). The Hall resistance $\mathscr{R}_{13,42}$ is determined by $T_{41}T_{23} - T_{43}T_{21}$, which is equal to N^2. Evaluation of the subdeterminant D in Eq. (3.5) yields $D = N^3$. All Hall resistances of the conductor of Fig. 8a are quantized and yield $\pm h/e^2 N$. The "longitudinal" resistances (for example $\mathscr{R}_{12,43}$) are zero. Thus the quantization is a consequence of the nature of the edge states: They provide a channel along which carriers propagate from one contact to the other with probability 1.

There is, of course, no lack of differing formulations of the quantum Hall effect. We have already mentioned the work of Kazarinov and Luryi (1982). Many discussions use a conventional linear response approach, which suffers from the difficulty that the local electric field is not known apriori. In contrast, in our approach it is only the chemical potentials of the electron reservoirs which appear. A number of discussions which use a scattering approach have been put forth. Streda et al. (1987) base their approach on the discussion of Büttiker et al. (1985) and study piled up charges to determine local potentials. Jain and Kivelson (1988) have obtained similar results in the limiting case of one Landau level only. Despite our criticism of this approach, (Büttiker 1986a; 1988a; 1989a) the discussion of Streda et al. (1987) has been defended by Sivan and Imry, (1989). None of these discussions does justice to the fact that a resistance depends on the detailed way probes couple to a conductor (Büttiker, 1988a; 1988b). Streda et al. (1987) in view of experiments have eventually adopted our point of view (see Haug et al. 1988; 1989). As mentioned above linear response can be recast to relate transport coefficients only to chemical potentials (Baranger and Stone, 1989b) leading again to Eqs. (3.3-3.7). Below, we consider a number of very simple situations which exemplify the flexibility of our approach (Büttiker, 1989b; 1989c).

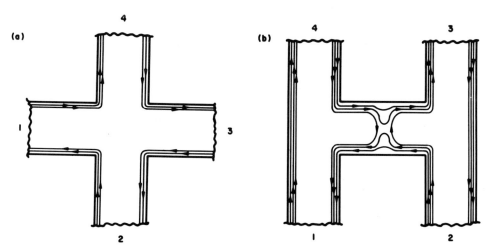

Figure 8. (a) Conductor with Hall probes. (After Büttiker, 1988a,b, 1989a,b) (b) Conductor with barrier reflecting K edge states. (After Washburn et al. 1988, Büttiker, 1988b, 1989b).

B. Anomalously Quantized Four-Terminal Resistances

The discussion given above can easily be applied to the case, where various current and voltage probes are interconnected with good metallic conductors. The result is that the two-terminal resistance is $\mathscr{R}_{12,12} = (h/e^2)(p/q)(1/N)$ with p and q integers as found experimentally by Fang and Stiles (1984). If, for instance, terminals 2 and 4 of the conductor in Fig. 8a are connected, the two-terminal resistance is proportional to 2/N. Next we analyze situations where the connection of edge states to various contacts is changed in a well controlled fashion. Fig. 8b shows a conductor where a gate creates a barrier to carrier flow. For a certain range of barrier height K edge states are reflected. Application of Eq. (2.5) predicts Hall resistances (Büttiker, 1988b)

$$\mathscr{R}_{13,42} = (\frac{h}{e^2}) \frac{1}{(N - K)} \tag{5.3}$$

$$\mathscr{R}_{42,13}(B) = \mathscr{R}_{13,42}(-B) = -(\frac{h}{e^2}) \frac{N - 2K}{N(N - K)} \tag{5.4}$$

and *quantized* longitudinal resistances which are symmetric in the field (Büttiker 1988b),

$$\mathscr{R}_{12,43}(B) = \mathscr{R}_{12,43}(-B) = \mathscr{R}_{43,12}(B) = (\frac{h}{e^2}) \frac{K}{N(N - K)} . \tag{5.5}$$

All other four-terminal resistance measurements on the conductor of Fig. 8b are zero. The plateaus predicted by Eqs. (5.3-5.5) have been observed in strikingly clear experiments by Washburn et al. (1988), and Haug et al. (1988; 1989). Interestingly, van Houten et al. (1988) found Eq. (5.5) to be a good approximation to the low field four-terminal magneto-resistance of a constriction, if there is some equilibration of the carriers between the constriction and the probes. In contrast, the derivation of Eqs. (5.3-5.5) assumes equilibration only in the reservoirs. Recently, Snell et al. (1989) have made a high-field experiment measuring all the resistances Eqs. (5.3-5.5) across a constriction.

So far we have assumed that carriers which reach a contact from the interior of the sample can escape into the reservoir with probability 1. This is called a contact without internal reflection (Büttiker 1988b). Correspondingly, if carriers approaching a contact have a probability of less than 1 to escape into the reservoir, we have a contact with internal reflection. (External reflection by carriers approaching the conductor from a reservoir is always assumed to be present: We have many quantum channels in the reservoirs and only a small number of states connecting contacts). A current source contact with internal reflection populates edge states in a non-equilibrium fashion, similar to the barrier discussed above. If

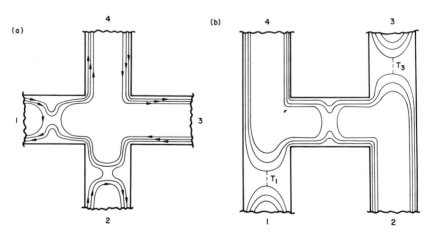

Figure 9.   (a) Conductor with contact 1 reflecting K edge states and contact 2 reflecting L edge states. (After van Wees, et al. 1989a). (b) Conductor with barrier and two weakly coupled contacts. (After Büttiker, 1989b).

contacts with no internal reflection and contacts with internal reflection *alternate* along the perimeter of the sample, all Hall resistances are still quantized (proportional to $1/N$) and all longitudinal resistances are zero. But if two contacts with internal reflection are adjacent, there is at least one Hall measurement which depends on the detailed scattering properties of the contacts. A clear demonstration of this has come with the work of van Wees et al. (1989a). They consider two contacts spaced closely compared to an equilibration length. The width of the contacts can be varied, and thus a barrier is created at the contacts which permits only a limited number of edge states to transmit. Fig. 9a shows a particular situation, where N-K edge states transmit at contact 1 and N-L edge states transmit at contact 2. The Hall resistance $\mathcal{R}_{13,42}$ is determined by the number N of bulk edge states, since carrier flow is from contact 1 to contact 3 via contact 4 which provides equilibration. But if carrier flow is from contact 2 to contact 4 contact 1 can not equilibrate the edge channels since it exhibits internal reflection. The Hall resistance is

$$\mathcal{R}_{24,13} = \left(\frac{h}{e^2} \right) \frac{1}{(N - M)} \tag{5.6}$$

where M=min(K,L). It is the contact which provides less reflection, i. e. which exhibits the larger conductance which determines the outcome of the measurement (van Wees et al. 1989a; van Houten et al. 1989).

Let us briefly return to the conductor of Fig. 8b. Clearly, the anomalous quantization found in the conductor of Fig. 8b also hinges on the properties of the contacts (at least as long as the contacts are within an equilibration length of the barrier (Büttiker, 1988b)). To show this, consider the conductor in Fig. 9b (Büttiker, 1989b, 1989c), where two of the contacts are separated by barriers from the main conducting channel. If this barrier forms a smooth saddle, the contacts interact only with the outermost edge state. Let the probability for transmission from this contact to the outermost edge be $T_1 < 1$ at contact 1 and $T_3 < 1$ at contact 3. The Hall resistances are anti-symmetric in the field and given by $(h/e^2)1/(N - K)$ independent of T_1 and T_3. *All longitudinal resistances are zero (!)* in contrast to the example discussed above. This is so, as seen by inspection of Fig. 9b. It is only the outermost edge state which is measured, and this state penetrates the barrier. These simple examples show the significance of the properties of contacts. This of course is only true as long as the contacts are close enough to the barrier to sense the differing population of the edge states. If the contacts are further than an equilibration length away from the barrier, then the outcome of the measurement is independent of the properties of the contacts and given by Eqs. (5.3-5.5). Nevertheless, the simple examples of Fig. 9 show that one cannot hope to provide a general theory of the quantum Hall effect without addressing the role of contacts.

Komiyama et al. (1989a; 1989b) have performed experiments on conductors with contacts which exhibit internal reflection. The contacts in these experiments do not have the simple properties shown in Fig. 9, where carriers in certain states are either totally transmitted or totally reflected, (except for carriers in one state which permits partial transmission and partial reflection). Instead, consideration of a completely general scattering matrix is needed to describe a contact. Komiyama and Hirai (1989) have extended the discussion of Büttiker (1988b) and have provided a quantitative analysis of the role of contacts in their experiments.

Perhaps the most profound consequence of the concern with the role of contacts is the discovery by Komiyama et al. (1989a; 1989b) and van Wees et al. (1989b) of the exceedingly long distance taken for equilibration of edge states which have initially been populated in a non-equilibrium manner (current injection through a disordered contact). Further these experiments demonstrate that an equilibration does not occur over distances of several hundred μm! Alphenaar et al. (1989) find that current injected into the outermost edge state is equlibrated over macroscopic distances (0.1 mm) with the remaining edge states, except the innermost edge state. A possible explanation of such long relaxation times is provided by the observation that edge states on the same side of the sample can be far apart if the confining potential is smooth. As shown by Martin and Feng (1989) both the elastic interedge scattering rate and the phonon interedge scattering rate acquire a factor $\exp{-1/2(\Delta y/l_B)^2}$. Here Δy is the distance between edge states (on the same side of the sample) and l_B is the magnetic length. Thus with increasing magnetic field both elastic scattering from one edge state to another as well as inelastic phonon scattering is exponentially suppressed.

C. Resonant Departures from the Hall Resistance

Transitions of carriers from one side of the sample to another side of the sample can occur not only because of impurities and deliberately introduced barriers but also for purely geometrical reasons. Consider the conductor shown in Fig. 10. The energy spectrum in the branches of the conductor consists of magneto-electric subbands as shown in Fig. 7a. The channels with low threshold energy might be determined by purely magnetic energies if the width of the wire is large compared to the magnetic length. The threshold energy of the higher lying modes is typically determined both by the confining potential and the magnetic energy. However, in the center of the cross, the electric confinement is relaxed, and therefore, it is possible to have states in the cross which have lower energies than would be predicted from consideration of the branches of the conductor alone. This was noticed by Peeters (1989) who calculates the bound state energies using a variational approach and by Schult et al. 1989, who present exact numerical computations for zero field. The effect on transport has been discussed by Ravenhalll et al. (1989) and Kirczenow (1989a). Kirczenow (1989a) has mapped these states as a function of magnetic field for a Hall conductor with quadratic confinement. The resonant states are somewhat be- low the threshold of the n-th magneto electric subband in the branches of the wire. In other words, as the magnetic field is lowered, the n+1-th magneto electric subband has a precursor which shows up as a resonant state in the center of the cross. It is clear that this is a universal phenomena which is to be expected whenever the magneto-electric confinement plays a role. (If the magnetic field is so strong that the spectrum in the center of the wire is independent of confinement, then of course this phenomena does not occur).

Büttiker (1988d) applied the Breit-Wigner formalism to describe resonant transmission in a Hall conductor. Here we are specifically interested in the case of four-fold symmetric conductors. The decay widths of the resonant states are then all equal, $\Gamma_i \equiv \Gamma$, i = 1,2,3,4. For a conductor which is four fold symmetric, only three transmission probabilities remain to be determined. (See Fig. 6c and Eq. (4.13)). For a four-fold symmetric conductor and a current pattern as in Fig. 10 Büttiker (1988d) predicts

$$T_+ = N - 3\Gamma^2/\Delta \tag{5.7a}$$

$$T_- = T_d = \Gamma^2/\Delta \tag{5.7b}$$

where $\Delta = (E_F - E_r)^2 + 4\Gamma^2$ and $E_r(B)$ is the energy of the resonant state. Therefore, at resonance, $T_+ = N - 3/4$, $T_- = T_d = 1/4$ independent of the confining potential and other physical parameters!

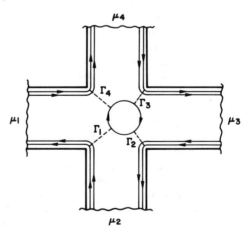

Figure 10.   Conductor with a resonant state coupled via decay widths Γ_i to the edge states. (After Büttiker, 1989d).

This is a muli-terminal analog of the well known fact that transmission through a single resonant state in a two-port conductor is equal to 1, if the decay rate to the right and to the left are equal. Using Eq. (5.7) and Eq. (4.13), the Hall resistance is predicted to decrease from a well quantized plateau at $\mathscr{R}_H = (h/e^2)(1/N)$ to a minimum value

$$\mathscr{R}_H = \mathscr{R}_{13,42} = \left(\frac{h}{e^2}\right)\frac{N-1}{N^2-N+1/2} \tag{5.8}$$

at resonance. Note that for N=1, Eq. (5.8) predicts a complete suppression of the Hall resistance. Such resonant features are clearly seen in the computations of Ravenhall et al. (1989) and Kirczenow (1989a). A more detailed comparison between these analytic predictions and the computational results is the subject of Büttiker and Kirczenow (1989). Interestingly, experiments on very narrow junctions do indeed reveal a sequence of resonant like suppressions of the Hall resistance (Ford et al. 1989b). However, the experimental data are considerably more complex than suggested by the discussion given here. It is possible that since real probes are not perfect, some localization near the threshold of the subbands leads to an even stronger confinement of states in the junction. With the transmission probabilities as specified by Eq. (5.7) we can similarly calculate the longitudinal resistance, (sometimes also referred to as bend resistance) which is zero, when the Hall resistance is quantized. At resonance, it is

$$\mathscr{R}_L = \mathscr{R}_{14,32} = \left(\frac{h}{e^2}\right)\left(\frac{1}{4}\right)\frac{(1-N)}{(N-1/2)(N^2-N+1/2)}. \tag{5.9}$$

which for N=1 is zero and for large N it is negative and proportional to $-1/(4N^2)$. A recent calculation of the bend resistance by Kirczenow (1989b) gives a slightly positive resistance for the one band case N=1 but perhaps is not at high enough fields to make the simple current pattern of Fig. 10 valid.

To summarize: We have shown, by discussing a few examples that an approach to electric conductance which emphasizes transmission from one electron reservoir to another, treating all probes on the same footing, contains a rich amount of physics. It is rather remarkable that the expressions Eqs. (3.4-3.7) apply over the entire range of magnetic fields as long as we deal with single electron transport. Treating all the probes, without apriori assumptions, is not a luxury, but an essential step to understanding transport in small systems. In the quantum Hall regime, where small means 0.1 mm, it is the definition of the Hall resistance with respect to the chemical potentials of the probes which leads to theoretical predictions which are in agreement with experiments.

APPENDIX A: MICROREVERSIBILITY AND CURRENT CONSERVATION

The resistances discussed above are formulated in terms of transmission probabilities. To understand some of the general properties of these transmission probabilities it is useful to consider not the probabilities but the probability amplitudes. Since scattering in the sample is elastic the scattering properties of the sample must be described by a scattering matrix S. The scattering matrix relates the incident current amplitude $\alpha_{j,l}$ in lead j and channel l to the outgoing current amplitudes $\alpha'_{i,k}$ in lead i and channel k. The linear relation between the in and outgoing current amplitudes is

$$\alpha'_{i,k} = \sum_{j=1}\sum_{l=1}^{l=M_j} s_{ij,kl}\alpha_{j,l}. \tag{A1}$$

The index j of the first sum runs over all the leads. Since current is conserved, it follows that S must be a unitary matrix, $S^\dagger = S^{-1}$. Here † denotes Hermitian conjugation. Time reversal (reversal of the momentum and the magnetic field) implies $S_*(-B) = S^{-1}(B)$. Here the asterisk denotes complex conjugation. Taken together these two conditions imply the reciprocity of the S matrix, $S^T(B) = S(-B)$, or, expressed in terms of the matrix elements,

$$s_{ij,kl}(B) = s_{ji,lk}(-B) \tag{A2}$$

The scattering amplitude $s_{ij,kl}(B)$ connecting incident carriers in lead j and channel l and outgoing carriers in lead i and channel k in in the presence of a magnetic field, is the same as the scattering amplitude connecting incident carriers in lead i and channel k and out-going carriers in lead j channel l in the presence of a magnetic field -B. The elements of the S matrix can be grouped into transmission and reflection matrices,

$$\begin{pmatrix} r_{11} & t_{12} & t_{13} \\ t_{21} & r_{22} & t_{32} \\ t_{31} & t_{32} & r_{33} \end{pmatrix} \tag{A3}$$

and similarly for a conductor with an arbitrary number of leads. In Eq. (A3) r_{11} is a matrix withhcthe elements $r_{ii,mn}$ and t_{ij} is a matrix with elemnts $t_{ij,mn}$. Thus the transmission and reflection probabilities, $T_{ij,mn} = |t_{ij,mn}|^2$, $R_{ij,mn} = |r_{ij,mn}|^2$ have the property,

$$T_{ij,kl}(B) = T_{ji,lk}(-B), \quad R_{ii,kl}(B) = R_{ii,lk}(-B). \tag{A4}$$

In terms of these transmission and reflection probabilities current conservation requires,

$$\sum_{l=1}^{l=M_i} R_{ii,kl} + \sum_{j \neq i} \sum_{l=1}^{l=M_j} T_{ij,kl} = 1, \tag{A5}$$

and

$$\sum_{k=1}^{k=M_i} R_{ii,kl} + \sum_{j \neq i} \sum_{k=1}^{k=M_j} T_{ji,kl} = 1. \tag{A6}$$

Mathematically, Eq. (A5) and (A6) state that the rows and colums of the matrix given by Eq. (A3) are normalized to 1. Physically, Eq. (A5) states that if all incident channels are full, then the out-going channels must also be full. Eq. (A6) states that the sum of all the transmitted and reflected currents due to an incident current of 1 in lead i and channel k must also be 1. Consider now the total reflection and transmission probabilities. Using the reciprocity of the reflection probabilities, we find

$$R_{ii}(B) = \sum_{k=1, l=1}^{k=M_i, l=M_i} R_{ii,kl}(B) = \sum_{k=1, l=1}^{k=M_i, l=M_i} R_{ii,lk}(-B) = R_{ii}(-B) \tag{A7}$$

and

$$T_{ij}(B) = \sum_{k=1, l=1}^{k=M_i, l=M_j} T_{ij,kl}(B) = \sum_{k=1, l=1}^{k=M_i, l=M_j} T_{ji,lk}(-B) = T_{ji}(-B). \tag{A8}$$

Thus the total reflection probability is symmetric in the magnetic field and the total transmission probability obeys a reciprocity relation. If we use the definition of the total transmission and reflection probabilities, Eq. (A5) gives,

$$M_i = R_{ii} + \sum_{j \neq i} T_{ij}, \tag{A9}$$

and from Eq. (A6),

$$M_i = R_{ii} + \sum_{j \neq i} T_{ji}. \tag{A10}$$

In a more formal way the total transmission probabilities in Eqs. (A7) and (A8) can be written as a trace of the matrices introduced in Eq. (A3);

$$T_{ij} = \mathrm{Tr}(t_{ij}^\dagger t_{ij}), \quad R_{ii} = \mathrm{Tr}(r_{ii}^\dagger r_{ii}). \tag{A11}$$

As is well known, the trace has certain invariance properties. In particular, there exists a unitary transformation $U = \Pi_i U_i$ such that incident and outgoing currents are separately conserved. Under this transformation the reflection and transmission matrices introduced in Eq. (A3) become $U_i^{-1} t_{ij} U_j$ and

70

$U_i^{-1} r_{ii} U_i$. The group of transformations U leaves the total transmission and reflection probabilities (defined by Eqs. A7, A8 and A11) invariant. There is an equivalence class of scattering matrices $U^{-1}SU$ characterized by the same resistance (Eq. 3.5). Most importantly, there exists a particular U such that the transformed transmission and reflection matrices are *diagonal*. Such a representation is useful, for instance, in the discussion of Breit-Wigner resonances (Büttiker, 1988c, 1988d).

REFERENCES

Alphenhaar, B. W., McEuen, P. L., Wheeler, R. G., and Sacks, R. N., 1989 (unpublished).

Anderson, P. W., Thouless, D. J., Abrahams, E., and Fisher, D. S., 1980, Phys. Rev. **B22**, 3519.

Aronov, A. G., and Sharvin, Yu. V., 1987 Rev. Mod. Phys. **59**, 755.

Avishai, Y., and Band, Y. B., 1989, Phys. Rev. **B62**, 2527.

Azbel, M. Ya, 1981, J. Phys. **C14**, L225.

Baranger, H. U., and Stone, A. D., 1989a, Phys. Rev. Lett. **63**, 414.

Baranger, H. U., and Stone, A. D., 1989b, Phys. Rev. **B**, (unpublished).

Baranger, H. U., Stone, A. D., and DiVincenzo, D. P., 1988, Phys. Rev. B **37**, 6521.

Beenakker, C. W. J., and van Houten, H., 1988, Phys. Rev. Lett. **60**, 2406.

Beenakker, C. W. J., and van Houten, H., 1989a, unpublished.

Beenakker, C. W. J., and van Houten, H., 1989b, Phys. Rev. **B39**, 10445.

Beenakker, C. W. J., van Houten, H., and van Wees, B. J., 1989, Supperlattices and
Microstructures **5**, 127.

Benoit, A. D., Washburn, S., Umbach, C. P., Laibowitz, R. P., and Webb, R. A., 1986,
Phys. Rev. Lett. **57** , 1765.

Beton, P. H., Snell, B. R., Main, P. C., Neves, A., Owers-Bradley, J. R., Eaves, L., Henini, M.,
Hughes, O. H., Beaumont, S. P., and Wilkinson, C. D. W., (1989), J. Phys. Cond. Matter (unpublished).

Büttiker, M., 1985, Phys. Rev. **B32**, 1846.

Büttiker, M., 1986a, Phys. Rev. Lett. **57**, 1761.

Büttiker, M., 1986b, Phys. Rev. **B33**, 3020.

Büttiker, M., 1986c, in *New Techniques and Ideas in Quantum Measurement Theory* ,
Annals of the New York Academy of Sciences, Vol. 480, page 194.

Büttiker, M., 1987, Phys. Rev. B **35** , 4123.

Büttiker, M., 1988a, IBM J. Res. Develop. **32**, 317.

Büttiker, M., 1988b, Phys. Rev. **B38**, 9375.

Büttiker, M., 1988c, IBM J. Res. Develop. **32**, 63.

Büttiker, M., 1988d, Phys. Rev. **B38**, 12724.

Büttiker, M., 1989a, Phys. Rev. Lett. **62**,229.

Büttiker, M., 1989b, in "Nanostructure Physics and Fabrication",
M. A. Read and W. P. Kirk, eds., Academic Press, Boston (in press).

Büttiker, M., 1989c, Surface Science, (unpublished).

Büttiker, M., 1989d, Phys. Rev. **B40**, 3409.
An extended account of this work is given in *Analogies in Optics and Micorelectronics*, W. van
Haeringen and D. Lenstra, eds., Kluwer Academic Publishers, Dordrecht.

Büttiker, M., and Kirczenow, G., 1989, (unpublished).

Büttiker, M., Imry, Y., Landauer, R., and Pinhas, S., 1985, Phys. Rev. **B31**, 6207.

Casimir, H. B. G., 1945, Rev. Mod. Phys. **17**, 343.

Chang, A. M., Timp, G., Cunningham, J. E., Mankiewich, P., M., Behringer, R., E., and
Howard, R. E., 1988, Solid State Commun. **76**, 769.

Chang, A. M., Chang, T. Y., and Baranger, H. U., 1989, Phys. Rev. Lett. **63**, 996.

Cheung, H. F., Gefen, Y., and Riedel, E. K., 1988, IBM J. Res. Develop. **32**, 359.

D'Amato, J. L., and Patawski, H. M., 1989, "Conductance of a Disordered Linear Chain
Including Inelastic Scattering Events", (unpublished).

D'Amato, J. L., Pastawski, H. M., and Weisz, J. F., 1989, Phys. Rev. **B39**, 3554.

Datta, S., 1989, Phys. Rev. **B40**, 5830.

DiVincenzo, D. P., and Kane, C. L., 1988, Phys. Rev. **B38**, 3006.

Duke, C. B., 1969, in *Solid State Physics,* Supplement 10, (Academic Press, New York).

Economou, E. N., and Soukoulis, C. M., 1981, Phys. Rev. Lett. **46**, 618.

Escapa, L., and Garcia, N., 1989, J. Phys. **C1**, 2125.

Fang, F. F., and Stiles, P. J., 1984, Phys. Rev. **B29**, 3749.

Fisher, D. S., and Lee, P. A., 1981, Phys. Rev. **B23**, 6851.

Ford, C. J. B., Washburn, S., Büttiker, M., Knoedler, C. M., and Hong, J. M., 1989a, Phys. Rev. Lett. **62**, 2724.

Ford, C., J., B., Washburn, S., Büttiker, M., Knoedler, C. M., and Hong, J. M., 1989b, Surface Science, (unpublished).

Frenkel, J., 1930, Phys. Rev. **36**, 1604.

Glazman, L. I., Lesovik, G. B., Khmel'nitskii, D. E., and Shekter, R. I., 1988, JETP Lett. **48**, 239.

Halperin, B. J., 1982, Phys. Rev. **B25**, 2185.

Haug, R. J., Kucera, J., Streda, P., and von Klitzing, K., 1989, Phys. Rev. **B39**, 10892.

Haug, R., J., MacDonald, A. H., Streda, P., and von Klitzing, K., 1988, Phys. Rev. Lett. **61**, 2797.

Imry, Y., 1986, in *Directions of Condensed Matter Physics*, G. Grinstein and G. Mazenko, eds., World Scientific, Singapore, 1986. Vol. 1. page 101.

Jain, J. K., and Kivelson, S. A., 1988, Phys. Rev. Lett. **60**, 1542.

Kane, C. L., Lee, P. A., DiVincenzo, D. P., 1988, Phys. Rev. B **38**, 2995.

Kazarinov, R. F., and Luryi, S., 1982, Phys. Rev. **B25**, 7626.

Kirczenow, G., 1988, Solid State Commun. **68**, 715.

Kirczenow, G., 1989a, Phys. Rev. Lett. **62**, 2993.

Kirczenow, G., 1989b, Solid State Commun. **71**, 469.

Komiyama, S., and Hirai, H., 1989, "Theory of Contacts in a Two Dimensional Electron Gas at High Magnetic Fields", (unpublished).

Komiyama, S., Hirai, H., Sasa, S., and Fuji, F., 1989a, "Influence of Disordered Contacts on the Four-Terminal Measurements of Integral Quantum Hall Effects", (unpublished).

Komiyama, S., Hirai, H., Sasa, S., and Fuji, F., 1989b, "Quantitative Analysis of the Role of Contacts in the Measurement of Integral Quantum Hall Effects", (unpublished).

Komiyama, S., Hirai, H., Sasa, S., and Hiyamizu, S., 1989, "Violation of the Integral Quantum Hall Effect: Influence of Back Scattering and A Role of Voltage Contacts", (unpublished).

Langreth, D. C., and Abrahams, E., 1981, Phys. Rev. **B24**, 2978.

Landauer, R., 1957, IBM J. Res. Dev. **1**, 223.

Landauer, R., 1970, Phil. Mag.**21**, 863.

Landauer, R., 1981, Phys. Lett. **85A**, 91.

Landauer, R., 1987, Z. Phys. **68**, 217.

Landauer, R., 1989, J. Phys. C, (unpublished).

MacDonald, A. H., and Streda, P., 1984, Phys. Rev. **B29**, 1616.

Martin, T., and Feng, S., 1989, "Suppression of Scattering in Electron Transport in Mesoscopic Quantum Hall Systems", (unpublished).

Peeters, F. M., 1988, Phys. Rev. Lett. **61**, 589.

Peeters, F. M., 1989, Supperlattices and Microstructures, **6**, 217.

Ravenhall, D. G., Wyld, H. W., and Schult, R. L., 1989, Phys. Rev. Lett. **62**, 1780.

Roukes, M. L., Scherer, A., Allen, Jr., S. J., Craighead, H. G., Ruthen, R. M., Beebe, E. D., and Harbison, J. P., 1987, Phys. Rev. Lett. **59**, 3011.

Schult, R. L., Ravenhall, D. G., and Wyld, H. W., 1988, Phys. Rev. **B39**, 5476.

Sharvin, Yu. V., 1965, Soviet Phys. JETP **21**, 655.

Sivan, U., Imry, Y., and Hartzstein, C., 1989, Phys. Rev. **B39**, 1242.

Snell, B. R., Beton, P. H., Main, P. C., Neves, A., Owers-Bradely, J. R., Eaves, L., Heini, M., Hughes, O. H., Beaumont, S. P., and Wilkinson, C. D. W., J. Phys. C (unpublished).

Sokolovski, D., 1988, Phys. Lett. **A123**, 381.

Syphers, D. A., and Stiles, P. J., 1985, Phys. Rev. **B32**, 6620.

Stone, A. D., and Szafer, A., 1988, IBM J. Res. Developm., **32**, 384.

Streda, P., Kucera, J., and MacDonald, A. H., 1987, Phys. Rev. Lett. **59**, 1973.

Streda, P., 1989, J. Phys. **C1**, L12025.

Szafer, A., and Stone, A. D., 1989, Phys. Rev. Lett. **62**, 300.

Takagaki, Y., Gamo, K., Namba, S., Ishida, S., Takaoka, S., Murase, K., Ishibashi, K., and Aoyagi, Y., 1988, Solid State Communic. **68**, 1051.

Takagaki, Y., Gamo, K., Namba, S., Takaoka, S., Murase, K., and Ishida S., 1989, Solid State Communic. **71**, 809.

Tekman, E., and Ciraci, S., 1989, Phys. Rev. **B39**, 8772.

Timp, G., Baranger, H. U., deVegar, P., Cunningham, J. E., Howard, R. E., Behringer, R., and Mankiewhich, P. M., 1988, Phys. Rev. Lett. **60**, 2081.

Thouless, D. J., 1981, Phys. Rev. Lett. **47**, 972.

Tsu, R., and Esaki, L., 1973, Appl. Phys. Lett. **22**, 562.

van der Marel, D., and Haanappel, E. G., 1989, Phys. Rev. **B39**, 7811.

van der Pauw, L. J., 1958, Phillips Res. Rep. **13**, 1.

van Houten, H., Beenakker, C. W. J., and van Loosdrecht, P. H. M., Thornton, T. J.,
Ahmed, H., and Pepper, M., Foxon, C. T., and Harris, J. J., 1988, Phys. Rev. **B37**, 8534.

van Houten, H., Beenakker, C. W. J., Williamson, J. G., Broekaart, M. E. I., and
van Loosdrecht, P. H. M., van Wees, B. J., Moij, J. E., Foxon, C. T., and Harris, J. J., 1989,
Phys. Rev. **B39**, 8556.

van Wees et al., van Houten, H., Beenakker, C. W. J., Williamson, J. G.,
Kouwenhouven, L. P., van der Marel, D., and Foxon, C. T., 1988, Phys. Rev. Lett. **60**, 848.

van Wees, B. J., Willems, E. M. M., Harmans, C. J. P. M., Bennakker, C. W. J.,
van Houten, H., Williamson, J. G., Foxon, C. T., and Harris, J. J., 1989, Phys. Rev. Lett. **62**,
1181.

van Wees, B. J., Willems, E. M. M., Kouwenhoven, L. P., Harmans, C. J. P. M.,
Williamson, H. G., Foxon, C. T., and Harris, J. J., 1989, Phys. Rev. **B39**, 8066.

von Klitzing, K., Dorda, G., and Pepper, M., 1988, Phys. Rev. Lett. **45**, 494.

Washburn, S., Fowler, A. B., Schmid, H., and Kern, D., 1988, Phys. Rev. Lett. **61**, 2801.

Webb, R. A., and Washburn, S., 1988, Physics Today **41**, Dec., 46.

Wharam, D. A., Thornton, T. J., Newbury, R., Pepper, M., Ahmed, H., Frost, J. E. F.,
Hasko, D. G., Peacock, D. C., Ritchie, D. A., and Jones, G. A. C., 1988, J. Phys. **C21**, L209.

Yacoby, A., and Imry, Y., 1989, (unpublished).

SEMI-CLASSICAL THEORY OF MAGNETORESISTANCE ANOMALIES

IN BALLISTIC MULTI-PROBE CONDUCTORS

C.W.J. Beenakker and H. van Houten

Philips Research Laboratories
5600 JA Eindhoven
The Netherlands

1. INTRODUCTION

The regime of ballistic transport in a two-dimensional electron gas (2DEG) was opened up a few years ago, when it became possible technically to reduce the dimensions of a conductor to below a mean free path. In this regime the resistance is determined by the geometry of the conductor, to the extent that impurity scattering can be neglected. In the usual regime of diffusive transport, the Hall bar geometry (a straight current-carrying channel with small side contacts for voltage drop measurements) is most convenient to determine the various components of the resistivity tensor separately. A down-scaled Hall bar was therefore the natural first choice as a geometry to study ballistic transport in a 2DEG (Timp et al., 1987; Roukes et al., 1987; Takagaki et al., 1988; Simmons et al., 1988; Chang et al., 1988; Ford et al., 1988). The point contact geometry (a short and narrow constriction) was an alternative choice (Van Wees et al., 1988; Wharam et al., 1988; Van Houten et al., 1988a). As it turns out, it is much easier to understand ballistic transport through a point contact than through a narrow Hall bar. The reason is that the resistance of a point contact is determined mainly by the number of occupied 1-dimensional subbands at the narrowest point of the constriction, and not so much by its shape (cf. the very similar results of Van Wees et al. (1988) and Wharam et al. (1988) on the quantized resistance of point contacts of a rather different design). The resistances measured in a narrow channel geometry, in contrast, are mainly determined by scattering at the junction with the side probes (Timp et al., 1988), which is different for junctions of different shape. The strong dependence of the low-field Hall resistance on the junction shape was demonstrated theoretically by Baranger and Stone (1989), and experimentally by Ford et al. (1989a) and Chang et al. (1989). These results superseded many earlier attempts (including one of our own) to explain the discovery by Roukes et al. (1987) of the *quenching of the Hall effect* without modelling the shape of the junction realistically (Beenakker and Van Houten, 1988; Peeters, 1988; Phillips, 1988; Akera and Ando, 1989; Srivastava, 1989; Johnston and Schweitzer, 1989; Isawa, 1989). Baranger and Stone (1989) argued that the rounded corners (present in a realistic situation) at the junction between the main channel and the side branches lead to a suppression (quenching) of the Hall resistance at low magnetic fields as a consequence of the *horn collimation effect* (an effect first proposed in the context of the point contact geometry (Beenakker and Van Houten, 1989a)). A Hall bar with straight corners, in contrast, does not show a generic suppression of the Hall resistance (Ravenhall et al., 1989; Kirczenow, 1989a), although quenching can occur for special parameter values if only a few subbands are occupied in the channel.

Electronic Properties of Multilayers and Low-Dimensional Semiconductors Structures
Edited by J. M. Chamberlain *et al.*, Plenum Press, New York, 1990

75

The quenched Hall effect (Roukes et al., 1987; Ford et al., 1988; 1989a; Chang et al., 1989) is just one of a whole variety of magnetoresistance anomalies observed in narrow Hall bars. Other anomalies are: the *last Hall plateau* (Roukes et al., 1987; Timp et al., 1987; Simmons et al., 1988; Ford et al., 1988; 1989a; Chang et al., 1988; 1989), reminiscent of quantum Hall plateaus, but occurring at much lower fields; the *negative Hall resistance* (Ford et al., 1989a), as if the carriers were holes rather than electrons; the *bend resistance* (Takagaki et al., 1988; 1989a; 1989b; Timp et al., 1988; 1989), a longitudinal resistance associated with a current bend, which is negative at small magnetic fields and zero at large fields, with an overshoot to a positive value at intermediate fields; and more.

In a recent paper (Beenakker and Van Houten, 1989b) we have shown that, at least qualitatively, all these phenomena can be explained in terms of a few simple semi-classical mechanisms. Since the magnetoresistance anomalies could be reproduced by a numerical simulation of the trajectories of electrons at the Fermi energy, it could be concluded that these are essentially classical rather than quantum size effects. In the present paper several aspects of our theory of junction scattering are discussed in more detail. In addition we present a direct comparison between the simulation and representative experiments, in order to determine to what extent a quantitative description of the magnetoresistance anomalies can be obtained with a semi-classical model in which quantum interference effects and lateral quantisation are not taken into account.

The outline of this paper is as follows. In Sec. 2 the classical mechanisms responsible for the magnetoresistance anomalies are presented. Our method of simulation is described in Sec. 3, and the results are compared with experiment in Sec. 4. We conclude in Sec. 5 with a critical discussion of the merits and limitations of our approach, and of the various alternative mechanisms proposed for the quenching of the Hall effect. In that section we also describe the modification of electron focusing in a narrow channel geometry and discuss the possible role of chaos in ballistic transport.

2. MECHANISMS

The variety of magnetoresistance anomalies mentioned above can be understood in terms of a few simple characteristics of the curved trajectories in a classical billiard in the presence of a perpendicular magnetic field (Beenakker and Van Houten, 1989b). At very small magnetic fields, *collimation* and *scrambling* are the key concepts. The gradual widening of the channel on approaching the junction reduces the injection/acceptance cone, which is the cone of angles with the channel axis within which an electron is injected into the junction, or within which an electron can enter the channel coming from the junction. This is the horn collimation effect (Beenakker and Van Houten, 1989a). An experimental demonstration of this effect using two opposite point contacts as injector and collector of the collimated beam has been given by Molenkamp et al. (1989). If the injection/acceptance cone is smaller than 90°, then the cones of two channels at right angles do not overlap. That means that an electron approaching the side probe coming from the main channel will be reflected (Fig. 1a), and will then typically undergo multiple reflections in the junction region (Fig. 1b). The trajectory is thus scrambled, whereby the probability for the electron to enter the left or right side probe in a weak magnetic field is equalized. This suppresses the Hall voltage. Our "scrambling" mechanism for the quenching of the Hall effect requires a weaker collimation than the "nozzle" mechanism put forward by Baranger and Stone (1989) (we return to both these mechanisms in Sec. 5). Scrambling is not effective in the geometry shown in Fig. 1c, in which a large portion of the boundary in the junction is oriented at approximately 45° with the channel axis. An electron reflected from a side probe at this boundary has a large probability of entering the opposite side probe. This is the "rebound" mechanism for a negative Hall resistance proposed by Ford et al. (1989a).

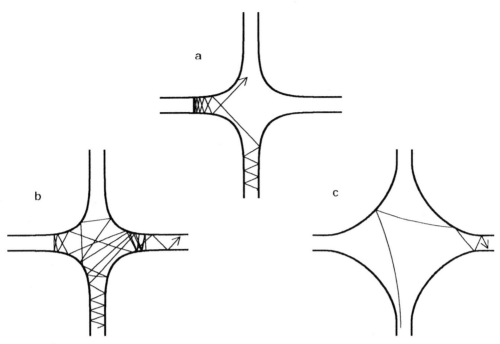

Fig. 1. Classical trajectories in an electron billiard, illustrating the collimation (a), scrambling (b), and rebound (c) effects.

At somewhat larger magnetic fields, *guiding* takes over. As illustrated in Fig. 2a, the electron is guided by the magnetic field along equipotentials around the corner. Guiding is fully effective when the cyclotron radius $l_{cycl} \equiv \hbar k_F / eB$ (with k_F the Fermi wave vector) becomes smaller than the minimal radius of curvature r_{min} of the corner, that is for magnetic fields greater than the guiding field $B_g \equiv \hbar k_F / er_{min}$. In the regime $B \gtrsim B_g$ the junction can not scatter the electron back into the channel from which it came. The absence of backscattering is characteristic for the quantum Hall effect regime (Büttiker, 1988), but is in this case an entirely classical, weak-field phenomenon (Van Houten et al., 1988b). Because of the absence of backscattering, the longitudinal resistance vanishes, and the Hall resistance R_{II} becomes equal to the contact resistance of the channel, just as in the quantum Hall effect — but without quantisation of R_{II}. The contact resistance $R_{contact} \approx (h/2e^2)(\pi/k_F W)$ is approximately independent of the magnetic field for fields such that the cyclotron diameter $2l_{cycl}$ is greater than the channel width W, that is for fields below $B_{crit} \equiv 2\hbar k_F / eW$ (Van Houten et al., 1989). This explains the occurrence of the socalled "last plateau" in R_{II} for $B_g \lesssim B \lesssim B_{crit}$ as a classical effect. At the low-field end of the plateau, the Hall resistance is sensitive to *geometrical resonances* which increase the fraction of electrons guided around the corner into the side probe. Fig. 2b illustrates the occurrence of such a geometrical resonance as a result of the magnetic focusing of electrons into the side probe, at magnetic fields such that the separation D of the two perpendicular channels (measured along the junction boundary, see Fig. 2b) is an integer multiple of the cyclotron diameter. This is in direct analogy with electron focusing in a double point contact geometry (Van Houten et al., 1988a; 1989). As shown in Fig. 3, electron focusing thus leads to periodic oscillations superimposed on the Hall plateau with maxima at approximately integer multiples of the focusing field $B_{focus} \equiv 2\hbar k_F / eD$ (indicated by arrows). In this geometry the oscillations become very large because the boundary at the junction has no curvature.

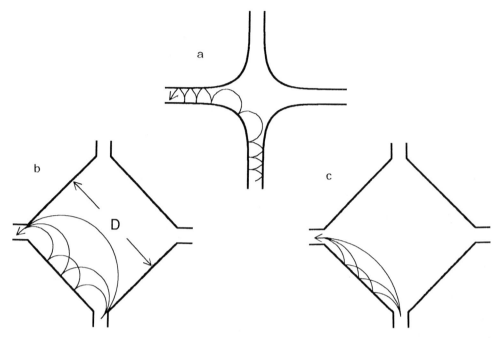

Fig. 2. Illustration of the guiding effect (a), and of two mechanisms leading to geometrical resonances (b,c). In (b) the trajectories are shown at three multiples of the focusing field B_{focus}, in (c) at three multiples of $B_{focus} / \sqrt{2}$.

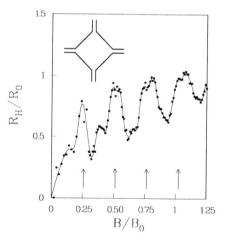

Fig. 3. Hall resistance in the hard-wall geometry shown in the inset. The curve drawn through the calculated data points is a guide to the eye. The arrows indicate magnetic fields at multiples of $B_{focus} \equiv (2W/D) B_0$ at which electron focusing in the junction occurs (cf. Fig. 2b). The parameters B_0 and R_0 are defined in Eq. (5).

Electron focusing in a Hall cross requires that electrons can be injected into the junction at 90° with the side wall connecting the injection probe with the side probe. The collimation effect, however, favors injection at 45° with this boundary, and thus may suppress electron focusing in favor of another geometrical resonance, illustrated in Fig. 2c. A collimated beam is bent by the magnetic field into the side probe (without focusing) when D is an integer multiple of the chord length $2l_{cycl}/\sqrt{2}$ of the electrons in the beam. This leads to oscillations in R_{H} with periodicity $B_{focus}/\sqrt{2}$. In Fig. 3 these more rapid oscillations are not clearly visible because of the absence of appreciable collimation in this particular geometry. In realistically smooth geometries both mechanisms discussed above, as well as additional geometrical resonances, may play a role. Note that these mechanisms for oscillations in the resistance depend on a commensurability between the cyclotron radius and a characteristic dimension of the junction, but do not involve the wave length of the electrons as an independent length scale. This distinguishes these geometrical resonances conceptually from the quantum resonances due to bound states in the junction considered by others (Avishai and Band, 1989; Ravenhall et al., 1989; Kirczenow, 1989a; 1989b; Peeters, 1989).

3. MODEL

To calculate the resistances in the semi-classical limit we use the Landauer-Büttiker formalism (Landauer, 1957; Büttiker, 1986), by which the resistances can be expressed as rational functions of transmission probabilities for electrons with the Fermi energy. The central equations, in a form suitable for a semi-classical calculation, are

$$I_i = G_i(1 - t_{i \to i})V_i - \sum_{j \neq i} G_j t_{j \to i} V_j, \tag{1}$$

where I_i is the current in channel i, eV_i is the chemical potential of a reservoir in equilibrium connected to channel i, $G_i \equiv (2e^2/h)N_i$ is the contact conductance of channel i, and $t_{j \to i}$ is the fraction of the current injected into channel j which leaves the system via channel i. The number N_i in the definition of the contact conductance is the number of transverse waveguide modes at the Fermi energy in channel i. In the semi-classical limit, N is treated as a continuous variable, e.g. at zero magnetic field $N \equiv k_F W/\pi$ for a channel defined by a square well confining potential with width W. The transmission probabilities in a magnetic field B satisfy the symmetry relation (Büttiker, 1986)

$$G_j(B) t_{j \to i}(B) = G_i(B) t_{i \to j}(-B), \tag{2}$$

(note that G is symmetric in B), and obey the normalization

$$\sum_j t_{i \to j} = 1. \tag{3}$$

Eqs. (2) and (3) together imply that

$$\sum_j G_j t_{j \to i} = G_i. \tag{4}$$

Once the coefficients in Eq. (1) are known, the voltages V_i can be obtained by solving the set of linear equations for given currents I_i. The four-terminal resistances are defined as $R_{ij,kl} \equiv (V_k - V_l)/I$, where $I_i = -I_j \equiv I$ and $I_m = 0$ for $m \neq i,j$.

The algorithm used to calculate the coefficients in Eq. (1) is straightforward, except for one point (the injection distribution). We simulate the injection of a large number (from 10^3 to 4×10^4, depending on the accuracy required) of electrons towards

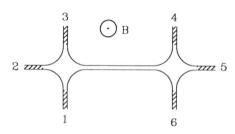

Fig. 4. Double junction geometry, with the hard-wall, field-free leads shown shaded.

the junction through channel j, and integrate Newton's equations of motion numerically to determine the fraction $t_{j \to i}$ of electrons leaving the junction via channel i. The injection distribution has to be chosen such that the current injected in the channel is uniformly distributed among the modes, to simulate injection by a reservoir in thermal equilibrium. In a hard-wall channel (defined by a square well confining potential), in zero magnetic field, this is realized by injecting the electrons uniformly over the channel width W, with Fermi velocity v_F, and angular distribution $P(\alpha) = (\cos \alpha)/2$ (α in the interval $(-\pi/2, \pi/2)$ being the angle with the channel axis). The contact conductance is then given by $G = (2e^2/h)(k_F W/\pi)$, and all coefficients in Eq. (1) can be obtained. For other confining potentials, or for $B \neq 0$, both the injection distribution and the contact conductances are different, and not easily calculated. We circumvent this difficulty by attaching to each channel of the structure a hard-wall lead in which $B = 0$ (shaded in Fig. 4). This trick does not change the resistances in the semi-classical limit (see below), while it permits us to use the simple expressions for $P(\alpha)$ and G given above.

It remains to prove the correctness of the injection trick. Consider attaching to channel i a hard-wall, field-free lead (one of the shaded leads in Fig. 4). First note that trajectories which leave the junction through channel i are not reflected at the interface with the shaded lead. Moreover, if the shaded lead is attached at more than a few channel widths from the junction it has no influence on the dynamics in the junction itself. The transmission probabilities $t_{j \to j}$ and $t_{j \to i}$ for $j \neq i$ are therefore unaffected. The contact conductances G_j for $j \neq i$ are also unchanged, of course. This holds for all B, so that in view of the symmetry relation (2) the product $G_i t_{i \to j}$ for $j \neq i$ is unchanged as well. Finally, also the term $G_i(1 - t_{i \to i})$ remains the same, since by virtue of Eq. (4) this quantity is given by

$$G_i(1 - t_{i \to i}) = \sum_{j \neq i} G_j t_{j \to i}.$$

All the coefficients in Eq. (1), which determines the resistances, are therefore the same as they were before the shaded lead was attached to channel i. We can now repeat the argument and attach a hard-wall, field-free lead to each of the channels without any effect on the resistances. Note that this injection trick is correct in the semi-classical limit only. In the quantum-mechanical problem, spurious reflections will occur on approaching the shaded lead from the unshaded part of the channel. These may be eliminated by a gradual (adiabatic) transition to a hard-wall, field-free lead (Baranger and Stone, 1989). By our method the properties of the reservoirs, on which the Landauer-Büttiker formalism is based, are taken into account properly. For a discussion of the extent to which actual ohmic contacts approximate the theoretical reservoirs, see e.g. the paper by Komiyama and Hirai (1989).

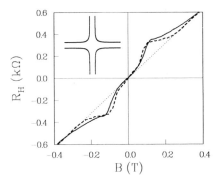

Fig. 5. Hall resistance as measured (solid curve) by Simmons et al. (1988), and as calculated (dashed curve) for the hard-wall geometry in the inset ($W = 0.8\,\mu$m and $E_F = 14$ meV). The dotted line is R_H in a bulk 2DEG.

4. THEORY VERSUS EXPERIMENT

An overview of the magnetoresistance anomalies exhibited by the semi-classical theory has been given in our earlier paper (Beenakker and Van Houten, 1989b). Here we present a direct comparison between theory and representative experiments on laterally confined two-dimensional electron gases in high-mobility GaAs-AlGaAs heterostructures. In the calculation we chose a parabolic confining potential for the narrowest channels (of width around 100 nm), and a square well confining potential for wider channels. In the junction the equipotentials are segments of the curve $|x|^p + |y|^p = $ constant, with the power $p > 1$ parameterizing the smoothness of the corners (the larger p, the sharper the corners). We first discuss the Hall resistance R_H.

Fig. 5 shows the precursor of the classical Hall plateau (the "last plateau") in a relatively wide Hall cross. The experimental data* (solid curve) is from a paper by Simmons et al. (1988). Our calculation (dashed curve) is for a square well confining potential of channel width $W = 0.8\,\mu$m (as estimated in the experimental paper), and with the relatively sharp corners shown in the inset (corresponding to $p = 8$, $r_{min} \approx 0.8W$). The Fermi energy used in the calculation is $E_F = 14$ meV, which corresponds (via $n_s = E_F\,m/\pi\hbar^2$) to a sheet density in the channel of $n_s = 3.9 \times 10^{15}$ m^{-2}, somewhat below the value of 4.9×10^{15} m^{-2} of the bulk material in the experiment. Good agreement between theory and experiment is seen in Fig. 5. Near zero magnetic field, the Hall resistance in this geometry is close to the linear result $R_H = B/en_s$ for a bulk 2DEG (dotted line). The corners are sufficiently smooth to generate a Hall

---

*The experimental curves in Figs. 5 and 7 are (anti-)symmetrized resistances,

$$R^{\wedge}_{ij,kl} = \frac{1}{2}\,[R_{ij,kl}(B) - R_{kl,ij}(-B)]\,, \quad R^{S}_{ij,kl} = \frac{1}{2}\,[R_{ij,kl}(B) + R_{kl,ij}(-B)]\,,$$

reported in this way in many of the experimental papers.

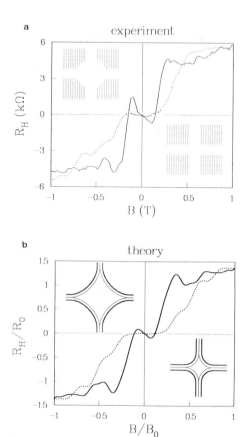

Fig. 6. Hall resistance as measured (a) by Ford et al. (1989a), and as calculated (b). In (a) as well as in (b), the solid curve corresponds to the geometry in the upper left inset, and the dotted curve to the geometry in the lower right inset. The insets in (a) indicate the shape of the gates, not the actual confining potential. The insets in (b) show equipotentials of the confining potential at E_F (thick contour) and 0 (thin contour). The potential rises parabolically from 0 to E_F, and vanishes in the diamond-shaped region at the center of the junction.

plateau[†] via the guiding mechanism discussed in Sec. 2. The horn collimation effect, however, is not sufficiently large to suppress R_H at small B. Indeed, the injection/acceptance cone for this junction is considerably wider than the maximal angular opening of 90° required for quenching of the Hall effect via the scrambling mechanism of Sec. 2 (see the angular injection distribution given in Fig. 3, dashed curve, of our earlier paper (Beenakker and Van Houten, 1989b), which shows an injection/acceptance cone of about 115° for this geometry).

The low-field Hall resistance changes drastically if the channel width becomes smaller, relative to the radius of curvature of the corners. Fig. 6a shows experimental data by Ford et al. (1989a). The solid and dotted curves are for the geometries shown respectively in the upper left and lower right insets of Fig. 6a. Note that these insets indicate the gates with which the Hall crosses are defined electrostatically. The equipotentials in the 2DEG will be smoother than the contours of the gates. The experiment shows a well developed Hall plateau, with superimposed fine structure. At small positive fields R_H is either quenched or negative, depending on the geometry. The geometry is seen to affect also the width of the Hall plateau — but not the height. In Fig. 6b we give our numerical results for the two geometries in the insets, which we believe to be reasonable representations of the confining potential induced by the gates in the experiment (although no attempt was made to actually solve the electrostatic problem). We used a parabolic confining potential in the channel and equipotentials at the corners defined by $p = 1.7$ and $p = 2$ in the upper left and lower right inset, respectively[*]. In the theoretical plot the resistance and the magnetic field are given in units of

$$R_0 \equiv \frac{h}{2e^2} \frac{\pi}{k_F W} \quad , \quad B_0 \equiv \frac{\hbar k_F}{eW} \quad , \tag{5}$$

where the channel width W for the parabolic confinement is defined as the separation of the equipotentials at the Fermi energy. The experimental estimates $W \approx 90$ nm, $n_s \approx 1.2 \times 10^{15}$ m^{-2} imply $R_0 = 5.2$ kΩ, $B_0 = 0.64$ T. With these parameters the calculated resistance and field scales do not agree well with the experiment, which may be due in part to the uncertainties in our modelling of the shape of the experimental confining potential. The $\pm B$ asymmetry in the experimental plot is undoubtedly due to asymmetries in the cross geometry (in the calculation the geometry has four-fold symmetry, which leads automatically to $R_H(B) = -R_H(-B)$). Apart from these differences, there is agreement in all the important features: the appearance of quenched and negative Hall resistances, the independence of the height of the last Hall plateau on the smoothness of the corners, and the shift of the onset of the last plateau to lower fields for smoother corners. The oscillations on the last plateau in the calculation (which as we discussed in Sec. 2 are due to geometrical resonances) are also quite similar to those in the experiment, in support of our claim that these are classical rather than quantum resonances.

[†] In this junction with a rapidly varying curvature, the guiding field $B_g \approx 0.16$ T of Sec. 2 is somewhat too large an estimate of the low-field onset of the Hall plateau. The upper limit of the plateau is accurately given by $B_{crit} \approx 0.26$ T.

[*] The geometry in the lower right inset (with an approximately constant curvature of the corners) has $r_{min} = 2W$. The resulting guiding field $B_g = 0.5B_0$ is seen to correspond accurately to the low-field onset of the Hall plateau (cf. the dashed curve in Fig. 6b). The angular opening of the injection/acceptance cone in this case is below 90°, consistent with the appearance of a quenched R_H in the calculation.

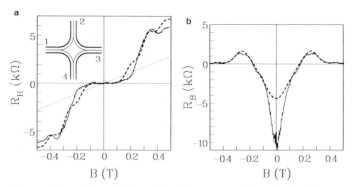

Fig. 7. Hall resistance $R_H \equiv R_{13,24}$ (a) and bend resistance $R_B \equiv R_{12,43}$ (b), as measured (solid curves) by Timp et al. (1989), and as calculated (dashed curves) for the geometry in the inset (consisting of a parabolic confining potential with the equipotentials at E_F and 0 shown respectively as thick and thin contours; the parameters are $W = 100$ nm and $E_F = 3.9$ meV). The dotted line in (a) is R_H in a bulk 2DEG.

We now turn to the bend resistance R_B. In Fig. 7 we show experimental data[†] by Timp et al. (1989) (solid curves) on $R_B \equiv R_{12,43}$ and $R_H \equiv R_{13,24}$ measured in the same Hall cross (defined by gates of a shape similar to that in the lower right inset of Fig. 6a; see the inset of Fig. 7a for the numbering of the channels). The dashed curves are calculated for a parabolic confining potential in the channels (with the experimental values $W = 100$ nm, $E_F = 3.9$ meV), and with corners as shown in the inset of Fig. 7a (defined by $p = 2$). The calculated quenching of the Hall resistance and the onset of the last plateau are in good agreement with the experiment, and also the observed overshoot of the bend resistance around 0.2 T as well as the width of the negative peak in R_B around zero field are well described by the calculation. The calculated height of the negative peak, however, is too small by more than a factor of two. We consider this disagreement to be significant in view of the quantitative agreement with the other features in both R_B and R_H. The negative peak in R_B is due to the fact that the collimation effect couples the current source 1 more strongly to voltage probe 3 than to voltage probe 4, so that $R_B \propto V_4 - V_3$ is negative for small magnetic fields (at larger fields the Lorentz force destroys collimation by bending the trajectories, so that R_B shoots up to a positive value[*], until guiding takes over and brings R_B down to zero by eliminating backscattering at the junction). The discrepancy in Fig. 7b thus seems to indicate that the semi-classical calculation underestimates the collimation effect in this geometry.

As we showed earlier (Beenakker and Van Houten, 1989b), the positive resistance peak in the bend resistance coincides in magnetic field range with a peak of enhanced longitudinal resistance R_L measured along the current-carrying channel ($R_L \equiv R_{25,34}$ in the geometry of Fig. 4). The peak in the longitudinal resistance has the same origin as in the case of the bend resistance discussed above, viz. the destruction of the collimation effect by the magnetic field. A collimated beam propagating along the channel will not be scattered very much by the side branches, and thus corresponds to a low R_L. A weak

---

[†]The (anti-)symmetrized resistances are plotted in Fig. 7.

[*]We only find the positive overshoot in R_B for rounded corners. This explains the near absence of the effect in the calculation of Kirczenow (1989b) for a junction with straight corners.

magnetic field destroys the collimation effect, thereby increasing the backscattering by the side branches and thus increasing R_L. Magnetic guiding at larger fields reduces R_L, so that our calculations show a "camel-back" B–dependence. Backscattering by channel wall irregularities leads to a similar non-monotonic magnetoresistance, an effect discovered in sodium wires fourty years ago by MacDonald (1949). This effect (not included in our calculations in which a smooth, straight channel is assumed) is actually the dominant mechanism for the camel-back magnetoresistance in the systems studied thus far, as has been demonstrated convincingly in a set of experiments by Thornton et al. (1989).

We conclude this section with a discussion of the temperature dependence of the magnetoresistance anomalies. The theoretical results in Figs. 5-7 are for $T = 0$. The resistance $R(T, E_F)$ at temperature T and chemical potential E_F follows from Eq. (1) with thermally averaged coefficients

$$I_i = <G_i(1 - t_{i \rightarrow i})> V_i - \sum_{j \neq i} <G_j t_{j \rightarrow i}> V_j, \tag{6}$$

Here $< ... >$ denotes the thermal average

$$<G t> \equiv \int dE\, G(E)\, t(E)\, \frac{d}{dE_F}\, f(E - E_F), \tag{7}$$

where f is the Fermi function

$$f(E - E_F) = \left(1 + \exp \frac{E - E_F}{k_B T} \right)^{-1}. \tag{8}$$

The resistance R which follows from Eq. (6) is a rational function of the thermally averaged transmission probabilities. In a first approximation we can interchange the evaluation of the rational function and the average, and write $R(T, E_F) \approx < R(0, E_F) >$. For $k_B T$ small compared to E_F, the thermal average can be approximated by the average of $R(0, E)$ over an energy interval $\Delta E = 3.5 k_B T$ around E_F (corresponding to the width of the derivative of the Fermi function). For the following considerations we assume a hard-wall confining potential, so that the geometry of the equipotentials is the same at each energy. (The conclusions hold also for a smooth potential, provided that the geometry does not change significantly if the energy of the equipotentials varies by ΔE around E_F.) For a fixed geometry, $R(0, E)$ depends on E only via the scaling variables R_0 and B_0 defined in Eq. (5), according to $R(0, E) \equiv R_0 \rho(\beta)$ with $\beta \equiv B/B_0$. The dimensionless resistance ρ is the quantity plotted e.g. in Fig. 6b. Since $d\beta/dE = - \beta/2E$ and R_0 are both only weakly dependent on E, the energy average of R over ΔE corresponds approximately to the magnetic field average of ρ over the interval $\Delta \beta = \Delta E \beta/2E_F$. The finest details in our magnetoresistance plots (cf. Fig. 6b) occur for $\beta \lesssim 1$ and require a resolution $\Delta \beta \gtrsim 0.1$, so that at temperatures $T \sim 0.1 E_F/k_B \sim 10$ K these features are still resolved. Note that the energy separation of the subbands does not enter in our criterion for the temperature dependence, since the wave length is not an independent variable in the semi-classical theory.

The experiments shown above were carried out at temperatures around 1 K, for which we expect our zero-temperature semi-classical calculation to be appropriate. At lower temperatures the effects of quantum mechanical phase coherence which we have neglected will become more important (Ford et al., 1989b). At higher temperatures the thermal average smears out the magnetoresistance anomalies, and eventually inelastic scattering causes a transition to the diffusive transport regime in which the resistances have their normal B–dependence. Takagaki et al. (1989b) find that the bend resistance is almost independent of temperature below 10 K, which is consistent with the above considerations.

5. CONCLUDING REMARKS

Merits and limitations

The overall agreement between the experiments and the semi-classical calculations demonstrated in this paper is remarkable in view of the fact that the channel width in the narrowest structures considered is comparable to the Fermi wave length. When the first experiments on these "electron waveguides" appeared, it was expected that the presence of only a small number of occupied transverse waveguide modes would fundamentally alter the nature of electron transport (Timp et al., 1987). Our results show instead that the modal structure plays only a minor role, and that the magnetoresistance anomalies observed are characteristic for the *classical* ballistic transport regime. The reason that a phenomenon such as the quenching of the Hall effect has been observed only in Hall crosses with narrow channels is simply that the radius of curvature of the corners at the junction is too small compared to the channel width in wider structures. This is not an essential limitation, and the various magnetoresistance anomalies discussed here should be observable in macroscopic Hall bars with artificially smoothed corners — provided of course that the dimensions of the junction remain well below the mean free path. Ballistic transport is essential, but a small number of occupied modes is not.

Although we believe that the characteristic features of the magnetoresistance anomalies are now understood, several interesting points of disagreement between theory and experiment remain which merit further investigation. One of these is the discrepancy in the magnitude of the negative bend resistance at zero magnetic field, which we discussed in Sec. 4. The disappearance of a region of quenched Hall resistance at low electron density is another unexpected observation by Chang et al. (1989) and Roukes et al. (1989). The semi-classical theory discussed in this paper predicts a universal behavior (for a given geometry) if the resistance and magnetic field are scaled by R_0 and B_0 defined in Eq. (5). For a square well confining potential the channel width W is the same at each energy, and since $B_0 \propto k_F$ one would expect the field region of quenched Hall resistance to vary with the electron density as $\sqrt{n_s}$. For a more realistic smooth confining potential, W depends on E_F and thus on n_s as well, in a way which is difficult to estimate reliably. In any case, the experiments point to a systematic disappearance of the quench at the lowest densities, which is not accounted for by the present theory (and has been attributed by Chang et al. (1989) to enhanced diffraction at low electron density as a result of the increase in the Fermi wave length). As a third point, we mention the curious density dependence of the quenching observed in approximately straight junctions by Roukes et al. (1989), who find a low-field suppression of R_{II} which occurs only at or near certain specific values of the electron density. Our semi-classical model applied to a straight Hall cross (either defined by a square well or by a parabolic confining potential) gives a low-field slope of R_{II} close to its bulk 2D value. The fully quantum mechanical calculations for a straight junction (Ravenhall et al., 1989; Kirczenow, 1989a) do give quenching at special parameter values, but not for the many-mode channels in this experiment (in which quenching occurs with as many as 10 modes occupied, whereas in the calculations a straight cross with more than 3 occupied modes in the channel does not show a quench).

In addition to the points of disagreement discussed above, there are fine details in the measured magnetoresistances, expecially at the lowest temperatures (below 100 mK), which are not obtained in the semi-classical approximation. The quantum mechanical calculations (Ravenhall et al., 1989; Kirczenow, 1989a; 1989b; Baranger and Stone, 1989) show a great deal of fine structure due to interference of the waves scattered by the junction. The fine structure in most experiments is not quite as pronounced as in the calculations, presumably partly as a result of a loss of phase coherence after many multiple scatterings in the junction (more than 10 boundary collisions in the junction before an electron escapes into one of the channels are common in our simulation, and it could well be that phase coherence is not maintained for the correspondingly long trapping times). The limited degree of phase coherence in the experiments, and the smoothing effect of a finite temperature, help to make the semi-classical model work so well even for the narrowest channels.

Some of the most pronounced features in the quantum mechanical calculations are due to transmission resonances which result from the presence of bound states in the junction (Avishai and Band, 1989; Ravenhall et al., 1989; Kirczenow, 1989a; 1989b; Peeters, 1989). In Sec. 2 of this paper we have emphasized a different mechanism for transmission resonances which has a classical rather than a quantum mechanical origin. As we have shown in Sec. 4, the oscillations on the last Hall plateau observed experimentally are quite well accounted for by these geometrical resonances. One way to distinguish experimentally between these resonance mechanisms is by means of the temperature dependence, which should be much weaker for the classical than for the quantum effect (cf. Sec. 4). One would thus conclude that the fluctuations in Fig. 6a, measured by Ford et al. (1989a) at 4.2 K, have a classical origin — while the fine structure which Ford et al. (1989b) observe only at mK temperatures is intrinsically quantum mechanical.

Routes to quenching

Among the magnetoresistance anomalies observed in the ballistic regime, the quenching of the Hall effect (Roukes et al., 1987) has the most subtle explanation, and is the most sensitive to the geometry. As we discussed in our earlier article (Beenakker and Van Houten, 1989b), and in Sec. 2 of the present paper, long trapping times in the junction play an essential role: the scrambling of the trajectories after multiple reflections suppresses the asymmetry between the transmission probabilities t_l and t_r to enter the left or right voltage probe, and without this transmission asymmetry there can be no Hall voltage. We emphasize that this *scrambling mechanism* is consistent with the original findings of Baranger and Stone (1989) that quenching requires collimation. The point is that the collimation effect leads to non-overlapping injection/acceptance cones of two perpendicular channels, which ensures that electrons can not enter the voltage probe from the current source directly — but only after multiple reflections (cf. Sec. 2). In this way a rather weak collimation to within an injection/acceptance cone of about 90° angular opening is sufficient to induce a suppression of the Hall resistance via the scrambling mechanism.

Collimation can also suppress R_H directly by strongly reducing t_l and t_r relative to t_s (the probability for transmission straight through the junction). This *nozzle mechanism*, introduced by Baranger and Stone (1989), requires a strong collimation of the injected beam in order to affect R_H appreciably. In the geometries considered here, we find that quenching of R_H is due predominantly to scrambling and not to the nozzle mechanism (t_l and t_r each remain more than 30% of t_s), but data by Baranger and Stone (1989) shows that both mechanisms can play an important role.

There is a third proposed mechanism for the quenching of the Hall effect (Ravenhall, 1989; Kirczenow, 1989a), which is the reduction of the transmission asymmetry due to a bound state in the junction. The *bound state mechanism* is purely quantum mechanical and does not require collimation (in contrast to the classical scrambling and nozzle mechanisms). Numerical calculations have shown that it is only effective in straight Hall crosses with very narrow channels (not more than 3 modes occupied), and even then for special values of the Fermi energy only. Although this mechanism can not account for the experiments performed thus far, it may become of

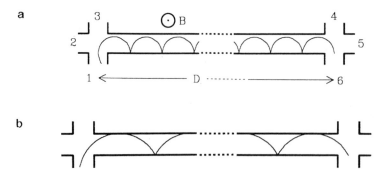

Fig. 8. Classical trajectories in a narrow channel, illustrating two types of geometrical resonances: electron focusing of skipping orbits (a), and quasi-electron focusing of traversing trajectories (b).

importance in future work. We note that while the bound state and the scrambling
mechanism for quenching have a different origin, what unifies the two is that in both
cases long trapping times in the junction are involved, suppressing R_H by eliminating
the transmission asymmetry which the Lorentz force tries to impose. The nozzle mechanism, to the contrary, deals with trajectories which move straight through the junction
with minimal time delay, so that it is distinct from the other two mechanisms in this
respect.

Narrow-channel electron focusing

In Sec. 2 we showed how electron focusing in a junction (from current to voltage
probe) can lead to large periodic oscillations in the Hall resistance in special geometries.
In this paragraph we wish to describe a similar effect in a narrow channel. We consider
the geometry shown in Fig. 8, defined by a hard-wall potential with *straight* rather than
rounded corners. The resistance $R_T \equiv R_{12,65}$ is an example of what M.L. Roukes has
termed a "transfer resistance" at this Conference. The net current flows entirely in one
junction (from lead 1 to 2), while the voltage difference is measured between two side
probes 6 and 5 in the other junction where no net current flows. Fig. 9 shows the result
of our semi-classical calculation of R_T in this geometry. For one field direction R_T decreases smoothly with B, while for the other field direction a striking oscillatory pattern
is superimposed.

The oscillations for $|B| > B_0$ are due to magnetic focusing of skipping orbits
along the boundary from lead 1 to 6, as in the electron focusing experiment with point
contacts in metals (Tsoi, 1974) or in a wide 2DEG (Van Houten et al., 1988a; 1989).
The focusing periodicity is (cf. Sec. 2)

$$B_{focus} \equiv 2\hbar k_F / eD \equiv (2W/D) B_0 , \qquad (9)$$

which is 0.075 B_0 for the center-to-center separation $D = 26.66 \, W$ of side branches 1 and
6 used in the calculation. This agrees well with the periodicity of the oscillations for
$|B| > B_0$ in Fig. 9. The oscillations die out as $|B|$ approaches $B_{crit} \equiv 2B_0$, since the
voltage difference $V_6 - V_5$ vanishes if the cyclotron diameter $2l_{cycl}$ becomes less than
W (cf. Sec. 2).

Electron focusing is not possible for $|B| < B_0$, since skipping orbits with the
maximum chord length $2l_{cycl}$ shown in Fig. 8a require a channel of width at least l_{cycl}
to prevent collisions with the opposite channel wall. Collisions with both the channel

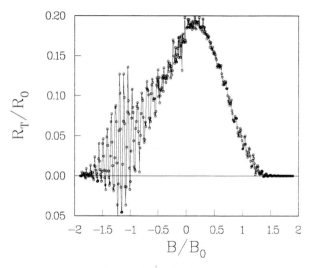

Fig. 9. Transfer resistance $R_{12,65}$ calculated for the hard-wall geometry of Fig. 8. The curve drawn through the data points is a guide to the eye.

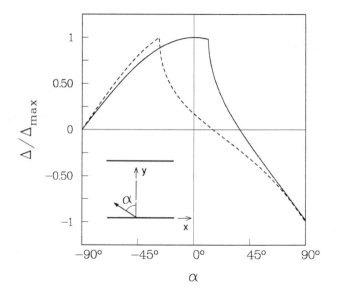

Fig. 10. Dependence on the injection angle α of the separation Δ between two subsequent collisions with the same boundary in the channel indicated in the inset ($\Delta > 0$ corresponds to motion in the positive x–direction). Results for two field values are plotted, corresponding to electron focusing (solid curve, $B = -1.2B_0$, as in Fig. 8a) and to quasi-electron focusing (dashed curve, $B = -0.53B_0$, as in Fig. 8b).

walls transform the skipping orbit into a traversing trajectory (Fig. 8b), and this destroys the focusing effect: a narrow flux tube containing traversing trajectories leaving probe 1 cannot be focused to a point at probe 6 by the magnetic field. Fig. 10 shows why: In this figure we have plotted for the two magnetic fields corresponding to Fig. 8a (solid curve, $B = -1.2B_0$) and 8b (dashed curve, $B = -0.53B_0$) the dependence on the injection angle α (indicated in the inset) of the distance Δ between two subsequent collisions with the same channel boundary (Δ is normalized by its maximum Δ_{max} as a function of α). For $|B| > B_0$ the curve $\Delta(\alpha)$ has a *smooth* maximum, while for $|B| < B_0$ the maximum is a *cusp*. In the former case Δ is stationary at Δ_{max} for $\alpha \approx 0$, and thus the skipping orbits injected nearly perpendicular to the x −axis are focused at multiples of Δ_{max}. There is no point of stationary Δ in the latter case, so that the traversing trajectories can not be focused.

Although focusing can not occur for $|B| < B_0$, one sees from Fig. 9 that oscillations in R_T persist almost down to zero field, albeit with decreasing amplitude and with a spacing ΔB which is not constant but is gradually reduced as $B \to 0$. We refer to these low-field oscillations in a narrow channel as *quasi-electron focusing*, since they result from a geometrical resonance involving traversing trajectories which is the analogue of focusing of skipping orbits in higher fields, or in wider channels. In both field regimes a peak in R_T occurs whenever $D = p\Delta_{max}$, with p an integer. If $W > l_{cyc}$, then $\Delta_{max} = 2l_{cyc}$ is the maximum chord length of a skipping orbit, and the above criterion is the usual electron focusing condition

$$|B| = pB_{focus}, \quad W > l_{cyc}. \tag{10}$$

If, on the other hand, $W < l_{cyc}$, then one has $\Delta_{max} = 2W(2l_{cyc}/W - 1)^{1/2}$, so that one obtains the condition

$$|B| = pB_{focus}\frac{4pWD}{D^2 + (2pW)^2}, \quad W < l_{cyc}. \tag{11}$$

In agreement with the numerical results in Fig. 9, the oscillations in the resistance determined by Eq. (11) become more rapid at lower fields (corresponding to smaller values of the integer p), although the periodicity remains approximately equal to the focusing periodicity B_{focus} as long as W is not much smaller than l_{cyc}. The fields B_p defined in Eq. (11) are such that for $|B| < B_p$ electrons can be transmitted from probe 1 to probe 6 after $2p - 1$ specular reflections with the channel walls. The contribution of these trajectories to the transmission probability increases with B until at B_p this contribution drops abruptly to zero, leading to a sequence of oscillations in R_T. A similar effect in thin metal films has been discussed by Korzh (1975).

Chaotic scattering

The many multiple reflections in a junction with rounded corners lead to a strong sensitivity of the choice of exit channel (through which the electron leaves the junction) on the injection parameters. A plot of exit channel versus injection angle α shows an irregularly fluctuating "chaotic" behavior, as shown in Fig. 11a for zero magnetic field, and in Figs. 11b and c for $B = 0.2B_0$ and $B = B_0$ respectively. Chaotic scattering in similar geometries in the absence of a magnetic field has received considerable attention recently (Bleher et al., 1989; and references therein). We find that intervals in α of irregular dependence of the exit channel on the injection angle are separated by intervals in which one particular exit channel is favored. These "islands" of regular scattering grow with B, until for $B > B_{crit} \equiv 2B_0$ all electrons are guided into one particular channel (number 1 in the case of Fig. 11).

The chaotic behavior in Fig. 11 is a manifestation of the scrambling mechanism for the quenching of the Hall effect discussed above, but is not directly visible in the resistances considered thus far. The reason is that the average over the injection parameters which determines the transmission probabilities smears out most of the irregular fluctuations apparent in Fig. 11. A way in which one may be able to study the chaotic scattering in a resistance measurement is by measuring the voltage difference over a region through which no net current flows. The transfer resistance R_T considered above is such a quantity. In Fig. 12 we show our calculations of the $B -$ dependence of R_T in the hard-wall geometry with rounded corners of Fig. 4. Because of the multiple

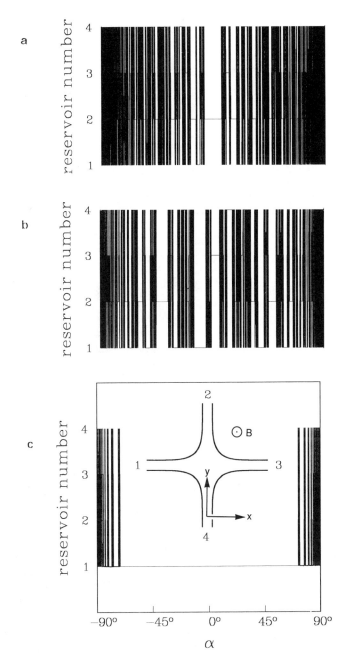

Fig. 11. Dependence of the choice of exit channel (numbered 1 through 4) on the injection angle, for $B = 0$ (a), $B = 0.2B_0$ (b), and $B = B_0$ (c). The electrons are injected into the junction shown in the inset, starting from the origin of the coordinate system at an angle α with the positive y-axis.

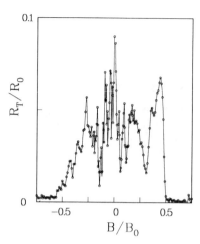

Fig. 12. Transfer resistance $R_{12,65}$ calculated for the hard-wall geometry of Fig. 4. The curve drawn through the data points is a guide to the eye.

scattering in the junction, R_T looks very different from the result in Fig. 9 for a junction with straight corners (in which multiple scattering is not possible). The (quasi-)electron focusing oscillations for negative B are not apparent in Fig. 12, but irregular fluctuations occur for both field directions. The fine details of these fluctuations are masked by numerical noise (which we estimate at $0.01 - 0.02R_0$ in these simulations, consisting of an average over 4×10^4 electrons). Voltage fluctuations in R_T have been observed by Takagaki et al. (1989a), who attributed them to a quantum interference effect. Our simulation shows that similar fluctuations can result from classical chaotic scattering.

Essentially, what we are doing in this simulation is to use one junction as an injector of ballistic electrons, and the other junction as a detector. Provided that the transport remains ballistic over the distance separating the two junctions (which may be difficult to realize experimentally because of the presence of a small amount of diffuse boundary scattering (Thornton et al., 1989)) this measurement is more sensitive to details of the junction scattering than the usual Hall or bend resistance measurements. Such an experiment would provide a rare opportunity to study chaotic scattering in the solid state, in a regime of unusual length scales and magnetic fields.

It is a pleasure to acknowledge stimulating discussions on this subject with the participants of the NATO ASI, in particular with M. Büttiker and M.L. Roukes.

REFERENCES

Akera, H., and Ando, T., 1989, *Phys.Rev.B*, 39:5508.
Avishai, Y., and Band, Y.B., 1989, *Phys.Rev.Lett.*, 62:2527.
Baranger, H.U., and Stone, A.D., 1989, *Phys.Rev.Lett.*, 63:414;
    also *in*: "Science and Engineering of 1- and 0-Dimensional Semiconductors", S.P.
    Beaumont and C.M. Sotomayor-Torres, eds., Plenum, London, to be published.
Beenakker, C.W.J., and Van Houten, H., 1988, *Phys.Rev.Lett.*, 60:2406;
    see also, Beenakker, C.W.J., Van Houten, H., and Van Wees, B.J., 1989,
    Superlattices and Microstructures, 5:127.

Beenakker, C.W.J., and Van Houten, H., 1989a, *Phys.Rev.B*, 39:10445;
see also, Van Houten, H., and Beenakker, C.W.J., *in:* "Nanostructure Physics and Fabrication", M.A. Reed and W.P. Kirk, eds., Academic Press, New York, to be published.

Beenakker, C.W.J., and Van Houten, H., 1989b, *Phys.Rev.Lett.*, 63:1857.

Bleher, S., Ott, E., and Grebogi, C., 1989, *Phys.Rev.Lett.*, 63:919.

Büttiker, M., 1986, *Phys.Rev.Lett.*, 57:1761; 1988, *IBM J.Res.Dev.*, 32:317.

Büttiker, M., 1988, *Phys.Rev.B*, 38:9375.

Chang, A.M., Timp, G., Howard, R.E., Behringer, R.E., Mankiewich, P.M., Cunningham, J.E., Chang, T.Y., and Chelluri, B., 1988, *Superlattices and Microstructures*, 4:515.

Chang, A.M., Chang, T.Y., and Baranger, H.U., 1989, *Phys.Rev.Lett.*, 63:996.

Ford, C.J.B., Thornton, T.J., Newbury, R., Pepper, M., Ahmed, H., Peacock, D.C., Ritchie, D.A., Frost, J.E.F., and Jones, G.A.C., 1988, *Phys.Rev.B*, 38:8518.

Ford, C.J.B., Washburn, S., Büttiker, M., Knoedler, C.M., and Hong, J.M., 1989a, *Phys.Rev.Lett.*, 62:2724.

Ford, C.J.B., Washburn, S., Büttiker, M., Knoedler, C.M., and Hong, J.M., 1989b, *Surf.Sci.*, to be published.

Isawa, Y., 1989, preprint.

Johnston, R., and Schweitzer, L., 1989, *J.Phys.Condensed Matter*, 1:4465.

Kirczenow, G., 1989a, *Phys.Rev.Lett.*, 62:2993.

Kirczenow, G., 1989b, *Solid State Comm.*, 71:469.

Komiyama, S., and Hirai, H., 1989, preprint.

Korzh, S.A., 1975, *Sov.Phys.JETP*, 41:70.

Landauer, R., 1957, *IBM J.Res.Dev.*, 1:223; 1988, 32:306.

MacDonald, D.K.C., 1949, *Nature*, 163:637;
see also, Pippard, A.B., 1989, "Magnetoresistance in Metals", Cambridge University Press, Cambridge.

Molenkamp, L.W., Staring, A.A.M., Beenakker, C.W.J., Eppenga, R., Timmering, C.E., Williamson, J.G., Harmans, C.J.P.M., and Foxon, C.T., 1989, *Phys.Rev.B*, to be published.

Peeters, F.M., 1988, *Phys.Rev.Lett.*, 61:589;
see also, 1989, *Superlattices and Microstructures*, 6:217.

Peeters, F.M., 1989, *in:* "Science and Engineering of 1- and 0-Dimensional Semiconductors", S.P. Beaumont and C.M. Sotomayor-Torres, eds., Plenum, London, to be published.

Phillips, J.C., 1988, *Phil.Mag.B*, 58:361.

Ravenhall, D.G., Wyld, H.W., Schult, R.L., 1989, *Phys.Rev.Lett.*, 62:1780.

Roukes, M.L., Scherer, A., Allen, S.J., Craighead, H.G., Ruthen, R.M., Beebe, E.D., and Harbison, J.P., 1987, *Phys.Rev.Lett.*, 59:3011.

Roukes, M.L., Thornton, T.J., Scherer, A., Simmons, J.A., Van der Gaag, B.P., and Beebe, E.D., 1989, *in:* "Science and Engineering of 1- and 0-Dimensional Semiconductors", S.P. Beaumont and C.M. Sotomayor-Torres, eds., Plenum, London, to be published.

Simmons, J.A., Tsui, D.C., and Weimann, G., 1988, *Surf.Sci.*, 196:81.

Srivastava, V., 1989, *J.Phys.Condensed Matter*, 1:1919; 1:2025;
Srivastava, V., and Srinivasan, V., 1989, *J.Phys.Condensed Matter*, 1:3281.

Takagaki, Y., Gamo, K., Namba, S., Ishida, S., Takaoka, S., Murase, K., Ishibashi, K., and Aoyagi, Y., 1988, *Solid State Comm.*, 68:1051.

Takagaki, Y., Gamo, K., Namba, S., Takaoka, S., Murase, K., Ishida, S., Ishibashi, K., and Aoyagi, Y., 1989a, *Solid State Comm.*, 69:811.

Takagaki, Y., Gamo, K., Namba, S., Takaoka, S., Murase, K., Ishida, S., 1989b, *Solid State Comm.*, 71:809.

Thornton, T.J., Roukes, M.L., Scherer, A., Van Der Gaag, B., 1989, preprint.

Timp, G., Chang, A.M., Mankiewich, P., Behringer, R., Cunningham, J.E., Chang, T.Y., and Howard, R.E., 1987, *Phys.Rev.Lett.*, 59:732.

Timp, G., Baranger, H.U., deVegvar, P., Cunningham, J.E., Howard, R.E., Behringer, R., and Mankiewich, P.M., 1988, *Phys.Rev.Lett.*, 60:2081.

Timp, G., Behringer, R., Sampere, S., Cunningham, J.E., and Howard, R.E., 1989, *in*: "Nanostructure Physics and Fabrication", M.A. Reed and W.P. Kirk, eds., Academic Press, New York, to be published.

Tsoi, V.S., 1974, *JETP Lett.*, 19:70.

Van Houten, H., Van Wees, B.J., Mooij, J.E., Beenakker, C.W.J., Williamson, J.G., and Foxon, C.T., 1988a, *Europhys.Lett.*, 5:721;
Beenakker, C.W.J., Van Houten, H., and Van Wees, B.J., 1988, *Europhys.Lett.*, 7:359.

Van Houten, H., Beenakker, C.W.J., Van Loosdrecht, P.H.M., Thornton, T.J., Ahmed, H., Pepper, M., Foxon, C.T., and Harris, J.J., 1988b, *Phys.Rev.B*, 37:8534.

Van Houten, H., Beenakker, C.W.J., Williamson, J.G., Broekaart, M.E.I., Van Loosdrecht, P.H.M., Van Wees, B.J., Mooij, J.E., Foxon, C.T., and Harris, J.J., 1989, *Phys.Rev.B*, 39:8556;
for a review of electron focusing in a 2DEG, see: Beenakker, C.W.J., Van Houten, H., and Van Wees, B.J., 1989, *Festkörperprobleme*, 29:299.

Van Wees, B.J., Van Houten, H., Beenakker, C.W.J., Williamson, J.G., Kouwenhoven, L.P., Van der Marel, D., and Foxon, C.T., 1988, *Phys.Rev.Lett.*, 60:848.

Wharam, D.A., Thornton, T.J., Newbury, R., Pepper, M., Ahmed, H., Frost, J.E.F., Hasko, D.G., Peacock, D.C., Ritchie, D., and Jones, G.A.C., 1988, *J.Phys.C*, 21:L209.

ELECTRON-BOUNDARY SCATTERING IN QUANTUM WIRES

M.L. Roukes, T.J. Thornton, A. Scherer, and B.P. Van der Gaag

Bellcore
Red Bank, New Jersey 07701 USA

INTRODUCTION.

To reduce the dimensionality of an electronic system it must be confined within artificially imposed *boundaries*. In the most idealized consideration of the problem, the properties of confined electrons depend solely upon the volume containing them. In all real systems, however, the characteristics of the boundaries themselves play a significant role in the physics observed. Recent advances in epitaxial growth techniques now permit nearly ideal heterointerfaces to be created over appreciable areas. But even for these, the crystallographic (and therefore the electronic) properties at the surfaces are quite different that those of the bulk. Recent dramatic demonstrations of surface reconstructions obtained through scanning tunneling microscopy provide a particularly striking example. However the more general situation is even far from this ideal. Any real boundary, when viewed over a large enough area, always reveals randomness. In the case of the best epitaxially-grown interfaces this will be manifested as a finite domain size for the last few atomic layers, as schematically depicted in Fig. 1 *(left)*. The edges of these domains delineate random patches of surfaces in registry with the underlying crystal. A quantum well between two such interfaces would be characterized by a thickness which varies stochastically across the growth plane.

Surface roughness causes small quantum systems to depart from the behavior expected in the case of ideal confinement. Kubo (1962) first considered the nature of the eigenvalue spectrum of an ensemble of small metal particles, using random matrix theory to assess the effect of random boundary conditions. Irregular surfaces within an individual quantum system can act as a source of level broadening, and may be viewed as reducing the lifetimes of the confined states. Such considerations apparently troubled Schreiffer (1957) enough to dismiss the likelihood that his prediction of two-dimensional confinement beneath the gate of a *Si*-MOSFET might ever be experimentally realized. Now, in certain highly optimized two-dimensional electron gas (2DEG) systems, such as those formed at and epitaxial *GaAs/AlGaAs* heterojunction, "interface roughness scattering" can be dramatically reduced – to the point where it becomes insignificant compared to other intrinsic sources of scattering within the crystal.

The actual boundaries which confine electrons in any structure can always be viewed as being electrostatic in nature. For both *Si*-MOSFETS and III-V heterojunctions, the boundaries confining the electrons are created by the potentials at very smooth surfaces. In the first case it occurs between *Si* and its oxide, in the second it involves an epitaxially-grown interface. In a quantum wires, by contrast, the electrostatic boundaries are far less smooth and homogeneous. We illustrate this by considering a typical cross-section of a quantum wire formed between "pinched"- ("split"-) gate electrodes (*Fig. 1, right*) These are metal films patterned with well-defined gaps, deposited upon the surface of a typical *GaAs/AlGaAs* heterojunction (Thornton *et al.*, 1986; Zheng *et al.*, 1986). Upon application of sufficient negative bias potential, the 2DEG immediately beneath the gates becomes depleted of carriers. A narrow conducting region remains below the gap in the electrodes which consititutes the quantum wire. The "edges" of these wires are delineated by the depletion regions controlled by imposed potential on the remote gates. The donor layer intervenes between this imposed gate potential and the 2DEG. Since this is comprised of a very large number of randomly-located

Electronic Properties of Multilayers and Low-Dimensional Semiconductors Structures
Edited by J. M. Chamberlain *et al.*, Plenum Press, New York, 1990

95

ionized donor cores, it generates an additional fluctuating component to the total potential felt by electrons at the heterointerface. It is this same fluctuating potential which determines the edge of the depleted region and thus, the boundaries of the quantum wire. This results in edge roughness on length scales of order $\sim 1000\text{Å}$. To contrast, in the case of randomness at an epitaxial interface (*Fig. 1, left*) the height of the interface fluctuates on the scale of one monolayer. Here, the roughness has a characteristic length of order a lattice constant, typically a few Å.

This randomness at the edges of a narrow wire can have a strong effect upon its transport properties. Understanding the anomalies which develop gives us new insight into the nature of electron interactions at boundaries. In what follows, two separate investigations we have recently carried out are described (Thornton et al., 1989b; Roukes et al., 1990). They employ two particular anomalies: observed in the magnetoresistance of narrow wires, and in we have termed the "transfer resistance" of narrow wires with junctions.

BOUNDARY REFLECTION OR BOUNDARY SCATTERING?

The question of the *specularity* of a boundary has a long history. The term refers to the degree to which a surface causes mirror-like reflection (angle of incidence = angle of reflection) of a flux of particles incident upon it. Fuchs (1938) first introduced a single parameter, p, to describe the degree of specularity of boundary collisions. He used it as a simple phenomenological means of incorporating boundary scattering into the Boltzmann equation. With this parameter he imposed a boundary condition upon the non-equilibrium electron distribution function, $g(\mathbf{x},\mathbf{p})$. This may be written as $g(\mathbf{x}_s,\mathbf{p}_i) = p\, g(\mathbf{x}_s,\mathbf{p}_f) + (1-p)\, g_0(\mathbf{x}_s,\mathbf{p})$. In this simple picture of boundary scattering, only the fraction p of the incident flux of electrons striking the surface are specularly reflected, and have their normal momentum reversed: $\mathbf{p}_f = \mathbf{p}_i - [2\mathbf{p}_i\cdot\hat{\mathbf{n}}(\mathbf{x}_s)]\hat{\mathbf{n}}(\mathbf{x}_s)$. Here, $\hat{\mathbf{n}}(\mathbf{x}_s)$ is a unit vector normal to the surface described by the $\mathbf{x}_s$. On the other hand, a fraction $(1-p)$ suffer "catastrophic scattering", and are randomly ejected after the collision with uniform probability over the hemisphere of possible emergent angles. Thus, $p=0$ describes the limit of *completely diffuse* scattering, where a single boundary collision is sufficient to restore the local distribution function to equilibrium, $g_0(\mathbf{x},\mathbf{p})$.

Early investigations of the physics of metals were interpreted as indicating that, in general, it was safe to assume that boundary scattering was essentially *diffuse*: $p \sim 0$. Evidence substantiating this viewpoint came from many studies of the resistance of thin metal films (see Chambers, 1969). Particularly strong evidence came from Pippard's (1957) experiments investigating the anomalous skin effect in *Cu*. Amidst all these early experiments suggesting that scattering from metals surfaces is completely diffuse ($p \sim 0$), some evidence did appear suggesting otherwise. Of note, in particular, is data on carefully prepared thin, single crystal *Bi* films (Friedman and Koenig, 1960) in which the contribution to the resistance attributed to boundary scattering appeared to be smaller than expected by assuming $p = 0$. We will comment on this further below.

The first strong experimental evidence for a high degree of *specular* electron-boundary scattering ($p \sim 1$) came from microwave surface impedance resonances (Khaikin, 1960; Koch and Kip, 1965). These resonances were explained by Prange and Nee (1968) as arising from surface Landau levels. The resonances were ascribed to electronic transitions between levels bound to the surface by the magnetic field. Their classical analogs are electron trajectories which skip along the surface. These authors considered a semiclassical picture for the broadening of these resonances due to diffuse scattering at the boundaries. After each bounce it was supposed that a fraction $(1-p)$ of the skipping electrons are randomly scattered away from the trajectories. From their arguments an average degree of specularity of the collisions at the surface could be deduced. Analyzing existing experimental data, they inferred that an exceedingly high degree of specularity was operative, $(1-p) \sim 10^{-2}$. Collisions at *glancing* incidence to the surface, involved in the skipping trajectories of the experiments, were argued as being those most likely to experience specular reflection. This is perhaps consistent with intuition based on the phenomenon of total internal reflection in optics. The concept of the angular dependence of p has been further developed by Soffer (1967), but might be too literal an interpretation of such a simple classical picture for electron-boundary collisions in real systems.

Tsoi was first to lay claim to investigating boundary scattering at nearly *normal* incidence to a metal surface (Tsoi, 1974b). He embarked upon these experiments as the first application of his discovery of transverse electron focusing (Tsoi, 1974a). The technique involved touching a carefully prepared metal surface with fine metallic points serving as injector and collector electrodes, with the

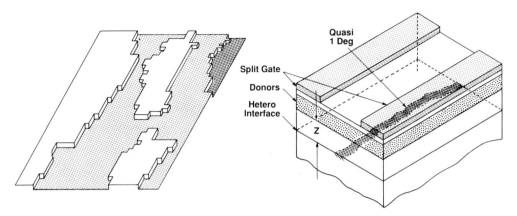

Figure 1. *(left)* Representation of a surface roughness at an epitaxially-grown interface. Terraced monolayer domains cause the position of the interface to vary randomly across the plane. *(right)* Edge roughness in a quantum wire arises from charge fluctuations within the donor layer located above the heterointerface. The undoped spacer layer of thickness z intervenes.

metal foil serving as a common ground electrode. Focussing peaks, maxima in the detected voltage, were found the applied magnetic field, B, was swept upward. (A uniform **B** was directed in the plane of the film and perpendicular to the line between electrodes.) These peaks occurred when an integral number of cyclotron diameters, N, fit between the current and voltage contacts. For $N = 1$, the focussed electrons follow a circular trajectory (half a cyclotron orbit) in traversing from injector to collector. For $N > 1$, the trajectory involves $N - 1$ *boundary collisions*, as the injected electrons hop along a cycloidal path to the collector. Tsoi observed that amplitude of the peaks progressively decreased with N, and argued that this results from the finite probability of scattering specularly after each bounce. From this he deduced a value of $p \sim 0.8$ for his very carefully prepared, single crystal *Bi* samples. It is very important to note however, that Pippard (1989) has recently shown that electron focusing peaks can persist well into the "hopping regime" ($N > 1$, where boundary collisions are required for traversal from injector to collector) even in the limit of *completely diffuse scattering, $p = 0$*. In the absence of a full theory for electron focusing experiments, he cautions against taking values of p obtained from focusing traces too literally.

Careful sample preparation was crucial to the success of these early experiments suggesting that p can be finite. They involved single crystals of very high purity metals, having crystallograpically-oriented surfaces prepared with extreme care (see, e.g., Koch and Kuo, 1966; Doezema and Koch, 1972). Epitaxial growth techniques now routinely provide the experimentalist with interfaces of very high quality. Investigations carried out so far indicate that a significant degree of specularity can be obtained with these systems. Hensel *et al.* (1985) investigated the resistivity of thin $CoSi_2$ and $NiSi_2$ epilayers. They observed a much smaller thickness dependence than expected for their thin layers and interpreted this as reflecting a large component of specular scattering, $p \sim 0.8 - 0.9$, at the epitaxial $Si/CoSi_2$ interface. In experiments involving another highly ideal interface, that formed by Si and its oxide, Hartstein *et al.* (1976) investigated scattering in Si inversion layers. They were able to experimentally isolate the component arising from surface roughness from the other predominant scattering mechanism, involving ionized impurities in the oxide layer. Their analysis of the data was roughly consistent with a more refined interpretation subsequently presented by Ando (1977), who argued the data was consistent with rms surface fluctuations of height of $\Delta \sim 4.3$Å and average extent (correlation length) $\Lambda \sim 15$Å.

In a remotely-doped $GaAs/AlGaAs$ heterojunction, a spacer layer intervenes between the ionized donors and the heterointerface where the 2DEG is formed (Fig 1., *right*). The spatial separation of the charged donors has the important effect of decreasing coulomb scattering from ionized cores. This has enabled very high in-plane electron mobilities to be attained. The remoteness of this region containing the randomly located charges has another important effect. Because of the separation, the short wavelength (high-q) spatial Fourier components of the fluctuating potential are felt only weakly by electrons at the heterointerface. These components become exponentially suppressed as spacer thickness, z, is increased (*Fig. 1, right*). The resulting effective potential felt by electrons at the heterointerface predominantly contains *long range* (wavelength) *fluctuations*. This has very important consequences for a narrow pinched gate wire (Nixon and Davies, 1989). The combined effect of the gate potential and the randomly-located donor cores cause the lateral depletion profiles delineating the edges of the wires to be jagged, especially at low densities near pinch-off. These irregular profiles define wires with edge roughness on a typical length scale of 1000Å. This is a striking demonstration of the decay of the high-q components of the potential, given that the average distance between donor cores is ~ 100Å. The length scale of fluctuations in quantum wires is more than two orders of magnitude larger than that felt by electrons in the in the *Si* inversion layers of Harstein *et al.*. This illustrates the fundamentally different nature of boundary scattering in quantum wires.

MAGNETORESISTANCE IN A QUANTUM WIRE

When a two-dimensional electron gas is transformed into a narrow conducting path, a strong increase in the magnetoresistance at low-B is observed. In general, the increase in the resistance at zero field is greater than expected from geometrical reduction alone — the (2D) *sheet resistivity* is seen to markedly increase as the conducting path width decreases. This results in a *negative* magnetoresistance with a maximum value at *zero* magnetic field. The magnitude of this peak at zero field grows as the conducting path width decreases. This effect has been studied by several groups (Choi *et al.*, 1986; Zheng *et al.*, 1986; Simmons *et al.*, 1988; van Houten *et al.*, 1988) — all have noted that field scale of this negative magnetoresistance is *not* characteristic of weak localization. Elsewhere we shall describe our recent investigation of this phenomenon (Thornton *et al.*, 1990). Recently, however, we have reported opposite behavior in narrow high mobility wires: a *positive* zero-field magnetoresistance (Roukes *et al.*, 1987). This is characterized by a local minimum at $B = 0$ leading to an anomalous maximum at low but *finite* field. This unusual behavior is clearly displayed in Fig. 2 for wires patterned by ion exposure (Thornton *et al.*, 1989b). As shown it can become very prominent for wires narrower than $\sim 1\mu m$. This low-B maximum occurs at a field scaling inversely with wire width, W. We observe this phenomenon most clearly in narrow wires patterned using a highly optimized low energy ion exposure technique — this produces narrow channels with small edge depletion lengths and carrier density which is essentially unchanged from that of the unpatterned 2DEG (Scherer and Roukes, 1989). The inset, displaying similar peaks from a 12 μm long pinched-gate wire at different gate voltages, however, shows that this phenomena is not specific to a particular method of confinement. A similar positive magnetoresistance and low-B maximum is evident in some of the data of Timp *et al.* (1988), and Ford *et al.* (1988), in which conventional etching and pinched gating techniques, respectively, were employed for confinement.

Our studies of narrow wires patterned by ion exposure show that increasing the ion dose beyond an optimum value eliminates the postive magnetoresistance by increasing the zero field resistance until it overwhelms the finite-B peak. (In Fig. 6a similar behavior is demonstrated in a gated sample as the mean free path, ℓ_0, decreases with electron density.) The data we obtain in this situation is similar to that obtained from wires defined by conventional etching, such as that recently reported by several groups (van Houten *et al.* 1986; Timp *et al.*, 1988; Simmons *et al.*, 1988). We observe that, concomitant with this loss of the positive magnetoresistance, there is a strong increase in the amplitude of resistance fluctuations and the appearance of random offsets in the Hall resistance at zero field, $R_H(0)$. We interpret these as general trends indicating an overall increase in disorder.

Quantum interference phenomena, such as Ahronov-Bohm oscillations in ring structures, are strongly damped with increasing temperature as phonon scattering begins to destroy phase coherence. In general, temperatures of $\sim 1K$ or below are required for their observation. By contrast, we have found that this low-B peak survives to moderately high temperatures, $40-80K$, dependent upon wire width. This robust phenomenon remained unexplained for several years after its observation. Kirczenow (1989a, 1989b) first presented quantum mechanical calculations demonstrating the effect of junction scattering upon the (series) magnetoresistance. His results showed peaks at finite-B arising from junction resonances which, at the outset, appeared remarkably similar to those observed experimentally. Our attempts to link these theoretical results with the experimental observations proved unsuccessful. Subsequently, we established that the finite-B peak was not a junction scattering phenomenon but actually an intrinsic property of the wires. It was found to linearly scale with the length, L, of a wire segment between two junctions, for $L > 2\mu m$. We shall discuss this scaling in more

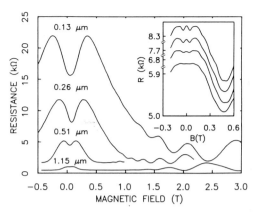

Figure 2. The 4.2K magnetoresistance of 12μm long wires of various widths patterned by ion exposure. *(inset)* Similar data from a 12μm pinched-gate wire for gate voltages V_g = -0.78, -0.83, -0.86 and -0.89 V.

detail below. This conclusion was further corroborated by experiments with long wires connecting two broad 2DEG regions — in these experiments undertaken *without* junctions, the peak at finite-B was still prominently displayed. Recently, purely classical simulations of magnetotransport through ballistic junctions (Beenakker and van Houten, 1989a) also predict similar low-B structure in the magnetoresistance. These appear to arise from resonance-like behavior due to the temporary trapping of electrons within multiply-scattered classical paths in the junction region. Only the peak at finite-B now attributed to boundary scattering has clearly been observed in experiments thus far. The unusual features predicted to arise from quantum mechanical and classical junction resonances have not yet been reported.

Subsequent experiments, and comparison with previous investigations of boundary scattering at metal sufaces, have now led us to explanations consistent with experiments. Below, we briefly review the relevant past work in metals before describing our recent studies.

BOUNDARY SCATTERING IN METALS

The subject of boundary scattering began in 1898 when Isabella Stone observed that very thin metal films have a higher resistivity than the bulk (Stone, 1898). Thomson was first to formulate a qualitative theory of the phenomenon in terms of increased scattering because of rough surfaces (Thomson, 1901). The first complete theory of classical scattering from diffuse boundaries was advanced by Fuchs (1938). The field was infused with a resurgence of interest in the 1960's when purification techniques were developed capable of yielding metals with very long mean free paths: $\ell_0 > 1mm$, in some cases. This greatly reduced scattering in the bulk (internal scattering) to strongly enhance the importance of interactions at the surfaces. Several excellent reviews of this early work are available (Chambers, 1969; Pippard, 1989).

Important for our investigations in quantum wires is the early work of MacDonald and Sarginson (1950). They were first to consider the transverse magnetoresistance of thin plates of metal, in what has become known as the "MacDonald geometry". This involves a applied magnetic field and an induced current flow, both in the plane but mutually perpendicular (Fig. 3a, top inset). Hereafter, "transverse" magnetoresistance refers to the situation where field and current density vectors are orthogonal. In their treatment of the problem, scattering at the metal surfaces was assumed to be completely diffuse, $p = 0$. The classical theory will prove directly relevant to our investigations of the transverse magnetoresistance of a "quantum" wire having many occupied Q1D modes. Ditlefsen and

Lothe (1966) improved upon this theory, explicitly avoiding several key assumptions that had been made. Their results were qualitatively similar results to those obtained earlier, but numerical differences were noted. Data from their improved theory of the transverse magnetoresistance, $R(B)$, is shown in Fig. 3a. *(solid curve)*. A positive magnetoresistance is seen, leading from zero field up to a *maximum* at a field, $B_{max} = 0.55B_0$. Here, $B_0 = p_F/eW$ is the field where the cyclotron radius, $r_c = p_F/eB$, equals the thickness of the plate, W. p_F is the Fermi momentum. Beyond the maximum in $R(B)$ the resistivity falls abruptly, closely approaching the bulk value at fields where the cyclotron diameter becomes smaller than the thickness ($B > 2B_0$). The theory showed that the magnitude of the resistivity at $B = 0$, the relative height of the peak, and the ultimate resistance for fields $B > 2B_0$, *all* depend upon the ratio of the thickness to the mean free path. The solid curve of Fig. 3a depicts these results of Ditlefson and Lothe for $W/\ell_0 = 1/36$. This particular choice of parameters was intended to closely approximate the experimental conditions of Forsvoll and Holwech (1964), whose data from thin *Al* plates is shown in Fig. 3b. The theoretical data qualitatively, but not quantitatively, agrees with these experimental findings.

Pippard (1989) has recently solved the simpler problem of boundary scattering in a 2D geometry (Fig. 3a, bottom inset) in the limit of infinite mean free path, also assuming that boundary scattering is completely diffuse, $p = 0$. His results, depicted as the dashed curve of Fig. 3a, closely follow the 3D predictions of Ditlefsen and Lothe. At $B = 0$ and for $B > 2B_0$ differences arise because of his assumption that ℓ_0 is infinite. Important for the interpretation of our results, however, is the fact that the position of the maximum, B_{max}, appears insensitive to both dimensionality and the ratio W/ℓ_0.

These qualitative features of the magnetoresistance, $R(B)$, may be understood by considering trajectories for the electrons within the metal. Consider the schematic (top) view of a "MacDonald geometry" conductor as displayed in Fig. 3c. At zero and very low magnetic fields the few trajectories (*e.g., path no. 1*) which do not involve diffuse collisions with the surfaces will dominate the conductivity. These act to partially "short out" the resistance from boundary scattering. The value of the resistivity measured at zero field depends on the bulk mean free path, ℓ_0, and thickness, W. (The thickness dependence arises from the effect of the rough surfaces (see Ziman, 1960).) The application of a modest magnetic field trajectories causes the formerly straight to curve and, thus, terminate in catastrophic collisions with the boundaries (*path no. 2*). With the loss of these straight trajectories, the short-circuiting of the boundary scattering resistivity no longer occurs and the total measured resistivity increases above its zero-field value. At high magnetic fields, however, the cyclotron diameter becomes smaller than W. In this regime the walls become unimportant and the resistivity drops once again to the bulk value expected at these fields. This picture, valid for the case of catastropic boundary scattering ($p = 0$), is easily extended to the situation of finite specularity ($p > 0$).

BOUNDARY SCATTERING REVISITED: QUANTUM WIRES

The existing classical theory predicts the appearance and location of the low-B maximum. The agreement between the 2D and 3D theories indicates that the peak location, B_{max}, should be relatively insensitive to dimensionality and to the precise value of the mean free path, ℓ_0. The amplitude of the maximum, however, may not exhibit a similar lack of dependence upon these parameters. In Fig. 4 we test these classical predictions for B_{max}. The cyclotron diameter at the maximum, $r_c(B_{max}) = p_F/eB_{max} = \sqrt{2\pi\hbar^2 n_s}/eB_{max}$, is calculated for eighteen different wires using measured values of the electron density and the peak position. Fig. 4a displays the remarkable agreement with the theory which is obtained.

The predicted location of the peak, $B_{max} = 0.55B_0 = 0.55(p_F/eW)$ depends upon the square of the electron density through the Fermi momentum, $p_F = \sqrt{2\pi\hbar^2 n_s}$. This is verified in Fig. 4b where the $\sqrt{n_s}$ dependence is seen for three quantum wires with self-aligned gates having different widths. These samples were patterned using very low energy ion exposure (Scherer and Roukes, 1989). This technique is unique in the fact that it produces wires which retain a width remaining essentially constant as the electron density is varied (Thornton *et al.*, 1989a; Roukes *et al.*, 1989). This property makes the test of scaling shown in Fig. 4b possible; it has also recently enabled us (Roukes *et al.*, 1989) to quantitatively test the predictions of a classical model for transport in ballistic multiprobe conductors (Beenakker and van Houten, 1989). An end-view schematic comparing quantum wires, patterned by pinched gates *(left)* and by low energy ion exposure with self-aligned gates *(right)*, is shown in Fig. 5. In the latter technique, the unprotected regions are exposed to a very low energy ion

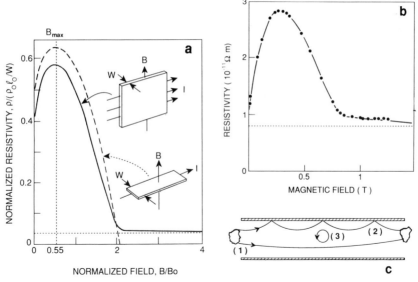

Figure 3. (a) Normalized magnetoresistivity curves from the theory of Ditlefsen and Lothe (1966) *(solid)* for the a 3D metal with thickness $W/\ell_0 = 1/36$ in the "MacDonald geometry". The classical 2D predictions of Pippard (1989) *(dashed)* in the limit of infinite ℓ_0 are also shown. **(b)** Experimental data of Forsvoll and Holwech (1964) from rolled aluminum plates of thickness $W = 25\mu m$. **(c)** Conductance at low-B is dominated by electrons following straight trajectories *(1)* which are not scattered by diffuse boundaries, but travel ballistically between impurity collisions within the wire's interior. These "short-circuit" the boundary resistance. Upon application of a modest field, B_{max}, all straight trajectories have been converted to curved paths *(2)* which terminate in collisions with the diffuse boundaries. The resistance rises to include the full effect of boundary scattering. At higher fields, a complete cyclotron orbit fits within the boundaries *(3)* and diffusive transport characteristic of the bulk is recovered.

beam ($\sim 100 eV$) resulting in only a neglible amount of material removal. This technique relies upon physical alteration of the region close to the surface to cause carrier depletion from the heterointerface (Scherer and Roukes, 1989), and is far gentler than methods which actually etch the crystal.

The classical theory assumes the boundaries are completely diffuse. We have carried out experiments to actually probe the degree of specularity. In Fig. 6 we explore the density dependence of the low-field magnetoresistance in a 0.27 μm wide self-aligned gated wire in more detail. In the raw data traces *(Fig. 6a)* the peak at B_{max} is seen in the low resistance (i.e. high density) traces on either side of zero field. As density is decreased the overall resistance rises and a peak in the resistance emerges at $B = 0$ (negative magnetoresistance). At low n_s this zero field peak ultimately overwhelms the weaker maximum at finite-B. We can qualitatively characterize disappearance of the peak at finite-B with decreasing electron density by plotting resistances ratios: $R(B_{max})$ divided by the value at zero field, $R(0)$. As seen *(Fig. 6b)*, this normalization dramatically enhances the prominence of the effect at high densities. We attribute this peak's disappearance to the reduction in the transport mean free path, ℓ_0, at low densities. This is consistent with our observation that controlled introduction of disorder during fabrication acts to reduce ℓ_0 and obfuscate the effect in similar fashion. The inset of Fig. 6b shows the amplitude of the effect plotted against the transport mean free path, $\ell_0 = 2\pi\hbar^2/(ep_FR_0)$, calculated from the measured density and zero field sheet resistivity, R_0. For these samples the finite-B peak disappears when the ℓ_0 is smaller than $\sim 2\mu m$. We interpret this critical value of ℓ_0 as the *boundary scattering length*, ℓ_B, the average distance an electron travels down the wire before the probability of catastrophic boundary scattering becomes approximately unity. Its precise value depends upon wire width. Our picture is that the peak at finite-B peak disappears when $\ell_0 < \ell_B$ and bulk scattering becomes more likely than a boundary scattering event.

This same logic can be applied to the disappearance of the finite-B peak at high temperatures. As temperature is increased phonon scattering will reduces of ℓ_0. Observing the reduction of this feature with temperature, we (somewhat arbitrarily) assume that phonon scattering has reduced ℓ_0 to become comparable with ℓ_B when $R(B_{max})/R(0) = 1$. We denote the temperature where this occurs as T'. In a separate experiment we extract the temperature dependence of ℓ_0 from the temperature dependence of the zero field resistance in a wide wire made from the same heterojunction. From this we obtain $\ell_0(T')$. This procedure leads us to a second qualitative measure of the boundary scattering length in these wires: $\ell_B \sim \ell_0(T')$. Here it is implicitly assumed that the ℓ_B itself is independent of temperature over the range measured, which we believe is reasonable. This yields values of ℓ_B of 1.2- and 5-μm for wires of width $W \sim 0.13$- and 0.51-μm respectively.

A third qualitative measure of ℓ_B in narrow wires can be obtained by the scaling of finite-B peak with length. We investigated how the maximum depends upon the length of a narrow wire link between two junctions, in early attempts to establish whether the peak arises from junction scattering. We found that the effect linearly scales with length, remaining a constant fraction of the resistivity for wires having lengths from 160- down to 2-μm. Neither we, nor other groups, observe this effect in studies of quantum point contacts. This and our scaling experiments indicate that lengths shorter than $2\mu m$ are of particular interest — in this range the effect disappears. Fig. 7 depicts results from our investigations of this length dependence. The inset shows that the amplitude of the finite-B maximum is a constant fraction of the zero field resistance for wires having lengths, $L > 2\mu m$. The solid lines are a fit to a function of the form $R(B_{max})/R(0) = A[1 - \exp(-L/\ell_B)]$, used in absence of a formal theory as a concrete means of extracting an effective boundary scattering length from the data. Note that the formula saturates (approaches unity) for lengths greater than ℓ_B; this correctly describes the scaling observed experimentally in the large-L regime. The data shows that both $R(B_{max})$ and $R(0)$ increase linearly with length beyond a critical distance; asymptotically this formula has their ratio tend to the constant A. The inset shows that A depends upon width. This analysis leads us to values of ℓ_B which are 0.5- and 1-μm for $W \sim 0.13$- and 0.30-μm, respectively.

From the values of of the boundary scattering length deduced as described above, the specularity of scattering at the edges of these quantum wires can be estimated. We may characterize a boundary either by Fuch's parameter, p, or equivalently, by N_{av}, the average number of boundary collisions required to restore the distribution function to a local equilibrium. On average, this equilibration is supposed to occur over a length ℓ_B. In a wire of width W, a very approximate estimate for the average number of boundary collisions suffered during traversal of a wire segment of length ℓ_B is $N_{av} \sim \ell_B/W$. If after *each* boundary scattering event the electron's momentum is completely randomized, then $N_{av} \sim 1$. This means that in this limit of catastrophic boundary scattering $\ell_B \sim W$. If, on the other hand, scattering at the boundaries is partially specular, an electron will have *finite* probability of surviving a collision, thus $\ell_B > W$ and correspondingly, $N_{av} > 1$. We conclude that the probability of specular scattering is $p = 1 - 1/N_{av}$ or, very approximately, $p = 1 - W/\ell_B$. Our expectation is that since p is a property of the boundary, wires of different width fabricated under identical conditions should display similar values. The data obtained from our experiments to date appears consistent with this view. The three qualitative approaches to extracting the boundary scattering length from the data lead to values of p ranging from 0.73-0.9 for narrow wires patterned by ion exposure. This means that there are, very approximately, $N_{av} \sim 4 - 10$ boundary collisions before momentum memory is lost.

It is of interest to compare the quality of the boundaries obtained through various fabrication techniques available. As previously discussed, we have amassed a large amount of data on boundary scattering for the case of confinement via ion-exposure. In our wires defined using pinched gate confinement, however, the peak at finite-B is much less pronounced. This has made it difficult to obtain as much systematic information for this case. A further complication is that channel width and density in a pinched-gate wire are both strongly dependent upon gate voltage and the width of the gap between electrodes (Ford et al., 1988; Wharam *et al.*, 1989). A careful comparison of confinement boundaries using the peak at finite-B requires samples having identical widths, mobilities and carrier densities. This is extremely difficult to achieve. A preliminary comparison, however, can be obtained through the following arguments in conjunction with data such as in the inset of Fig. 2. In a pinched-gate wire, an increasingly negative gate potential reduces both the width *and* the density of the carriers remaining beneath the gap between electrodes. The mobility and, correspondingly, the transport mean free path, ℓ_0, decrease with the decreasing density — as mentioned, this causes the the boundary scattering peak to be suppressed, while the zero field resistance increases. We note the value of gate voltage, V_g', where $R(B_{max})$ decreases to become comparable with the zero-field resistance, $R(0)$. As we've done in previous **cases,** we make the rough assumption that this indicates that $\ell_B \sim \ell_0$. Separate

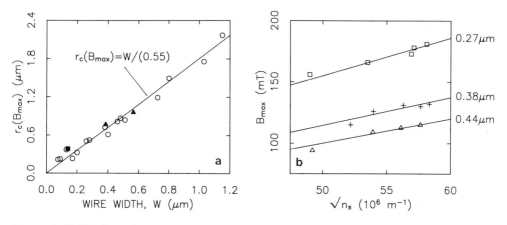

Figure 4. Verification of the classical scaling of the anomalous magnetoresistance peak. **(a)** For pinched and self-aligned gated wires of different widths, W, the cyclotron radius at B_{max} follows the prediction of the classical theory, $W/r_c = (0.55)$. The solid triangles are from the data of Ford *et al.* (1988), the solid square is from the data of Timp *et al.* (1988). **(b)** Density dependence of the postion of the low-B magnetoresistance maximum, B_{max}, for three wire widths. The data are obtained from ion exposed self-aligned gated wires which retain constant width as n_s is varied.

determination $\ell_0(V_g')$ then yields a qualitative measure of ℓ_B. Deducing $\ell_0(V_g')$ from the zero field resistivity in turn requires knowledge of $W(V_g')$ and $n_s(V_g')$. These are extracted from magnetic depopulation measurements (Berggren *et al.*, 1986) and from high-B Shubnikov-de Haas oscillations, respectively. Given this information we arrive at an estimate of the specularity of our electrostatic boundaries, $p = 1 - W/\ell_0(V_g')$. For the curve in the inset displaying the a resistance ratio ~ 1, we deduce values $\ell_0 \sim 1.85\mu m$ for a wire width $W \sim 0.1\mu m$. This leads to a specularity of $p \sim 0.95$, meaning catastrophic scattering only occurs after $N_{av} \sim 1/(1-p) \sim 20$ boundary reflections. Again, this serves only as a very rough estimate and should not be taken too literally.

Ostensibly, the preliminary conclusion we arrive at from this comparison is that split-gate electrostatic confinement ($p \sim 0.95$, $N_{av} \sim 20$) yields smoother boundaries than those obtained from ion-exposure ($p \sim 0.73 - 0.90$, $N_{av} \sim 4 - 10$). We stress that the prescriptions we have used to deduce values of p, however, are quite approximate. Additionally, by necessity they are not identical for the two cases investigated. Future work may provide a more quantitative comparison. The recent modelling of a split gate wire indicates that near pinch-off the low electron density in the channel results in weak screening of potential fluctuations (Nixon and Davies, 1989). This causes a large

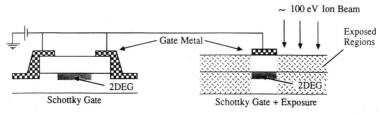

Figure 5. Sectional (end) views of narrow wires. *(left)* Electrostatic depletion of carriers in the 2DEG under pinched (split) gates leaves narrow channels between ungated regions. *(right)* In the technique of self-aligned gates, the conductivity of the 2DEG is reduced to zero except beneath masked regions. The metal masks which remain are subsequently used in experiments as gate electrodes capable of altering the electron density in the narrow conducting paths below them. If the mask's pattern is transferred into the 2DEG by an optimized low energy ion exposure step (Scherer and Roukes, 1989), rather than by conventional etching methods, edge depletion is minimized and the conducting path width remains essentially constant as n_s is varied.

103

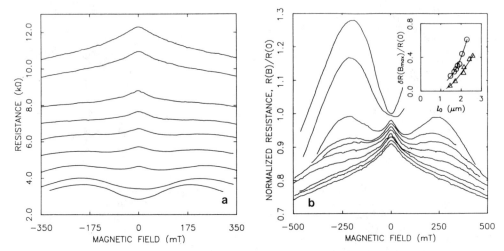

Figure 6. Density dependence of the low-B maximum in a $8\mu m$ long self-aligned gated wire of constant width, $W \sim 270nm$. *(left)* Raw data from $n_s \sim 2 - 5 \times 10^{11} cm^{-2}$. *(right)* Magnetoresistance traces normalized to their zero field value. The curves are offset by .01 units along the vertical for clarity. The positive magnetoresistance leading up to the low-B peak seen at low densities decays at higher n_s and ultimately disappears into the dominant negative magnetoresistance peak at $B = 0$. *(inset)* The amplitude of the maximum is plotted against the transport mean free path, ℓ_0, for 0.27- (triangles) and 0.10-μm (circles) wires.

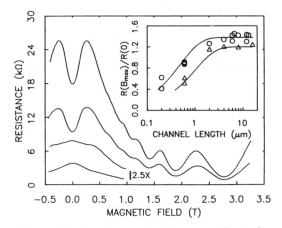

Figure 7. Length dependence of the low-B anomaly. Traces are shown from $0.13\mu m$ wires having lengths of 0.2, 0.6, 8.0 and 16.0 μm. *(inset)* The size of the anomaly; qualitatively characterized by the resistance at the maximum, $R(B_{max})$, normalized to R at zero field; decays for short channels. Data from two sets of wires of width 0.13- (circles) and 0.30-μm (triangles) are shown.

increase in edge roughness as the wire is pinched down to narrow widths. (This situation is schematically depicted in Fig. 1). At low channel densities close to pinch-off, screening of these fluctuations is very feeble and the remaining electrons begin to form "puddles". Ultimately these become disconnected and a metal-insulator transition occurs. These authors note that in a long split-gate wire the jagged depletion profiles result in edge roughness and random variations in width. They speculate that the latter may account for the absence of conductance quantization data noted in investigations of pinched-gate wires *longer* that $\sim 1 \mu m$. Note that the quantization phenomenon disappears at a length comparable to ℓ_B. Conversely, we have found that the boundary scattering peak disappears for lengths *shorter* than ℓ_B. It seems extremely likely that the same mechanism, boundary scattering, controls both.

From the results of these calculations, it appears that the edge roughness is not constant in a pinched gate wire. As V_g and, with it, n_s and W are varied, the parameter p may change drastically, especially near pinch-off. By contrast, for a wire defined by ion-exposure the situation may be quite different. Through optimization of this method, we have fabricated narrow wires which, as lithographic (mask) width is decreased down to $W > 50nm$, show negligible decrease in electron density (Scherer and Roukes, 1989). This suggests that the edges of these wires are delineated by potential contours much steeper than those in pinched gates wires of comparable width. Further confirming this is the fact that the channel width remains essentially constant while electron density is varied (through V_g) in ion-exposed wires with self-aligned gates. In our analysis of data presented earlier, we have implicitly assumed that p is a property of the ion-exposed boundary depending most strongly on fabrication conditions. These considerations appear to substantiate this idea.

BOUNDARY SCATTERING INVESTIGATIONS WITH MULTIPROBE CONDUCTORS

In the previous section we have described investigations focusing on an anomaly in the magnetoresistance of narrow quasi-ballistic wires. The experimental data presented thus far has reflected properties intrinsic to the wires themselves. With the addition of strongly coupled probes to these samples, new transport anomalies are manifested (Roukes *et al.*, 1987; Timp *et al.*, 1988; Takagaki et al., 1988). Elsewhere we have shown that random scattering obscures the observation of phenomena arising from the added junctions (Roukes *et al.*, 1989b). Viewing this from a more positive perspective, this suppression offers the possibility of a new experimental approach to investigate the nature of scattering within these samples. One anomaly, the *transfer resistance*, is particularly useful in this context. As shown in what follows, recent studies of scattering in narrow wires using this phenomenon agreem with the results we have presented above. Additionally, this new technique is easily employed over a wide range of densities and for various methods of confinement. Below, we shall first explore how this effect is developed before describing its application.

THE BALLISTIC TRANSFER RESISTANCE, $R_T(L,B)$

The Bend Resistances.

Takagaki and coworkers (Takagaki *et al.*, 1988) first noted that an unexpected, *negative*, four-probe resistance could be induced across a miniature ballistic "cross" junction at low temperatures. In their experiments a probe configuration was used which forced current to flow in a bent path through adjacent leads, while voltage was probed through the remaining pair diagonally across the junction (*Fig. 8(a)*). Use of these authors' polarity convention for the voltage probes, $V^{(+)}$ and $V^{(-)}$, as shown in the figure, gives a *positive* four-probe resistance of the junction region as expected at elevated temperatures. At low temperatures, however, what they termed a *negative bend resistance*, $R_B = [V^{(+)} - V^{(-)}]/I$, was observed. Here, I is the magnitude of the imposed (constant) current. In these first experiments the negative resistance was found to be easily suppressed by application of a small magnetic field (*see Fig. 10a*). R_B has also been referred to as the local bend resistance.

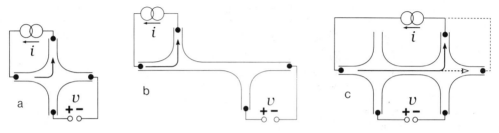

Figure 8. (a) Probe configuration for the negative bend resistance, $R_B(B)$. It is one example of a larger class of what may be termed *transfer resistances*, $R_T(L,B)$, representing the specific case of zero probe separation, $L = 0$. As depicted in configuration **(b)**, transfer resistances are generalized "resistances" in that they involve voltage measurements out of the *net* path of the current (which we depict by solid lines with arrows). In the case of R_B, at low temperatures carriers are ballistically transfered straight through the junction region into the $V^{(-)}$ probe; this results in a negative induced voltage at $B = 0$. At higher temperatures when diffusive transport is recovered, the (positive) resistance of the junction is observed for the same probe configuration. For the more general case of finite probe separations, $L > W$, the transfer resistance at zero field, $R_T(L,B = 0)$, vanishes in the diffusive limit. The probe configuration of sketch **(c)** is employed to measure the non-local bend resistance, ΔR_s. This is the difference between two four-probe resistances: $R_s^{(B)}$ measured with bent net current path (solid line), and $R_s^{(S)}$ with a straight path (dashed line). Here, the conventional probe arrangement yields the series resistance of the wire. This can be enhanced when a bent current path causes increased backscattering at the junction.

It was conjectured that the negative bend resistance is a ballistic effect, arising when carriers are not scattered within the junction, but instead pass straight through, directly into the negative $(V^{(-)})$ voltage probe. This would act to raise the chemical potential of the $V^{(-)}$ probe above that of the $V^{(+)}$ probe, causing a "negative" *four*-terminal resistance to appear in the measurements. The two-terminal resistances, meanwhile, always remain positive. The effect disappears at elevated temperatures when electrons scatter from phonons many times within the junction region. Transport at higher temperatures becomes diffusive – electrons *drift* from the "injecting" to the "collecting" current probe and a finite potential drop of conventional polarity is established across the junction. When the voltage probes and their attached reservoirs come to equilibrium with this local potential the "ordinary" resistance of the junction is measured.

Theoretical investigations of this phenomenon have recently been carried out by modelling electron transport through narrow junctions. Büttiker's equations (Büttiker, 1986) are employed to generate resistances from the transmission probablities, T_{ij}, for an electron which enters the junction from lead j and ultimately leaves via lead i. Calculations of these T_{ij}'s have followed two distinct routes. The first approach is *quantum mechanical*: Schrödinger's equation is solved for simple potentials approximating those at the junctions of real devices. In this picture, within each lead, and far from the junction, the electron eigenstates are plane waves in the longitudinal direction and confined (e.g. particle-in-a-box) states in the transverse direction. In the vicinity of the junction, the local potential acts to scatter electrons between these states and, consequently, from one lead to another. The calculations yield transmission probabilities at the Fermi energy, $T_{ij,mn}(E_F)$, describing transferral of an electron from mode n in lead j, to mode m in lead i. Traces over the mode indices for each lead generate the T_{ij}'s used to calculate of R_B. Avishai and Band (1989) were first to report calculations of R_B at zero temperature in absence of a magnetic field; for square-cornered crosses delineated by an infinite square-well confining potential. R_B was found to decrease with increasing mode occupancy, $4 > N_{occ} > 16$, a cusp being displayed in R_B as each new mode became occupied. Kirczenow (1989c) subsequently reported much more extensive calculations at finite magnetic field for $0 > N_{occ} > 5$, modelling square crosses with parabolic transverse potentials at $T = 0$. Similar overall behavior with mode occupancy was found, but strong additional features at finite-B were demonstrated. This behavior was shown to arise from quantum mechanical resonances at the junctions. The second theoretical approach is *classical*: the trajectories of electrons constrained to a potential energy surface (modelling the shape of a multiprobe conductor) are calculated. In this picture, the electrons are classical particles which, as if billiard balls, move according to Newton's equations and the Lorentz force (Beenakker and van Houten, 1989). In the limit of infinite-wall potentials, the problem reduces to free motion within a set of boundary conditions describing the conductor's geometry. The calculations are employed to generate classical transmission probabilities, T_{ij}, analogous to the quantum mechanical traces described above. These in conjunction with Büttiker's model generate simulated magnetoresistance curves. For the probe configuration of Fig. 8(a) a negative bend resistance is reproduced. Its magnitude is found to depend upon the Fermi momentum, p_F, which, in the classical limit, is a completely smooth, monotonic, and continuous variable. R_B is predicted to increase with decreasing p_F, similar to quantum mechanical predictions for dependence upon N_{occ}. In the quantum case, however, R_B has a characteristic dependence upon k_F for *each* subband and cusps occur each time the *discrete* variable N_{occ} changes. The fact that a classical model successfully reproduces the shape of this low-field anomaly is important. It demonstrates that the essential physics involves ballistic transport within the multiprobe geometry, rather than the few-subband aspect of transport in the samples as initially conjectured.

Both the quantum mechanical and the classical approaches to the problem indicate that R_B is strongly dependent upon the shape of the junction involved. Recent arguments of Beenakker and van Houten (1989b) lead us to expect this, since rounding of the junction corners acts to *collimate* the electron beam. Through this collimation, a greater proportion of the flux injected into the junction is directed straight ahead and into the $V^{(-)}$ probe. In samples where the fabrication technique causes the electron density to decrease in narrow wires compared to that of the wider junction regions, the collimation effect can also be enhanced by the larger Fermi velocity within the junction. This mechanism is probabably operative in samples with edge depletion lengths (see Roukes *et al.*, 1989a) which are comparable to the wire width, W, as is the case in many recent experiments involving multiprobe conductors (e.g. Timp *et al.*, 1988, 1989; Takagaki *et al.*, 1988; Ford *et al.*, 1989; Chang and Chang, 1989). Quenching of the Hall resistance at low-B (Roukes *et al.*, 1987) is another transport anomaly known to be extremely sensitive to the precise geometry of the junction (Ford *et al.*, 1989). Together, both of these anomalies constitute, in some sense, a characterization of a particular junction. The same T_{ij}'s, albeit in different combinations, generate the transport coefficients obtained from these separate measurements. Worthy of note is the fact that neither quantum nor classical modelling have yet produced results which simultaneously agree with experimental data for R_H and R_B from a given junction.

The classical model predicts that $R_B(B)$ depends upon Fermi momentum, p_F, probe width, W, and the shape of the junction. As mentioned earlier, ion-exposed wires with self-aligned gates have the unique property that their electrical geometry remains essentially fixed while n_s is varied. This can be put to an interesting use in the present context. For samples exhibiting this property, R_B should monotonically *scale* with the single parameter p_F. Quantitative tests of this prediction (Roukes et al., 1989b) fail to display the expected scaling. This experimental observation is attributed to the fact that several assumptions of the simple model are not valid for real samples. First, the boundary collisions are not purely specular; in real samples diffuse boundary scattering eliminates contributions to R_B arising from trajectories requiring many reflections. Second, the mean free path is not infinite; in actuality ℓ_0 monotonically decreases as electron density is lowered. Complicated, multiply scattered trajectories, involving effective path-lengths within the junction that are *longer* than ℓ_0 do not contribute to R_B. This has the effect of reducing the importance of collimation in real devices. As ℓ_0 continues to decrease as density is reduced, fewer and fewer of the collimation trajectories are relevant. Ultimately, only the "straight shot" trajectories remain to contribute to the development of a negative-R_B; these paths lead ballistic electrons from the injecting probe directly into the $V^{(-)}$ probe without collisions with the edges. For junctions with extremely rounded corners (radius of curvature $r_j > W$) the "cone" describing the angular range of such trajectories becomes extremely narrow, $\Delta\alpha = 2\tan^{-1}[1/(1 + 2\hat{r}_j)] \sim 1/\hat{r}_j$, where $\hat{r}_j = r_j/W$. In this situation a precipitous decay of R_B with decreasing density is manifested (Roukes *et al.*, 1990). These considerations also have important implications regarding the possibility of observing deterministic chaos in these structures. We shall return to this below.

Recently, in experiments with junctions defined electrostatically by four point contacts, cusps in R_B have been reported as gate voltage is varied (Timp *et al.*, 1989). In this special case modal structure is developed, despite the fact that *many* modes are occupied within the relatively broad "junction" region encircled by the point contacts. This interesting phenomenon apparently arises because the point contacts selectively inject current only into the lowest-lying modes, thereby inducing a non-equilibrium momentum distribution within the junction. This enhances the contribution to R_B from specific modes. The regime explored is, most likely, not analogous to that occuring in narrow

Figure 9. Samples for investigation of R_T. *(top left)* Arrays of $0.95 \times 0.8mm$ devices. From the twelve $100 \times 100\mu m$ optically-defined ohmic contacts at the edge of each device, a lead frame converges with wires terminating near the center. Within this field-of-view the electron beam lithography is carried out. *(right)* The center of the device showing three four-probe conductors used to measure $R_T(L)$ with (left to right) $L = 0$-, 1.0-, and 0.5-μm. In this sample the $200nm$ wide wires flare out to $\sim 2\ \mu m$ leads which connect to the tips of the optically-defined lead frame. *(bottom left)* Detailed view of junctions, comprising two separate samples to measure $R_T(L = 0)$ (via the cross geometry) and $R_T(L = 2\mu m)$ (via the twin-T junctions).

junctions, e.g. as modelled Kirczenow (1989c), for which resonances might be expected. Nor will the decay of the non-equilibrium distribution function in this multimode regime be similar to that of a narrow junction in which, by contrast, the level spacing becomes appreciable compared to E_F and very few modes are occupied. Within a broad junction, the large number of occupied modes results in scattering which is effectively two-dimensional.

Prior to the experiments of Takagaki *et al.* at a cross junction, Timp *et al.* (1988) observed resistance increases resulting from bends placed in the current path of small multiprobe conductors. The four-terminal (series) resistance of a segment of narrow wire increased by an increment $\Delta R_s = [R_s^{(B)} - R_s^{(S)}]$ when current was forced to turn a corner at a junction. This measurement is depicted in Fig. 8(c). Here, $R_s^{(B)}$ and $R_s^{(S)}$ refer to the bent *(dotted)* and straight-through *(solid)* current paths shown. This effect clearly demonstrates the non-local nature of transport in these quasi-ballistic samples; in the diffusive limit one expects that the potential distribution sensed by the voltage probes would be negligibly affected by such a change in the current probes. This increase in resistance, ΔR_s, often called the non-local bend resistance, was ascribed to fact that electrons in the lowest-lying quasi one-dimensional modes should have small probability of transmission into the side probes. Quantum mechanical calculations of junction scattering have confirmed this idea, for transport in systems with only a *few* occupied subbands (Ravenhall *et al.*, 1989; Kirczenow, 1989a, 1989b, 1989c; Baranger and Stone, 1989). As briefly mentioned above, however, recent simulations and experiments indicate that the physics responsible for most anomalies seen in experiments is, in essence, *classical*. We expect this is also true for the finite-ΔR_s phenomenon observed in these experiments. Reanalysis of the early data for, e.g., the position of the boundary scattering peak at B_{max} (Fig. 4) and the resistance of the last plateau of the Hall resistance (see Roukes *et al.* 1989b), suggests the experiments were carried out on wider wires than initial estimates indicated. This would lead to estimated values for N_{occ} which are too low; thus, explanations for transport with *many* occupied modes (> 5) may be most relevant. In the multi-mode regime at finite temperature, it is now clear that *modal* structure in transport coefficients is generally suppressed (except in the special case mentioned above). Nonetheless, ballistic "size effects" are manifested, taking the form of the observed transport anomalies. Baranger and Stone (1989a) have reported simulations of ΔR_s in the *few*-subband regime (this involves a minimum configuration of five probes and a specific model for random scattering).

To summarize, classical modelling of transport in a ballistic junction qualitatively reproduces the negative bend resistance observed in smooth-cornered junctions with many occupied modes. To obtain quantitative agreement with experiments further refinements will be required. In particular, the consideration of scattering mechanisms resulting in a finite momentum-memory for the electrons appears essential (Roukes *et al.*, 1989). For "abrupt" junctions bounded by sharp corners, however, non-monotonic density dependence emerges which is completely beyond the scope of the simple classical model. Future experiments with even smaller, more abruptly defined junctions may permit the exploration of new phenomena, such as resonant junction scattering, predicted by quantum simulations in the few-mode regime. Even though theory and experiment have not yet completely converged in these matters, it is possible to employ these experimentally observed phenomena to investigate electron scattering at boundaries. This is described in what follows.

Transfer Resistances.

The negative bend resistance, R_B, is one example of a larger class of what may be termed *transfer resistances*, $R_T(L,B)$. These are *generalized* resistances — the voltage probes may be placed *outside* the path of the current. The transfer resistance geometry, as depicted in the inset of Fig. 4(b) for example, can be envisioned as being formed from a symmetric cross junction. Stretching Fig. 4(a) along the horizontal yields adjacent T-junctions separated by length, L, with probe assignments remaining fixed. The current path (shown as solid lines with arrows in these figures) is outside the region between voltage probes. Given the configuration of the probes $R_T(L)$ vanishes for $L > W$ in the diffusive limit which is approached at elevated temperatures. For clean samples at low temperatures, however, $R_T(L)$ for $L > W$ becomes finite and negative when *ballistic transfer* of carriers out of the *net* path of the current is possible. In the limit of zero separation, the bend resistance is recovered: $R_T(L = 0, B) = R_B(B)$. For finite separation, $L > 0$, electrons must traverse a segment of narrow conductor, which we call a "propagation region", before they are actually transferred to the $V^{(-)}$ probe.

We can conceptualize the development of the transfer resistance in terms of the local electron distribution function, $g(\mathbf{x},\mathbf{p})$, in the vicinity of the junctions. Our discussion is an heuristic extension of the model envisioned by Büttiker (1986) to the situation where scattering *within the leads* is explicitly considered. Both the applied field and collimation (resulting from the "injecting" T-junction) act to induce a non-equilibrium momentum distribution in the forward direction. The "propagation region" and the $V^{(-)}$ probe of the second T-junction are most strongly affected. Within the wires this relaxes to local equilibrium, an isotropic distribution over the Fermi disc, by momentum scattering processes. This elevates the steady-state average of the local distribution function in these regions. An isotropically-directed (quasi diffusive) backflow of carriers from these regions is thereby established, leaking carriers back to the net current path. This satisfies the conditions of charge conservation and insures that *net* current flows only into the current probes. But most important in the present context is that the injected non-equilibrium momentum distribution decays within the "propagation region" *as a direct consequence of random scattering*. This provides us with a unique means to explore the scattering mechanisms operative in this segment of narrow wire.

SPATIAL DECAY OF THE BALLISTIC TRANSFER RESISTANCE

The electron micrographs of Fig. 9 display samples fabricated to observed the spatial decay of R_T. In perfecting these measurements, several generations of devices with slightly different geometries were fabricated — those shown represent one of the later generations. A given measurement involves fabrication of a *set* of samples. This is a family of four-terminal devices, all fabricated simultaneously, configured to measure $R_T(L)$ with different probe separations, L. Each particular device within this set is comprised of two T-junctions: an "injector" and "collector" junction, separated by the propagation region of fixed length L. (For $L = 0$, the two junctions merge together to become a straight cross junction.) The injector T-junction acts to induce an non-equilibrium local distribution, $g(\mathbf{x},\mathbf{p})$, within the propagation region which is anisotropic in $\mathbf{p}$-space, being peaked in the forward direction (pointing down the propagation region towards the collector). By reciprocity, we may view the collector as serving to sense the $\mathbf{p}$-space anisotropy of $g(\mathbf{x},\mathbf{p})$ in the locale of this second junction. The precise shape of the injector junction, i.e. the degree of rounding of the corners, strongly affects the collimation of emerging electrons. We may picture there being an angular distribution function characterized by an average cone of injection $\Delta\alpha$, for the injector junctions; similarly, the collector may be described by its cone of acceptance (cf. Beenakker and van Houten, 1989). In the experiments, these effective $\Delta\alpha$ are kept constant within each set by defining all of the devices simultaneously in one ion exposure. Thus, to a very high degree, only the length of the propagation region, L, varies between the devices of each set.

In Fig. 10 data from a set of samples with probe width, $W \sim 300nm$, are shown. This family of devices provided probe separations $L < 1\mu m$. An additional trace at $L = 6\mu m$ was also obtained. The negative resistance at zero field is seen to decay with increasing L; a small vestigial signal is still detected even after the injected carriers have propagated down a $6\mu m$ long wire (*Fig. 10a, inset*). This graphically demonstrates ballistic transport surviving over appreciable distances in these samples at $B = 0$. We explore this in Fig. 10b by plotting the zero field value, $R_T(0)$, against the length of the propagation region, L. The results plotted semi-logarithmically suggest that the spatial decay involves two exponential lengths. For this set of samples we find a rapid decay for $L < 0.5\mu m$, characterized by an exponential decay length $\ell_d \sim 0.2\mu m$. This appears to cross over to a decay more than an order of magnitude slower, $\ell_d \sim 2.4\mu m$, for $L > 0.5\mu m$. This family of devices was designed anticipating much more rapid exponential decays. To confirm that two lengths scales are operative, new sets of devices were fabricated, specifically exploring the range $0.5\mu m < L < 6\mu m$. Data from these show similar behavior. In Fig. 11, data from a set of samples having probe width $W \sim 150nm$ reveal a decay at short separations $\ell_d \sim 0.35\mu m$, which ultimately settles to a much slower decay, $\ell_d \sim 1.3\mu m$, for $L > 1\mu m$.

The exponential spatial decay of R_T at large L occurs with a characteristic length, ℓ_d, comparable to boundary scattering lengths, ℓ_B, discussed previously deduced from the low-B peak in the magnetoresistance. Both ℓ_B and ℓ_d are shorter than the transport mean free path, ℓ_0. In a wide conductor patterned from the same heterojunction as the narrow samples, $\ell_0 \sim 8\mu m$ at the equilibrium ($V_g = 0$) density, $n_s \sim 3.0 \times 10^{11} cm^{-2}$. The rapid spatial decay of the transfer resistance for short separations appears to reflect an interaction between closely-spaced junctions. In part, this arises from a loss of collimation with increasing L, as the cross junction ($L = 0$) distorts to ultimately emerge as a pair of well-defined T-junctions ($L > W$). At large probe separations the spatial decay crosses over to a slower exponential decay reflecting scattering processes intrinsic to the propagation region itself. The precise crossover length, the absolute magnitude of the signal, and the initial (rapid) decay length all depend upon several experimental parameters: the probe width, W; the shape of the junctions; and the scattering rates within the wires. The slow exponential decay seen for large probe separations, $L > W$, appears most sensitive to the properties of the boundaries themselves. This leads us to interpret this *larger* decay length as reflecting the total rate of momentum scattering, $\tau_p^{-1} \sim \ell_d/v_F$ within the narrow wire segment forming the propagation region. For devices of similar width, fabricated under the same

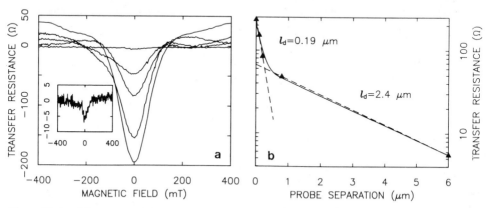

Figure 10. Spatial decay of the transfer resistance of a $\sim 300nm$ wire. *(left)* $R_T(L)$ at low-B shown for probe separations $L = 0$- (cross junction), 0.1-, 0.2, 0.8-, and 6-μm. *(inset)* 20X magnification of the resistance axis for $6\mu m$ probe separation. A vestige of the ballistic signal persists. *(right)* The signal appears decay with two characteristic exponential decrements. For $L < 0.5\mu m$ the transfer resistance quickly decays with an exponential length $\ell_d = 0.19\mu m$. For $L > 0.5\mu m$ however, the spatial decay is an order of magnitude slower.

conditions, we note that the boundary scattering length extracted from the finite-B peak in the magnetoresistance, ℓ_B, is comparable to this larger spatial decay length, ℓ_d, extracted from $R_T(L)$ at large probe separations. We interpret this as indicating that momentum coherence in these narrow wires is primarily determined by the quality of the boundaries. Further discussion of these experiments is presented elsewhere (Roukes *et al.*, 1990).

The local angular distribution of $g(\mathbf{x}, \mathbf{p})$ at point $\mathbf{x}$ within a narrow wire, can be described by a function $P(\mathbf{x}, \alpha)$. Here we take α as the angle between $\mathbf{p}$ and the channel axis. In the limit of perfect specularity, $p = 1$, the shape of this distribution is conserved after each boundary collision. In this situation electrons can propagate an arbitrary length down the wire while retaining their initially-prepared angular distribution; the angular distribution has no $\mathbf{x}$-dependence. The situation $p < 1$ is quite different. In the simplest approximation, where p is a constant independent of angle, $P(\alpha)$ is rapidly restored to isotropy. Here, after propagating a distance of order $\ell_B \sim W/(1-p)$, memory of the initial distribution is lost.† Further complication arises if p depends upon angle, as has been suggested from experiments investigating the scattering-induced broadening of surface Landau levels. If this dependence occurs, different segments of the angular distribution, $P(\alpha)$, will relax at their own characteristic rate. In real samples, however, the degree of specularity will always be less than unity for *all* angles: $p(\alpha) < 1$. This means that memory of anisotropic initial conditions is lost in a long enough wire segment and a uniform distribution, $P(\alpha) \sim 1/\pi$, ultimately emerges as the distribution tends toward isotropy. In this light, the exponential length, ℓ_d, obtained from the spatial decay of R_T is actually a weighted average of equilibration lengths for different segments of the angular distribution. The weighting is dependent upon the angular acceptance function of the collector junction, determined to a large degree by geometry.

In the present context we note that Beenakker and van Houten (1989) have employed an *anisotropic* angular distribution, $P(\alpha) = \cos(\alpha)/2$, in their recent classical modelling of ballistic multiprobe conductors. Given their assumption that boundary reflections are perfectly specular, this distribution is conserved along the injector probe until it is altered by junction scattering. The sensitivity of their numerical results to the precise form of the initial distribution has not been reported.

As mentioned earlier, the particular feature distinguishing a transfer resistance is that it is obtained from a configuration where the voltage probes are spatially separated from the path of *net* current flow. For this reason it provides a very sensitive means of investigating the return to equilibrium of the injected momentum distribution. In ideal samples, for probe separations greater than the wire width, $L > W$, the ballistic effect is observed without *any* background resistance signal. For $L < W$ a small positive contribution to the total resistance arises from residual random scattering within the junction region (i.e. the ordinary junction resistance). This is easily dominated at low temperatures by the negative ballistic effect. At high temperatures, however, this positive resistance becomes apparent since the negative ballistic signal has been suppressed. We have described the decay of the ballistic signal at low temperatures when the propagation region is lengthened. The number of decades over which this spatial decay can be followed is, in practice, most strongly determined by the inter-lead coupling between the voltage and current paths. The largest voltage drop in the sample occurs along the net path of the current. The voltage probes, in close proximity, are used to detect a signal of much weaker amplitude. Parasitic coupling mechanisms ultimately dominate the weak residual signal which persists at large probe separations, L. In the experiments, sources of such undesired coupling arise from both the intrinsic properties of the sample and the external leads and

---

†This simple picture is approximately true for the nearly specular case, but needs substantial refinement for small p. In the nearly diffuse limit, the estimate above becomes $\ell_B \sim W$. The real situation is somewhat more complicated. At such short distances the angular weight of trajectories which do not interact with the boundaries is substantial: $\Delta\alpha/\pi \sim (2/\pi)\tan^{-1}(W/L)$. For $L = W$, this implies $\Delta\alpha/\pi \sim 0.5$. In this situation the anisotropy in $g(\mathbf{x}, \mathbf{p})$ decays over lengths $L > W$ at first largely due to the *geometrical* reduction of the contribution from these "straight shot" trajectories. Subsequently, catastrophic boundary collisions and bulk momentum scattering processes in the interior of the wire dominate the return to isotropy as in the nearly specular case. These considerations point out ambiguities involved in defining a boundary scattering length, ℓ_B.

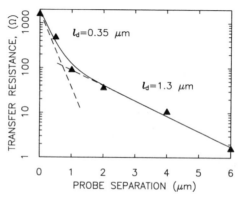

Figure 11. Spatial decay of $R_T(L)$ for a $\sim 150nm$ wide wire. The initial rapid decay, characterized by a decay length $\ell_d \sim .35\mu m$, gives way to a much slower decay $(\ell_d \sim 1.3\mu m)$ for probe separations, L, greater than $\sim 1 \mu m$.

circuitry. A check on the validity of measurements in the large-L regime is available, however, through the symmetry relation discovered by Büttiker (1986). The ballistic signal is symmetric under exchange of current and voltage leads: $R_{ij,kl}(B) = R_{kl,ij}(-B)$, where the first *(second)* pair of indices are the current *(voltage)* probes. The parasitic coupling between leads depends, in general, upon their exact configuration. As a result, this undesired portion of the detected signal does not generally satisfy the symmetry relation. The samples depicted in Fig. 9 each have *only* the four requisite leads involved in measurement of R_T — this greatly minimizes undesired coupling.

In their simulations of a transfer resistance involving two cross junctions, Beenakker and van Houten (1989) noted strong fluctuations in R_T vs. B. This suggests the emergence of *deterministic chaos* in this model; it is known to occur in other related classical systems. Their interpretation is strengthened by a subsequent investigation of the dependence of exit probe upon the initial angle of entry into the junction. Very striking sensitivity to this injection angle is found. To date we have not observed similar irregular features in our experiments, instead generally finding very smooth behavior in $R_T(B)$ when, with optimal fabrication, disorder in the channels is minimized (*e.g.*, Fig. 10a). With particular junction geometries electron focusing peaks can become pronounced (Roukes *et al.*, 1990). Our recent tests of the classically-predicted scaling of transport anomalies (Roukes *et al.*, 1989b) have bearing on this problem. As described earlier, if an anomaly arises from complicated trajectories involving long multiply-scattered paths within the junction, generally it will fail to satisfy the predictions of scaling. This suggests that scattering, both at diffuse boundaries and in the channel interior, eliminate these deterministic contributions to the measured transport coefficients. It is conceivable that in real systems, the relatively *low* number of boundary reflections sustained before complete loss of "momentum memory" occurs may preclude the observation of chaotic behavior. The important role of random scattering in establishing a finite momentum-memory has not been included in models investigated to date. A second problem is that all experimentally measurable transport properties involve *angular averages* of the distribution function. This is true in case of transfer resistances as well, where the voltages developed depend upon the collector junction's angular acceptance function. It appears that innovative sample configurations will be required before chaotic features emerge from transport experiments such as these.

Prior to our work involving the transfer resistance, Timp *et al.* (1988) reported a preliminary investigation of the decay of the non-local bend resistance, ΔR_s. Note that in Fig. 8(c) the bend in the current path occurs at a cross junction which also forms the end of the wire "segment" between voltage probes. In their study, a third junction was introduced, permitting the bend in the current path to be separated and located downstream from this "end" voltage probe. The amplitude of the resistance difference, ΔR_s, measured in this three-junction configuration was found to decrease as the distance of the remote bend from the region between voltage probes was lengthened. Two samples were studied, within each device $\Delta R_s(L)$ was measured at four separations, $L = 0$-, 0.7-, 0.9- and 2.2-

μm. The values of $\Delta R_s(L)$ extracted approximately followed an exponential decay with L in both devices. From a fit to data at these four separations, decay lengths $\ell_d \sim 360$- and 420-nm were extracted for the two samples. In wide channels of the same material a transport mean free path, ℓ_0, of $\sim 5\mu m$ was deduced. It was conjectured that the extremely rapid decay of ΔR_s, and the large disparity between ℓ_0 and ℓ_d, arise because of different rates existing for *small*- and *large*-angle scattering. Note that the transport mean free path derives its name from the fact that it is extracted from the mobility, $\ell_0 = p_F\mu/e$. Contributions from *large*-angle scattering are dominant in determining μ, since they are most effective in degrading the forward momentum. But even though scattering through *small* angles is ineffectual in this regard, (by definition) the electron still suffers a change of its momentum state. This will act to suppress any mode-dependent phenomenon. Quantum mechanical simulations of ΔR_s show a strong effect, however, only for low mode occupancies. In this extreme quasi one-dimensional limit small-angle scattering is actually *suppressed*. In this regime, level spacings are large compared to E_F and, as a consequence, all scattering events involve correspondingly large changes in forward momentum. Only with many occupied modes, i.e. in the 2D limit, can this *intermode* scattering proceed with arbitrarily small transfer of forward momentum and does the concept of *small*-angle scattering apply.

We mention three factors which may possibly account for the large difference between the decay lengths we have recently measured via the transfer resistance, and those previously obtained from the non-local bend resistance. First, in measurement of $\Delta R_s(L)$ at zero separation, $L = 0$, a configuration involving *two* junctions is employed. A *third* junction is introduced to obtain $\Delta R_s(L)$ finite lengths, $L > 0$. Büttiker (1985) first showed that probes themselves can act as inelastic scatterers; in the experiments considered here, the change from a two- to a three-junction configuration could result in an enhanced decay. Second, the raw traces obtained in the study of ΔR_s were very asymmetric. (These traces were subsequently "symmetrized" to permit extraction of ΔR_s.) Asymmetries in transport data are actually reflective of asymmetries intrinsic to the electrical geometry of the device. If a narrow sample, e.g. having the configuration of Fig. 8(c), contains *perfectly* symmetric cross junctions, the Hall resistance and the magnetoresistance will be antisymmetic and symmetric, respectively, in the absence of disorder. It has been shown experimentally that disorder introduced during fabrication can strongly increase deviations from these expected symmetries. This appears concomitant with increased scattering within the narrow channels (Roukes et al., 1989a). Third, at large separations when the non-local bend resistance has decayed to a small value, its extraction from the large background signal becomes difficult (Roukes *et al.*, 1990). In this regime, ΔR_s has become the small difference between two much larger resistances ($R_s^{(B)}$ and $R_s^{(S)}$), both comprised almost entirely of the resistance of the wire segment near the voltage probes. This adversely affects the accuracy of measurements at large-L.

Intermode scattering is important in quantum wires since it sets an upper limit on the length scale over which coherent transport, without mode-mixing, can be sustained. To insure that mode-dependent quantum effects can be observed, this length must be sufficient to span the active region of the device probed. These studies appear to indicate that this coherence *can* be maintained over relatively long distances — long by "nanofabrication" standards, anyway. The offers the possibility of future exploration mode-coherent effects in a wide variety of new structures.

BOUNDARY SCATTERING: CLASSICAL vs. QUANTUM REGIMES

For the experiments we have discussed, classical phenomenology has shown a high degree of success in explaining the observations. The low-B peak in the magnetoresistance scales as predicted by the classical theory of boundary scattering in metals. Transport anomalies such as the transfer resistances are qualitatively modelled through analyses of classical trajectories, even in situations where modal behavior begins to weakly modulate the overall classical features. The question thus arises: under what conditions might we expect a breakdown of classical approaches to boundary scattering?

There appear to be three distinct regimes of boundary scattering which we may loosely categorize as *classical, diffractive,* and *quantum*. In the *classical* regime Fuch's specularity parameter, p, is sufficient to describe the scattering properties of a surface. Characterizing a boundary by a constant p is a tacit assumption that some fraction of the distribution function is restored to equilibrium after each collision, and that this fraction is fixed, irrespective of the incident angle and momentum (wavelength). This can hold only for a perfectly random interface. For electrons to be scattered uniformly over all angles, the surface must be comprised of randomly-shaped asperities. Furthermore, for this behavior to be exhibited at all wavelengths, the spectrum of correlations for the roughnesss must be essentially uniform over all length scales. Obviously, this behavior is not to be expected from boundaries such as depicted in Fig. 1. Ando (1977) modelled surface roughness at a Si/SiO_2 interface by assuming a Gaussian correlation spectrum $<\Delta(\mathbf{r})\Delta(\mathbf{r}')> = \Delta^2 \exp[(\mathbf{r}-\mathbf{r}')^2/\Lambda^2]$. As mentioned, applying this model to the data of Hartstein et al. (1976) gave agreement for an average displacement of the interface, $\Delta \sim 4.3\text{Å}$ and average range of its spatial variation in the direction parallel to the surface, $\Lambda \sim 15\text{Å}$. Since the electron wavelength, $\lambda_F = 2\pi/k_F = \sqrt{2\pi/n_s}$, was a few hundred Å in the experiment, it appears that only the tail of the spectrum of correlations was "sensed" by the electrons.

This example involving a Si/SiO_2 interface illustrates a general point. When the electron wavelength is much longer than the dominant length scale (correlation length) of the roughness the degree of specularity can be quite high (Ziman, 1960). In a 2DEG the electron wavelength can become nearly three orders of magnitude longer than a typical lattice constant and, hence, much larger than atomic-scale surface roughness at interfaces or surfaces. A similar situation also holds in pure Bi; this may account for early observations of high specularity in resistance measurements in thin plates, and Tsoi's electron focusing experiments. In metals, by contrast, the much higher electron density yields much shorter values of λ_F and an analogous regime does not exist. In this situation the surface asperities strongly scatter electrons.

The possibility of a *diffractive* regime, can occur if there exists a characteristic length scale, ξ, for asperities at the surface. In this situation, quasi regular features at the surface result in a spectrum of correlations which is not uniform, but instead contains structure near $q \sim \xi^{-1}$. This is the case for the interface schematically depicted in Fig. 1, and for the model of the Si/SiO_2 interface used by Ando. When the wavelength of electrons incident upon such a surface is comparable to ξ, patently non-specular scattering can occur. This is analogous to the situation in optics where the wave nature of light is manifested by diffraction from small features. In this regime, geometric optics and ray tracing are no longer valid – a strong decrease in the apparent value of p would be expected. Although electron diffraction from surfaces is well known, it appears that there are no analogous manifestations yet reported from electron-boundary scattering within conductors.

Finally, the *quantum* regime occurs in the limit of only a few occupied modes. Here, similar to the case of diffraction, electrons can no longer be treated as particles colliding with the surface. In this situation the classical picture breaks down because spatially compact wavepackets cannot be constructed given the few modes available. Additionally, when the level spacings become appreciable compared to E_F the phase space for scattering is severely restricted. We have described the suppression of small-angle scattering in this limit; similar arguments have led to the prediction of mobility enhancement in 1D (Sakaki, 1980). Tesanovic, Jaric, and Maekawa (1987) have made the interesting prediction that in the 2D limit, quantum enhancement of the resistance due to surface scattering should arise. We may understand this by drawing an analogy to a classical situation discussed previously. We argued that electrons in trajectories moving parallel to the plane of the sample at zero magnetic field "short out" the enhanced resistance due to diffuse boundary scattering. Application of a small magnetic field eliminates their contributions to the conductivity by bending them and causing them to terminate in catastrophic collisions with the surface. The resistivity then rises above its zero-field, shorted-circuited, value. In the quantum mechanical case, a somewhat similar effect arises – zero point motion excludes classically allowed states of zero perpendicular momentum. In the few (2D) mode limit, the loss of these straight trajectories is predicted to result in a finite contribution from boundary scattering, even at $B = 0$.

CONCLUSIONS AND PROJECTIONS

We have explored boundary scattering in "quantum" wires in the multimode regime. Electron transport in a narrow conductor strongly depends upon the quality of the boundaries imposed by the fabrication process. If the edges of the conductor are rough, level broadening and asymmetries in the conducting geometry can obscure the effects of confinement. Classical explanations of the transport phenomena recently observed with these devices have provided a great deal of new insight into the nature of electron interactions at the surfaces. Optimal fabrication techniques have been shown to yield boundaries of high quality. This indicates that fabrication of extremely narrow channels with long momentum coherence lengths is an achievable goal.

Future experiments will allow investigations of boundary scattering in the extreme quasi one-dimensional limit, where strong corrections to classical behavior should occur. For this regime, it will be necessary to revise existing quantum theories of surface scattering to lower dimensionality (e.g. Ando, 1977; Tesanovic et al., 1987; Trivedi and Ashcroft, 1988) Even in the multi-mode limit, however, interesting experiments remain to be done. As an example, it should be possible to perform a "spectroscopy" of the spectrum of correlations of surface roughness in devices where electron wavelength is "tunable" while the imposed potential remains essentially fixed. With current advanced fabrication techniques, it may even be possible to "engineer" this spectrum of correlations.

We gratefully acknowledge helpful discussions on these topics with O. Alerhand, S.J. Allen, Jr., H.U. Baranger, C.W.J. Beenakker, and G. Kirczenow.

REFERENCES

Ando, T., 1977, J. Phys. Soc. Japan, **43**, 1616.
Avishai, Y., and Band, Y.B., 1989, Phys. Rev. Lett. **62**, 2527.
Baranger, H.U., and Stone, A.D., 1989a, in *Science and Technology of 1- and 0-Dimensional Semiconductors*, S.P. Beaumont and C.M. Sotomayor-Torres, eds., Plenum, London.
− 1989b, Phys. Rev. Lett. **63**, 414.
Beenakker, C.W.J., and van Houten, H., 1989a, Phys. Rev. Lett. **63**, 1857.; also in this book.
− 1989b, Phys. Rev. **B39**, 10445.
Berggren et al., 1986, Phys. Rev. Lett. **57**, 1769.
Büttiker, M., 1985, Phys. Rev. **B32**, 1846.
− 1986, Phys. Rev. Lett. **57**, 1761.
Chambers, 1969, in *The Physics of Metals*, J.M. Ziman, ed. Cambridge Univ. Press, U.K.
Chang, A.M., and Chang, T.Y., 1989, Phys. Rev. Lett. **63**, 996.
Choi, K.K. et al., 1986, Phys. Rev. **B33**, 8216.
Ditlefsen, E., and Lothe, J., 1966, Philos. Mag. **14**, 759.
Doezema, R., and Koch, J.R., 1972, Phys. Rev. **B5**, 3866.
Ford, C.J.B., et al., 1988, Phys. Rev. **B38**, 8518.
− 1989, Phys. Rev. Lett. **62**, 2724.
Forsvoll, K., and Holwech, I., 1964, Philos. Mag. **9**, 435.
Friedman, A.N., and Koenig, S.H., 1960, I.B.M. J. Res. Dev., **4**, 158.
Fuchs, K., 1938, Proc. Camb. Philos. Soc. **34**, 100.
Hartstein, A., et al., 1976, Surf. Sci. **58**, 178.
Hensel, J.C., et al., 1985, Phys. Rev. Lett. **54**, 1840.
Khaikin, M.S., 1960, Sov. Phys. J.E.T.P. **12**, 152.
Kirczenow, G., 1989a, Phys. Rev. Lett. **62**, 1920.
− 1989b, Phys. Rev. Lett. **62**, 2993.
− 1989c, Solid State Comm. **71**, 469.
Koch, J.F., and Kip, G., 1965, in *Low Temperature Physics, LT9*, edited by J.G. Daunt, et al., Plenum Press, Inc., New York, p. B818
− and Kuo, C.C., 1966, Phys. Rev. **143**, 470.
Kubo, R., 1962, J. Phys. Soc. Japan **17**, 975.
MacDonald, D., and Sarginson, K., 1950, Proc. Roy. Soc. London **A203**, 223.
Nixon, J.A., and Davies, J.H., 1989, unpublished.
Pippard, A.B., 1957, Phil. Trans. Roy. Soc. **A 250**, 325.
− 1989, *Magnetoresistance in Metals*, Cambridge Univ. Press, U.K.
Prange, R.E., and Nee, T.W., 1968, Phys. Rev. **168**, 779.
Ravenhall, D.G., et al., 1989, Phys. Rev. Lett. **62**, 1920.
Roukes, M.L., et al., 1987, Phys. Rev. Lett. **57**, 3011.
− 1989a, in *Science and Technology of 1- and 0-Dimensional Semiconductors* S.P. Beaumont and C.M. Sotomayor-Torres, eds., Plenum, London.
− 1989b, submitted to Phys. Rev. Lett.
− 1990, unpublished.
Sakaki, H., 1980, Japan. J. Appl. Phys. **19**, 1735.
Scherer, A., and Roukes, M.L., 1989, Appl. Phys. Lett. **55**, 377.

Schreiffer, J.R., 1957, in *Semiconductor Surface Physics*, edited by R. H. Kingston, Univ. of Penn. Press, Philadelphia, p.55

Simmons, J.A., *et al.*, 1988, Surf. Sci. **196**, 81.

Soffer, S.B., 1967, J. Appl. Phys. **28**, 1710.

Stone, I., 1898, Phys. Rev. **6**, 1.

Takagaki, Y., *et al.*, 1988, Solid State Comm. **68**, 1051.

Tesanovic, Z., *et al.*, 1987, Phys. Rev. Lett. **57**, 2760.

Thomson, J.J., 1901, Proc. Camb. Phil. Soc. **11**, 120.

Thornton, T.J., *et al.*, 1986, Phys. Rev. Lett. **56**, 1181.

- 1989a, in *Science and Technology of 1- and 0-Dimensional Semiconductors* S.P. Beaumont and C.M. Sotomayor-Torres, eds., Plenum, London.

- 1989b, Phys. Rev. Lett. **63**, 2128.

- 1990, unpublished.

Timp, G., *et al.*, 1988, Phys. Rev. Lett. **60**, 2081.

- 1989, Phys. Rev. Lett. **63**, 2268.

Trivedi, N., and Ashcroft, N.W., 1988, Phys. Rev. **B38**, 12298.

Tsoi, V.S., 1974a, J.E.T.P. Letts. **19**, 70.

- 1974b, Sov. Phys. J.E.T.P. **41**, 927.

van Houten, H., *et al.*, 1986, Appl. Phys. Lett. **49**, 1781.

van Wees, B., *et al.*, 1988, Phys. Rev. Lett. **60**, 848.

Wharam, D., *et al.*, 1988, J. Phys. C, **21**, L209.

- 1989, Phys. Rev. B., **B39**, 6283.

Zheng, H.Z., *et al.*, 1986, Phys. Rev. **B34**, 5635.

Ziman, J.M., 1960, *Electrons and Phonons*, Oxford. Univ. Press, U.K., Chapter XI.

VERTICAL TRANSPORT, TUNNELING CYCLOTRON RESONANCE, AND

SATURATED MINI-BAND TRANSPORT IN SEMICONDUCTOR SUPERLATTICES

S. James Allen, Jr., R. Bhat, G. Brozak[*],
E.A. de Andrada e Silva[**], F. DeRosa, L.T. Florez,
P. Grabbe, J.P. Harbison, D.M. Hwang, M. Koza,
P.F. Miceli, S.A. Schwarz, L.J. Sham[**] and M.C. Tamargo

Bell Communications Research, Inc.
Redbank, NJ 07701, USA

[*] Physics Department, Northeastern University
Boston, MA 02115

[**] Physics Department, University of California
San Diego, La Jolla CA 92093

INTRODUCTION

The focus of the early vision of Esaki and Tsu (1970), on vertical transport in artificial semiconductor superlattices, foresaw engineered bandstuctures with narrow mini-bands and forbidden mini-gaps that would lead to novel transport physics and devices. The phenomenon that captured their attention was Bloch oscillation, a phenomenon that can be readily understood in terms of saturated band transport in a narrow conduction band (Ktitorov, 1972). The key role played by the superlattice is the production of narrow mini-bands that can be saturated by modest electric fields and current density, which in turn would lead to differential negative resistance, gain and eventually Bloch oscillations. Saturated transport is the key phenomena and is a unique feature of miniband transport. Here we describe cyclotron resonance experiments that directly probe mini-band transport. Saturated band transport manifests itself as a saturation of the cyclotron resonance frequency in strong magnetic fields.

Cyclotron resonance is a classic band structure probe that is able to measure without ambiguity the transport mass and scattering rate in any system that is sufficiently clean that the carrier is able to circle the applied magnetic field at least a few times before scattering. As a result, in a superlattice, when the magnetic field is oriented perpendicular to the growth direction, the observation of cyclotron resonance has the power to determine the vertical transport mass as well as the scattering rate. It is important to point out that cyclotron resonance requires that the electron execute coherent transport around the magnetic field and the probe is specific to quantum, coherent tunneling transport in the vertical direction. Phonon assisted hopping

Electronic Properties of Multilayers and Low-Dimensional Semiconductors Structures
Edited by J. M. Chamberlain *et al.*, Plenum Press, New York, 1990

117

through barriers or thermally activated transport over barriers cannot contribute to the resonance since they are basically scattering processes. The simple observation of cyclotron resonance in the tunneling direction is an unambiguous proof of quantum or coherent transport along the superlattice direction.

One should recognize some features that are unique to cyclotron resonance in superlattices in the tunneling direction. Implicit in the discussion of cyclotron resonance in normal solids is the assumption that the cyclotron diameter is large compared to the period of the periodic potential, and that the cyclotron frequency is much less than the bandwidth. Unlike normal solids the length scale of the periodic potential in a semiconductor superlattice can become comparable to the magnetic length or cyclotron diameter and more important the magnetic energy can exceed the mini-bandwidth. This is rarely, if ever, encountered in conventional solids, but under these conditions the cyclotron resonance will begin to "sense" the "graininess" or microscopic character of the periodic potential. These effects were anticipated as early as 1955 by Harper (1955,1967) but not observed until 1985 by Maan and coworkers by interband magneto-luminescence (Maan, 1984, 1987; Belle et al.,1985a, 1985b) and by Duffield and co-workers (1987, 1988) in tunneling cyclotron resonace. Interband magneto-luminescence experimants revealed that the Landau level fan of interband transitions disappeared when the cyclotron energy exceeded the mini-band width. In cyclotron resonance, the low field cyclotron resonance evolves at high magnetic fields into a field independent absorption controlled by the single barrier tunneling rate. All of these features are unique to magneto-transport in superlattice systems, in which one can control both the period of the periodic potential or the mini-band width compared with the magnetic length or energy, respectively. They will be demonstrated and explored below.

SUPERLATTICES, MINI-BANDS AND MINI-GAPS

Before describing cylotron resonance in semiconductor superlattices we will briefly review mini-band structure in superlattices and the material parameters that control it. Following Esaki and Tsu (1971) we consider a superlattice shown in Figure 1, comprised of GaAs quantum wells and AlGaAs barriers. The barriers may be characterized by their height, V, which is a measure of the conduction band offset (Throughout, we shall discuss only electron transport.) and their width, b. The periodic potential is assumed to have a period z_o. This potential is the classic Kronig-Penney potential and can be solved exactly (Kittel,1986). However here we simply sketch the resulting band structure and relate the band widths and bandgaps to the results one would obtain if the potential were weak and nearly free electron like or were strong and described by a tight binding model.

The dashed curve in Figure 2 is the free electron dispersion relation relating energy to momentum. At a momentum, $k = \pi/z_o$, the electron is perfectly reflected by the superlattice and an energy gap opens in which no propagating states can be supported by the superlattice.

If we approximate the barrier by a delta-function with strength, Vb, and Vb/z_o is much less than the kinetic energy at the zone boundary, ie. $Vb/z_o \ll \epsilon_o = \hbar^2/2m \cdot (\pi/z_o)^2$, we are in the nearly free electron limit. Here the energy gap is given by $2Vb/z_o$. The ground state mini-band width is approximately

$$2\Delta = \epsilon_o - Vb/z_o \tag{1}$$

Clearly if the barrier is strong compared with the kinetic energy this approximation fails and one uses a tight binding model in which the starting basis is not the free electron states but the quantum well state shown in Fig. 1. In the tight binding approximation the mini-bandwidth is given by

$$2\Delta = (8/\pi^2) \cdot \epsilon_o{}^2/(Vb/z_o). \tag{2}$$

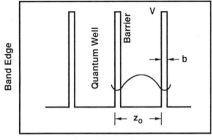

Fig. 1. Conduction band edge versus distance
along the growth direction. Quantum well
state confined by barriers of strength V
and width b separated by a distance z_o.

It proves to be the case that for much of the discussion that follows we can approximate the miniband dispersion along the growth direction by a simple cosinusoidal function

$$E(k) = - \Delta\cos(kz_o). \tag{3}$$

The effective mass at the bottom of the mini-band is simply

$$m_s{}^{-1} = \Delta z_o{}^2/\hbar^2. \tag{4}$$

This straight forward parameterization of the miniband dispersion should be sufficiently accurate to simply extract from measurements of the mini-band mass at the bottom of the band a measure of the total mini-band width. From the mini-bandwidth, we can predict saturation of the cyclotron resonance frequency at high magnetic fields and compare with experiment on a phenomenological level, unencumbered by the microscopic description of the barriers. On another level we can then take these phenomenological parameters and relate them to real material parameters.

It should be recognized that there is no real gap in the density of states since the dispersion relation shown in Fig. 2 does not account for the in-plane motion of the electrons. The in plane motion is largely unperturbed by the presence of the barriers and the in plane mass, m_0, is close to that of the quantum well material (GaAs in this case). It may suffer weak corrections due non-parabolicity and the fact that the electron drags the tails of its wavefunction in the barrier even though it is not required to tunnel through it. (Brozak, to be published)

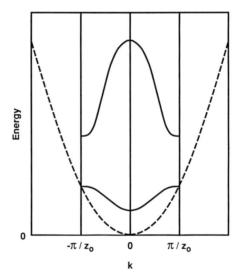

Fig. 2. Energy versus momentum along the growth direction of the superlattice. Dashed curve is the free electron dispersion relation. An energy gap occurs at the reciprocal superlattice boundary at $k = \pi/z_0$.

CYCLOTRON RESONANCE

In Fig. 3 we show the electron cyclotron motion for the magnetic field oriented perpendicular to the growth direction. For small magnetic fields or small superlattice periods, the cyclotron orbit encompasses several superlattice periods (orbit A in Fig. 3), the resonance condition is insensitive to the position of the orbit center

120

with respect to the barriers and quantum wells, and the resonance is determined by the superlattice band structure. In particular, the cyclotron frequency is given by

$$\omega_c = eB/(m_o m_s)^{1/2}.$$ (5)

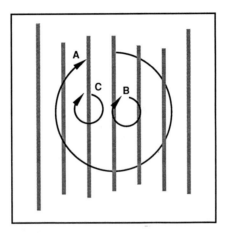

Fig. 3. Schematic drawing of cyclotron orbits in a plane including the growth direction. "A" encompasses several periods and its resonance is insensitive to orbit center whereas "B" and "C" will have different resonace conditions.

In the presence of large magnetic fields or large superlattice periods the resonance frequency will depend on the position of the orbit center with respect to the barriers and quantum wells (orbits B and C) and the resonance absorption will be inhomogeneously broadened. In particular, it might be expected that the lowest frequency feature in the inhomogeneously broadened line will correspond to orbits impaled on the barrier. Under these conditions the resonance condition is determined by the tunneling rate through the barrier and less by the time it takes the electron to go around the magnetic field. In this limit the frequency is expected to saturate and be a direct measure of barrier tunneling rates.

All of the above discussion is couched in classical terms. A quantitative description must be quantum mechanical since the resonance is usually observed in the quantum limit, that is to say only a few Landau levels occupied. Nonetheless, the essential physics and phenomena can be understood by these arguments.

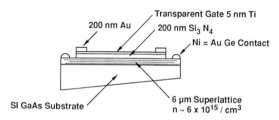

Fig.4. Sample geometry showing transparent gate and source and drain
   contacts. Substrate is wedged to eliminate interference effects.

EXPERIMENT

   Cyclotron resonance experiments were carried out on superlattice
samples comprised of GaAs quantum wells and AlGaAs barriers grown by
either MBE or organo-metallic chemical vapor deposition (OMCVD) on
semi-insulating substrates of GaAs. It was necessary to keep the doping
levels low for two reasons. First, it is desirable to minimize charged
impurity scattering. Second, the plasma resonance must be kept
substantially below the cyclotron resonance otherwise no cyclotron
motion is detected. To this end, dopant levels were maintained in the
$n \sim 10^{15}/cm^3$ range. In the MBE material this was done by
controlling Si dopant whereas in the OMCVD material, nominally undoped
material was caused to emerge n type at the requisite doping levels by
controlling the growth conditions. Since the depletion layer thickness
at $n \sim 10^{15}/cm^3$ is of the order of a micron, the superlattice
epitaxial layers had to be several microns thick and were typically 5-6
microns.

   Cyclotron resonance experiments were carried out in magnetic fields
as high as 15 Tesla using a Fourier transform interferometer that
covered the frequency range from 5 to 250 cm^{-1}, the beginning of the
restrahl band in GaAs. Sample temperature was maintained above 75 K to
ensure that most of the carriers were thermally activated from the
donors.

   In some cases to enhance the signal to noise, large area field
effect devices were fabricated with transparent gates. (Figure 4.) In
this way the electrons in the superlattice could be turned on and off by
application of a gate voltage. This is an effective way of obtaining
background free absorption of far-infrared radiation by the carriers in
the superlattice. A less effective alternative ratios the transmission
data with and without magnetic field. (Duffield et al., 1986)

   Experiments were performed with the magnetic field oriented both
parallel and perpendicular to the growth direction. Parallel to the
growth direction, one recovers the in plane effective mass. With the
magnetic field oriented perpendicular to the growth direction, the
electron is forced to execute a cyclotron motion as shown in Fig. 3. If
a resonance can be observed, then the electrons are executing coherent
cyclotron motion. Since the motion requires coherent tunneling
transport, these observations are an unambiguous measure of coherent
tunneling transport in the vertical direction.

MINI-BAND EFFECTIVE MASS

   When the cyclotron diameter is larger than the period or the
cyclotron energy is less than the mini-band width, the resonance

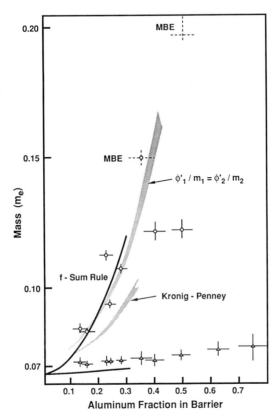

Fig. 5. Mini-band effective mass along the growth direction of
superlattices with 10 nm period, 2nm barrier width and varying
barrier Al content.   Triangles denote the in-plane effective
mass while circles denote the mini-band effective mass.   The
shaded curves are the calculated mass for a simple Kronig-Penney
model and a model including the mass discontinuity at the
quantum well barrier interface.   The solid curve is from Johnson
et al. (1987).

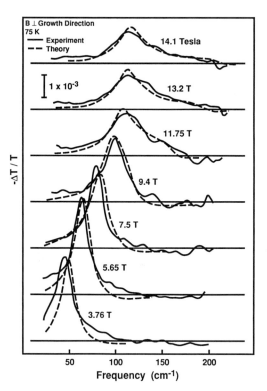

Fig. 6   Fractional change in transmission as a function of frequency
with magnetic field, oriented perpendicular to the growth
direction, as a parameter.   Dashed curves are the result of the
model calculation.

condition determines the effective mini-band mass. A set of experiments were performed on superlattices with a period of 10 nm, a barrier width of 2 nm but with varying amounts of Al in the AlGaAs barrier. In this way the transport mass along the growth direction could be determined as a function of barrier height. The results are shown in Fig 5.

There are a number of conclusions to be drawn from this result. First, the mini-band mass increases with barrier height as expected. Although there is a fair amount of scatter, the mini-band mass is in rough accord with a calculation that takes into account the mass discontinuity at the quantum well barrier interface. This requires continuity of the wavefunction and its derivative normalized by the mass. (Ben Daniel and Duke, 1961; White and Sham, 1981; Bastard, 1982) The results are also in good agreement with the approach of Johnson et al. based on the f-sum rule. (Johnson et al., 1987) Remarkably, it was also found that the scattering rate was the same in the superlattice direction as it was in the plane of the quantum wells and that the electrons propagated by coherent tunneling through tens of barriers without scattering. (Duffield, 1986)

The scatter in the data especially for large Al content reflects the sensitivity of the tunneling rate and mini-band mass to the barrier and superlattice parameters. Quantitative comparison with theoretical models requires precise knowledge of the material parameters. We will return to this issue later when we focus on tunneling through pure AlAs barriers.

MINI-BAND BREAKDOWN

From the low magnetic field data shown in From Fig. 5., we can estimate that a superlattice with 10nm period, 2nm wide barriers and 25% Al in the barrier will have a miniband mass of approximately .1 m_e, where m_e is the free electron mass. Using equation (4) we estimate the corresponding mini-band width to be ~ 120 cm^{-1}. We expect that when the cyclotron energy approaches the mini-bandwidth that it will saturate.

In Fig. 6 the cyclotron resonance is displayed for magnetic fields up to 14 Tesla. At low magnetic fields a nearly linear increase in resonance position with magnetic field is seen. This is used to deduce the effective mini-band mass shown in Fig. 5. As the resonance exceeds ~ 100 cm^{-1} it saturates and broadens, developing a tail to high frequencies.

This behavior can be successfully modeled by diagonalizing Schroedinger's equation in a magnetic field in a one dimensional periodic potential. Following Maan (1984) we calculate the eigen states of the following Hamiltonian

$$H = - \frac{\hbar^2}{2m} \frac{\partial^2}{\partial z^2} + \frac{1}{2}m\omega_c^2 z^2 + V(z+z_1) \qquad (6)$$

where $\hbar$ and m are Planck's constant and the electron mass in GaAs, ω_c is the cyclotron frequency and V(z) is the periodic potential. The origin has been chosen to be the orbit center which is displaced by z_1 from the origin of the periodic potential. We approximate the 2nm wide barriers by delta functions with strength given by the barrier height-width product.

The eigenvalues that result from this calculation are shown in Fig. 7. In the Landau gauge the orbit center or center of the wave function remains a good quantum number and Fig. 7 displays the Landau levels as a function of the orbit center with respect to the periodic potential. In a uniform system these would be a series of straight lines separated by $\hbar\omega_c$.

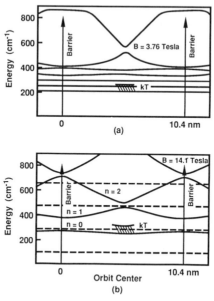

Fig. 7. Landau levels versus orbit position with respect to the barriers for a magnetic field of 3.76 Tesla, (a) and 14.1 Tesla, (b). The dashed lines indicate where the levels would be without the barriers . The cross hatched area indicates the scale of kT for the experiments.

At modest magnetic fields, ~ 4 T, the lowest levels still retain the appearance of a conventional Landau level diagram with the energy independent of orbit center. The energy level separation is given by the geometric mean of the in-plane and mini-band mass, (5). As we rise in Landau level index the levels develop considerable dispersion when the energy exceeds the mini-band width.

At this point it is useful to refine the conditions for no Landau level dispersion. Earlier we suggested that the relevant parameter could be taken to be either the ratio magnetic length to superlattice period or cyclotron frequency to miniband width. These two ratios are effectively the same measure only for the lowest Landau levels. The higher Landau levels have in fact larger cyclotron diameter and hence have larger ratio of cyclotron diameter to superlattice period yet show more dispersion. Although the extent of the wavefunction is larger for the higher Landau levels the wavefunctions contain a larger number of nodes reflecting the higher kinetic energy of the particle and it is the oscillatory features of the higher Landau levels that sense the relative position of the wavefunction and the periodic potential. As a result the critical parameter is in fact the energy relative to the mini-band width. This fact is readily apparent in the original interband magneto-luminescence of Maan (1984). There, the high lying Landau levels disappeared as they passed through the the the top of the mini-band. In the experiments we are discussing here, however, we are always probing the lowest Landau levels and the ratio of magnetic length/period is essentially equivalent to magnetic energy/mini-bandwidth in controlling the dispersion of the low lying levels.

In Fig. 7b, at ~14 Tesla, all Landau levels suffer dispersion due to the superlattice periodic potential. The resonance will be given by a superposition of vertical transitions from the ground state to the first excited state. It is apparent that the lowest transition frequency is found at the barrier. At that point the electron may be thought, in classical terms, to execute the cyclotron motion shown by C in Fig. 3. and tunnel twice per period through the barrier. The highest transition frequency is found for orbits centered on the quantum well. Here the electron is restrained not only by the magnetic field but the barriers on either side of the quantum well.

The full conductivity tensor may be derived for this system and the resulting far-infrared absorption calculated. The results are shown as a dashed lines in Fig. 6. It is apparent that the model reproduces in detail the saturating behavior of the cyclotron resonance accompanied by a broadening to high frequencies. The strongest feature in the inhomogeneously broadened line is the low frequency shoulder which corresponds to the barrier bound resonance, that is to say, resonance between states impaled on the barrier. For these orbits the time the electron takes to circle the magnetic field is determined predominantly by tunneling through the barrier and not by the magnetic field.

The relation of this frequency to the mini-bandwidth can also be readily understood. The mini-bandwidth is controlled by the tunneling rate between adjacent quantum wells. Since the saturated cyclotron resonance is also controlled by this tunneling rate it is not surprising that cyclotron resonance saturates at this same value.

TUNNELING THROUGH AlAs

There are two important issues in tunneling through pure AlAs barriers that can be quantitatively addressed by tunneling cyclotron resonance. First, in AlAs the lowest conduction band state is not at Γ but at X. Electrons in the Γ valley of the GaAs quantum well may couple to the X states in the AlAs barrier and take advantage of the substantially lower barrier. Second, even if there is little coupling to the X states and the tunneling is primarily through Γ, the GaAs conduction band minimum is a full volt below the AlAs conduction band minimum at Γ and it is doubtful that one can use the barrier band edge states at Γ to quantitatively describe the tunneling. Since it is clear that tunneling cyclotron resonance is a sensitive, quantitative and specific probe of barrier tunneling in semiconductor superlattices, these issues were addressed quantitatively in AlAs by performing tunneling cyclotron resonance in a suitable GaAs/AlAs superlattice by Brozak et al. (Brozak et al., to be published).

The superlattice was grown by MBE and consisted of 625 repetitions of nominally 7nm quantum wells and 1 nm barriers. It is apparent that in order to have a quantitative confrontation between experiment and theory it is critical to have an accurate measure of the material

parameters. The period was determined by x-ray diffraction to be
8.44±.01 nm. The average barrier thickness was determined by
measuring the average Al content by secondary ion mass spectroscopy,
(SIMS), (Schwarz, S.A. et al. to be published). This inferred a barrier
width of 1.05±.06 nm

The results are summarized in Fig. 8 where the absorption peak
frequency is plotted versus magnetic field for magnetic field oriented
both parallel to and perpendicular to the growth direction. The
saturating behavior discussed above is clearly seen in this superlattice
as well.

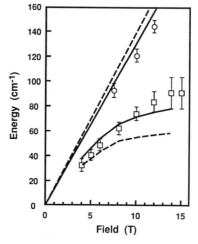

Fig. 8. Resonance frequency versus magnetic field for
a GaAs/AlAs, 7/1nm superlattice. Circles
correspond to the magnetic field along the
growth direction while the squares to it
being perpendicular. The dashed lines model
the results with the barrier   band edge
effective mass of .15 m_e. The solid curve
uses a renormalized value of .09 m_e.

The dashed curve is a prediction based on tunneling through AlAs
with a mass given by the Γ conduction band edge, .15 m_e. It is
apparent that the tunneling rate is substantially faster than predicted
with the band edge mass, of .15 m_e. Indeed, a tunneling mass of only
.09 m_e, represented by the solid curve, gives substantially better
agreement.

The actual value of the tunneling mass used was derived from a four
band k·P model of the complex barrier band structure related to the

Γ point. It should be remarked here that the most recent and
sophisticated treatments of the optical properties of short period
superlattice structures use quantum well and barrier band structure
paramters that automatically include these corrections (Moore, K.J. et
al., 1988). However, the optical spectra are controlled primarily by
the character of the states in the quantum well and suffer excitonic
corrections that make it unlikely that these measurements are critically
sensitive to tunneling rates through or into the barriers. Tunneling
cyclotron resonance on the other hand is entirely controlled by
tunneling and therefore is an unambiguous measure of the tunneling rates
and miniband dispersion.

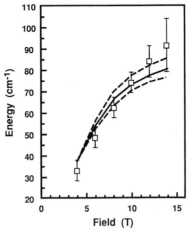

Fig. 9. Cylotron resonance versus magnetic field
        with the field perpendicular to the growth
        direction compared with the model
        calculation. The dashed lines vary the
        barrier thickness by   .06nm, 1/4
        monolayer.

It is interesting and important to ascertain how sensitive the model
and fit is to the material parameters. In Fig. 9. we show the
experimental data as well as the fit shown in Fig 8. varying the
thickness of the barrier by as little as .06nm, approximately 1/4 of a
monolayer. As sensitive as this measurement is it can test the theory
only as well as the material parameters of the superlattice are known.

We can project these results on the complex band structure (Schulman
and Chang, 1985). This is done in Fig. 10 where we show the imaginary
and real propagation constants connected to the Γ point of the

conduction band minima in the AlAs barrier. Here we have taken the barrier mass of .09 m_e that gives a good fit and translated it into an imaginary propagation constant which is indicated in Fig. 10.

There is no evidence of tunneling through the barrier X point although the barrier for these states is only several tenths of a volt. Recent experiments that directly probe Γ-X mixing indicate that appreciable mixing will occur only if the states are closer than 10 meV (Pulsford et al., 1989; Meynadier et al., 1988).

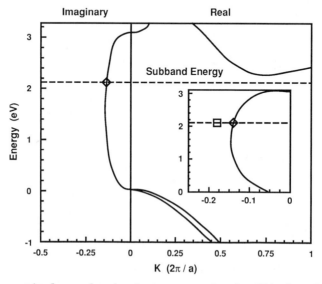

Fig. 10 Theoretical complex band structure in the AlAs barrier after Shulman and Chang (1985). Inset is an expanded view of the complex band structure. The diamond represents the result for m = .09m_e while the square indicates the imaginary wavevector extrapolated from the band edge mass of .15m_e.

SUMMARY

Tunneling cyclotron resonance has been shown to be a sensitive, specific and quantitative probe of barrier transport in semiconductor superlattices. As an equilibrium probe of superlattice mini-band transport, it is sine quo non. In low magnetic fields conventional cyclotron resonance is obtained with a cyclotron mass directly related to the mini-band mass. In large magnetic fields or long period superlattices the cyclotron resonance saturates as the cyclotron frequency exceeds the mini-band. This is a unique feature of cyclotron resonance in superlattices with narrow mini-bands and is a direct manifestation of saturated band transport. The fact that one can directly observe saturated band transport in these superlattice systems indicate that they have all the attributes required to observe miniband transport saturated by strong electric fields and Bloch oscillation (Esaki and Tsu, 1970).

REFERENCES

Bastard, G., 1982, Theoretical investigation of superlattice band structure in the envelope approximation, Phys. Rev., 25:7584.

Belle, G., Maan, J.C. and Weimann, G., 1985, Measurement of the miniband width in a superlattice with interband absorption in a magnetic field parallel to the layers, Solid State Commun., 56:65.

Belle, G., Maan, J.C. and Weimann, G., 1986, Observation of magnetic levels in a superlattice with a magnetic field parallel to the layers, Surf. Sci., 170:611.

Ben Daniel, D.J. and Duke, C.B., 1966, Space charge effects on electron tunneling, Phys. Rev. 152:683.

Brozak, G., DeRosa, F., Hwang, D.M., Micelli, P., Schwarz, S.A., Harbison, J.P., Florez, L.T. and Allen, S.J, Jr., to be published, Measurement of the effective mass along the growth direction of a thin barrier GaAs-AlAs superlattice, Surf. Sci.

Brozak, G., de Andrada e Silva, E.A., Sham, L.J., DeRosa, F., Miceli, P., Schwarz, S.A., Harbison, J.P., Florez, L.T. and Allen, S.J., Jr., to be published, Tunneling cyclotron resonance and the renormalized effective mass in semiconductor barriers.

Esaki, L. and Tsu, R., 1970, Superlattice and negative differential conductivity in semiconductors, I.B.M. J. Res. Develop., 14:61.

Duffield, T., Bhat, R., Koza, M., DeRosa, Hwang, D.M., Grabbe, P.and Allen, S.J., Jr., 1986, Electron mass tunneling along the growth direction of (Al,Ga)As/GaAs semiconductor superlattices, Phys. Rev. Lett., 56:2724.

Duffield, T., Bhat, R., Koza, M., DeRosa, F., Rush, K.M. and Allen, S.J., Jr., 1987, Barrier bound resonances in semiconductor superlattices in strong magnetic fields, Phys. Rev. Lett., 59:2693.

Duffield, T., Bhat, R., Koza, M., Hwang, D.M., DeRosa, F., Grabbe, P. and Allen, S.J., Jr., 1988, Breakdown of cyclotron resonance in semiconductor superlattices, Solid State Commun., 65:1483.

Harper, P.G., 1955, Proc. Phys. Soc. London, A58:879.

Harper, P.G., 1967, Quantum states of crystal electrons in a uniform magnetic field, J. Phys. Chem. Solids, 82:495.

Johnson, N.F., Ehrenreich, H., Hass, K.C. and McGill, T.C., 1987, f-Sum rule and effective masses in superlattices, Phys. Rev. Lett., 59:2352.

Kittel, C., 1986 "Introduction to Solid State Physics", John Wiley & Sons, Inc., New York.

Ktitorov, S.A, Simin, G.S. and Sindalovski, V.Ya., 1972, Bragg reflections and the high frequency conductivity of an electronic solid state plasma, Sov. Phys.-Sol. St., 13:1872.

Maan, J.C.,, 1984, Combined electric and magnetic field effects in semiconductor heterostructures, Springer Series in Solid State Sciences, 53:183.

Maan, J.C., 1987, Magnetic quantization in superlattices, Festkorperprobleme, 27:137.

Meynadier, M.H., Nahory, R.E., Worlock, J.M., Tamargo, M.C.,
de Miguel, J.L. and Sturge, M.D., 1988, Indirect-direct anticrossing
in GaAs-AlAs superlattices induced by an electric field: Evidence
of Γ-X mixing, Phys. Rev. Lett., 60:1338.

Moore, K.J., Duggan, G., Dawson, P. and Foxon, C.T., 1988, Short period
superlattices: Optical properties and electronic structure, Phys.
Rev., B38: 5535.

Pulsford, N.J., Nicholas, R.J., Dawson, Moore, K.J., Duggan, G. and C.T.
Foxon, to be published, Γ and X miniband structure in GaAs-AlAs
short period superlattices, Proceedings of the 4th International
Conference on Modulated Semiconductor Structures.

Schulman, J.N. and Chang, Y.-C., 1985, Band mixing in semiconductor
superlattices, Phys. Rev., B31:2056.

Schwarz, S.A., Schwartz, C.L., Harbison, J.P. and Florez, L.T., to be
published, Applications of $Al_xGa_{1-x}As$ Stoichiometry Measurement
by SIMS, in: "SIMS VII", John Wiley & Sons, Inc., New York

White, S.R. and Sham, L.J., 1981, Electronic properties of flat-band
semiconductor heterostructures, Phys. Rev. Lett., 47:879.

MAGNETOQUANTUM OSCILLATIONS IN A LATERAL SUPERLATTICE

Dieter Weiss

Max-Planck-Institut für Festkörperforschung
Heisenbergstr. 1, D-7000 Stuttgart 80, FRG

The low temperature magnetoresistance of a high mobility two-dimensional electron gas is dominated by Shubnikov-de Haas oscillations, reflecting the discrete nature
of the electron energy spectrum. When a weak one- or two-dimensional periodic potential is superimposed on the two-dimensional electron gas a novel type of oscillations
occurs which reflects the commensurability of the relevant lengths in these systems
– the cyclotron orbit diameter at the Fermi energy and the period a of the periodic
potential. In addition the electron mean free path l_e also plays a role since the effect is
observable only in mesoscopic systems where l_e is significantly longer than the period a
of the potential. The essential aspects of these novel commensurability oscillations are
discussed here in detail, starting from the discussion of same basic magnetotransport
properties in an unmodulated two-dimensional electron gas (2-DEG).

INTRODUCTION: MAGNETOTRANSPORT IN A 2-DEG

At the interface of GaAs-AlGaAs heterojunctions the electrons are localized in the
direction perpendicular to the layers (z-direction) and form a two-dimensional electron gas where the electrons can move freely in the x-y plane. The confinement in
z-direction leads to the formation of subbands. We consider the situation where only
the lowest subband (E_0) is occupied. In such a system the density of states is constant
(see e.g. Aoki, 1987). In the presence of a magnetic field, however, the electron motion
is completely quantized with energy eigenvalues (with respect to E_0) given by

$$E_n = (n + \frac{1}{2})\hbar\omega_c, \qquad n = 0, 1, ..., \tag{1}$$

with the Landau quantum number n and the cyclotron frequency $\omega_c = eB/m^*$ where B
is the magnetic field, and m* is the effective mass in GaAs equal to 0.067 m_0. The density of states (DOS) is a series of δ-functions which are, however, collision broadened
due to scattering processes characterized by a linewidth Γ. Within the selfconsistent
Born approximation (SCBA) (Ando and Uemura, 1974; Gerhardts, 1975) assuming
only short-range scatterers (with δ-function like scattering potentials) the DOS is a
series of semi-ellipses with linewidth $\Gamma = \sqrt{\frac{2}{\pi}\hbar\omega_c\frac{\hbar}{\tau}}$ where τ is the transport scattering
time (time between two scattering events) and is connected to the electron mobility
$\mu = \frac{e}{m^*}\tau$. Each of these Landau levels (LL) is occupied by N_L electrons per unit area

Electronic Properties of Multilayers and Low-Dimensional Semiconductors Structures   **133**
Edited by J. M. Chamberlain *et al.*, Plenum Press, New York, 1990

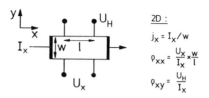

Figure 1. Determination of ρ_{xx} and ρ_{xy} by measuring the Hall voltage U_H and the voltage drop U_x along the applied current I_x in a Hall bar geometry. B is perpendicular to the plane.

given by $N_L = \frac{eB}{2\pi\hbar}$ (here spin splitting is included). The filling factor is defined as ratio between carrier density N_S and $N_L(\nu = N_s/N_L)$ and describes the occupancy of the Landau levels. For a homogeneous two-dimensional electron gas the wavefunction $|n, x_0\rangle$ of the electrons can be described as a plane wave in y-direction (Landau gauge $\vec{A} = (0, Bx)$) with wavenumber k_y ($\sim e^{ik_y y}$), and an eigenstate of a linear harmonic oscillator in the x-direction centered at $x_0 = -l^2 k_y$ (x_0 = center coordinate, l = magnetic length = $\sqrt{\hbar/eB}$). The extent of the wavefunction in x-direction is given by $2l\sqrt{2n+1}$ which for high quantum numbers n is the classical cyclotron diameter $2R_c = 2v_F/\omega_c = 2l^2 k_F$ where $k_F = \sqrt{2\pi N_S}$, the Fermi wavenumber.

A typical magnetoresistance experiment is sketched in Fig. 1. A constant current I_x is applied and the voltages U_x and U_H (Hallvoltage) are measured as a function of the magnetic field in z-directon. In a magnetic field the electric field $\vec{E}$ and the current density $\vec{j}$ are connected by the resistivity and conductivity tensor $\underline{\rho}$ and $\underline{\sigma}$, respectively:

$$\vec{E} = \underline{\rho}\vec{j} \text{ and } \vec{j} = \underline{\sigma}\vec{E}. \tag{2}$$

which are connected by

$$\underline{\rho} = \begin{pmatrix} \rho_{xx} & \rho_{xy} \\ \rho_{yx} & \rho_{yy} \end{pmatrix} = \frac{1}{\sigma_{xx}\sigma_{yy} - \sigma_{yx}\sigma_{xy}} \begin{pmatrix} \sigma_{yy} & -\sigma_{xy} \\ -\sigma_{yx} & \sigma_{xx} \end{pmatrix} \tag{3}$$

In a homogeneous 2DEG system the Onsager relation (Landau and Lifshitz, 1976) with $\sigma_{xx} = \sigma_{yy}$ and $\sigma_{yx} = -\sigma_{xy}$ holds. If $\sigma_{xx} = 0, \rho_{xx} = \sigma_{xx}/(\sigma_{xy}^2 + \sigma_{xx}^2)$ implies that $\rho_{xx} = 0$, simultaneously, which may sound strange but is true as long as $\sigma_{xy} \neq 0$ holds. Classically the conductivity components are given by the Drude formulas within the constant relaxation time approximation (τ = constant independent from DOS):

$$\sigma_{xx} = \frac{N_s e^2 \tau}{m^*} \frac{1}{1 + (\omega_c \tau)^2} \text{ and } \sigma_{xy} = -\frac{N_s e}{B} \frac{(\omega_c \tau)^2}{1 + (\omega_c \tau)^2}. \tag{4}$$

From this follows that the resistivity is independent of B, $\rho_{xx} = m^*/e^2 N_s \tau$. The magnetoresistance in a degenerate 2DEG, however, displays Shubnikov-de Haas (SdH) oscillations, reflecting the discrete nature of the degenerate Landau energy spectrum. Theoretically one has to go beyond the constant relaxation time approximation and consider a scattering rate τ which depends on the DOS. Within the SCBA one now obtains an oscillating conductivity $\sigma_{xx} \propto e^2 D(E_F)^2$ (Ando and Uemura, 1974) which vanishes when the Fermi energy is between two LL's.

In the following it will be shown that a novel type of magnetoresistance oscillations arises when a 1-dimensional or 2-dimensional periodic potential is superimposed on a 2-DEG. The system under consideration is mesoscopic in the sense, that the period of the modulation potential is small compared to the mean free path of the electrons so that ballistic transport over several hills and valleys of the potential can take place.

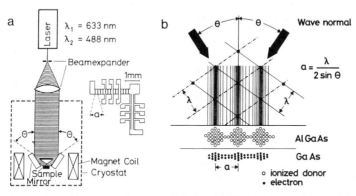

Figure 2. Schematic experimental set up (left hand side) and top view of the L shaped sample geometry where the interference pattern is sketched (a). Sketch of the spatial modulation of the concentration of ionized donors in the AlGaAs layer and of electrons in the 2-DEG produced by holographic illumination using two interfering laser beams with wavelength λ. The interference pattern created is shown schematically (b).

CREATION OF A LATERAL SUPERLATTICE: HOLOGRAPHIC ILLUMINATION

In selectively doped GaAs-AlGaAs heterostructures a persistent increase in the two-dimensional electron density is observed at temperatures below T = 150 K if the device is illuminated with infrared or visible light. This phenomenon is usually explained on the basis of the properties of DX-centers (persistent photoconductivity effect: PPC) which seem to be related to a deep Si donor. The increase in the electron density depends on the photon flux absorbed in the semiconductor so that a spatially modulated photon flux generates a modulation in the carrier density. In our measurements a holographic illumination of the heterostructure at liquid helium temperatures is used to produce a periodic potential with a period on the order of the wavelength of the interfering beams. In Fig.2b the interference of two plane light waves which create a periodic modulation of the ionized donors and therefore of the carrier density is shown schematically.

The existence of a periodically modulated carrier density in such structures has been demonstrated by Tsubaki et al. (1984) who measured the anisotropy of the resistivity parallel and perpendicular to the interference fringes at 90 K. In the presence of a magnetic field, a new oscillatory phenomenon in such modulated systems is observed if one goes to lower temperatures, demonstrating clearly that a periodic modulation of the 2-DEG is present. The potential modulation obtained by this technique is on the order of 1meV where the Fermi energy E_F in our samples is typically 10 meV.

The samples used in the experiment were conventional GaAs-AlGaAs heterostructures grown by molecular beam epitaxy with carrier densities between $1.5 \cdot 10^{11} cm^{-2}$ and $4.3 \cdot 10^{11} cm^{-2}$ and low temperature mobilities ranging from $0.23 \cdot 10^{6} cm^{2}/Vs$ to $1 \cdot 10^{6} cm^{2}/Vs$. Illumination of the samples increases both the carrier density and the mobility at low temperatures. The heterojunctions discussed in the following sections consist of a semi-insulating GaAs substrate, followed by a $1\mu m$-$4\mu m$ thick undoped GaAs buffer layer, an undoped AlGaAs spacer (6nm-33nm), Si-doped AlGaAs (33nm-84nm), and an undoped GaAs toplayer ($\approx$ 22nm). We have chosen an L-shaped geometry (sketched on the right hand side of Fig.2a) to investigate the magnetotransport properties parallel and perpendicular to the interference fringes. Such a mesa structure was produced using standard photolithographic and etch techniques. Ohmic contacts to the 2-DEG were formed by alloying AuGe/Ni layers at 450°C. Some of the samples investigated have an evaporated semi-transparent NiCr front gate (thickness $\approx$ 8nm) or a back gate, respectively, in order to vary the carrier density after holographic illumination. Such a semi-transparent front gate is also essential for performing magnetocapacitance measurements after holographic illumination.

The experiment was carried out using either a 5mW HeNe laser (λ = 633 nm) or a 3 mW Argon-Ion Laser (λ = 488 nm) both linearly polarized. The experimental realization of the holographic illumination is shown schematically in Fig.2a. The laser system was mounted on top of the sample holder which was immersed in liquid helium (4.2 K) within a 10-Tesla magnet system. The laser beam which was expanded to a diameter of 40 mm entered the sample holder through a quartz window and a shutter. The shutter ensures well defined illumination times of the sample down to 25 ms. Short exposure times are important to prevent jumping of the fringes; therefore exposure times between 25 ms and 100 ms were typically chosen. The mirrors which split the laser beam into two coherent waves, are located close to the device and are arranged in such a way that an interference pattern with a period a is generated at the surface of the device. An aperture mounted above covers the sample from direct illumination. The period a of the fringes created in this way depends on the wavelength λ of the laser and the incident angle Θ (see Fig.2): $a = \lambda/2sin\Theta$. With the wavelengths used in this experiment we have realized periods between 282 and 382 nm for the interference pattern.

The advantage of this kind of "microstructure engineering" is its simplicity and the achieved high mobility of the microstructured sample due to the absence of defects introduced by the usual pattern transfer techniques (see e.g. Mauterndorf, 1988).

MAGNETORESISTANCE OSCILLATIONS: EXPERIMENTS

After holographic illumination of the sample which was carried out between 1.5 and 4.2 K, we have measured the resistivities and Hall resistances parallel and perpendicular to the grating using the L-shaped sample geometry (see Fig.2). We have chosen

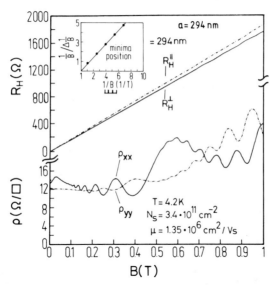

Figure 3. Magnetoresistivity ρ parallel (ρ_{yy}) and perpendicular (ρ_{xx}) to the interference fringes. The inset displays the 1/B dependence of the additional oscillations where the points correspond to minima in ρ_{xx}.

the x-axis to point in the direction perpendicular to the grating so that ρ_{xx} describes the transport perpendicular and ρ_{yy} parallel to the grating.

The magnetic field was perpendicular to the plane of the 2-DEG. The resistivities and Hall resistances are measured applying a constant current ($1\mu A$ - $10\mu A$) and measuring the voltage drop between potential probes along and perpendicular to the direction of current flow, respectively. A typical result obtained for a fringe period of $a = 294nm$ is shown in Fig.3. At magnetic fields above 0.5 T both ρ_{xx} and ρ_{yy} show the well known SdH-oscillations with a periodicity $\Delta(1/B)$ inversely proportional to the 2-DEG carrier density N_s. Below 0.5 T pronounced additional oscillations dominate ρ_{xx} while weaker oscillations with a phase shift of 180° relative to the ρ_{xx} data are visible in the ρ_{yy} measurements. The oscillations in ρ_{xx} are accompanied by a remarkable positive magnetoresistance at very low magnetic fields not present in ρ_{yy}. The Hall resistances, however, show no additional structure at magnetic fields below 0.5 T. Deviations from the clear linear behaviour of ρ_{xy} and ρ_{yx} occur only at magnetic fields above 0.5 T, and are connected to the SdH oscillations starting to be resolved. The slightly different slopes of ρ_{xy} and ρ_{yx} are due to a small difference in the carrier density of about 4% in the two branches of the sample.

A striking feature of this novel oscillations is their weak temperature dependence compared to SdH oscillations as is shown in Fig.4. At 13 K the additional oscillations in ρ_{xx} are still observable whereas the SdH oscillations are completely washed out. The strong temperature dependence of the SdH oscillations is due to the fact that their amplitudes depend on the Landau level separation $\hbar\omega_c$ as compared to the thermal energy $k_B T$. Therefore the additional oscillations do not originate from the Landau level separation $\hbar\omega_c$ as one would expect for electrons in higher subbands, e.g..

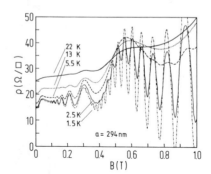

Figure 4. ρ_{xx} vs. B in the temperature range between 1.5 and 22K. The commensurability oscillations are less temperature dependent compared to the SdH-oscillations.

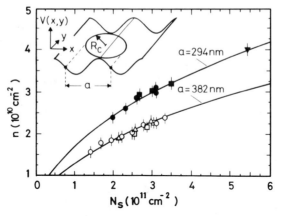

Figure 5. $n = \frac{e}{\pi\hbar}(\Delta\frac{1}{B})^{-1}$ versus N_s. Full symbols correspond to a laser wavelength λ=488nm, open symbols to λ=633nm and different symbols represent different samples. The solid lines are calculated using the condition that the cyclotron orbit diameter $2R_c$ is equal to an integer multiple of the interference period a, as is sketched in the inset.

The novel oscillations in ρ_{xx} and ρ_{yy} are perfectly periodic in $1/B$ as is demonstrated in the inset of Fig.3. The periodicity is obtained from the minima of ρ_{xx} which can be characterized by the commensurability condition

$$2R_c = (\lambda - \frac{1}{4})a, \qquad \lambda = 1, 2, 3, ..., \tag{5}$$

between the cyclotron diameter at the Fermi level, $2R_c$ and the period a of the modulation. For magnetic field values satisfying Eq.(5) minima are observed in ρ_{xx}. The periodicity $\Delta(1/B)$ can easily be deduced from Eq.(5) using the formulas given in the introduction:

$$\Delta\frac{1}{B} = e\frac{a}{2\hbar k_F} \tag{6}$$

where $\Delta(1/B)$ is the difference between adjacent minima (or maxima) of ρ_{xx} or ρ_{yy} on a $1/B$ scale.

The validity of Eq.(5) has been confirmed by performing these experiments on different samples, by changing the carrier density with an applied gate voltage, and by using two laser wavelengths in order to vary the period a. This is demonstrated in Fig.5 where the periodicity $\Delta(1/B)$ (displayed as carrier density $n = \frac{e}{\pi\hbar}(\Delta\frac{1}{B})^{-1}$) is plotted as a function of the carrier density N_s and the period a. The solid lines correspond to Eq.(6).

This simple picture is in excellent agreement with the experimental data displayed in Fig.5. To resolve an oscillation an elastic mean free path l_e at least as long as the perimeter of the cyclotron orbit is required. This agrees with our finding that the number of oscillation periods resolved depends on the mobility of the sample: the higher the mobility the more oscillations are observable since the electrons can traverse more periods of the potential ballistically. Similar experimental results have been observed recently by Winkler et al. (1989) using conventionally microstructured samples. The key for the explanation of this novel oscillatory behaviour of ρ_{xx} and ρ_{yy} in such periodically modulated 2-DEG's is a modification of the Landau level spectrum discussed in the next section.

LANDAU LEVELS IN A PERIODIC POTENTIAL: THEORY

The energy spectrum of electrons subjected to both a magnetic field and a periodic one-dimensional potential has been calculated by several authors (Aizin and Volkov, 1984; Kelly, 1985; Chaplik, 1985) using first order perturbation theory. Starting point is a Hamiltonian of the form

$$H = \frac{1}{2m^*}\left[-\hbar^2\frac{\partial^2}{\partial x^2} + \left(\frac{\hbar}{i}\frac{\partial}{\partial y} + \frac{e}{c}Bx\right)^2\right] + V_0 cos(Kx) \tag{7}$$

containing a periodic potential in x-direction $V(x) = V_0 cos(Kx)$ with period $a = 2\pi/K$. The energy spectrum can be taken in first order perturbation theory in V and is given

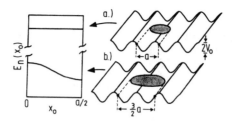

Figure 6. Simplified picture explaining the oscillating Landau level width.

by:

$$E_n(x_0) \approx (n + \frac{1}{2})\hbar\omega_c + \langle nx_0 \mid V(x) \mid nx_0 \rangle. \tag{8}$$

The right hand side matrix element can be regarded as effective potential acting on an electron averaged over the spatial extent of the wavefunction $\mid nx_0 \rangle$ given by $2l\sqrt{2n+1}$ which is equal to the classical cyclotron diameter $2R_c$ for high quantum numbers n. Two extremal situations can be considered as is sketched in Fig.6. Assuming, for sake of simplicity, a stepfunction like wavefunction, the matrix element $\langle nx_0 \mid V(x) \mid nx_0 \rangle$ at the Fermi energy vanishes when the cyclotron diameter equals an integer of the period a (Fig. 7a) leading to a flat Landau band, independent of the center coordinate x_0. On the other hand, a maximum contribution of the matrix element is expected for a cyclotron diameter equal to an odd integer of half the period a leading to Landau bands with strong curvature with respect to x_0 (Fig.6b). More precisely, the matrix element $\langle nx_0 \mid V(x) \mid nx_0 \rangle$ can be calculated analytically giving

$$E_n(x_0) \approx (n + \frac{1}{2})\hbar\omega_c + U_n \cos Kx_0 \tag{9}$$

with $U_n = V_0 \exp(-\frac{1}{2}X)L_n(X)$ where $X = \frac{1}{2}K^2l^2$ and $L_n(X)$ stands for the n-th Laguerre polynominal. $L_n(X)$ is an oscillating function of both its index n and its argument X where the flat band situation (LL independent of x_0) is given by $L_n(X) = 0$. This flat band condition $L_n(X) = 0$ can be expressed in terms of the cyclotron radius R_c and is given by Eq.(5) (Gerhardts et al., 1989). A typical energy spectrum – calculated in first order perturbation theory – for $B = 0.8T, V_o = 1.5meV$ and $a = 100nm$ is plotted in Fig.7. The corresponding DOS is sketched on the right hand side in Fig.7. Note the double peak structure of the DOS at the band edges which one typically expects for a 1-D bandstructure due to the van Hove singularities. Near the Fermi energy E_F and for the parameter values of Fig.7 the first order approximation is excellent for $B > 0.1T$ but it breaks down for $B \to 0$ and has to be calculated numerically (Gerhardts, 1989).

The most important point here to understand is that a one-dimensional periodic potential lifts the degeneracy of the Landau levels and leads to Landau bands of finite width. The bandwidth ($\approx 2U_n$) depends on the band index n in an oscillatory manner. The experimental verification of such a modified energy spectrum is the subject of the next section.

MODIFIED LANDAU LEVEL SPECTRUM: EXPERIMENTAL PROOF

In a two-dimensional electron gas the thermodynamic density of states $D_T(B)$ at the Fermi energy given by

$$D_T(B) = \frac{\partial N_s}{\partial \mu} = \int dE D(E) \frac{df(E - \mu)}{d\mu}, \tag{10}$$

where $f(x) = [\exp(x/kT) + 1]^{-1}$ is the Fermi function and the chemical potential μ is determined for a given electron density from $N_s = \int dE D(E) f(E - \mu)$. $D_T(B)$ is directly connected to magnetocapacitance $C(B)$ which is measured between gate and 2-DEG of the GaAs-AlGaAs heterojunction (Smith et al., 1985; Mosser et al., 1986):

$$\frac{1}{C(B)} = \frac{1}{C(0)} - \frac{1}{e^2 D_T(0)} + \frac{1}{e^2 D_T(B)}. \tag{11}$$

Here $D_T(0) = m^*/\pi\hbar^2$, the thermodynamic DOS at B = 0.

In a 2-DEG, $C(B)$ oscillates as a function of the magnetic field and the height of the magnetocapacitance maxima directly reflects the maximum value of the thermodynamic DOS at the Fermi energy which gives information about the Landau level width Γ. This has been used previously in homogeneous systems to investigate systematically the LL width Γ as a function of the electron mobility in such samples (Weiss and von Klitzing, 1987).

The active region of the sample used for the magnetocapacitance experiment was defined by etching a Hall-bar geometry and evaporating an 8 nm thick semi-transparent

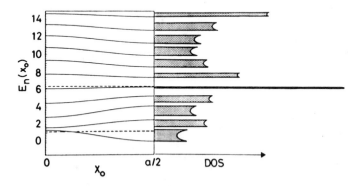

Figure 7. Calculated energy spectrum (first order perturbation theory) for B=0.8T, $V_0 = 1.5meV$ and a=100nm. The corresponding DOS is sketched. Flat parts of $E_n(x_0)$ lead to singularities in the DOS. The dashed lines correspond to the flat band situation determined by Eq.(5).

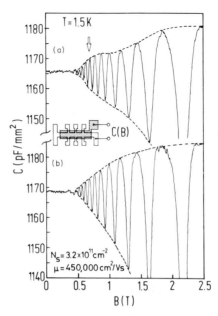

Figure 8. Measured magnetocapacitance (a) of a modulated sample compared to the capacitance of an essentially unmodulated sample (b). The arrow corresponds to the magnetic field value fulfilling Eq.5 for $\lambda = 1$. The inset sketches the measurement.

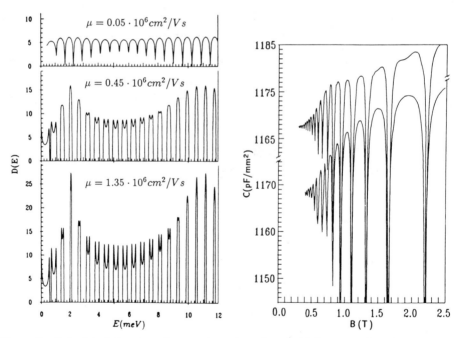

Figure 9. Calculated density of states in units of $m^*/\pi\hbar^2$ for several collision broadenings $\Gamma = \frac{e\hbar}{m^*}\sqrt{2B/\pi\mu}$ at B=0.35T and V_0=0.35meV.

Figure 10. Calculated magnetocapacitance versus magnetic field for $N_S = 3.2\cdot10^{11}cm^{-2}$ and $a = 365nm$. A B-independent linewidth is chosen to be $\Gamma = 0.3meV$. The upper curve is for $V_0 = 0.7meV$, and the lower one for the weak modulation $V_0 = 0.1meV$.

NiCr film as gate electrode. Prepared in such a way the sample shows a low temperature mobility of 240,000 cm^2/Vs and a carrier density of $1.5 \cdot 10^{11} cm^{-2}$ as has been
determined by SdH oscillations and low field Hall measurements. After holographic
illumination (duration 90 ms) using a 5mW HeNe laser which creates a periodic modulation with period $a = 365$ nm the carrier density was $3.2 \cdot 10^{11} cm^{-2}$ with a mobility of
$450,000 cm^2/Vs$. The capacitance between the semi-transparent gate and the 2-DEG
is measured with an ac-technique as sketched in Fig.8 (Smith et al., 1985; Mosser et.
al., 1986). In a homogeneous 2-DEG it has been shown experimentally that the LL
linewidth Γ has a magnetic field dependence of the form $\Gamma \propto B^\alpha$ with $0 \leq \alpha \leq 0.5$
(Weiss and von Klitzing, 1987; Gornik et al., 1985; Eisenstein et al., 1985), whereas
theoretically $\Gamma \propto \sqrt{B}$ is expected for short range scatterers and Γ independent of B for
long range scatterers (Ando and Uemura, 1974; Gerhardts, 1975). Since the Landau
level degeneracy is proportional to B, the peak values of the DOS in the individual
LL's and, as a consequence, the peak values of the capacitance are expected to increase monotonically with a structureless envelope with increasing magnetic field. On the
other hand the envelope of the magnetocapacitance minima decreases monotonically
with B due to the increasing LL separation $\hbar\omega_c$. In Fig.8 the magnetocapacitance data
after an initial holographic illumination (a) is compared with the capacitance measured
after an additional illumination which essentially smears out the periodic modulation
(b). The carrier density in (b) has been adjusted to the same value as before the additional illumination using a negative gate voltage. In contrast to Fig.8b, where the
magnetocapacitance behaves as usually observed in a 2-DEG, the capacitance oscillations in Fig.8a display a pronounced modulation of both the minima and maxima
which is easily explained from the energy spectrum plotted in Fig.7. At about 0.69 T
(marked by an arrow) the cyclotron diameter at the Fermi energy equals three quarter
of the period a and corresponds to the last flat band situation ($\lambda = 1$). Therefore, the
magnetocapacitance values near 0.69 T are approximately equal in Fig.8a and Fig.8b.
If now the magnetic field is increased, broader Landau bands are swept through the
Fermi level, and cause the nonmonotonic behaviour visible in Fig.8a. At higher magnetic fields the modulation broadening saturates and the usual LL degeneracy again
raises the DOS in a LL with increasing field. It should be mentioned that in the magnetocapacitance the same modulation effect is observed for different angles between
the one-dimensional modulation and the long axis of the Hall bar as is expected for a
thermodynamic quantity in contrast, of course, to the magnetoresistivity. Note that
below 0.69 T no oscillations comparable to those one observes in magnetotransport
experiments are resolved. This behaviour will be considered in the next section.

The discussion of the magnetocapacitance up to now uses qualitative arguments
rather than quantitative ones. In order to check the magnetocapacitance data theoretically, microscopic calculations of the DOS based on a generalisation of the well known
selfconsistent Born approximation have been performed by Zhang (Weiss et al., 1989).
In these calculations the energy spectrum of the homogeneous 2-DEG (e.g. Ando et
al., 1982) – which leads to the semi-elliptic, n-independent shape of the density of
states with a linewidth $\Gamma = \sqrt{\frac{2}{\pi}\hbar\omega_c \frac{\hbar}{\tau}}$ – is replaced by the energy spectrum obtained
from Eq.(7) in first order perturbation theory where for each modulation broadened
Landau band an average selfenergy is calculated. The resulting density of states is
plotted in Fig.9 for different mobilities which characterize the collision induced linewidth broadening. In this approximation the double peak structure of the LL's due
to the van Hove singularities is resolved if the collision broadening Γ is sufficiently
small compared to the modulation broadening U_n. The oscillatory n-dependence of
the modulation broadened LL's leads to an oscillation of the peak values of the DOS

with maxima for the narrowest levels. Once the DOS is calculated the data can be easily converted into magnetocapacitance data using Eq.(10) and Eq.(11). Calculating the magnetocapacitance for a modulation amplitude $V_0 = 0.7meV$ (Fig.10) reproduces the experimentally observed modulation of the envelope of the capacitance maxima (Fig.8a). A weaker amplitude of the periodic potential $V_0 = 0.1meV$ leads to the behaviour of the envelope usually observed in homogeneous 2-DEG's. The results of Fig.10 are obtained for a B-independent value $\Gamma = 0.3meV$, which yields a much better agreement with the experimental capacitance results than a linewidth calculated from the mobility μ according to the relation $\Gamma \propto (B/\mu)^{1/2}$ valid for short range scatterers. This phenomenon is well known for unmodulated samples of comparable mobility (Weiss and von Klitzing, 1987) and indicates the importance of long range scatterers (Ando and Uemura, 1974).

The van Hove singularities, not resolved experimentally in Fig.8a, are visible in the calculated data where a splitting of the Landau level is observed in the upper curve ($V_0 = 0.7meV$) at about 2 T.

The modified energy spectrum which has been proven experimentally in this section is the key for the explanation of the periodic potential induced oscillations given in the next section.

MAGNETORESISTANCE OSCILLATIONS: THEORY

The theory presented here follows closely the calculations of Gerhardts et al. (1989). The oscillations in ρ_{xx} (σ_{yy}) can be understood within a simple damping theory which means that electron scattering is described by a constant relaxation time τ. The k_y dispersion of the Landau energy spectrum leads to an additional contribution to the conductivity σ_{yy} which is within the framework of Kubo's formula (e.g. Kubo et al., 1965) given by

$$\Delta\sigma_{yy} = \frac{2e^2\hbar}{2\pi l^2} \int_0^a \frac{dx_0}{a} \sum_n \left(-\frac{1}{\gamma}\frac{df}{dE}(E_n(x_0)) \mid \langle nx_0 \mid v_y \mid nx_0 \rangle \mid^2 \right) \tag{12}$$

where $\gamma = \hbar/\tau$, f is the Fermi function, $\mid nx_0 \rangle$ are the eigenstates of Eq.(7) and v_y is the velocity operator in y direction. These eigenstates carry current in the y direction

$$\langle nx_0 \mid v_y \mid nx_0 \rangle = -\frac{1}{m^*\omega_c}\frac{dE_n}{dx_0} = \frac{1}{\hbar}\frac{dE_n}{dk_y} \tag{13}$$

but not in x-direction,

$$\langle nx_o \mid v_x \mid nx_o \rangle = 0, \tag{14}$$

which is the reason for the anisotropic behaviour of σ_{xx} (corresponding to ρ_{yy}) and σ_{yy} (corresponding to ρ_{xx}). In Eq.(13) the modified energy spectrum comes into play. The matrix element $\langle x_0 n \mid v_y \mid x_0 n \rangle$ vanishes always then when flat Landau bands (see e.g. Fig.7) are located at the Fermi energy, and $\Delta\sigma_{yy} = 0$. Consequently also $\Delta\rho_{xx}$ – the extra contribution to the resistivity ρ_{xx} – vanishes since $\Delta\rho_{xx} \approx \Delta\sigma_{yy}/\sigma_{xy}^2$ if $\omega_c\tau > 1$. On the other hand dE_n/dx_o displays a maximum value when the Fermi energy is located within the Landau band with the strongest dispersion and therefore $\Delta\sigma_{yy}$ ($\Delta\rho_{xx}$) is at maximum. A calculation based on the evaluation of Eq.12 which is compared to experimental magnetoresistivity data is shown in Fig. 11.

Assuming a modulation potential of 0.3 meV in the calculations reproduces nicely the oscillations of ρ_{xx} (solid lines in Fig.11). Note that the weak temperature de-

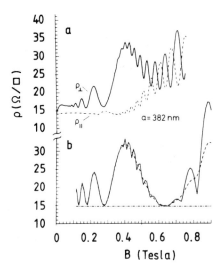

Figure 11. Magnetoresistivities for current perpendicular (ρ_{xx}, solid lines) and parallel (ρ_{yy}, dash-dotted lines) to the interference fringes, for a sample with $N_S = 3.16 \cdot 10^{11} cm^{-2}$, $\mu = 1.3 \cdot 10^6 cm^2/Vs$, and $a = 382nm$ — (a) measured at temperature $T = 2.2K$; (b) calculated for $T = 2.2K$ and for $4.2K$ (ρ_{xx} dashed line, ρ_{yy} shows no temperature dependence), using $V_0 = 0.3meV$ (Gerhardts et al., 1989).

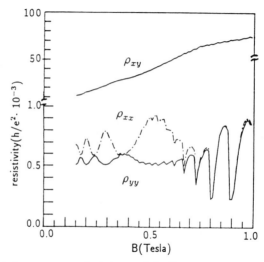

Figure 12. Calculated magnetoresistivity and Hall resistance for T=4.2K, a=294nm, $N_s = 3.14 \cdot 10^{11} cm^{-2}$, $\hbar/\tau$=0.0128meV and V_0=0.25meV. The oscillations in ρ_{yy} are due to a DOS dependent scattering rate so that $\rho_{yy} \propto e^2 D(E_F)^2$ (Gerhardts and Zhang, 1989).

pendence of the commensurability oscillations in ρ_{xx} (dashed line in Fig.11) is given correctly by the calculation in agreement with the experimental findings in Fig.4. The temperature dependence of these oscillations is much weaker than that of SdH oscillations, since the relevant energy is the distance between flat bands, which is much larger than the mean distance between adjacent bands. Winkler et al. (1989) have explained the oscillations in ρ_{xx} – measured in samples with microstructured gates – along very similar theoretical lines. Using the quasi-classical large-n limit, which means

$$ e^{-\frac{1}{2}X} L_n(X) \approx \pi^{-\frac{1}{2}} (nX)^{-\frac{1}{4}} \cos\left(2\sqrt{nX} - \frac{\pi}{4}\right), \tag{15} $$

and assuming that $\hbar\omega_c < kT$ (high temperature limit) one can deduce their result from Eq.(12)

$$ \Delta\sigma_{yy} \approx \frac{e^2}{2\pi\hbar} \frac{V_0^2}{\gamma\hbar\omega_c} \frac{4}{ak_F} \cos^2\left(2\pi\frac{R_c}{a} - \frac{\pi}{4}\right). \tag{16} $$

Eq.(16) can be rewritten to give $\Delta\rho_{xx}$ making use of the fact that $\Delta\rho_{xx} \approx \sigma_{yy}/\sigma_{xy}^2$:

$$ \Delta\rho_{xx} \approx \frac{1}{2\pi\hbar} \frac{V_0^2}{\gamma\hbar\omega_c} \frac{4}{ak_F} \frac{B^2}{N_s^2} \cos^2\left(2\pi\frac{R_c}{a} - \frac{\pi}{4}\right). \tag{17} $$

Equation (17) may be used to estimate from the amplitudes $\Delta\rho_{xx}^{max}$ of the commensurability oscillations the amplitude of the superimposed periodic potential. From the maximum of ρ_{xx} at 0.41T (Fig.11a) one estimates $V_0 = 0.28meV$ in good agreement with the calculation in Fig.11b.

While the low field oscillations of ρ_{xx} are nicely reproduced by the calculation, the calculated ρ_{yy}-data (dashed-dotted line in Fig.11b) display simply the magnetic field independent Drude result in contrast to the experiment which shows maxima when the Landau bands are flat (DOS maximum at E_F). This is not too surprising since one cannot describe the usual SdH oscillations of a homogeneous 2-DEG within the

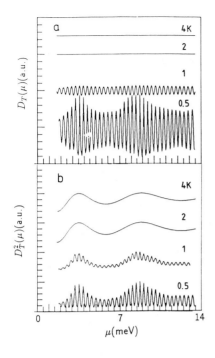

Figure 13. Thermal average of (a) DOS $D(E)$, and (b) $D(E)^2$, versus chemical potential for different temperatures. The curves are normalized and shifted so that (a) the average value is 1, (b) the maximum value is 1. $B=0.2$T, $V_0=0.25$meV (Zhang and Gerhardts, 1989).

constant relaxation time approximation; you end up with the simple Drude result. The same result has been obtained by Beenakker (1989) using a semiclassical model (guiding center drift resonance). Beenakker noticed that the mean square Hall-drift velocity has an oscillating behaviour according Eq.(5) which, expressed as one-dimensional diffusion constant, leads via Einstein's law to the novel oscillations of ρ_{xx} while ρ_{yy} shows no magnetic field dependence. In order to understand the experimentally observed ρ_{yy} oscillations one has to go beyond this approximation – in analogy to the description of the SdH oscillations – and consider a density-of-state-dependent scattering rate. In the calculations one has to go through the formalism of the selfconsistent Born approximation (Ando and Uemura, 1974; Gerhardts, 1975) using the solutions of Eq.7. A detailed description of this theory has been given by Zhang and Gerhardts (1989). In analogy to the theory of SdH oscillations they find that

$$\sigma_{xx} \propto D_T^2(\mu) = \int dE \frac{df(E-\mu)}{d\mu} D(E)^2 \tag{18}$$

where $D_T^2(\mu)$ is the thermal average of the square of the DOS, not to be confused with the square of the thermodynamic DOS $D_T(\mu)$ defined in Eq.(10). Therefore, the weak antiphase oscillations in ρ_{yy} are in phase with the density of states oscillations and maxima in ρ_{yy} are always observed when the DOS at the Fermi-energy is at maximum, in contrast to ρ_{xx} which displays minima when the Landau bands are flat since $dE_n/dx_0 = 0$. Calculated curves of ρ_{xx} and ρ_{yy} are shown in Fig.12 demonstrating this behaviour in agreement with the experiment (Fig.3 and 11a). Eq.(18) also explains why the quantization of the Landau levels is essential for the explanation of the oscillations at least of ρ_{yy} while the quantization is not resolved in magnetocapacitance measurements at low magnetic fields. The reason for this is visualized in Fig.13 where

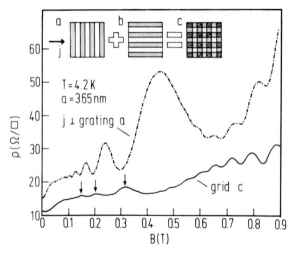

Figure 14. Magnetoresistance in a grating (j ⊥ grating) and grid. The creation of the holographically defined pattern is shown schematically.

the thermodynamic density of states $D_T(\mu)$ (Eq.10) which determines the magneto-capacitance signal is compared to $D_T^2(\mu)$, the quantity which is responsible for the behaviour of ρ_{yy} (Eq.18). Up to 1K the Landau level quantization is resolved in $D_T(\mu)$ as well as in $D_T^2(\mu)$. At higher temperatures all the structure in D_T (which determines the magnetocapacitance) is washed out, $D_T^2(\mu)$, however, which determines ρ_{yy} shows still pronounced oscillations with maxima at flat band energies given by $2R_c = (\lambda - \frac{1}{4})a$.

MAGNETORESISTANCE IN A TWO-DIMENSIONAL PERIODIC POTENTIAL

In this last section some preliminary results of low field magnetotransport experiments in a two-dimensional periodic potential are presented. In such a potential grid the commensurability problem becomes more severe as compared to the 2-D case and results in a complicated energy spectrum (Hofstadter, 1976), and the shape of the DOS is not clear. The two-dimensional periodic potential ($V_0 \ll E_F$) with $a = 365nm$ is created by successively illuminating holographically a high mobility heterostructure ($\mu = 1.2 \cdot 10^6 cm^2/Vs$). Holographic illumination of type (a) in Fig.14 produces additional oscillations in the magnetoresistance as mentioned above (ρ_{xx}, dash-dotted line in Fig.14). An additional holographic illumination where the sample has been rotated by 90° results then in a grid potential sketched in Fig.14c. The magnetoresistance obtained under such conditions (solid line in Fig.14) displays a weak oscillating behaviour also corresponding to the commensurability condition Eq.(5), with maxima where ρ_{xx} –measured for situation (a)– shows minima. If one starts with an illumination of type (b) followed by (a) one ends up with the same result. The result obtained for the magnetoresistance in a two-dimensional periodic potential is therefore very close to the result one gets when the current flows parallel to the potential grating, discussed as additional oscillations in ρ_{yy} above (see e.g. Fig.3 and Fig.11). Therefore one may speculate that the DOS in a weak grid potential is similar to those in a grating and that the oscillating magnetoresistance (apart from SdH-oscillations) in a two-dimensional potential also reflects the oscillating scattering rate due to corresponding oscillations in the DOS according Eq.13.

148

ACKNOWLEDGEMENTS

I would like to thank K. v.Klitzing , R. R. Gerhardts, C. Zhang and D. Heitmann for valuable discussions and I am grateful to K. Ploog and G. Weimann for providing me with high quality samples.

REFERENCES

Aizin, G. R., and Volkov, V. A., Sov. Phys. JETP **60**, 844 (1984) [Zh. Eksp. Teor. Fiz **87**, 1469 (1984)]

Ando, T., and Uemura, Y., J. Phys. Soc. Jpn. **36**, 959 (1974)

Ando, T., Fowler, A. B., and Stern, F., Rev. Mod. Phys. **54**, 437 (1982)

Aoki, H., Rep. Prog. Phys. **50**, 655 (1987)

Beenakker, C. W. J., Phys. Rev. Lett. **62**, 2020 (1989)

Chaplik, A. V., Solid State Commun. **53**, 539 (1985)

Eisenstein, J. P., Störmer, H. L., Narayanamurti, V., Cho, A. Y., and Gossard, A. C., Phys. Rev. Lett. **55**, 1820 (1985), 539 (1985)

Gerhardts, R. R., Z. Physik **B21**, 285 (1975)

Gerhardts, R. R., Weiss, D., and von Klitzing, K., Phys. Rev. Lett. **62**, 1173 (1989)

Gerhardts, R. R., in "Science and Engineering of 1- and 0-Dimensional Semicondcutors", edited by S.P. Beaumont and C.M. Sotomayor Torres (Plenum, London), Proc. NATO ARW, 29 March - 1 April 1989, Cadiz, Spain, to be published

Gerhardts, R. R., and Zhang, C., Proc. Eighth Intern. Conf. Electronic Properties of Two-Dimensinal Systems, Grenoble, France, 4-8 Sept. 1989, to be published in Surf. Sci.

Gornik, E., Lassnig, R., Strasser, G., Störmer, H. L., Gossard, A. C., Wiegmann, W., Phys. Rev. Lett. **54**, 1820 (1985)

Gudmundsson, V., and Gerhardts, R. R., Phys. Rev. **B35**, 8005 (1987)

Hofstadter, D. R., Phys. Rev. **B14**, 2239 (1976)

Kelly, H. J., J. Phys. C. **18**, 6341 (1985)

Kubo, R., Miyake, S. J., and Hashitsume, N., Solid State Physics **17**, 239 (1965)

Landau, L. D., and Lifshitz, E. M., Course of Theoretical Physics vol. 5 (Oxford: Pergamon) 1976

Mauterndorf, 1988, Physics and Technology of Submicron Structures, Vol. 83 of Springer Series in Solid-State Sciences, ed. by H. Heinrich, G. Bauer, and F. Kuchar (Springer, Berlin, 1988)

Mosser, V., Weiss, D., von Klitzing, K., Ploog, K., and Weimann, G., Solid State Commun. **58**, 5 (1986)

Smith, T. P., Goldberg, B. B., Stiles, P. J., and Heilblum, M., Phys. Rev. **B32**, 2696 (1985)

Tsubaki, T., Sakaki, H., Yoshino, J., and Sekiguchi, Y., Appl. Phys. Lett. **45**, 663 (1984)

Weiss, D., and von Klitzing, K., in "High Magnetic Fields in Semiconductors Physics", Vol. 71 of Springer Series in Solid State Science, edited by G. Landwehr (Springer, Berlin, 1987), p. 57

Weiss, D., von Klitzing, K., Ploog, K., and Weimann, G., Europhys. Lett. **8**, 179 (1989); also in "The Application of High Magnetic Fields in Semiconductor Physics", ed. G. Landwehr, Springer Series in Solid State Sciences (Berlin), to be published

Weiss, D., Zhang, C., Gerhardts, R. R., von Klitzing, K., and Weimann, G., Phys. Rev. **B39**, 13030 (1989)

Weiss, D., von Klitzing, K., Ploog, K., and Weimann, G., Proc. Eighth Intern. Conf.

Electronic Properties of Two-Dimensional Systems, Grenoble, France 4-8 Sept. 1989, to be published in Surf. Sci.

Winkler, R. W., Kotthaus, J. P., and Ploog, K., Phys. Lett. **62**, 1177 (1989)

Zhang, C., and Gerhardts, R. R., submitted to Phys. Rev. B

TECHNIQUES FOR LATERAL SUPERLATTICES

Detlef Heitmann

Max-Planck-Institut für Festkörperforschung
Heisenbergstr. 1, 7000 Stuttgart 80, West-Germany

We review the fabrication of laterally microstructured semiconductor systems by holographic lithography and dry etching techniques. We discuss applications to electron systems of low dimensions, i.e., two-, one- and zero-dimensional systems (2DES, 1DES, 0DES), in particular grating coupler induced 2D plasmon resonances, intersubband resonances, minigaps and local plasmons in charge density modulated systems, dc transport and far infrared response in 1DES and 0DES as well as quantum well exciton polaritons in the photoluminescence of microstructured quantum well systems.

INTRODUCTION

Two-dimensional electronic systems (2DES),[1] which can be realized in metal-insulator-semiconductor (MIS), semiconductor heterostructures and quantum well systems, are very important for many applications and are ideally suited for fundamental research. The reason for the latter is that the most important parameters of the electronic system, e.g., the electron density, the subband spacing and the density of states can be easily and reproducibly varied by electric fields via a gate voltage or by external magnetic fields. Moreover, in the novel heterostructure and superlattice systems, which are grown with highly sophisticated growth techniques such as molecular beam epitaxy (MBE), it is possible to tailor even the bandstructure of the system (bandstructure engineering). It is thus possible to optimize the system in such a way that certain physical phenomena can be investigated under ideal conditions. This has initialized a broad spectrum of research during the last two decades and given rise to an incredible spectrum of novel, unique results. Many of these results will be addressed also in other contributions in this book.

The MIS-, heterostructure- and superlattice systems are all layered structures. They are thus ideally suited for an additional lateral structuration to reduce the dimensionality of the system even further. In the limit of very small dimensions it is possible to realize quantum confined systems, i.e., 1D quantum wires and 0D quantum dots.[2-29] Lateral structures of small dimensions can be prepared by different techniques, e.g.; conventional contact mask technology, e-beam and ion-beam lithography, ion-beam bombardment through narrow masks, or growth on tilted high-index surfaces of the semiconductor material.[27]

Electronic Properties of Multilayers and Low-Dimensional Semiconductors Structures
Edited by J. M. Chamberlain *et al.,* Plenum Press, New York, 1990

151

In these lecture notes I will concentrate on lateral structures which have been prepared with holographic lithography. With this technique it is possible to realize periodically structured samples with small periodicities, a, ($a \approx 100nm$) which have an extremely high homogeneity over large areas ($3x3mm^2$ and more). The large areas of these systems makes them ideally suited, as we will show in this paper, for optical experiments in the visible (VI), near infrared (NIR) and far infrared (FIR) spectral range, where large area samples are necessary to achieve a sufficient signal to noise ratio. The periodicity of these structures can be used, if chosen in an appropriate manner, to study interesting lateral superlattice effects, e.g., minigaps in the excitation spectrum. The smallness of the lateral structures leads into the regime of quantum confinement and novel transport phenomena.

I will give in these lecture notes examples that periodic grating structures can be applied in different ways. In some experiments they are used in a 'passive' way, i.e., the electronic system is not directly affected by the microstructure. The periodic structure acts as a 'grating coupler' which couples radiation with elementary excitations in the system, e.g. 2D plasmons and intersubband resonances in modulation-doped systems or polaritons in quantum well systems. In a next step, periodic structures can be used in an 'active' way, i.e., to produce electron systems with a spatially modulated electron density, $N_s = N_s(x)$. (Throughout this paper z denotes the direction normal to the layered structures, x in a linear grating structure the direction perpendicular to the grating rules and y along the wires.) The density modulated systems show interesting lateral superlattice effects e.g., for the 2D plasmon dispersion. An ultimate goal is that the lateral structures give rise to an additional quantum confinement, e.g. 1D quantum wires and 0D quantum dots. We will discuss examples for such systems.

I would like to use these lecture notes to give an extended description of the preparation steps and experimental techniques. I will then discuss a series of different experiments in which periodically structured systems have been applied. Within the available space I can only concentrate on the basic messages of these experiments. For more details, a full explanation and a more complete list of references I refer the reader to the original papers.

PREPARATION OF LATERAL MICROSTRUCTURES
BY HOLOGRAPHIC LITHOGRAPHY AND ETCHING TECHNIQUES

A set up for holographic lithography is sketched in Fig. 1. An expanded parallel laser beam is realized by a combination of a microscope lens, a pinhole and a collimating optic. A beamsplitter splits the beam into two components. The superposition of these two beams in the plane of the sample surface leads to a sinusoidal modulation of the light intensity, which is used to expose a light sensitive photoresist layer on top of the sample. The periodicity is $a = \lambda/(2sin\delta)$ and can be controlled via the angle of incidence of the beams, δ. λ is the wavelength of the laser. In our current set up we use an Ar^+ laser at $\lambda = 457.9nm$ with a power of about $1.8W$. The spatial filter consists of a 20X microscope lens, a 10 or 5 μm high power pinhole and a 250mm double lens collimator corrected for minimal spatial aberration. The intensity in the expanded beam is $5 - 10mWcm^{-2}$. The interferometric part of the set up is optimized for a short optical pathlength ($< 200mm$) to increase stability. It is strictly symmetric to achieve coherence over an sample area of at least 10mm in diameter with a laser of relatively large band width ($5GHz$). Thus no complicated intracavity etalon is necessary. Working with the blue laser line has the advantage that one can use conventional and thus high quality optics, and, very conveniently, one can visually optimize the alignment and homogeneity of the beams.

For $\lambda = 457.9nm$ the smallest periodicity is in air limited to $a \approx 230nm$. It can be further decreased if one uses prism arrangements or a miniaturised set up in a fluid thus that the effective wavelengths λ/n is decreased by the index of refraction n. In this way we have produced gratings with periodicities down to $a \approx 180nm$. For smaller periodicities one can of course also work with other laser lines; Ar^+-UV-lines

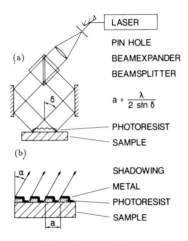

LASER

PIN HOLE

(a)

BEAMEXPANDER

BEAMSPLITTER

$$a = \frac{\lambda}{2 \sin \delta}$$

PHOTORESIST

SAMPLE

(b)

SHADOWING

METAL

PHOTORESIST

SAMPLE

Fig. 1 (a) Schematical set-up for holographic lithography to produce structures with small periodicities. (b) Preparation of metal stripes by tilted angle evaporation (shadowing).

($\approx 350nm$), frequency doubled $525nm\,Ar^{+}$-laser lines, $HgCd$ laser at $\lambda = 325nm$ or even excimer lasers (e.g. KrF or ArF at $\lambda = 249$ or $193nm$, respectively). Optics and alignment are a little bit more complicated, nevertheless, excellent results have been reported.[30,31] An additional reduction of the period can be achieved by tricky arrangements, e.g. spatial frequency doubling in the near-field of a master grating.[32]

For the photoresist we normally use Shipley AZ type lacquer (e.g. S-1400-17). It is spin-coated onto the sample surface. Prebaking (80°C, 20min) and sometimes post baking is used to increase adhesion and selectivity for following dry etching processes. With a power density of $5mW\,cm^{-2}$ per laser beam we need typical exposure times of $10s$ for $1min$ development in 5:1 diluted Shipley developer. The width of the photoresist stripes, t, can be varied via the exposure and development time, practically between $t \approx 0.2a$ to $0.8a$. (We denote in the following a 'geometrical' width by ' t ', in contrast to an 'electrical' width, ' w '. The latter characterizes the extension of the electron wavefunctions.) Sometimes we also apply a controlled very short dry etching in an O_2-plasma (see below) for an additional definition of the photoresist profile. For highly reflecting sample surfaces one also gets a standing wave pattern in the z-direction normal to the surface. In an unfortunate case there is a plane of low intensity (nodes) at a certain distance from the sample surface, which stops a further development of the photoresist into deeper layers. Thus, if ever possible from the selectivity of the following process steps, one tries to work with thin photoresist layers $d_{PR} < \lambda/2n_r$ (n_r = index of refraction for the photoresist). Typical thicknesses are $d_{PR} \approx 100 - 300nm$.

With the holographic technique one can not only produce linear structures but also cross-grating structures by use of a double exposure with an inbetween 90° rotation of the sample in the plane of the sample. Depending on the exposure time, t_e, one either achieves a photoresist grid with dots of free sample surface or, for larger t_e, isolated photoresist dots. Such structures are the starting point for 0DES. Moreover, one can use instead of parallel laser beams a superposition of divergent and/or convergent beams. Then it is possible to fabricate curved and 'chirped' gratings, i.e. gratings with a varying periodicity $a = a(x,y)$. Such structures are useful for focussing grating couplers.[33] For more details on holographic lithography see e.g. Ref.. [34]

The photoresist masks are the starting point for additional processes. For the excitation of 2D plasmons or intersubband resonances, which will be discussed below, one needs periodic highly conducting metal stripes. These can be fabricated by the

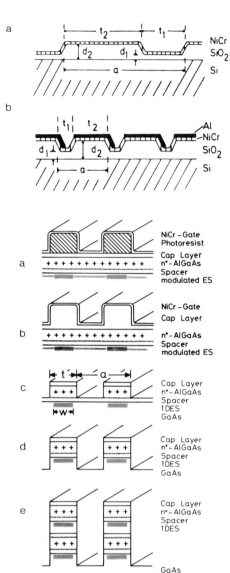

Fig. 2   Sketch of Si-MOS systems with modulated oxide thickness for gate voltage induced density modulated and isolated wire systems. In (a) the oxide is covered with a continuous semitransparent $NiCr$ gate. In (b) additional metal stripes are shadowed onto the grating to enhance the grating coupler efficiency.

Fig. 3   Some examples of realized periodic arrays of quantum wire structures in the $AlGaAs/GaAs$ system. (a) and (b) are split-gate configurations where via a gate voltage and a varying distance between gate and channel carriers are depleted leaving isolated quantum wires. In (a) the gate-distance modulation is achieved via a modulated photoresist, in (b) via an etched cap layer. (c)-(e) sketch examples for mesa etched quantum wire structures: (c) shows schematically a 'shallow'-mesa etched structure, (d) a 'deep'-mesa etched single-layer structure, (e) a 'deep'-mesa etched double-layer structure. (from Ref.[43])

method of shadowing.[35–38] The metal (e.g. Al, Au, Ag) is evaporated onto the sample at an angle α. Thus one side of the photoresist profile is left uncovered and can be lifted off (see Fig. 1b). The width of the metal stripes for a given height of the photoresist stripe can be varied via the shadowing angle α ($\alpha \approx 30° - 80°$). This is important to optimize the efficiency of the grating coupler as will be discussed below. Similar results can also be achieved by a special technique where the uppermost photoresist layer is hardened and an undercut in the photoresist profile is produced after the development.[39] The metal is evaporated normally onto the sample and lift-off is possible due to the uncovered undercut side walls of the resist profile. This latter process has the advantage that it also can be applied to cross-grating structures.

A first step into the direction of electron systems with dimensions smaller than 2 are systems with a spatially modulated charge density[40,41] $N_s = N_s(x)$. A straightforward way to realize such systems is to produce a photoresist grating on top of the sample and to cover the periodic profile with a continuous metal gate as sketched in Fig. 3a.[42,9,13,43] For optical spectroscopy one uses a thin layer of $NiCr(5 - 10nm)$ which is semitransparent both in the FIR and VI spectral range. It also has a high sticking coefficient on Si, SiO_2 and $GaAs$ and does not form islands even for very thin films.

The photoresist has good isolating properties. Thus, if a gate voltage V_g is applied between the gate and the channel, a spatially modulated electron gas is induced with $N_s(x) = \epsilon\epsilon_0(V_g - V_t(x)/ed(x)$, where ϵ is the static DK of the insulating layer, $V_t(x)$ and $d(x)$ are respectively the spatially varying threshold voltage and distance between gate and channel. This expression is actually only true for small $d <\approx 0.1a$. For larger d the density modulation is smeared out, since the higher Fourier components of the static gate field distribution decay stronger in the z-direction. Photoresist gate modulation has been successfully demonstrated both on the Si-MOS and on the $GaAs$ heterostructure system.[42,9,13,43] However, sometimes one encounters difficulties, e.g. sticking problems and hysteresis effects in the gate voltage characteristic, in particular after several cool down cycles to He-temperatures. Here, instead of the photoresist, it is also possible to etch profiles into the SiO_2 (Fig. 2)[40] or into a thicker $GaAs$ cap layer (Fig. 3b) on top of an $AlGaAs/GaAs$ heterostructure.[5,43] This technique with a continuous gate on a thickness modulated insulating layer has the advantage that the whole gate area is 'two-dimensionally' connected. We have also prepared metal stripes by shadowing directly on the sample surface and connecting the stripes at the edge of the gate area to apply the gate voltage. Here one has to be very careful that as few as possible of the $3mm$ long and perhaps only $100nm$ wide stripes are interrupted. These wires are then electrically dead and the charge density under them is not modulated, giving, e.g. on 'normally-on' $AlGaAs/GaAs$ heterostructures, an unwanted 2D signal in the experiments.

It is well known that the carrier density in $AlGaAs/GaAs$ heterostructures can be increased by illumination with bandgap radiation which ionizes deep donors. The electrons are transferred into the channel and stay there, at $T < 70K$, practically forever (persistent photoeffect). If this illumination is performed through a periodical mask, e.g. highly reflecting metal stripes, one achieves a periodically modulated persistent photoeffect.[44] It is even possible to perform in-situ holography directly on a $AlGaAs$-structure to produce a charge density modulation.[47,48] It was found that the spatial resolution is about the same as the distance of the remote donors in the doped $AlGaAs$ from the channel, i.e., typically $20nm$.[47]

The structures with the varying gate distance described above (Fig. 2, Fig. 3a+b) are not only suited for density modulated systems. If $V_g > V_t(x)$ in a certain region of the period and vice versa in the other, isolated wires are formed. For very small dimensions of these wires, e.g. $w < 300nm$ in $GaAs$, it is possible to achieve quantum confined 1DES. That is that, in addition to the original 2D confinement in the z-direction, the electron motion in the x-direction is also quantized in discrete energy levels. We will discuss several aspects of such 1DES, in particular the FIR response, below.

Gated structures have the advantage, that the width of the wires and thus the confinement energies and number of 1D subbands can be tuned continuously. In particular, if conducting substrates are used, very high confinement energies $> 5meV$ can be achieved. On semi-insulating $GaAs$-substrates the minimal width and the confinement in gated structures is limited due to pinch-off effects in very small wires. Semi-insulated substrates are required for FIR spectroscopy since doped substrates are non-transparent. We are currently working with semitransparent δ-doped layers acting as a backgate contact close to the channel to overcome this problem. Another disadvantage of gated structures is that due to photoexcited leakage current they normally do not work under permanent illumination (e.g. in Raman spectroscopy or photoluminescence). Here and also for other reasons, etched structures are very interesting.

Actually, a system prepared by 'deep mesa etching' as sketched in Fig. 3d, seems to be the most straightforward way to produce a quantum confined 1DES.[14] Here, starting from a photoresist mask, wires are fabricated by etching all the way into the active $GaAs$ layer, leading ideally to a direct 'geometrical' confinement of the electrons. In reality it turns out that deep mesa etching is a very ambitious technological task to prepare such structures. The reason is that the free etched surfaces create surface states which trap electrons. Only with optimized processes (see below) was it possible to realize 1DES with deep mesa etching. For the $AlGaAs$-system it was found that structures with geometrical dimensions t as small as $550nm$ could be prepared. It was found that the actual width, w, of the electron channel was about $300nm$, thus lateral depletion lengths $d_e \approx 100nm$ are formed on either side of the structure. This is qualitatively in agreement with the well known fact that at a clean $GaAs$ surface the Fermi level pinning occurs near midgap.[49] Quantitatively, however, there are still open questions, in particular, it is quite possible that our special etching process leads to a passivation of the surface and helps to keep the lateral depletion lengths small. Other authors have found much larger lateral depletion lengths $(d_e \approx 1000nm)$ which make it practically impossible to realize 1DES.[50] It is also possible that for our process some of the electrons from the donors in the $AlGaAs$ layers are transferred to surface states in the side walls. These negatively charged surface states would help to confine the carriers in the 1D channel. For $GaAs$ we have thus the situation that the electrons are more 'electrically' confined by the remote donors in the $AlGaAs$ and possibly by charged surface states in the sidewalls.

For the $InGaAs$-system it is known that the Fermi level pinning occurs near the conduction band edge.[49] Thus this system seems to be a very good candidate for the deep mesa etching technique. Indeed, we have realized 1DES in deep mesa etched modulation-doped $InGaAs$ wires with geometrical width $t = 300nm$.[51] It is found that the electrical width w of the channel is nearly the same as the geometrical width t, indicating a more 'geometrical' confinement as compared to the $AlGaAs/GaAs$ system.

One of the interesting aspects of the deep mesa etched structures is that one can not only prepare single-layered wire structure, but, as sketched in Fig. 3e, one can start from a multi-layered modulation-doped $AlGaAs$ structure and etch through all the different layers. These structures are very interesting for possible technical applications of 1DES. The proposed advantages of 1DES-devices, the inherent increase in the mobility due to the reduced scattering in the 1D-k-space,[52] and the phase coherence effect in the ballistic transport regime only work efficiently if only the lowest 1D subband is occupied. This implies that the current density is very small. This disadvantage can be overcome by using many parallel wires, integrated, not only in the x-direction, but also in the z-direction. The multi-layered structures also show interesting physical phenomena which will be treated below.

For one-layered 1DES one can also use the technique of 'shallow' mesa etching,[7,14] where, as sketched in Fig. 3c, only the doped $AlGaAs$ layer is etched. This has the advantage that no surface states are created in the active region and enables the fabrication of very small structures (our experience so far $w \approx 160nm$). The important requirement for this technique is a very well defined etching process to stop exactly at the desired distance above the active channel.

ETCHING TECHNIQUES

There are many wet etching solutions known for $SiO_2, Si, GaAs, AlGaAs$ etc. Very important for wet etching is that tails of photoresist are stripped before the etching. Nevertheless, for very small dimensions that are of interest here, dry etching seems to give better, in particular more reproducible results. So we concentrate on these processes here.

For the etching of photoresist (stripping and profile definition) we use an isotropic plasma etching process in an O_2 plasma which is induced in an inductively coupled barrel reactor. A so called 'etch tunnel', consisting of a metal mesh with $\approx 1mm$ diameter holes, assures that only uncharged radicals hit the sample surface. This prevents radiation damage. The typical pressure is 0.3 Torr and the etching rate of AZ-photoresist is $40nm/min$.

For $GaAs$ quantum wire structures we start from high mobility modulation-doped $Al_xGa_{1-x}As/GaAs(x = 0.3)$ heterostructures and multi-quantum well systems, grown

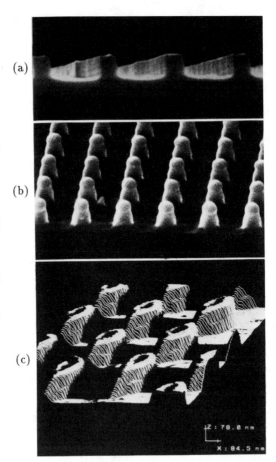

(a)

(b)

(c)

Fig. 4 (a) Electron micrograph of an etched two-layered quantum well wire structure with periodicity $a = 1100nm$, geometrical wire width $t = 400nm$ and etching depth $500nm$. (b) Mesa etched cross-grating structure with a periodicity $a = 500nm$ and a depth of $d = 150nm$. The substrate is semi-insulating $GaAs$. The photoresist dots, serving as the etching mask, are still not removed. (c) shows the same structure after removing the photoresist and sputtering with gold in a tunneling electron microscope contour map (from Refs.[14,57]).

by molecular beam epitaxy (MBE). The doped $AlGaAs$ regions in these structures are typically $50nm$ thick and the spacer layers $10nm$. The width and separation of the quantum wells as shown in Fig. 3d are $50nm$ and $130nm$, respectively. The essential requirements for the fabrication of the quantum wires are, (a) to minimize all processes that cause radiation damage or create surface states and thus act as electron traps in the narrow wires and (b), to keep the number of process steps to a minimum, again to minimize possible damage and to increase reproducebility. For the deep mesa etching we have selected a $SiCl_4$ reactive ion etching (RIE) process. [53] As it is generally known, the RIE process[54] has two components, the physical etching and the chemical etching. Whereas the physical component etches anisotropically and gives thus steep side walls, it has less selectivity with respect to the mask, and, important for our demands, creates damages. Chemical etching has a high selectivity but etches isotropically. We have thus balanced these two aspects. On the one hand we want enough physical etching to achieve an accurate transfer of the mask and steep vertical sidewalls. This is particularly important for the multi-layered quantum wires (Fig. 3e). On the other hand the physical component should be small, to achieve enough selectivity for our one-layered photoresist technique, and to make any damage as small as possible.[55]

We use a two-plate reactor where the lower electrode ($280mm$ diameter) is powered by a $13.56MHz$ generator. The upper grounded plate is in a distance of $55mm$ from the lower one and is 1.6 times larger in diameter. A very low power density of $0.24W/cm^2$ keeps the self-excited dc bias voltage, which essentially determines the physical etching, as low as $150V$. The samples are placed on the lower electrode and are directly exposed to the plasma. They are inserted into a hole of suitable size in a $GaAs$ wafer to avoid inhomogeneities from increased electric fields at the edges of a separated sample. In our system we can control the pressure, p, and the flow, f, independently. The characteristic property of high flow is an increased number of reactive ions reaching the sample surface and increasing the chemical etching. A low pressure and correspondingly high bias voltage results in a pronounced physical etching with steep walls. The parameters for our optimized standard process are a power density of $0.24W/cm^2$, a pressure $p = 30mTorr$ and a gas flow $f = 50sccm$. The etching rates for this process are $100nm/min$ and $20nm/min$ for the $GaAs$ and for the photoresist, respectively. We have successfully prepared multi-layered quantum wire structures with up to 5 layers.[16] Here an etching depth of about $1000nm$ was required. For the actual structure shown in Fig. 4a with a geometrical width of $t = 400nm$ we found that after the processing no mobile electrons were left in the systems. All electrons were trapped in defect and surface states. However, for slightly wider structures, $t = 550nm$, we had mobile electrons in the channels and could prove the formation of 1D subbands,[14] as will be discussed below.

For the etching of SiO_2, as needed for structures in Fig. 2, there exist a variety of different etching processes. We have made good experiences with a CHF_3 RIE process in a parallel-plate reactor. Optimized conditions, i.e., low radiation damage, no redeposition of polymers from reaction products from the photoresist were found for a pressure of $6 \cdot 10^{-2}Torr$ and a CHF_3 flow of 65 standard liters per hour. The self-excited bias voltage was $300V$ under these conditions. We achieved an etching rate of about $10nm/min$ for SiO_2, the etching rate of AZ-photoresist was $3nm$. We found that in particular during the RIE process (and not so strongly in the barrel reactor) the UV radiation causes a significant reduction of the breakdown voltage of the SiO_2. These radiation damages could be nearly completely removed by an annealing process in forming gas (10 standard liters per hour gas flow, 95% Ar and 5% H_2 at $400°C$). Annealing time was $45min$, cool down time $6h$. More details for the SiO_2 processing are included in Ref.[56]

In Fig. 4 we show some examples of etched structures. Figure 4a depicts an etched two-layered quantum well wire structure with periodicity $a = 1100nm$, and geometrical wire width $t = 400nm$. Figure 4b shows a mesa etched $GaAs$ dot structure.[57] The photoresist dots, which served as an etching mask, are still present on the top of the $GaAs$ dots. Figure 4c shows a tunneling electron microscope contour map of the same structure. I would like to note here that it is the easier part to produce small structures and show beautiful, impressive pictures. The much harder part of the preparation is the optimisation of all the detailed process parameters such that the electronic properties of the system fulfill the desired and necessary requirements. This very often requires many many trys, extreme accuracy and much patience.

GRATING COUPLER INDUCED EXCITATIONS IN 2DES

The elementary dynamic excitations of a homogeneous 2DES are 2D plasmons (collective intraband excitations), intersubband resonances (resonant transitions between the 2D subbands of the 2DES) and, in a magnetic field B, cyclotron resonances (CR), (transitions between Landau levels). Under typical experimental conditions these resonances occur in the FIR frequency range (in particular regarding the resonance condition $\omega\tau > 1$). Thus FIR spectroscopy (transmission, reflection and emission[58]) is a very powerful tool to study these systems (for reviews on spectroscopy see e.g. Refs.[59,60,43,45,46]). In transmission spectroscopy FIR radiation is transmitted through the sample and the transmission T is measured. To eliminate the frequency dependent sensitivity of the spectrometer the spectra are normalized to a reference spectrum. It is very convenient if one can deplete the carrier density via a gate voltage. Then one can perform a differential spectroscopy and evaluate $\Delta T = (T(N_s) - T(0))/T(0)$. This expression is, for small signals, proportional to the real part of the dynamic conductivity and can thus be directly related to the microscopic properties of the electronic system. For sharp resonances, e.g. CR, a spectrum taken in a magnetic field B_0 can also be used as a reference. For more details see e.g. Refs.[59,60]

CR can be directly observed in normal transmission. However, intersubband resonances in highly symmetrical systems microscopically correspond to oscillations of

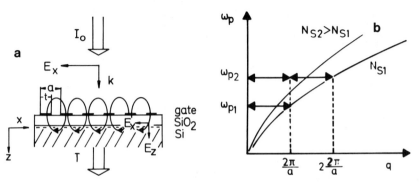

Fig. 5   (a) Grating coupler effect of a periodical structure. For normally incident FIR radiation spatially modulated parallel ($e_x(\omega, q)$) and perpendicular ($e_z(\omega, q)$) electric field components are induced.   (b) Coupling of FIR radiation to 2D plasmons of wavevectors $q_1 = 2\pi/a$ and $q_2 = 2 \cdot 2\pi/a$.

electrons in the z-direction perpendicular to the interface. Since, due to the high index of refraction of semiconductor materials, even at non-normal incidence the e_z-component of the incident electric field is very low, special tricks are required to excite intersubband resonance effeciently with FIR radiation. One of these tricks is to use a metal stripe grating coupler as will be discussed below.

2D plasmons were first observed for electrons on the surface of liquid He,[61] for semiconductor systems in the $Si(100)$ system. [35–38] The first calculations of the 2D plasmon dispersion were performed by Ritchie, [62] Stern[63] and Chaplik. [64] (For recent extended reviews on 2D plasmons see Refs.[38,65] (experiments) and Ref.[66] (theory).) 2D plasmons have the dispersion

$$\omega_p^2 = \frac{N_s e^2}{2\bar{\epsilon}\epsilon m^*} \cdot q \qquad (1)$$

where N_s is the 2D charge density, m^* the effective mass, $\bar{\epsilon} = \bar{\epsilon}(\omega, q)$ is the effective dielectric function (which also depends on the geometrical surrounding) and q is the wave vector in the x-y-plane.

2D plasmons are nonradiative modes in the sense that $q > \omega/c$ (taking into account retardation, also at small q and ω). Thus no direct coupling with transmitted radiation is possible. An effective way to excite these nonradiative modes is with a grating coupler. A grating coupler may consist of periodic stripes (periodicity a) of high and low conductivity materials (Fig. 5a). (In principle, any periodic variation of the dielectric properties in the vicinity of the 2DES, e.g., a modulation of the 2DES itself, causes a grating coupler effect). Then the incident field, which has a homogeneous e_x-component only, is short-circuited in the high conductivity regime. Thus, in the near-field of the grating, the electromagnetic fields are spatially modulated and consist of a series of Fourier components

$$\vec{e} = \sum_{n=-\infty}^{+\infty} (e_x^n, 0, e_z^n) \cdot \exp(i(k_x^n x - \omega t)) \qquad (2)$$

with $k_x^n = n \cdot 2\pi/a$. Fourier components with wavevector k_x^n will couple with 2D plasmons if $q = k^n$ and ω satisfies the dispersion relation (Fig. 5b). It is also important

159

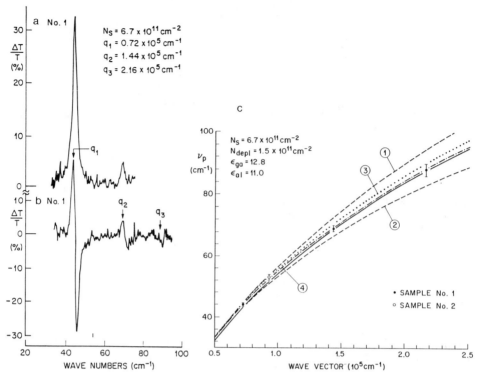

Fig. 6   Experimental plasmon resonances for a space-charge layer in $GaAs$ with $N_s = 6.7 \cdot 10^{11} cm^{-2}$. The arrows mark plasmon resonance positions. For the upper trace (a) transmissions at $B = 0T$ and $B \geq 8T$ have been rationed to determine $\Delta T/T$. The lower trace (b), qualitatively the derivative of the upper trace, is obtained by changing N_s slightly via the persistent photoeffect. (c) Theoretical and experimental plasmon dispersions. The solid line is the classical local plasmon dispersion. The curves marked 1-4 are defined as follows: curve 1, plasmon dispersion including nonlocal correction; curve 2, plasmon dispersion including finite-thickness effect; curve 3, plasmon dispersion including nonlocal and finite-thickness corrections combined; and curve 4, plasmon dispersion including all correction terms (from Ref.[70]).

that in the near-field the grating induces e_z-components of the electric field. Via these field components it is possible to excite intersubband resonances. [67,68] A full general calculation of the dielectric response in the presence of a grating is a complicated numerical problem. Approximate treatments for the conditions here are given in Refs.. [38,69] The effectiveness of the grating coupler depends on its design. The nth Fourier components of the fields decay with $\exp(-\mid k_x^n \mid \cdot z)$ for $\mid k_x \mid >> \omega/c$. Thus the distance d_z of the 2DES from the grating should be small compared with the periodicity ($d_z <\approx 0.1a/n$).

High mobility modulation-doped $AlGaAs/GaAs$ heterostructures are an ideal system to study 2D plasmons. In Fig. 6a we show grating coupler induced 2D plasmon excitations from Ref.. [70] The periodicity of the grating coupler is $a = 880nm$. The transmission spectrum, T, of the FIR radiation is normalized to a spectrum taken at $B = 0T$. A magnetic field shifts the plasmon resonance, such that the effective conductivity at $B = 8T$ is low in the frequency range of Fig. 6a. Two resonances are observed which are excited via the first and second Fourier components of the grating and correspond to wavevectors $q_1 = 2\pi/a$ and $q_2 = 2 \cdot 2\pi/a$. In the differential spectrum in Fig. 6b (here $\Delta T/T$ is evaluated for two slightly different densities) three plasmon resonances are resolved. This allows a detailed study of the q-dependence of the dispersion, as depicted in Fig. 6c. It is shown in Ref.[70] that several corrections to the simple formula (2) become effective at high wavevectors q. In particular effects three effects are important. They arise (a) from nonlocal properties of the 2DES, i.e., the finite compressibility of the Fermi gas (Fermi pressure), (b) from virtual coupling to intersubband transitions and (c) from the finite thickness of the 2DES. The finite thickness becomes important if q approaches $1/d_e$, the inverse spatial extension of the electron wavefunctions in the z-direction. It is shown that these effects nearly cancel each other in agreement with the experimental findings.

2D plasmons have also been studied for the Si-MOS system. An interesting aspect of the plasmon excitation is that one can measure the dispersion for different directions in the surface. For $Si(110)$ the plasmon dispersion is found to be anisotropic for different direction $\vec{q}$ with respect the [001] direction. [71] This arises from the anisotropic bandstructure $E(\vec{k})$ of a $Si(110)$ surface. Also 2D plasmons in hole space charge layers of $Si(110)$ have been investigated. [72] Here the plasmon dispersion reflects the nonparabolicity of the hole surface bandstructure and the anisotropy of the $Si(110)$ surface.

GRATING COUPLER INDUCED INTERSUBBAND RESONANCES

Besides 2D plasmons, intersubband resonances (ISR) are characteristic dynamic excitations in heterostructures and quantum wells. ISR represent oscillations of the carriers perpendicular to the interface. Thus in highly symmetrical systems, e.g., electrons in $Si(100)$-MOS systems or in $GaAs$-heterostructures, an e_z-component of the exciting electric field is necessary to excite these transitions. Strip-line and prism coupler arrangements have been used to study ISR on $Si(100)$. [73,74] Another very powerful method is the grating coupler technique. [67,68] As we have discussed above, a grating coupler excites in the near field e_z-components of the electric field (Fig. 5a). Experimental spectra measured on $Si(100)$ samples are shown in Fig. 7a. The grating period is $a = 1800 nm$. For $N_s = 3.3 \cdot 10^{12} cm^{-2}$ two resonances, $\tilde{E}_{01}$ and $\tilde{E}_{02}$, are observed which correspond, respectively, to resonant transitions from the lowest subband to the first and second excited subbands. With increasing N_s the resonances shift to higher frequencies, corresponding to a larger subband separation in the steeper potential well at larger surface electric fields. For $N_s > 8 \cdot 10^{12} cm^{-2}$ additional resonances $\tilde{E}'_{01}$ are observed which can be attributed to resonant transitions in the primed subband system. [1] The primed subband system arises from the projection of four volume energy ellipsoids of Si onto the $Si(100)$ surface and is separated in k-space by $0.86 \cdot 2\pi/A$ (A= crystal lattice constant) in [001] and equivalent directions. It is known that these subbands are occupied for $N_s > 7.5 \cdot 10^{12} cm^{-2}$ (Ref. 1).

The resonance energy measured in an ISR spectrum is not directly the subband spacing $E_{01} = E_1 - E_0$. Two effects shift the observed resonance with respect to the subband spacing, $\tilde{E}_{01} = E_{01} \cdot \sqrt{1 + \alpha - \beta}$. The first, the so-called exciton shift, [75] results from the energy renormalization when an electron is transferred to the first excited subband E_1 leaving a 'hole' in the E_0 subband. The exciton effect, characterized in a two-band model by β, shifts the resonance energy to smaller energies. A second effect, characterized by α, is the so-called depolarization shift, which increases the resonance energy. It arises from the resonant screening of the microscopic one particle dipole excitation by the collective effect of all other electrons in the potential well. [76] To include this effect in a calculation it is required to take into account the full time dependence in the Hartree potentials[77] beyond a purely static calculation of subband energies.

For $Si(111)$ and $Si(110)$, the surface bandstructure results from the projection of volume energy ellipsoids which are tilted with respect to the surface. [1] Due to the anisotropic energy contours, the x- and z-components of the dynamic surface current are coupled and thus ISR can be excited with a parallel field component e_x.[68] Whereas parallel excited ISR (labeled $\bar{E}_{01}$) is not affected by the depolarization shift, $\bar{E}_{01} = E_{01} \cdot \sqrt{1 - \beta}$, the perpendicular excited resonance ($\tilde{E}_{01}$) is affected by both effects, $\tilde{E}_{01} = E_{01} \cdot \sqrt{1 + \alpha - \beta}$. Thus, if a grating coupler is used on $Si(111)$, (Fig. 7b) both resonances, directly parallel excited resonances and perpendicular grating cou-

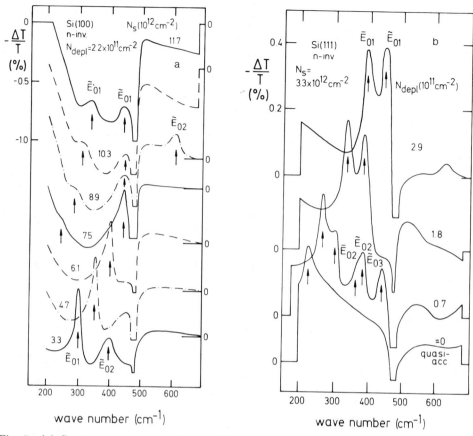

Fig. 7 (a) Grating coupler induced intersubband resonances for $Si(100)$ at different charge densities N_s. Resonances $\tilde{E}_{01}$ and $\tilde{E}_{02}$ in the lower subband system and $\tilde{E}_{01}$ in the second subband system are observed. In the regime of the optical phonon frequency of SiO_2 (about $480 cm^{-1}$), a resonant coupling to polaritons is measured, which is not fully shown here for clarity. (b) Excitation of intersubband resonances on a $Si(111)$ sample with a grating coupler for different N_{depl}. Directly parallel excited $\bar{E}_{oi}$ resonances and grating coupler induced depolarization shifted $\tilde{E}_{oi}$ resonances are present (from Ref.[68]).

pler excited (and thus depolarization shifted) resonances are observed. We find in the spectra in Fig. 7b that the depolarization shift slightly increases with the depletion field that is characterized by N_{depl}. It becomes much smaller for transitions to higher subbands (E_{02} and E_{03}). Very detailed investigations of grating coupler induced ISR for different surface orientation on Si are reported in Ref.. [68] Grating couplers have also been used to investigate in a very tricky experiment ($Si(100)$-MOS structure under external stress) a resonant coupling of intersubband resonances and 2D plasmons, leading to an anti-crossing of the dispersions. [79] Recently grating couplers have also been applied to investigate the $GaAs$ system.[80]

CHARGE DENSITY MODULATED SYSTEMS

A first step in the direction of lower dimensional electronic systems is a density modulated system.[40,41] Fig. 8 shows measurements on $Si(100)$ -MOS systems with periodically modulated oxide thickness as discussed above. The thickness of the oxide is d_1 in the region t_1 and is (except for the small region of the slopes) d_2 for the rest $t_2 = a - t_1$ of the period a. A continuous layer of NiCr ($3nm$) is evaporated with varying angles onto the structured oxide. If a gate voltage V_g is applied between this gate and the substrate, an electron gas with a low charge density N_{s2} in the region t_2

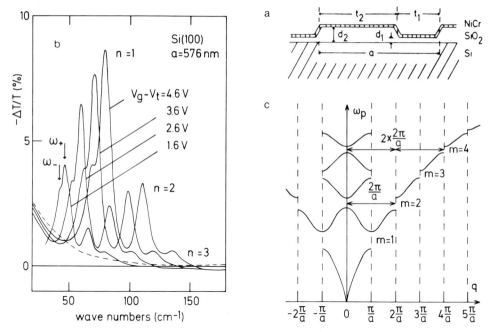

Fig. 8 Schematical geometry of a MOS sample with modulated oxide thickness (a), excitation of 2D plasmons with split resonances (b) due to the superlattice effect of the charge density modulation on the 2D plasmon dispersion (c) (from Refs.[40,41]).

and with high densities N_{s1} in the region t_1 is induced. d_1 and d_2 are, respectively, typically 20 and $50nm$.

In Fig. 8b FIR transmission spectra are shown. Superimposed on the Drude background, well pronounced plasmon resonances are observed which shift with increasing gate voltage and corresponding charge densities to higher wavenumbers, roughly as is expected from the plasmon dispersion (equation (1)). Resonances ω_1 and ω_2 are indicated. They are excited, respectively, via n=1 and n=2 reciprocal grating vectors $q_n = n(2\pi/a)$. Characteristic for these charge density modulated systems is that the plasmon excitations are split into two resonances, ω_{n-} and ω_{n+}. To discuss the origin of this splitting, we show in Fig. 8c the plasmon dispersion in a charge density modulated system. The superlattice effect of the periodically modulated charge density creates Brillouin zones with boundaries $q = m\pi/a$, $m = \pm1, \pm2, \ldots$ If we fold the plasmon dispersion back into the first Brillouin zone, $-\pi/a < q < \pi/a$ we expect a splitting of the plasmon dispersion at the zone boundaries and for the center of the Brilluoin zone at $q = 0$. The plasmon dispersion forms bands with minigaps at $q = 0$ and $q = \pi/a$. Since we use the same grating that produces the Brillouin zones also for the coupling process $(q_n = n(2\pi/a))$, we can only observe the gaps at $m=2,4,\ldots$ The resonances observed in Fig. 8b are thus the lower and upper branches of this dispersion at $q = 0$.

The plasmon dispersion in a charge density modulated system has been calculated in Ref.. [81] In this perturbation theory approach it is found that the splitting of the m-th gap is proportional to N_{sm} where N_{sm} is the Fourier coefficient of the charge density Fourier series:

$$N_s(x) = \sum_{m=-\infty}^{+\infty} N_{sm} e^{i2\pi m x/a} . \qquad (3)$$

Thus to observe a large splitting of the plasmon resonance for $q = 2\pi/a$, a system

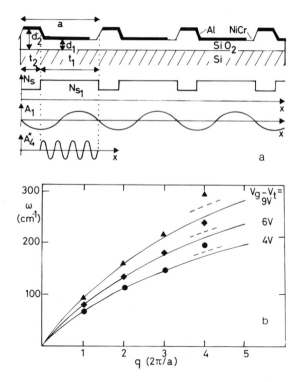

Fig. 9 Schematical sample configuration of an oxide thickness modulated MOS system. The oxide
is covered with a continuous semi-
transparent $NiCr$ gate and periodic
non-transparent Al-stripes. Via a
gate voltage a modulated charge
density $N_s = N_s(x)$ is induced. A_1
and A_4^* show schematically the amplitude of the plasmon oscillations
of extended plasmons and of plasmons which are localized in the region t_1. (b) Experimental plasmon
resonance positions from Fig. 3
versus wavevector q. Up to $q =
3(2\pi/a)$ the experimental resonance
positions follow the plasmon dispersion calculated with the averaged
charge density (full line). The resonance plotted at the position of $q =
4(2\pi/a)$ has a strongly increased res-
onance frequency due to the localization of the plasmons in the region
t_1.(from Ref.[82])

with a large second Fourier component N_{s2} in the charge density distribution has to
be prepared. This explains a strong influence of the sample geometry on the amount
of the experimentally observed splitting and also the fact that the splitting observed
for higher gaps (m=4,6,...) is generally smaller. [40,41] Another characteristic feature of
plasmons in a periodically modulated system is that they have the character of standing
waves, where both branches have different symmetry. Because of the symmetry, the
upper branch ω_{1+} for the configuration in Fig. 8c has a radiative character, whereas
the ω_{1-} branch is less radiative. Thus FIR radiation can excite the ω_{1+} branch with
higher efficiency. This is observed in Fig. 8b.

LOCALIZATION OF PLASMONS

In Fig. 9 we show measurements[82] on an oxide-thickness modulated system with
additional metal stripe grating couplers and, what is important for this experiment, a
largely extended high density regime t_1. The gap between the metal stripes is made
very small leading to a very efficient excitation of higher Fourier components. Indeed
resonances up to wave vectors $q_4 = 4(2\pi/a)$ are observed. The q_4-resonance shows
an unexpectedly high frequency which is caused by a localization of plasmons in the
following sense. At low wavevectors $q < 3(2\pi/a)$ the plasmon oscillations extend
over several superlattice periods and are thus governed by the average charge density
of the system. With increasing q the electromagnetic fields which accompany the
plasmon excitation in the surrounding media and which couple the different regions
decrease in both x- and z-direction with exp $(-| qx |)$ and exp $(-| qz |)$, respectively.
Thus, at high q, we can treat the regime t_1 independently and apply tentatively the
model of plasmons in a box of length t_1 (see Fig. 9a, amplitude A_4^*). Neglecting
any electromagnetic coupling and assuming ideal 'reflection' at the 'interface' t_1/t_2 we
would expect a plasmon wavevector of $q_4^* = 4(2\pi/t_1)$ for the mode ω_4^* considered here.
Using this value of q_4^* we obtain from Eq. (1) e.g. for $V_g - V_t = 9V$ with $d = d_2$
and $N_s = N_{s2}$ a plasmon frequency of $310 cm^{-1}$. This value is slightly higher than
the experimental value ($295 cm^{-1}$). However, the fact that the experimentally observed
resonance frequency ω_4^* lies between the fully localized frequency (box-model) and the
plasmon frequency that we calculate for 'free' plasmons with the parameters d_2 and

164

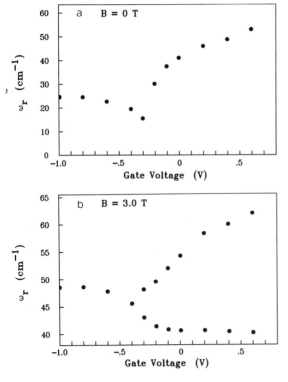

Fig. 10 Experimental FIR resonance positions measured on a split-gate sample (Fig.3a) at $B = 0T$ and $B = 3T$. For $V_g \geq -0.3V$ the system is density modulated, below isolated 1D wires are formed. (from Ref.[43])

N_{s2} (broken lines in Fig. 9) gives strong evidence that the frequency enhancement that we observe here is caused by a localization of plasmons in the region t_1. Such a frequency enhancement has recently been calculated.[83] Related effects also occur for wire structures and will be discussed below.

TRANSITION FROM 2DES TO 1DES

In Fig. 10 we show experimental FIR resonance positions[43] for a $AlGaAs/GaAs$ heterostructure with modulated gate distance as sketched in Fig. 3a. For $V_g = 0V$ and $B = 0T$ one observes the grating coupler induced 2D plasmon resonance at $\omega_p = 53cm^{-1}$, for $B = 3T$ the cyclotron resonance at $\omega_c = 40cm^{-1}$ and the magnetoplasmon resonance ω_{mp} at 63cm^{-1}. The latter is shifted with respect to ω_p according to the well known magnetoplasmon dispersion, [64,65] $\omega_{mp}^2 = \omega_p^2 + \omega_c^2$. With decreasing V_g we induce a density modulated system with a decreased average density. Correspondingly the plasmon frequency shifts to lower frequencies. At $V_g = -0.4V$ there is a sudden change in the slope and, for B=3T, ω_{mp} and ω_c merge. From dc transport one finds that this singular behaviour occurs in the range of gate voltages, where isolated wires are formed. Such a behaviour has first been observed in Ref.. [9] One finds that in the regime $V_g < -0.4$ the electron system forms a quantum confined 1DES, however the FIR response reflects only indirectly the 1D subband spacing. We will discuss the characterisation of 1DES and the FIR response in the next section.

ONE DIMENSIONAL ELECTRONIC SYSTEMS

1DES are used here for electronic systems with an energy spectrum that consists of a set of quantum confined 1D subbands (i,j= 0,1,2,3,..)

$$E^{ij}(k_y) = \frac{\hbar^2 k_y^2}{2m^*} + E_x^i + E_z^j \quad , \qquad (4)$$

where the electrons have a free dispersion only in the y-direction (k_y is the electron wave

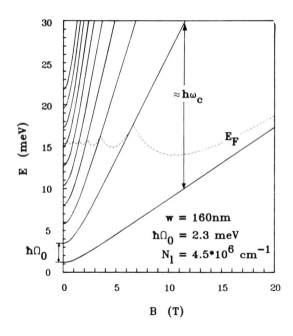

Fig. 11 Energy spectrum for electrons confined in a harmonic oscillator potential and in a magnetic field B. The dotted line shows the oscillations of the Fermi energy for the indicated linear charge density N_l. The hybrid 1D-subband- Landau-levels become successively depopulated with increasing B.

vector). E_z^j represents the quantized energy levels in the original 2DES with typical separations ranging from 10 to $100meV$. E_x^i represents the quantum confined energy states due to the additional lateral confinement acting in the x-direction. To achieve for instance a lateral quantization of $2meV$ in the $GaAs$ system with $m^* = 0.065m_o$, one has to confine the electrons, as we will see below, into a width w of some $100nm$.

A first approximation to calculate the energy levels in a 1DES, e.g., in a deep-mesa etched structure as sketched in Fig.3d, is to use a model with infinite potential walls. However this is only a very rough approximation. The actual potential depends on charged surface states at the side walls, in other words, on the Fermi level pinning at the surfaces. It depends on the electric field of the remote ionized donors in the $AlGaAs$ stripes and on the electrons in the channel itself due to self-consistent screening. The most sophisticated calculations for 1DES so far have been performed for the split-gate configuration (Fig.3b) by Laux et al. [84] It is found that for a small number of electrons in the channel the potential is nearly parabolic. However, with increasing 1D-charge density N_l, self-consistent effects become important very quickly. The potential flattens and the subband separation decreases drastically. Also experimentally it is not easy to characterize the 1DES, i.e., to determine subband spacing, electrical wire width and number of electrons per 1D subband.

DC MAGNETOTRANSPORT IN 1DES

In most studies so far 1DES are characterized by the magnetic depopulation of the 1D subbands which occurs if the 1DES is exposed to a perpendicular magnetic field B.[2] The underlying physics can be explained without loss of generality if one assumes a parabolic confinement potential[2] $V(x) = \frac{1}{2}m^*\Omega_0^2 x^2$. In this case the Schrödinger equation can be solved analytically. The magnetic field induces an additional potential $V_B(x) = \frac{1}{2}m^*\omega_c^2(x-x_0)^2$, where $\omega_c = eB/m^*$ is the cyclotron frequency. For this model the energy levels are given by

$$E^i(k_y, B) = \hbar\Omega(i + \frac{1}{2}) + \frac{\hbar^2 k_y^2}{2m_y^*(B)} \qquad (5)$$

with $\Omega^2 = \omega_c^2 + \Omega_0^2$, and $m_y^*(B) = m^*\Omega^2/\Omega_0^2$. This energy spectrum is shown in Fig.11. Since the 1D density of states, $D_{1D}(E, B)$, increases with increasing B the

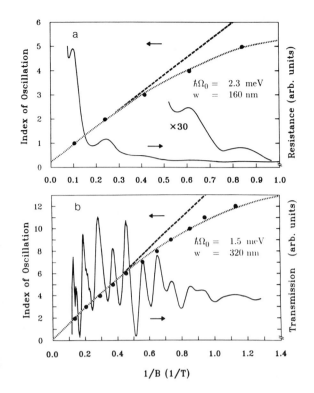

Fig. 12 Magnetotransport measurements (full lines, right scale) on 'shallow' mesa etched (a) ($a = 550nm, t = 250nm$) and in a 'deep' mesa etched two-layered (b) ($a = 1000nm, t = 550nm$) structure plotted versus $1/B$. A fan chart for the positions of the maxima in the magneto resistivity (full circles, left scale) exhibits deviations from a linear $1/B$ dependence (dashed lines) and indicates an one-dimensional energy structure in our samples. The dotted lines show the depopulation of 1D subbands within a harmonic oscillator model, calculated for the indicated values of the confining potentials $\hbar\Omega_0$ and the wire width w. (from Ref.[14])

1D subbands become successively depopulated (Fig.3b), giving rise to oscillations of the Fermi energy. In a transport measurement this leads to Shubnikov-de Haas (SdH) type of oscillations. In a 2DES the number of occupied Landau levels increases with decreasing B, leading ideally to an infinite number of SdH oscillations periodic in $1/B$. In a 1DES however, only a finite number of 1D subbands are occupied at $B = 0$, thus only a finite number of SdH oscillations occur, which are no longer linear in $1/B$.

We would like to demonstrate this behaviour for two types of our samples as shown in Fig. 3c and e. We performed magnetotransport measurements[14] at low temperatures ($T = 2.2K$) in perpendicular magnetic fields B. We defined on some of the one-layered quantum wire structures an active area of $2.5 \times 2.5mm^2$ by chemical mesa etching. Ohmic contacts of a Au/Ge alloy were aligned perpendicularly to the grating in order to measure the dc transport parallel to the stripes. On other samples a quasi-dc response was obtained by measuring the transmission of microwaves ($30 - 40GHz$) through the sample. Since for this measurement no contacts were needed, it was especially useful for samples with multi-quantum well (MQW) wires. Alloyed contacts would have short-circuited the channels of different layers.

In Fig. 12a we show measurements on a shallow-mesa etched sample (Fig.3c). The period of the wires was $a = 500nm$, the remaining width of the n-doped $AlGaAs$ was $t = 250nm$. Via Ohmic contacts the two-terminal resistance was measured as a function of the magnetic field. The dc conductivity shows well pronounced SdH-type oscillations. The important point is that the period of the oscillations is not constant in $1/B$, but exhibits distinct deviations at large values of $1/B$ and correspondingly small B. This clearly indicates the 1D character of our structure as was discussed above for the harmonic oscillator model. We calculated the depopulation of the 1D subbands by a magnetic field within this model using Ω_0 and the total 1D carrier density N_l as fitting parameters. The experimental fan chart in Fig. 12a is best described for $\hbar\Omega_0 = 2.3meV$ and $N_l = 4.5 \cdot 10^6 cm^{-1}$. For these values, six 1D subbands were occupied at $B = 0$. Defining the width w of the electron channel by the amplitude at the Fermi energy: $E_F(B = 0) = V(\frac{w}{2}) = \frac{1}{2}m^*\Omega_0^2(\frac{w}{2})^2$, we found $w = 160nm$ which was smaller than the

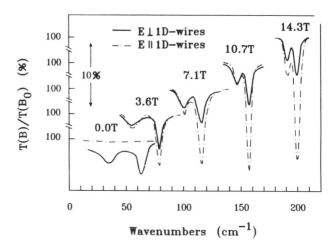

Fig. 13 Experimental FIR spectra, measured on a double-layered quantum wire structure at indicated magnetic fields B. Full lines and dash-dotted lines denote, respectively, polarization of the incident FIR radiation with the electric field vector perpendicular and parallel to the wires. (from Ref.[15])

geometrical width $t = 250nm$. Therefore we defined a 'lateral edge depletion region' on either side of length $w_{dl} = \frac{1}{2}(t - w) \approx 45nm$. As an example for deep mesa etched structures we show in Fig. 12b the quasi-dc conductivity of a two-layered quantum wire structure (Fig.3d), measured in microwave transmission as described above. SdH oscillations can be clearly resolved and show again distinct deviations from a linear $1/B$ behaviour at small B. An analysis within the harmonic oscillator potential model gives for the 1D confinement $\hbar\Omega_0 = 1.5meV$, $N_l = 15 \cdot 10^6 cm^{-1}$ and 16 occupied 1D subbands. From the channel width $w = 320nm$ we deduced that there is a lateral depletion length $w_{dl} = 100nm$ on either side of the wire.

FAR INFRARED SPECTROSCOPY ON QUANTUM WIRE STRUCTURES

We have seen above that the $GaAs$ quantum wire structures show typical subband separations of some meV. Thus FIR intersubband resonance spectroscopy appears to be a very promising method to determine the 1D bandstructure. However, it is found that the optical response exhibits a very complex behaviour, where the subband separation can be derived only indirectly. Here in particular the measurements on the multi-layered quantum wire structures are very helpful to understand the nature of the excitations.

In Fig.13 we show experimental FIR spectra for a two-layered quantum wire structure (Fig. 3e).[15] At $B = 0$ we observe two resonances at $\omega_r = 38cm^{-1}$ and $\omega_r = 63cm^{-1}$ if the incident electric field is polarized perpendicular to the wires. The resonances shift with increasing B to higher frequencies. With increasing B a resonance is observed also for parallel polarization with exactly the same resonance frequency as for perpendicular polarization. A very similar behaviour is also found for other quantum wire structures. However, the important point is that in n-layered systems up to n resonances are observed, e.g., for the one-layered system of Fig.3c or 3d, one resonance, and for a five-layered system, five resonances.[16] One finds that in a magnetic field B the experimental resonance positions for the quantum wire systems obey the relation $\omega_{ri}^2(B) = \omega_{ri}^2(B = 0) + \omega_c^2$.

The most surprising result is that these resonance frequencies ω_{ri} in the FIR spectra are significantly higher in energy than one would expect from the 1D subband separation $\hbar\Omega_0$, which was determined from the dc magnetotransport measurements. At $B = 0$ for a one-layered quantum wire system: $\hbar\Omega_0 = 1meV$, $\hbar\omega_r = 4meV$. For the two-layered system (Fig.13): $\hbar\Omega_0 = 1.5meV$, $\hbar\omega_{r1} = 8meV$, $\hbar\omega_{r2} = 4meV$. However, we have already seen from the discussion of the ISR in homogeneous 2DES that one has to be careful with the interpretation of optical spectra. In 2DES the intersubband resonance ω_r is shifted with respect to the one-particle transition due to the depolarization effect, i. e. the resonant screening effect of all electrons in the system. The effect was characterized by an effective plasma frequency $\omega_d = \sqrt{\alpha}\omega_{10}$ so that the observed resonance frequency could be written as $\omega_r^2 = \Omega_0^2 + \omega_d^2$. Using this model also for 1D intersubband resonance we find e.g. for a one-layered quantum wire structure[15] $\omega_d^2 = \omega_r^2 - \Omega_0^2 = (4meV)^2 - (1meV)^2 = (3.9meV)^2$. Thus the 1D-ISR here is strongly governed by the depolarization effect and without a very accurate model of the depolarization shift and detailed knowledge of the sample parameters it is not possible to extract the 1D subband separation from FIR spectra. This is also the case for the 1DES of $GaAs$ systems with split-gate configuration as discussed in Fig.10 and Ref.. [9,13] The best coincidence between 1D subband separation (measured with dc magnetotransport) and FIR resonance (they differ only by about 30%) has been found so far for quantum wires in $InSb$-MIS systems.[12] The reason for this is a much larger 1D subband spacing of $10meV$ which arises from the very small effective mass ($m^* = 0.014m_0$) of $InSb$. In addition a metal gate close to the 1D system and neighbouring grating stripes decreases the depolarization effect.[85,86]

It has been shown in Ref.[15] that for the conditions in $GaAs$ quantum wires with a strongly dominating collective contribution the FIR response can be much better explained in terms of a 'local' plasmon resonance. Indeed the magnetic field dependence of the FIR resonances and in particular the occurrence of two (n) resonances for two (n)-layered quantum wire structures resembles a plasmon type of excitation. For a two (n)-layered homogeneous 2DES it is known, that the collective excitation spectrum at small wavevectors q consists of two (n) branches (q is the plasmon wavevector in the plane).[87-89] Whereas for widely separated electron sheets ($qd \gg 1$, d = separation of the sheets) the plasmon branches of the individual layers are degenerated, the Coulomb coupling leads for small distances ($qd \approx 1$) to a splitting of the plasmon dispersion. The energetically highest plasmon branch represents an 'in-phase longitudinal oscillation' of all electron layers. The frequencies of the energetically lower plasmon branches are determined by the strength of the coupling between the electron sheets. The lower plasmon branches represent 'anti-phase' oscillations of the electrons in the different sheets and exhibit a linear q-dispersion at small q. Therefore they are called 'acoustical' plasmons in contrast to the energetically highest 'optical' branch. These type of optical and acoustic plasmons are observed in Fig. 13.

On the other hand, the resonances observed in the quantum wires of Fig. 13 differ significantly from excitations in a homogeneous system in the following points: (a) Besides the plasmon resonance one would expect to observe a cyclotron resonance in a homogeneous system. This resonance is completely quenched in our deep-mesa etched microstructured samples. All observed resonances ω_{ri} are shifted with respect to ω_c. (b) When we calculate the plasmon frequency of a homogeneous system (equation (1)) using the average dielectric constant for the microstructured region and the average 2D charge density $N_{s2D} = N_{s1D}/a$, we find that the experimentally observed resonances are significantly higher in energy than expected for the homogeneous system. We explain this frequency shift by 'localization' of plasmons in the following sense: Let us assume a single-layer '2DES' which is additionally confined in the x-direction to a width w. Then, in a very simple model, we can treat the 2D plasmon mode for the x-direction as a 'plasmon in a box'. The continuous 2D plasmon dispersion (equation (1)) of a homogeneous system with a free wavevector q is now quantized in fixed values of $q = \pi/w_e$ and correspondingly

$$\omega_{pl}^2 = \frac{N_s e^2}{2\bar{\epsilon}\epsilon_0 m^*} \frac{\pi}{w_e} \qquad (6).$$

Here $\bar{\epsilon}$ is the average dielectric constant. The effective width w_e is given by $w_e = w(1 + \alpha)$, where α takes account of the phase relation if the plasmon is 'reflected' at the walls of the box. This is of course a very rough model, which totally neglects Coulomb interaction with neighbouring electron stripes and leaves α so far undetermined. However, this model explains our experimentally observed upward shift of the resonance frequency with decreasing w.

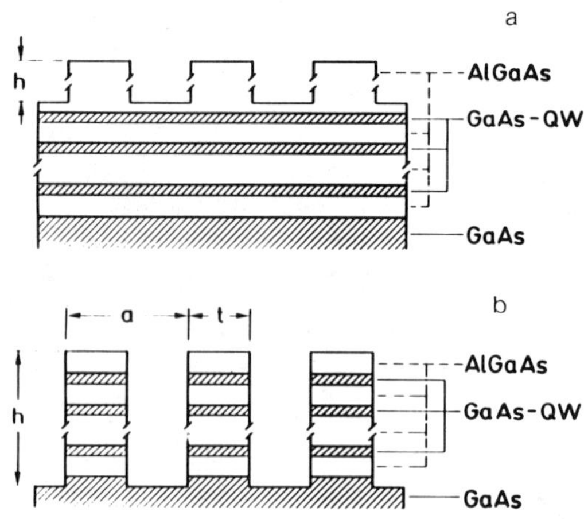

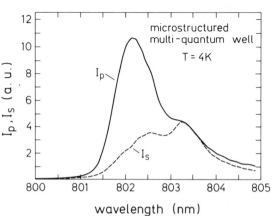

Fig. 14 Schematic configuration of the laterally microstructured three-layered undoped quantum well samples of type (a) and type (b). In (a) only the thick cap layer is structured and serves as a grating coupler. In (b) the etching is performed through the quantum wells.

Fig.15 Experimental luminescence spectra of a microstructured $GaAs$-MQW sample of type (b) with $a = 350nm$ and $t = 150nm$. I_p and I_s denote the p- and s- polarized luminescence intensity, respectively. The strong transition at $802.1nm$, which only occurs in p- polarization, arises from the radiative decay of QWEP. (From Ref.[25,26])

EXCITON-POLARITONS IN QUANTUM WELLS AND QUANTUM WIRES

So far we have considered modulation-doped systems. Very interesting physics also occurs in undoped quantum well systems. In particular, the photoluminescence spectra of such systems are strongly governed by intrinsic free excitonic effects, where the 2D-confined excitons show very interesting properties, e.g. a strongly enhanced oscillator strength.[90,91] Recently such studies have been extended to microstructured quantum well systems, in a search for additional confinement effects (e.g. Refs.[18-27]).

One of the interesting aspects here is that the microstructure itself leads to a grating coupler effect. In Fig. 14a we show a three-layered $GaAs$ quantum well system with a grating coupler etched into the thick cap layer. The photoluminescence spectra are found to be strongly dependent on the polarization of the exciting and luminescent radiation with respect to the grating rules. In particular one observes an increased p-polarized emission at the high energy side of the heavy-hole exciton transition. This emission arises from the radiative decay of quantum well exciton polaritons (QWEP).[25,26,92,93] QWEP are resonant electrodynamic excitations in the quantum well with wavevectors q in the plane of the quantum well and $|q| > \omega/c$ such that the fields decay exponentially in the positive and negative z-directions. They are thus normally not coupled with freely propagating, 'radiative' photons and do not contribute to the photoluminescence. Due to the grating coupler, however, QWEP are coupled with radiative photons of wavevector $q_{PH} = q_{PO} - n \cdot 2\pi/a < \omega/c$ which leads to a resonantly enhanced emission of radiation. Very similar effects are also observed[25] (see Fig. 15) in microstructured quantum well systems as shown in Fig. 14a with about $200nm$

narrow quantum well wire structures. Again the emission is strongly polarization dependent and governed by QWEP emission. Very recently we have succeeded in the fabrication of $70nm$ narrow quantum well wire systems.[94] It is found that the QWEP concepts break down at these narrow dimensions and instead of this one observes 1D excitons, i.e., excitons related to different energetically separated 1D electron subbands in the narrow wires. The 1D character of these excitons is manifestated in an increased exciton binding energy which was determined in magneto optical experiments.

SUMMARY

We have reviewed the preparation of lateral superlattices on semiconductor systems by use of holographic lithography and dry etching techniques. We have given a short survey on a variety of applications to study interesting physical phenomena in 2DES, 1DES and 0DES.

ACKNOWLEDGEMENTS

In these lecture notes I have reported on investigations which were only possible due to an excellent cooperation with many colleagues. I would like to thank very much all of my colleagues, as listed in the References, who have been working with me on these different subjects. I also acknowledge financial support from the BMFT.

REFERENCES

[1]T. Ando, A.B. Fowler, and F. Stern, Rev. Mod. Phys. **54**, 437 (1982)

[2]K.-F. Berggren, T.J. Thornton, D.J. Newson, and M. Pepper , Phys. Rev. Lett. **57**, 1769 (1986)

[3]H. van Houten, B.J. van Wees, M.G.J. Heijman, J.P. André, D. Andrews, and G.J. Davies , Appl. Phys. Lett. **49**, 1781 (1986)

[4]J. Cibert, P.M. Petroff, G.J. Dolan, S.J. Pearton, A.C. Gossard, and J.H. English , Appl. Phys. Lett. **49**, 1275 (1986)

[5]T.P. Smith, III., H. Arnot, J.M. Hong, C.M. Knoedler, S.E. Laux, and H. Schmid , Phys. Rev. Lett. **59**, 2802 (1987)

[6]M.L. Roukes, A. Scherer, S.J. Allen, Jr., H.G. Craighead, R.M. Ruthen, E.D. Beebe, and J.P. Harbison, Phys. Rev. Lett. **59**, 3011 (1987)

[7]H. van Houten, B.J. van Wees, J.E. Mooij, G. Roos, and K.-F. Berggren, Superlattices and Microstructures **3**, 497 (1987)

[8]G. Timp, A.M. Chang, P. Mankiewich, R. Behringer, J.E. Cunningham, T.Y. Chang, and R.E. Howard , Phys. Rev. Lett. **59**, 732 (1987)

[9]W. Hansen, M. Horst, J.P. Kotthaus, U. Merkt, Ch. Sikorski, and K. Ploog , Phys. Rev. Lett. **58**, 2586 (1987)

[10]B.J. van Wees, H. van Houten, C.W.J. Beenakker, J.G. Williamson, L.P. Kouwenhoven, D. van der Marel, and C.T. Foxon , Phys. Rev. Lett. **60**, 848 (1988)

[11]D.A. Wharam, T.J. Thornton, R. Newbury, M. Pepper, J.E.F. Frost, D.G. Hasko, D.C. Peacock, D.A. Ritchie, and G.A.C. Jones, J. Phys. C**21**, L209 (1988)

[12]J. Alsmeier, Ch. Sikorski, and U. Merkt , Phys. Rev. **B37**, 4314 (1988)

[13]F. Brinkop, W. Hansen, J.P. Kotthaus, and K. Ploog , Phys. Rev. **B37**, 6547 (1988)

[14]T. Demel, D. Heitmann, P. Grambow, and K. Ploog , Appl. Phys. Lett. **53**, 2176 (1988)

[15]T. Demel, D. Heitmann, P. Grambow, and K. Ploog , Phys. Rev. **B38**, 12732 (1988)

[16]T. Demel, D. Heitmann, P. Grambow, and K. Ploog, Superlattices and Microstructures **5**,287(1989)

[17]T.J. Thornton, M. Pepper, H. Ahmed, D. Andrews, and G.J. Davies, Phys. Rev. Lett. **56**, 1189 (1986)

[18]K. Kash, A. Scherer, J.M. Worlock, H.G. Craighead, and M.C. Tamargo, Appl. Phys. Lett. 49: 1043 (1986).

[19]J. Cibert, P.M. Petroff, G.J. Dolan, S.J. Pearton, A.C. Gossard, and J.H. English, Appl. Phys. Lett. **49**, 1275 (1986).

[20]M.A. Reed, R.T. Bate, K. Bradshaw, W.M. Duncan, W.R. Frensley, J.W. Lee, and M.D. Shih, J. Vac. Sci. Technol. **B4** 358 (1986).

[21]H. Temkin, G.J. Dolan, M.B. Panish, and S.N.G. Chu, Appl. Phys. Lett. **50**, 413 (1987).

[22]Y. Hirayama, S. Tarucha, Y. Suzuki, and H. Okamoto, Phys. Rev. **B37**, 2774 (1988).

[23]D. Gershoni, H. Temkin, G.J. Dolan, J. Dunsmuir, S.N.G. Chu, and M.B. Panish, Appl. Phys. Lett. **53**, 995 (1988).

[24]H.E.G. Arnot, M. Watt, C.M. Sotomayor-Torres, R. Glew, R. Cusco, J. Bates, and S.P. Beaumont, Superlattices and Microstr. **5**, 459 (1989)

[25]M. Kohl, D. Heitmann, P. Grambow, and K. Ploog, Phys. Rev. **B37**, 10927 (1988).

[26]M. Kohl, D. Heitmann, P. Grambow, and K. Ploog, Superlattices and Microstr. **5**, 235 (1989).

[27]M. Tsuchinya, J.M. Gaines, R.H. Yan, R. J. Simes, P.O. Holtz, L. A. Coldren, and P.M. Petroff , Phys. Rev. Lett. **62**, 4668 (1989)

[28]W. Hansen, T.P. Smith,III, K.Y.Lee,J.A. Brum, C.M. Knoedler, J.M. Hong, and D.P. Kern , Phys. Rev. Lett. **62**, 2168 (1989)

[29]Ch. Sikorski and U. Merkt , Phys. Rev. Lett. **62**, 2164 (1989)

[30]J.B. Brannon and J.F. Asmus, Appl. Phys. Lett. **38**, 299 (1981)

[31]T. Ahlhorn, H. Pohlmann, and J.P. Kotthaus, Proc. of SPIE 'Excimer Lasers and Applications', **1023**,231(1988)

[32]A.M. Hawryluk, H. I. Smith, and D. J. Ehrlich, J. Vac. Sci. Technol. **4**, 1200(1983)

[33]D. Heitmann and R.V. Pole, Appl. Phys. Lett. **37**, 585 (1980)

[34]G. Schmahl and D. Rudolf, Progr. in Optics **14**, (North Holland, Amsterdam) (1976)

[35]S.J. Allen Jr., D.C. Tsui, and R.A. Logan, Phys Rev. Lett. **38**, 980 (1977)

[36]T.N. Theis, J.P. Kotthaus, and P.J. Stiles, Solid State Commun. **24**, 273 (1977)

[37]T.N. Theis, J.P. Kotthaus, and J.P. Stiles, Solid State Commun. **26**, 603 (1978)

[38]T.N. Theis, Surf. Sci. **98**, 515 (1980)

[39]P. Grambow, E. Vasiliadou, T. Demel, K. Kern, D. Heitmann, and K. Ploog, submitted to Microelectronic Engineering

[40]U. Mackens, D. Heitmann, L. Prager, J.P. Kotthaus, and W. Beinvogl, Phys Rev. Lett. **53**, 1485 (1984)

[41]D. Heitmann, J.P. Kotthaus, U. Mackens, and W. Beinvogl, J. Superlattices and Microstructures **1**, 35 (1985)

[42]E Batke, W. Hansen, D. Heitmann, J.P. Kotthaus, U. Mackens, L. Prager, and W. Beinvogl, Proceed. of MRS, p. 155, Strassbourg (1985)

[43]T. Demel, D. Heitmann, and P. Grambow, Proceedings of Nato ARW on "Spectroscopy of semiconductor microstructures", Venice (1989), Eds. G. Fasol, A. Fasolino, and P. Lugli

[44]E. Batke, D. Heitmann, J.P. Kotthaus, and K. Ploog, Phys. Rev. Lett. **54**, 2367 (1985)

[45]U. Merkt, Ch. Sikorski, and J. Alsmeier, Proceedings of Nato ARW on "Spectroscopy of semiconductor microstructures", Venice (1989), Eds. G. Fasol, A. Fasolino, and P. Lugli

[46]D. Heitmann, T. Demel, P. Grambow, and K. Ploog, Advances in Solid State Physics **29**, Ed. U. Rössler (Vieweg, Braunschweig), p. 285

[47]K. Tsubaki, H. Sakaki, J. Yoshino, and Y. Sekiguchi, Appl. Phys. Letts. **45**, 663 (1984)

[48]D. Weiss , this volume

[49]L.G. Meiners and H.H. Wieders, Materials Science Report **3**,139,(1988)

[50]K.K. Choi, D.C. Tsui, and K. Alavi , Appl. Phys. Lett. **50**, 110 (1987)

[51]K. Kern, T. Demel, D. Heitmann, P. Grambow, K. Ploog, and M. Razeghi, Surf. Science, in press

[52]H. Sakaki, Jap. J. Appl. Phys. **19**, L735 (1980)

[53]M.B. Stern and P.F. Liao, J. Vac. Sci. Technol. **B1**, 1053 (1983)

[54]B. Chapman, Glow discharge processes, Wiley, New York (1980)

[55]P. Grambow, T. Demel, D. Heitmann, M. Kohl, R. Schüle, and K. Ploog, Micro-electronic Engineering **9**, 357 (1989)

[56]U. Mackens, Thesis, Hamburg (1986)

[57]J.M. Gomez-Rodriguez, L. Vazquez, A. Bartolome, A.M. Baro, P. Grambow, and D. Heitmann, Ultramicroscopy **6**, 355(1989)

[58]R.A. Höpfel and E. Gornik, Surf. Sci. **142**, 412 (1984)

[59]E. Batke and D. Heitmann, Infrared Phys. **24**, 189 (1984)

[60]D. Heitmann in "Physics and Application of Quantum Wells and Superlattices", eds. E.E. Mendez and K. von Klitzing, Plenum Press, New York, p. 317 (1987).

[61]C.C. Grimes and G. Adams, Phys. Rev. Lett. **36**, 145 (1976)

[62]R.H. Ritchie, Phys. Rev. **106**, 874 (1957)

[63]F. Stern, Phys. Rev. Lett. **18**, 546 (1967)

[64]A.V. Chaplik, Soviet Phys. JETP **35**, 395 (1972)

[65]D. Heitmann, Surf. Sci **170**, 332 (1986)

[66]A.V. Chaplik, Surf. Sci. Rep. **5**, 289 (1985)

[67]D. Heitmann, J.P. Kotthaus, and E.G. Mohr, Solid State Commun. **44**, 715 (1982)

[68]D. Heitmann and U. Mackens, Phys. Rev. **B33**, 8269 (1986)

[69]D.C. Tsui, S.J. Allen Jr., R.A. Logan, A. Kamgar, and S.N. Coppersmith, Surf. Sci. **73**, 419 (1978)

[70]E. Batke, D. Heitmann, and C.W. Tu, Phys. Rev. **B34**, 6951 (1986)

[71]E. Batke and D. Heitmann, Solid State Commun. **47**, 819 (1983)

[72]E. Batke, D. Heitmann, A.D. Wieck, and J.P. Kotthaus, Solid State Commun. **46**, 269 (1983)

[73]P. Kneschaurek, A. Kamgar, and J.F. Koch, Phys. Rev. **B14**, 1610 (1976)

[74]B.D. McCombe, R.T. Holm, and D.E. Schafer, Solid State Commun. **32**, 603 (1979)

[75]T. Ando, Z. Phys. **B26**, 263 (1977)

[76]W.P. Chen, Y.J. Chen, and E. Burstein, Surf. Sci. **58**, 263 (1976)

[77]S.J. Allen, D.C. Tsui, and B. Vinter, Solid State Commun. **20**, 425 (1976)

[78]T. Ando, T. Eda, and M. Nakayama, Solid State Commun. **23**, 751 (1977)

[79]S. Oelting, D. Heitmann, and J.P. Kotthaus, Phys. Rev. Lett. **56**, 1846 (1986)

[80]M. Helms, E. Colas, P. England, F. DeRosa, and S.J. Allen Jr., Appl. Phys. Letts. **53**, 1714(1989)

[81]M.V. Krasheninnikov and A.V. Chaplik, Sov. Phys. Semicond. **15**, 19 (1981)

[82]D. Heitmann and U. Mackens, Superlattices and Microstructures **4**,503(1988)

[83]V. Cataudella and V. Margliano Ramaglia, Phys. Rev. **B38**, 1828 (1988)

[84]S.E. Laux, D.J. Frank, and F. Stern, Surf. Sci. **196**,101(1988)

[85]W. Que and G. Kirczenow , Phys. Rev. **B37**, 7153 (1988) and , Phys. Rev. **B39**, 5998 (1989)

[86]A.V. Chaplik, Superlattices and Microstructures **6**, 329(1989)

[87]G. Fasol, N. Mestres, H.P. Hughes, A. Fischer, and K. Ploog , Phys. Rev. Lett. **56**, 2517 (1986)

[88]A. Pinczuk, M.G. Lamont, and A.C. Gossard , Phys. Rev. Lett. **56**, 2092 (1986)

[89]J.K. Jain and P.B. Allen , Phys. Rev. Lett. **54**, 2437 (1985)

[90]R. Dingle, W. Wiegmann, and C.H. Henry, Phys. Rev. Lett. **33**, 827 (1974).

[91]C. Weisbuch in "Physics and Application of Quantum Wells and Superlattices", eds. E.E. Mendez and K. von Klitzing, Plenum Press, New York, p. 261 (1987).

[92]M. Nakayama, Solid State Commun. **55**, 1053 (1985).

[93]D. Heitmann, M. Kohl, P. Grambow, and K. Ploog, Proceedings of Nato ARW on "Physics and Engineering of 1- and 0-Dimensional Semiconductors", Cadiz 1989. Eds. S. Beaumont and C.M. Sotomayor-Torres

[94]M. Kohl, D. Heitmann, P. Grambow, and K. Ploog, submitted to Phys. Rev. Letts.

MAGNETOCONDUCTANCE OSCILLATIONS LINEAR IN ΔB

IN SEMICONDUCTOR SURFACE SUPERLATTICES

D. K. Ferry, Jun Ma, R. A. Puechner, W.-P. Liu, R. Mezenner, A. M. Kriman, and G. N. Maracas

Department of Electrical and Computer Engineering, and
Center for Solid State Electronics Research
Arizona State University, Tempe AZ 85287-5706

INTRODUCTION

In the past decade, a number of experiments performed in thin metal and semiconductor films have demonstrated the possibility of observing quantum interference of electrons within condensed matter. Particularly convincing demonstrations have come from experiments in multiply-connected structures, for which the Aharonov-Bohm effect can be observed in magnetoresistance measurements of doubly-connected, one-dimensional rings whose characteristic lengths are shorter than the inelastic mean free path, or the diffusion length, of the electrons in metals (Webb *et al.*, 1985; Chandrasekhar *et al.*, 1985; Washburn and Webb, 1985) and semiconductors (Timp *et al.*, 1987a; Ford *et al.*, 1988; Ishibashi *et al.*, 1987b; Mankiewich *et al.*, 1988). Here, oscillations in the magnetoresistance are observed that arise from phase interference among topologically inequivalent paths (e.g. around opposite arms of a ring). The large majority of experiments demonstrating these effects have involved topologies with one or a relatively small number of individual rings (Umbach *et al.*, 1986).

In structures with extended topologies, such as a large two-dimensional lattice, magnetoconductance periodic in the magnetic field through a unit cell has been hypothesized for many years (Zak, 1964; Rauh *et al.*, 1974; Azbel, 1964). In this latter structure, the presence of two basic lengths, one the magnetic length $l_m=(\hbar/eB)^{1/2}$ and the second arising from the periodic potential, gives rise to two complementary limits that can be treated exactly. In one limit, the Landau regime, the periodic potential is regarded as a weak perturbation on the magnetic level structure, and this leads to oscillations periodic in reciprocal magnetic field, in a sense analogous to Shubnikov-de Haas oscillations. In the opposite regime, the Onsager regime, where the magnetic field is a weak perturbation on the periodic potential, the magnetoconductance is predicted to show oscillatory phenomena *periodic in the magnetic field,* in contrast to normal macroscopic magnetoconductance effects. While the required magnetic field is unreasonably high in normal semiconductor lattices, because of the small size of the unit cell, it should be an observable effect in multi-dimensional superlattices where the superlattice periodicity is hundreds of nanometers.

While several new phenomena have been observed in surface superlattices, both line meshes and full grid superlattices (Bernstein and Ferry, 1986; Gerhardts *et al.*, 1989; Winkler *et al.*, 1989), we know of no observations to date of this expected magnetoconductance effect. It should be remarked that these oscillations, while similar in character to the Aharonov-Bohm effect, are quite different. While the latter shows

Electronic Properties of Multilayers and Low-Dimensional Semiconductors Structures
Edited by J. M. Chamberlain *et al.*, Plenum Press, New York, 1990

175

diminished amplitudes in arrays of rings (Umbach *et al.*, 1986), the present oscillations are seen in large superlattices that extend over distances many times the inelastic mean free path.

Clearly, the observation of the above effects, which are linear in magnetic field, requires the phase coherence length of the electrons to be larger than the "lattice" period, whether the lattice is a single ring or an extended superlattice. If the phase breaking length is only of the order of a few times the lattice period *a*, then the current will be a relatively incoherent superposition of currents oscillating with the h/e period, but out of phase with one another. In a ring, one normally also expects to see the h/2e oscillation, characteristic of weak localization, become a dominant contributor to the overall magnetoconductance (Umbach *et al.*, 1986). It is not clear whether this latter oscillation should be expected in the superlattice. On the other hand, if the phase-breaking length is much larger than the lattice period, and comparable to the overall lattice size, then there is a question as to whether the interference of such a large number of unit cells will be in phase and observable in the macroscopic current. We examine these effects here.

THEORETICAL CONSIDERATIONS

The Aharonov-Bohm Effect

Now let us consider the Aharonov-Bohm effect as it is seen in quasi-two-dimensional semiconductor systems. The basic idea is illustrated in Fig. 1. A quasi-one-dimensional conducting channel is fabricated on the surface of a semiconductor or insulator. This channel can be metal, or in a high-electron mobility semiconductor heterostructure in which the inversion channel is defined by dry-etching of a mesa or by electrostatic confinement. In any case, it is preferable to have the waveguide sufficiently small that only one or a few discrete electron modes are allowed. The incident electrons, from the left in Fig. 1, have their wave split at the entrance to the ring. The waves recombine at the exit to the ring. The overall transmission through the ring from the left electrode to the right electrode depends on the relative size of the ring circumference in comparison with the electron's wavelength. In the Aharonov-Bohm effect, however, a magnetic field is threaded through the center of the ring. The non-conservative part of the vector potential $\mathbf{A}$ is then azimuthal, so that electrons passing through one side of the ring have their phase increased, while those passing through the other side have their phase decreased. Thus, varying the magnetic field can modulate the phase and produce conductance oscillations for the transport from one terminal to the other.

Now, we can estimate this effect by considering the phase of the electron through the above discussion, which leads to

$$\phi = \frac{1}{\hbar} (\mathbf{p} + e\mathbf{A}) \cdot \mathbf{x} , \tag{1}$$

and the change in phase introduced by the magnetic field is

$$\delta\phi = \frac{e}{\hbar} \oint \mathbf{A} \cdot d\mathbf{l} = \frac{e}{\hbar} \int_{ring} (\nabla \times \mathbf{A}) \cdot \mathbf{n} dS$$

$$= \frac{e}{\hbar} \int_{ring} \mathbf{B} \cdot \mathbf{n} \, dS = 2\pi \frac{\Phi}{\Phi_0} , \tag{2}$$

where $\Phi_0 = h/e$ is the quantum flux unit. Thus, the phase goes through a complete oscillation for the coupling of each unit of flux through the ring. This produces a modulation in the conductance that is periodic in flux, with a period of h/e. This periodic behavior is the Aharonov-Bohm effect.

When the mobility is high, and transport through the ring can occur ballistically without inelastic scattering, we get the above effect. There is another possibility that has been seen in samples with relatively low mobility. When there is significant elastic scattering, such as in a low mobility structure, the transport is more diffusive in nature. There can still be an interference, but it occurs not at the output end, but at the input end where the two counter-propagating parts of the electron wave both make a complete circuit. Since the two sides are almost time-reversed paths, they will automatically produce destructive interference at the input end. In this case, twice the flux is enclosed, and the conductance oscillates with a period of h/2e in flux. While it may seem that this is an unusual event, in fact it is quite likely to occur, and this coherent back-scattering is the primary source of weak localization phenomena.

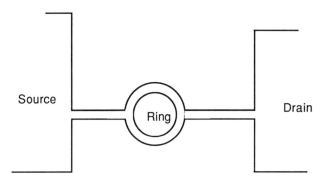

Fig. 1   A quasi-one-dimensional channel is formed, with a ring in it. When a magnetic field is passed through the center of the ring, the conductance can be modulated.

Superlattices

Consider the general Hamiltonian for electrons moving under the influence of a periodic potential $V(\mathbf{r})$ and a homogeneous magnetic field $\mathbf{B}=\nabla\times\mathbf{A}$, for which

$$H = \frac{1}{2m}(\mathbf{p} + e\mathbf{A})^2 + V(\mathbf{r}) , \qquad V(\mathbf{r} + \mathbf{L}) = V(\mathbf{r}) , \qquad (3)$$

where $\mathbf{L}=N\mathbf{a}_x+M\mathbf{a}_y$ defines the two-dimensional lattice and the other symbols have their usual meaning (in MKS units). As is well known, the group-theoretical considerations imply the existence of a magnetic translation operator, which generates the structural relationship for motion around the basic unit cell, defined by unit vectors $\mathbf{a}_x$ and $\mathbf{a}_y$ (Zak, 1964)

$$T(L_1\mathbf{a}_x)T(L_2\mathbf{a}_y)T(-L_1\mathbf{a}_x)T(-L_2\mathbf{a}_y) =$$

$$= T(L_1\mathbf{a}_x+L_2\mathbf{a}_y-L_1\mathbf{a}_x-L_2\mathbf{a}_y)\exp[iL_1L_2\alpha\bullet(\mathbf{a}_x\times\mathbf{a}_y)]$$

$$= \exp(2\pi ieBL_1L_2/h) , \qquad (4)$$

where $\alpha=e\mathbf{B}/h$. It is apparent that, although we are talking about a very different geometrical structure than that found in Aharonov-Bohm rings, the phase factor in (4) is precisely the phase integral around the ring which occurs in these latter structures to account for the interference of the electron. The question here is whether currents at different parts of the structure will exhibit the overall phase coherence necessary to observe the h/e oscillations in the magnetoconductance.

We can illustrate the above concepts of fluctuations by studying the global transport properties of the quantum boxes in the extended state regime by considering the miniband nature of the lateral superlattice (Reich et al., 1983). If the electrons become fully localized in the quantum boxes, with no long range order, we can study the conductance fluctuations mentioned above on a superlattice basis. That is, we can construct a truly realistic Anderson model with our superlattice basis. Localization phenomena can arise either from energy fluctuations on an ordered lattice or from disorder in the lattice. The Anderson model treats the localized regime by a tight-binding Hamiltonian in which the individual site energies ε_i are assumed to be randomly distributed over a range $[-W,W]$, and the sites are coupled to nearest neighbors by an overlap energy V_{ij}. This leads to the Hamiltonian

$$H = \sum_i \varepsilon_i a_i^+ a_i + \sum_{i,j} V_{ij} a_i^+ a_j , \tag{5}$$

where a^+ and a are the creation and annihilation operators for an electron at site i or j. We consider the usual case of diagonal disorder with $V_{ij}=V$. The ratio of the range of site energies W to the overlap energy V (W/V) gives a measure of the degree of disorder in the system. A numerical evaluation can be achieved by evaluating the Green's functions in the site representation and using these to evaluate the Kubo formula (Thouless and Kirkpatrick, 1981; Lee and Fisher, 1981). We display the latter, as it gives a straight-forward method to evaluate the universal conductance fluctuations and the results are expressible in terms of the Landauer multi-channel conductance formula as well. We consider a two-dimensional localized region of volume $\Omega = n_x n_y a^2$, where a is the spacing of the sites and x is the direction of current flow. Then, the conductance is

$$G = \frac{4e^2 V^2 n_x}{n_y h} \mathrm{Tr}\{G''(j'-1,j-1)G''(j,j') + G''(j',j)G''(j-1,j'-1)$$
$$- G''(j'-1,j)G''(j-1,j') - G''(j',j-1)G''(j,j'-1)\} , \tag{6}$$

where G'' is the imaginary part of the Green's function. To be able to apply this formalism to a stripe of localized region, the stripe is bounded with two perfect conductors that act as contacts. We then need to be able to compute the Green's functions anywhere in the sample. The stripe is composed of a finite disordered region and infinite ordered regions as leads on the left and right. We first calculate the Green's functions $G^L(j)$ for the left and G^R for the right semi-infinite perfect lattices. The Green's function at any other point is then built up by using a recursive method which constructs one column at a time from either the left or the right, as

$$G^L(j) = [G^0(j)^{-1} - |V|^2 G^L(j-1)]^{-1} , \quad G^R(j) = [G^0(j)^{-1} - |V|^2 G^R(j+1)]^{-1} . \tag{7}$$

These then lead to

$$G(j,j) = \{G^0(j)^{-1} - |V|^2 [G^L(j-1) + G^R(j+1)]\}^{-1}$$

$$G(j,j') = \left[\prod_{i=j}^{j'} G^L(i) V \right] G(j',j') \qquad \text{for } j' > j$$

$$G(j,j') = \left[\prod_{i=j}^{j'} G^R(i) V \right] G(j',j') \qquad \text{for } j' < j . \tag{8}$$

Results for magnetic field variation are shown in Fig. 2a. The apparent periodicity arises from the microscopic loops of area a^2 that exist in this lattice.

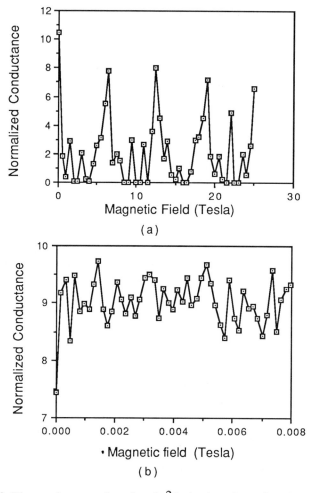

Fig. 2 The conductance, in units of e^2/h, is plotted as a function of the magnetic field, in units of $h/2\pi ea^2$. This plot is for values of W=2V, periodic boundary conditions, and a Fermi Energy E=V. In (a), the range is over several flux quanta per unit cell, while in (b), the low magnetic field range is expanded to illustrate the fractal nature of the conduction.

The amplitude of the oscillations and fluctuations shown in Fig. 2a are quite sensitive to the level of the Fermi energy, but seem to be independent of the exact details of the model. For example, the same results are obtained if the magnetic field is confined only to the localized region, and not to the contacts. We also note in Fig. 2, that the conductance goes to zero at several values of the magnetic field, which is usually an indication of the quantum Hall effect. Here, it appears that the structures are showing behavior similar to that calculated by Hofstadter (1976), with a magneto-conductance showing fractal behavior. It is important to note that the conductance fluctuations are truly fractal, and an expanded version of the region near the B=0 point is illustrated in Fig. 2b. The correlation function for these fluctuations has a magnetic correlation "length" which scales with sample size in a manner consistent with that observed in universal conductance fluctuations (Stone, 1985), which will be discussed further below.

EXPERIMENTAL RESULTS IN HIGH MOBILITY STRUCTURES

We report here on the observation of periodic magnetoconductance oscillations in large (both length and width are considerably larger than the inelastic mean free path) quasi-two-dimensional semiconductor structures in which an additional two-dimensional periodic superlattice potential has been applied. The magnetoconductance oscillations have a period given by a flux of h/e coupled through each unit cell of the superlattice potential. These oscillations are observed at relatively low magnetic fields B<1 T (clearly we are in the Onsager regime). In addition, there are conductance fluctuations which appear in the magnetoconductance and have an amplitude of the order of $0.1e^2/h$. This is larger than we expect from universal conductance fluctuations (discussed further below) in such large structures.

Samples are prepared by standard molecular beam epitaxy of a quantum well structure on an undoped GaAs substrate and buffer layer. We have studied both high-electron mobility inversion structures at the GaAs-AlGaAs interface, in which the carriers are produced by modulation doping the AlGaAs overlayer, and pseudomorphic InGaAs single quantum wells, in which the carriers are produced by modulation doping the GaAs overlayer. Here, we report only on the GaAs-InGaAs samples. The InGaAs layer, 13.5 nm thick with 20% In content, forms a quantum well and the top 40 nm thick GaAs layer contains Si doping (1×10^{18} cm^{-3}). An undoped GaAs cap layer 5 nm thick was grown last. The carrier density in the quantum well was about 2×10^{10} cm^{-2} at 5 K when -2.5V gate bias was applied.

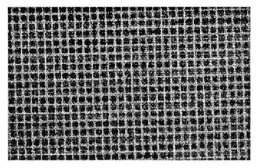

Fig. 3 The grid gate, pictured here, is placed over the 10 μm × 20 μm area at the cross of the mesa. It is composed of 40 nm lines on 160 nm centers.

Structures were prepared by etching a mesa in the sample, and writing the superlattice grid by electron-beam lithography and subsequent lift-off metallization. The individual lines of the grid gate were 40 nm wide on nominal center-to-center spacing of 160 nm. In Fig. 3, we illustrate the grid gate itself. The active area of the device structure measures 10 μm × 20 μm, and measurements are taken in both current directions. We estimate that the inelastic mean free path in these structures is of the order of 0.7 μm at 5 K, discussed below, which is considerably less than either dimension in the plane of the sample. In this structure, the electrons will be confined in three-dimensional quantum boxes lying below the open regions of the grid, and it is this total confinement which produces the full superlattice effects.

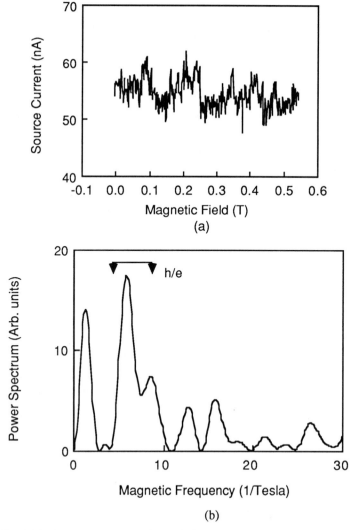

Fig. 4 (a) The source current through the sample (in the long direction)
for an applied voltage of 1 mV, as the magnetic field is varied.
(b) Fourier transform of the magnetoconductance data, after
taking out the d.c. value.

In Fig. 4a, we show the magnetoconductance σ_{xx}, in which the current is along the
long axis of the sample, at 5 K. It is apparent that there are significant fluctuations and a
weak periodicity of the conductance that is present in the magnetoconductance. The structure
is fully repeatable as long as the sample is maintained at low temperature, but does change
somewhat upon heating and recooling of the sample. The applied longitudinal voltage on the
sample was only 1 mV over the entire range, so that the amplitude of the fluctuations in
conductance that are also observable is about $0.1e^2/h$. We have Fourier transformed the
conductance in order to bring out the underlying periodicity, and this is shown in Fig. 4b.
The d.c. component has been removed to enhance the signal, but there is still a low
frequency component that arises from the weak overall magnetoresistance variations in the

sample. In this latter figure, we have marked the range expected from estimating the frequency that would arise from the fabricated superlattice periodicity, allowing for the possibility that the actual flux coupled to each well varies due to the finite width of the individual gate lines. A second set of weaker peaks is observed near the second harmonic. Whether these are related to h/2e oscillations seen in weak localization in rings or are simply the second harmonic has not been determined at this time.

In Fig. 4b, the dominant peak is approximately 7.5 T^{-1}, which corresponds to a unit cell whose side is 176 nm, while actual scanning electron microscopy measurement of the sample suggests a number closer to 168 nm. Considering the quality of the data, this agreement is quite good. A secondary peak is also observed which lies very close to the first, and within the range of the spread expected from the fabricated grid. This secondary peak could arise from a slightly different spacing over part of the grid, which could in turn arise from differences across the grid in the linearity of the electron beam sweep.

The source of the conductance fluctuations in the data, and the relatively large amplitude of these fluctuations compared to that expected for universal conductance fluctuations (UCF), is also quite interesting. A phenomenon related to the Aharonov-Bohm effect, UCF has been observed in quasi-one-dimensional metal (Umbach *et al.*, 1984, 1986; Benoit *et al.*, 1987) and semiconductor (Fowler *et al.*, 1982; Licini *et al.*, 1985; Skocpol *et al.*, 1986; Timp *et al.*, 1987b; Whittington *et al.*, 1986; Hiramoto *et al.*, 1987; Ishibashi *et al.*, 1987a; Brinkup *et al.*, 1988) wires. UCF is a general quantum-mechanical interference phenomenon that is sample specific, and is a direct consequence of the sensitivity of the conductivity to changes in the microscopic configuration (with varying electrochemical potential or magnetic field) of a mesoscopic system (Stone, 1985; Lee and Stone, 1985; Feng *et al.*, 1986; Lee *et al.*, 1987). UCF is regarded as arising from quantum interference of different modes, or paths, of the electrons as the chemical potential or the magnetic field is varied, so that interference effects appear in the end-to-end conductance, and are related to the Aharonov-Bohm effect. In general, the observations of these effects in the past have been confined to quasi-one-dimensional conductors. The structure we are investigating is considerably larger than the estimate of the inelastic mean free path. UCF has been found to decay faster than 1/L in quasi-one-dimensional wires (Umbach *et al.*, 1984; Benoit *et al.*, 1987) and faster than 1/N in a sequence of rings (Umbach *et al.*, 1986). The inelastic mean free path inferred below is such that the amplitude observed for the fluctuations is of this order of magnitude expected from these studies.

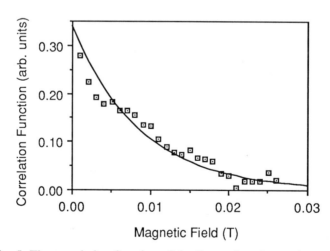

Fig. 5 The correlation function of the fluctuations in conductance. The solid line is a guide to the eye only, but exponential fitting with this type of behavior leads to a correlation "length" of 85 G.

In Fig. 5, we plot the correlation function calculated from the data of Fig. 4a, as a function of the field separation. This clearly evidences the expected exponential behavior, and has a correlation "length" of 85 G. This translates to a fundamental active area which is described by a inelastic mean free path of 0.7 µm. The size of this area also fits well the amplitude of the oscillations in terms of the number of basic areas that are being ensemble averaged. This averaging scales here exactly as that expected for UCF.

We suggest that the conductance fluctuation effect may be explained as well in terms of the fractal energy structures arising from application of a magnetic field to an electron gas in a two-dimensional periodic potential. It is well known that the Hamiltonian given in (1) can be solved for its eigenvalues only under certain conditions. As mentioned above, there are two fundamental lengths in the problem: one is the periodicity of the two-dimensional periodic potential, which here has $L_1=L_2=a$, and the second is the magnetic length $l_m=(\hbar/eB)^{1/2}$. The Hamiltonian can be solved for its eigenvalues when these two lengths are rationally related as $a/l_m=p/q$, where p and q are integers (Zak, 1964; Rauh et al., 1974; Azbel, 1964; Hofstadter, 1976; Thouless et al., 1982; Thouless, 1984). This property by itself leads to a magnetoconductance that exhibits fluctuations of the order of e^2/h, even in two-dimensional systems. This is probed with the theoretical calculations, which are similar to those introduced by Lee and Fisher (1981), that were discussed above. For both hard-wall and periodic boundary conditions, we find the existence of magnetoconductance fluctuations and periodicity in the flux coupled through each unit cell (Mezenner, 1988). (The details of these calculations will be published elsewhere.) These results are consistent with the interpretation expected from Hofstadter's work that the fractal nature of the eigenvalues for this system imply that small changes in the magnetic field produce significant changes in the eigenfunctions and therefore produce significant changes in the quantum interference in the structure. While the effect is essentially the same as that of UCF, the source of the effect has a different origin, arising here from the superlattice potentials. (The differences in the two theories are subtle, but can be explained for an Aharonov-Bohm ring. The UCF are generally assumed to arise from interference of modes in the multi-moded quasi-one-dimensional waveguides making up the arms of the rings. Thus, UCF is expected to disappear in a ring with truly single-mode waveguides. On the other hand, the ring still possesses two characteristic lengths – one geometric and one magnetic. Thus, the "fractal" theory suggests that conductance fluctuations will still exist even in a single-mode waveguide as infinitesimally small changes in the magnetic field will make macroscopic changes in the wavefunction for the electrons in the ring and should therefore macroscopically affect the resulting current.)

WEAK LOCALIZATION

We have also incorporated the LSSL gates into structures fabricated on material normally used for MESFETs. Here, the active layer is a 60 nm epitaxial layer grown by vapor phase epitaxy on a lightly doped substrate. The epitaxial layer is doped to 1.5×10^{18} cm^{-3}. It has been demonstrated previously that such layers will show a quasi-two dimensional behavior at low temperatures near pinchoff (Pepper, 1977). The structures measured here yield a variety of data, some of which is not fully understood at present. In particular, we note that at the highest magnetic fields used here, 0.9T, we still have $\omega_c\tau_{el}<1$, so that we do not expect to see magnetic quantum effects. In Fig. 6, we show the source conductance so that the presence of the negative magnetoresistance at low magnetic fields can be seen, which is clear evidence of weak localization. Other effects can also be seen in this latter figure. The source current shows sharp drops at regular values of the magnetic field, and sharp changes in the Hall voltage are often correlated with these. However, the sign of the change in the Hall voltage seems random in many samples. The source current drops occur at integral multiples of a flux quantum coupled through each unit cell of the surface superlattice, which we interpret to be periodic replicas of the negative magnetoresistance at zero magnetic field, and hence periodic replicas of the weak localization of the electrons!

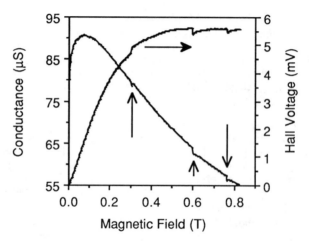

Fig. 6   Source current and Hall voltage for a MESFET at low
temperatures. We show the jumps in conductance and Hall
voltage that are seen in these samples (Ferry *et al.*, 1989;
Puechner *et al.*, 1989). Only the second, fourth, and fifth
drops are clearly seen (in addition to that at zero magnetic
field).

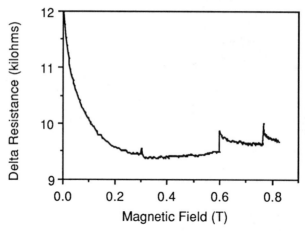

Fig . 7   Source magnetoresistance, with the linear part of the
magnetoresistance subtracted out, shows the localization
effects more clearly.

In contradistinction to the high electron mobility samples, where there was less than
one electron per quantum well, the electron density in these samples should be about 3-
5×10^{11} cm^{-2}, so that there may be as many as 100 electrons per quantum well. There are
certain to be complicated many-body corrections within each well. Nevertheless, it appears
from Fig. 6 that the basic periodicity in the conductance is about 1500 Gauss ($\pm$100 Gauss),

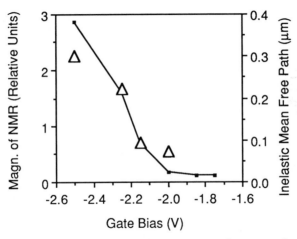

Fig. 8   The amplitude of the negative magnetoresistance peak at zero magnetic field (curve, in arbitrary units) and the inelastic mean free path (open triangles, in microns) are plotted as a function of the gate bias for a particular device. The values of inelastic mean free path measured vary from device to device.

which corresponds to about 165 nm periodicity, quite close to the periodicity in the grid. For comparison, the source (longitudinal) resistance is shown in Fig. 7, in which the linear magnetoresistance has been removed.

We can estimate the inelastic mean free path near these jumps from the negative magnetoresistance in the source current following the jump. In Fig. 8, the amplitude of the negative magnetoresistance at zero magnetic field is plotted for data on another sample, and the inelastic mean free path inferred from the basic digamma function weak localization behavior in a quasi-two-dimensional system (Altshuler *et al.*, 1980; Chakravarty and Schmid, 1986) is shown for comparison. It is clear that the electron system is becoming more two-dimensional as the device approaches pinch-off (negative gate bias). It should be pointed out that the value of the inelastic mean free path inferred from a data fit to several drops in the conductance at the various integer multiples of a flux quantum (coupled to each well) is nearly the same for a given sample and gate voltage. The quality of the fit to the first negative magnetoresistance peak, and to a higher conductance drop, are shown in Fig. 9 for one of the samples. The range of inelastic mean free paths obtained in the well-formed quasi-two-dimensional electron gas, for a range of samples, was 130-600 nm. In Fig. 10, the value of the inelastic mean free path computed for each of the drops (six) found in the data of Fig. 6 are shown for comparison, but are compared with the magnetoresistance data of Fig. 7. It is clear that the value found in this sample is about 0.2 μm, and the variation about this value is within the uncertainty of the fit obtained.

While we do not understand the jumps which occur in the Hall voltage, this could correspond to sweeping the Fermi level entirely through a miniband with the magnetic field. On the other hand, if the complicated band structure mentioned above is invoked, Thouless (1984) has speculated on the possibilities of jumps in the Hall conductance due to mixing of Landau levels and/or subbands, and this should occur at preferential values of the number of flux quantum per well. Moreover, the coupling of various orbits by Bragg scattering from the superlattice should occur at field values such that $\Phi = 2n\Phi_0$.

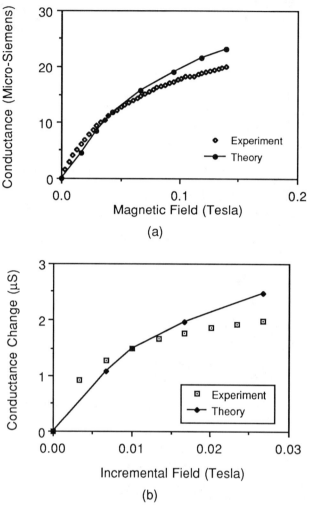

Fig. 9 The negative magnetoresistance from one of the MESFET samples is compared here with the theory of Altshuler *et al.* (1980). (a) The first NMR region, at zero magnetic field. The inelastic mean free path obtained from this fit is 223 nm. (b) The NMR region at the $\Phi=3\Phi_0$ (B=0.44 T). The inelastic mean free path for this last region is 0.185 μm. (The linear magnetoresistance, apparent in Fig. 6, has been subtracted out of the data.

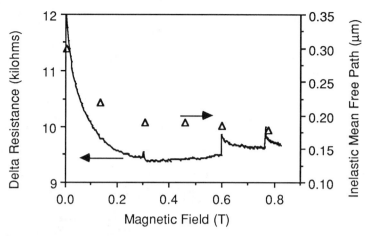

Fig. 10 The values obtained for the inelastic mean free path at the various jumps for the data of Fig. 6 and Fig. 7 are shown here. The variation in the values is within the uncertainty in calculating the individual values.

SUMMARY

In summary, we have measured the magnetoconductance in lateral surface superlattices, in which a two-dimensional electrostatic superlattice potential is imposed upon a quasi-two-dimensional electron gas. These measurements show the existence of periodic oscillations, with a period determined by the coupling of a quantized unit of magnetic flux h/e through each unit cell of the superlattice. In addition, we observe the presence of fluctuations in the magnetoconductance, for structures in which both lateral dimensions are large compared to the inelastic mean free path. The measurements have been carried out at relatively low magnetic fields (B< 1T), but it is only in the low field regime, where the magnetic energy is a perturbation on periodic potential, that the linear-in-B oscillations are expected (Zak, 1964; Azbel, 1964; Rauh *et al.*, 1974; Hofstadter, 1976; Thouless *et al.*, 1982; Thouless, 1984), while normal 1/B periodicity is expected at high magnetic fields.

The appearance of the oscillations differs in low mobility and high mobility samples. The latter do not exhibit weak localization with the superlattice present, and the oscillations are characterized by an oscillatory magnetoconductance. In the low mobility samples, however, weak localization dominates the overall conductance. The oscillations are characterized by the replication of the weak localization (at zero magnetic fields) at integer multiples of the fundamental flux quantum coupled into each unit cell of the superlattice. The sharp drop in conductance that accompanies this replicated localization effect is unexpected, and can be quite large. The oscillations are remarkably large considering that the active sample length is some 20 µm, and it would be expected that simple Aharonov-Bohm interference effects would have ensemble averaged for this situation in light of the relatively short inelastic mean free path (Umbach *et al.*, 1986). In addition, the conduction through the samples, in both the high mobility and low mobility cases, exhibits conductance fluctuations, which may be universal conductance fluctuations, but are more likely to be due to the fractal nature of the energy in systems with two characteristic lengths. This latter follows from the same ensemble averaging argument expected for the long sample length (compared to the inelastiv mean free path) mentioned above in connection with the oscillations. It is worth noting here, that the sample size, in both width and length, was many times the largest inelastic mean free path measured.

ACKNOWLEDGMENT

The authors are indebted to P. Newman, Fort Monmouth, P. Chu and H. H. Wieder, University of California at San Diego, and R. C. Curby, Avantek, for providing material used in this effort. This work was supported by the U. S. Office of Naval Research.

REFERENCES

Altshuler, B. L., Khmelnitskii, D. E., Larkin, A. I., and Lee, P. A., 1980, Magnetoresistance and Hall Effect in a Disordered Two-Dimensional Electron Gas, **Phys. Rev. B** 22: 5142.

Azbel, M. Ya., 1964, Energy Spectrum of a Conduction Electron in a Magnetic Field, **J. Exptl. Theoret. Phys.** 46: 929 (Translation in **Sov. Phys. - JETP** 46: 634).

Benoit, A., Umbach, C. P., Laibowitz, R. B., and Webb, R. A., 1987, Length Independent Voltage Fluctuations in Small Devices, **Phys. Rev. Lett.** 58: 2343.

Bernstein, G., and Ferry, D. K., 1986, Fabrication of Ultra-Short Gate MESFETs and BlochFETs by Electron-Beam Lithography, **Superlatt. Microstructures** 2: 373.

Brinkup, F., Hansen, W., Kotthaus, J. P., and Ploog, K., 1988, One-dimensional Subbands of Narrow electron Channels in Gated $Al_xGa_{1-x}As/GaAs$ Heterojunctions, **Phys. Rev. B** 37: 6547.

Chakravarty, S., and Schmid, A., 1986, Weak Localization: The Quasiclassical Theory of Electrons in a Random Potential, **Phys. Repts.** 140: 193.

Chandrasekhar, V., Rooks, M. J., Wind, S., and Prober, D. E., 1985, Observation of Aharonov-Bohm Interference Effects with Periods h/e and h/2e in Individual Micron-Size, Normal-Metal Rings, **Phys. Rev. Lett.** 55: 1610.

Feng, S., Lee, P. A., and Stone, A. D., 1986, Sensitivity of the Conductance of a Disordered Metal to the Motion of a Single Atom: Implications for 1/f Noise, **Phys. Rev. Lett.** 56: 1960.

Ferry, D. K., Bernstein, G., Puechner, R. A., Ma, J., Kriman, A. M., Mezenner, R., Liu, W.-P., Maracas, G. N., and Chamberlin, R., 1989, Magnetoconductance in Lateral Surface Superlattices, in *"High Magnetic Fields in Semiconductors II,"* Ed. by G. Landwehr (Springer-Verlag, Heidelberg, in press).

Ford, C. J. B., Thornton, T. J., Newbury, R., Pepper, M., Ahmed, H., Foxon, C. T., Harris, J. J., and Roberts, C., 1988, The Aharonov-Bohm effect in electrostatically defined heterojunction rings, **J. Phys. C** 21: L325.

Fowler, A. B., Hartstein, A., and Webb, R. A., 1982, Conductance in Restricted Dimensionality Accumulation Layers, **Phys. Rev. Lett.** 48: 196.

Gerhardts, R. R., Weiss, D., and v. Klitzing, K., 1989, Novel Magnetoresistance Oscillations in a Periodically Modulated Two-Dimensional Electron Gas, **Phys. Rev. Lett.** 62: 1173.

Hiramoto, T., Hirakawa, K., Iye, Y., and Ikoma, T., 1987, One-Dimensional GaAs wires fabricated by focused ion beam implantation, **Appl. Phys. Lett.** 51: 1620.

Hofstadter, D. R., 1976, Energy Levels and Wave Functions of Bloch Electrons in Rational and Irrational Magnetic Fields, **Phys. Rev. B** 14: 2239.

Ishibashi, K., Nagate, N., Gamo, K., Namba, S., Ishida, S., Murase, K., Kawabe, M., and Aoyagi, Y., 1987a, Universal Magnetoconductance Fluctuations in Narrow n^+GaAs Wires, **Sol. State Commun.** 61: 385.

Ishibashi, K., Takagaki, Y., Gamo, K., Namba, S., Ishida, S., Murase, K., Aoyagi, Y., and Kawabe, M., 1987b, Observation of Aharonov-Bohm Magnetoresistance Oscillations in Selectively Doped GaAs-AlGaAs Submicron Structures, **Sol. State Commun.** 64: 573.

Lee, P. A., and Fisher, D. S., 1981, Anderson Localization in Two Dimensions, **Phys. Rev. Lett.** 47: 882.

Lee, P. A., and Stone, A. D., 1985, Universal Conductance Fluctuations in Metals, **Phys. Rev. Lett.** 55: 1622.

Lee, P. A., Stone, A. D., and Fukuyama, H., 1987, Universal Conductance Fluctuations in Metals: Effects of Finite Temperature, Interactions, and Magnetic Field, **Phys. Rev. B** 35: 1039.

Licini, J. C., Bishop, D. J., Kastner, M. A., and Melngailis, J., 1985, Aperiodic Magnetoresistance Oscillations in Narrow Inversion Layers in Si, **Phys. Rev. Lett.** 55: 2987.

Mankiewich, P. M., Behringer, R. E., Howard, R. E., Chang, A. M., Chang, T. Y., Chelluri, B., Cunningham, J., and Timp, G., 1988, Observation of Aharonov-Bohm Effect in Quasi-one-dimensional GaAs/AlGaAs Rings, **J. Vac. Sci. Technol. B** 6:131.

Mezenner, R., 1988, Ph.D. Thesis, Arizona State University, unpublished.

Pepper, M., 1977, A Metal-Insulator Transition in the Impurity Band of n-Type GaAs Induced by Loss of Dimension, **J. Phys. C** 10: L173.

Puechner, R. A., Kriman, A. M., Bernstein, G., Liu, W.-P., Ma, J., Ferry, D. K., and Maracas, G.N., 1989, Conductance Oscillations in Lateral Surface Superlattices, in *"GaAs and Related Compounds 1988, "* (Amer. Inst. Phys., in press).

Reich, R. K., Grondin, R. O., and Ferry, D. K., 1983, Transport in Lateral Surface Superlattices, **Phys. Rev. B** 27: 3483.

Rauh, A., Wannier, G. H., and Obermair, G., 1974, Bloch Electrons in Irrational Magnetic Fields, **Phys. Stat. Sol. (b)** 63: 215.

Skocpol, W. J., Mankiewich, P. M., Howard, R. E., Jackel, L. D., Tennant, D. M., and Stone, A. D., 1986, Universal Conductance Fluctuations in Silicon Inversion-Layer Nanostructures, **Phys. Rev. Lett.** 56: 2865.

Stone, A. D., 1985, Magnetoresistance Fluctuations in Mesoscopic Wires and Rings, **Phys. Rev. Lett.** 54: 2692.

Thouless, D. J., 1984, Quantized Hall Effect in Two-Dimensional Periodic Potentials, **Phys. Repts.** 110: 279.

Thouless, D. J., and Kirkpatrick, S., 1981, Conductivity of the Disordered Linear Chain, **J. Phys. C** 14: 235.

Thouless, D. J., Kohmoto, M., Nightingale, M. P., and den Nijs, M., 1982, Quantized Hall Conductance in a Two-Dimensional Periodice Potential, **Phys. Rev. Lett.** 49: 405.

Timp, G., Chang, A. M., Cunningham, J. E., Chang, T. Y., Mankiewich, P., Behringer, R., and Howard, R. E., 1987a, Observation of the Aharonov-Bohm Effect for $\omega_c\tau>1$, **Phys. Rev. Lett.** 58: 2814.

Timp, G., Chang, A. M., Mankiewich, P., Behringer, R., Cunningham, J. E., Chang, T. Y., and Howard, R. E., 1987b, Quantum Transport in an Electron-Wave Guide, **Phys. Rev. Lett.** 59:732.

Umbach, C. P., Washburn, S., Laibowitz, R. B., and Webb, R. A., 1984, Magnetoresistance of Small, Quasi-One-Dimensional, Normal-Metal Rings and Lines, **Phys. Rev. B** 30: 4048.

Umbach, C. P., Van Haesendonck, C., Laibowitz, R. B., Washburn, S., and Webb, R. A., 1986, Direct Observation of Ensemble Averaging of the Aharonov-Bohm Effect in Normal-Metal Loops, **Phys. Rev. Lett.** 56: 386.

Washburn, S. and Webb, R. A., 1986, Aharonov-Bohm effect in normal metal Quantumcoherence and transport, **Adv. Phys.** 35: 375.

Webb, R. A., Washburn, S., Umbach, C., and Laibowitz, R. A., 1985, Observation of *h/e* Aharonov-Bohm Oscillations in Normal-Metal Rings, **Phys. Rev. Lett.** 54: 2696.

Winkler, R. W., Kotthaus, J. P., and Ploog, K., 1989, Landau-Band Conductivity in a Two-Dimensional Electron System Modulated by an Artificial One-Dimensional Superlattice Potential, **Phys. Rev. Lett.** 62: 1177.

Whittington, G. P., Main, P. C., Eaves, L., Taylor, R. P., Thoms, S., Beaumont, S. P., Wilkinson, C. D. W., Stanley, C. R., and Frost, J., 1986, Universal Conductance Fluctuations in the Mgnetoresistance of Submicron n$^+$GaAs Wires, **Superlatt. Microstruc.** 2: 381.

Zak, J., 1964, Magnetic Translation Group, **Phys. Rev.** 134: A1602; 134: A1607.

QUANTUM DEVICE MODELING WITH THE CONVOLUTION METHOD

T. P. Orlando, P. F. Bagwell, R. A. Ghanbari, and K. Ismail

Department of Electrical Engineering and Computer Science
Massachusetts Institute of Technology, Cambridge, MA 02139

1 Introduction

Recent advances in materials fabrication and nanolithography have made possible a generation of semiconducting structures whose conductance is governed by quantum mechanical phenomena. In particular, nanostructures on Si MOSFETs and GaAs MODFETs have shown modulations in their conductance versus gate voltage characteristics that have been attributed to quantum mechanical effects. In this paper, we review a modeling scheme which gives a unified way of understanding how these quantum effects are affected by temperature, mobility, voltage, and the structure of the device. This model provides not only a qualitative understanding of the various quantum phenomena, but also a basis for developing efficient computational algorithms for modeling specific devices. We have called this scheme the *convolution method* because most of the calculations can be written in terms of separate convolutions involving the individual phenomena of temperature, mobility, voltage, and structure.

In §2 specific quantum devices and their characteristics will be briefly described. The main part of the paper in §3 will present the convolution method and apply it to understanding the devices presented in §2.

2 Fabricated Devices

Quantum effects have been seen in the conductance of two principle types of semiconducting devices. In the first type, the electrons are electrostatically confined so that subbands are formed in one dimension, while the electrons remain free to move in the other dimension. The structure in the conductance of such quasi-one dimensional (Q1D) devices has been attributed to the filling of the Q1D subbands in both Si MOSFET and GaAs MODFET devices (Scott-Thomas et al., 1988; Ismail et al., 1989b; Warren et al, 1986). In the second type of structure, a periodic superlattice is imposed on the electrons in a two dimensional electron gas (2DEG). The modulation in the conductance of these devices is attributed mainly to Bragg diffraction from the potential induced by the lateral surface superlattice (LSSL) structure (Ismail

Electronic Properties of Multilayers and Low-Dimensional Semiconductors Structures
Edited by J. M. Chamberlain *et al.,* Plenum Press, New York, 1990

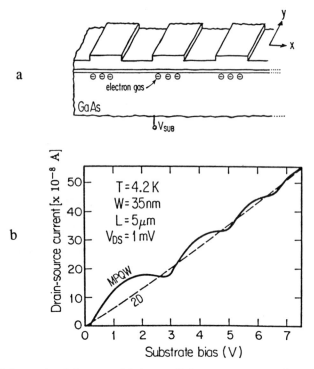

Figure 1. (a) Schematic of three multiple parallel quantum wires (MPQW) in a GaAs MODFET. (b) Drain-source current as a function of the substrate bias for a device with 100 parallel wires. The dashed line is for a scaled down 2D device. From Ismail et al. (1989b).

et al., 1988; Ismail et al., 1989a; Tokura and Tsubaki, 1987; Bernstein and Ferry, 1987; Warren et al., 1985). We now discuss three devices that illustrate the kinds of observed quantum effects.

Figure 1a depicts three parallel Q1D wires fabricated in a GaAs MODFET. The electrons are confined in the x-direction to a region about 40 nm wide, and are free to move in the y-direction. The actual device (Ismail et al., 1989b) consists of 100 parallel wires which were fabricated by ion milling a shallow grating into the doped AlGaAs layer through a mask produced by x-ray lithography. The period of the lines is 200 nm. The drain-source current along the y-direction is shown in Figure 1b as a function of back gate bias. As the back gate bias is increased the density of the electrons in the inversion layers increases, causing successive Q1D subbands to be occupied. The conductance decreases as electrons occupy each new subband. The structure in Figure 1b washes away with increased temperature and is less pronounced for devices with lower mobility.

The same x-ray mask used to make the Q1D conductors in Figure 1 can be rotated as shown in Figure 2a so that the electrons now flow perpendicular to the grating lines. In this device (Ismail et al., 1988) a 200 nm-period Schottky barrier grating gate of Ti/Au replaces the usual continuous gate in the MODFET configuration. As the gate-source potential is changed, the drain-source current along the x-direction is modulated as shown in Figure 2b. This modulation is believed to be caused by

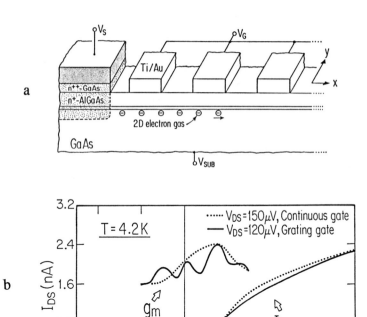

Figure 2. (a) Schematic view of a lateral surface superlattice (LSSL) device. (b) Drain-source current as a function of the gate-source voltage. The dotted curve is for a continuous gate device. The insert shows the transconductance g_m for the LSSL device. From Bagwell and Orlando (1989b) and Ismail et al. (1988).

Bragg diffraction of the electrons from the periodic potential. The structure in this device also washes away at higher temperatures and for devices with lower mobility. In addition, the structure diminishes with increasing drain-source voltage.

The structure in the conductance for the grating-gate LSSL is weak because the electrons experience a periodic potential in only one direction. A truly two dimensional periodic potential was fabricated in the grid-gate configuration as shown in Figure 3a. The period of the grid-gate is 200 nm and the linewidth is about 60 nm (Ismail et al., 1989a). A much larger modulation of the conductance is seen for this device in Figure 3b than in the grating-gate LSSL because the Bragg diffraction condition can be satisfied such that true minigaps are possible. As in the other two devices, the modulation of the conductances washes away as the temperature and drain-source voltage are increased, and as the mobility is decreased.

3 Device Modeling

There are three characteristic lengths that determine the type of electrical transport. The first is the length of the device L. The second is the average distance between elastic scattering events ℓ, which is known as the mean free path. The conduction electron has an average speed given by the Fermi velocity v_F, so that the

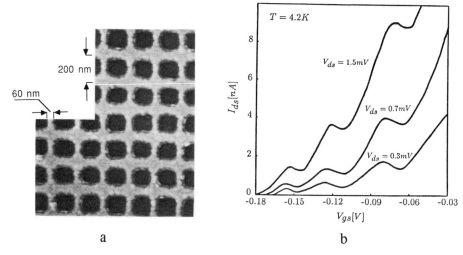

Figure 3. (a) SEM micrograph of a grid-gate with 200 nm period and a 60 nm line width. (b) Drain-source current as a function of the gate-source voltage for various drain-source voltages. From Ismail et al. (1989a).

mean free path can be related to a mean time τ between elastic scattering events by

$$\ell = v_F \tau \,. \tag{1}$$

The third characteristic length is L_ϕ, the phase breaking length. This is the length over which the phase of the quantum mechanical wavefunction of the conducting electron is correlated. The phase of the wavefunction is randomized, on average, each time τ_ϕ that the electron undergoes an inelastic scattering event which changes its energy. For this reason, we refer to L_ϕ as the inelastic scattering length. Depending on the relationship between these three lengths, the type of electrical conduction can range from diffusive transport where many scattering events occur, to ballistic transport where no scattering events occur. We now discuss these two limiting forms of electrical transport.

If $L \gg \ell$, then the electron will undergo many elastic scattering events before it traverses the device. Classically, the electron undergoes a random walk as it scatters, making its motion diffusive. The classical diffusion constant D is given by

$$D = \frac{1}{d} v_F^2 \tau = \frac{1}{d} \ell v_F, \tag{2}$$

where d is the dimensionality of the transport ($d = 2$ for a 2DEG). When electron motion is diffusive, its characteristic lengths and times are related via the diffusion constant. For example, the time τ_L for an electron to diffuse across the device and the phase breaking time τ_ϕ are given by

$$\tau_L = \frac{L^2}{D} \qquad \text{and} \qquad \tau_\phi = \frac{L_\phi^2}{D} \,, \tag{3}$$

as in Table 1.

The semiclassical Boltzmann transport equation gives a method for combining the concepts of classical diffusion with quantum mechanical properties. In the semiclassical method quantum mechanics is used to calculate the energy levels of the electrons.

Table 1. Length Scales and Time Constants

Diffusive $L \gg L_\phi \gg \ell$	Ballistic $L_\phi \gg \ell \gg L$
$\ell \approx \sqrt{D\tau} = v_F\tau$ $L_\phi = \sqrt{D\tau_\phi}$ $L = \sqrt{D\tau_L}$	$\ell = v_F\tau$ $L_\phi = v_F\tau_\phi$ $L = v_F\tau_L$

The electrons are then treated as wavepackets with their group velocities given by the dispersion relationship of the energy levels. This is the standard method used to calculate the conductivity of metals and semiconductors (Kittel, 1986; Ashcroft and Mermin, 1976). For the semiclassical method to be valid, there must be many inelastic scattering events in the sample such that $L \gg L_\phi \gg \ell$. Because the phase of the wavefunction is randomized at each inelastic event, the overall properties of the sample will be an ensemble average of all the possible scattering configurations. It is this averaging process that justifies the semiclassical method. Note that the wavefunction still has its phase correlated over the distance L_ϕ.

To see why there must be many inelastic scattering events to use the semiclassical method, let us suppose, on the contrary, that there are no inelastic scatterers in the sample. Quantum mechanically, the elastic scatterers could then be described by some scattering potential $V(\mathbf{r})$ which depends only on position. Therefore, the eigenstates for this potential could be found. These eigenstates have a discrete energy E so that the phase ϕ of the wavefunction would increase as $\phi = Et/\hbar$ and would be correlated for all times. Indeed, if $L_\phi \gg L$ the conduction electron would always have its phase correlated and the resulting conduction would be sensitive to the exact positions of the elastic scatterers. For such a regime of lengths $L_\phi \gg L \gg \ell$, the conduction would be dominated by unpredictable but repeatable fluctuations known as "universal conduction fluctuations" (Lee and Stone, 1985; Al'tshuler, 1985; Skocpol et al., 1986). To have diffusive transport, the phase of the wavefunction must be randomized many times during the transport across the sample, that is, $L \gg L_\phi \gg \ell$.

In the opposite limit to diffusive transport, there are no elastic scatterers in the sample so that the transport is ballistic ($\ell \gg L$). We also assume that the inelastic length is larger than the other two lengths so that $L_\phi \gg \ell \gg L$. In this case the electron moves without scattering at the Fermi velocity v_F. Therefore, the mean free path is given by Equation 1. The inelastic scattering length is related to the scattering time by

$$L_\phi = v_F\tau_\phi \tag{4}$$

and the time τ_L for the electron to ballistically traverse the sample is

$$\tau_L = \frac{L}{v_F}. \tag{5}$$

Table 1 lists these lengths and times.[†] We now discuss our method to calculate the transport in these two limiting regimes.

[†]In Table 1, L_ϕ, ℓ, τ_ϕ, and τ are defined as if the material between the contacts were infinite and disregards any influence of the contacts on the device. If the contacts are considered in the definitions

3.1 Diffusive transport

In the diffusive limit of transport where $L \gg L_\phi \gg \ell$, the calculation of the conductivity and the density of electrons can be done using the semiclassical method (Kittel, 1986; Ashcroft and Mermin, 1976). We will recast the usual expression for the density and the conductivity in terms of convolutions as the first example of our convolution method. Let $N(E)$ be the density of states for electrons in the material. The density of electrons n at a temperature T is given by

$$n(E_F, T) = \int_{-\infty}^{\infty} N(E')f(E' - E_F, T) \, dE', \tag{6}$$

where E_F is the chemical potential and $f(E, T)$ is the Fermi-Dirac distribution function. The thermodynamic density of states $N(E, T)$ is defined by

$$N(E, T) \equiv \frac{\partial n(E, T)}{\partial E} = \int_{-\infty}^{\infty} N(E')f'(E' - E, T) \, dE'. \tag{7}$$

Here $f'(E, T)$ is the negative derivative of $f(E, T)$ with respect to E and is given by

$$f'(E, T) = \frac{1}{4kT} \operatorname{sech}^2 \left(\frac{E}{2kT} \right). \tag{8}$$

This bell-shaped function has a full width at half maximum of $3.5kT$ and is symmetric in E. Due to this symmetry of $f'(E)$, the thermodynamic density of states in Equation 7 can also be written as

$$N(E, T) = N(E) \otimes f'(E, T). \tag{9}$$

The symbol $\otimes$ denotes a convolution which is given by

$$
\begin{aligned}
C(E) \otimes D(E) &= \int_{-\infty}^{\infty} C(E')D(E - E') \, dE' \\
&= \int_{-\infty}^{\infty} D(E')C(E - E') \, dE'
\end{aligned}
\tag{10}
$$

Because the convolution tends to broaden the first function by the width of the second function, $f'(E, T)$ will be referred to as the thermal broadening function.

In the semiclassical method the conductivity tensor is given by (Kittel, 1986; Ashcroft and Mermin, 1976).

$$\sigma^{ij}(E_F, T) = e^2 \int_{-\infty}^{\infty} v_i(E')v_j(E')\tau(E')N(E') \left[-\frac{\partial f(E' - E_F, T)}{\partial E'} \right] dE'. \tag{11}$$

At zero temperature the derivative of the Fermi function is a delta function so that the conductivity is simply

$$\sigma^{ij}(E) \equiv e^2 v_i(E)v_j(E)\tau(E)N(E). \tag{12}$$

Furthermore, the derivative of the Fermi function in Equation 11 is just $f'(E_F - E', T)$ since this function is symmetric about the Fermi energy. Therefore, the conductivity

---

of these lengths and times, it is then possible to interpret the contacts as a source of inelastic (phase-breaking) scattering such that $L_\phi = L$ and $\tau_\phi = \tau_L$. Nevertheless, ballistic transport still means that no scattering occurs in the material between the contacts.

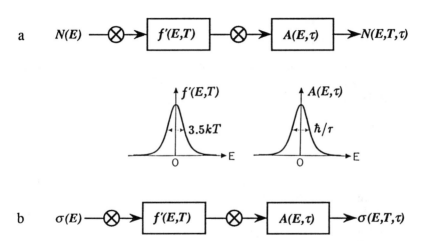

Figure 4. The convolution method for (a) the thermodynamic density of states $N(E,T,\tau)$ and (b) the conductivity $\sigma^{ij}(E,T,\tau)$. The thermal broadening function $f'(E,T)$ and the impurity broadening function $A(E,\tau)$ are shown graphically along with their widths. The symbol $\otimes$ denotes a convolution in energy.

as a function of temperature in Equation 11 can also be written as a convolution; namely,

$$\sigma^{ij}(E,T) = \sigma^{ij}(E) \otimes f'(E,T) . \tag{13}$$

We see that if the density of states and the conductivity are known at zero temperature, then their values at finite temperatures can be found by a simple convolution with the thermal broadening function $f'(E,T)$.

In the diffusive transport limit, the ensemble averaging (due to the phase randomization from inelastic scattering) leads to an additional energy level broadening (whose width is set by the elastic scattering). This additional broadening is characterized by the impurity broadening function $A(E,\tau)$ given by (Abrikosov et al., 1963; Rickayzen, 1980; Bagwell, 1988)

$$A(E,\tau) = \frac{1}{\pi} \frac{\hbar/2\tau}{E^2 + (\hbar/2\tau)^2} . \tag{14}$$

The impurity broadening function has a Lorentzian shape with a full width at half maximum of $\hbar/\tau$. The thermodynamic density of states including the broadening from both temperature and impurities is given by (Bagwell et al., 1989; Bagwell and Orlando, 1989b)

$$N(E,T,\tau) = N(E) \otimes f'(E,T) \otimes A(E,\tau) . \tag{15}$$

Likewise the conductivity is given by (Bagwell et al., 1989; Bagwell and Orlando, 1989b)

$$\sigma^{ij}(E,T,\tau) = \sigma^{ij}(E) \otimes f'(E,T) \otimes A(E,\tau) . \tag{16}$$

Figure 4 summarizes the convolution method for diffusive transport and shows the shapes of the two broadening function. Because the convolution with $f'(E,T)$ depends only on the temperature while the convolution with $A(E,\tau)$ depends only on the impurity scattering, we see that the broadening effects from temperature and impurities are independent effects.

This convolution method readily implies that, if the conductance has a feature which occurs over some energy range ΔE, then to see that structure the widths of the broadening functions have to be smaller than ΔE. For temperature and impurity broadening, this means that

$$\Delta E > 3.5kT \quad \text{and} \quad \Delta E > \frac{\hbar}{\tau}. \tag{17}$$

These two restrictions lead to the observability criteria (Antoniadis et al., 1985) that to see quantum effects, the electron should have a small effective mass, high mobility, low temperature, and be confined to a small length scale. Measurements clearly show the washing away of structure in device I-V characteristics as the temperature is increased and as the mobility is decreased (Ismail, 1989).

The convolution method clearly demonstrates why the structure in the conductivity tends to wash away at high temperatures and low mobilities. However, this method can also be used to understand device characteristics. In most field effect devices, the conductivity (of the source-drain) is measured as a function of gate-source voltage. We will assume here that the gate-source voltage is proportional to the density of electrons in the inversion layer. (If the density is related to the gate-source voltage in a more complicated fashion, then this more complicated relationship must be used.) Therefore, we seek in our convolution model to plot the conductivity versus density n. From the definition of the thermodynamic density of states in Equation 7, the density is

$$n(E_F, T) = \int_{-\infty}^{E_F} N(E, T, \tau) \, dE. \tag{18}$$

Another piece of information needed to model a specific device is the scattering time τ that enters the conductivity $\sigma^{ij}(E)$ and the impurity broadening function $A(E, \tau)$. This scattering time can be calculated from the quantum mechanical expression for the scattering rate τ^{-1} given by Fermi's golden rule; namely,

$$\frac{1}{\tau} = \frac{2\pi n_{\mathrm{imp}} |V|^2}{\hbar} N(E). \tag{19}$$

Here n_{imp} is the density of impurities and $|V|^2$ is the square of the matrix element between scattering states, which is assumed constant for the average of random impurities (Madelung, 1978). Hence, we see that τ is inversely proportional to the density of states. This is plausible because more states to scatter into implies a higher scattering rate and therefore a lower mobility. The constant of proportionality between τ and $N(E)$ can be found if the mobility $\mu(E) = e\tau(E)/m$ is known for a given energy E_o (or equivalent density n_o). Then,

$$\tau(E) = \frac{m\mu(E_o)}{e} \frac{N(E_o)}{N(E)} = \tau_o \frac{N(E_o)}{N(E)}. \tag{20}$$

Equations 15, 16, 18, and 19 are the central results of the convolution method needed to calculate the diffusive conductivity for a given device. We will now use these results to find the conductivity for a few devices. The first example will be for a one dimensional (1D) conductor. We will then show how the conductivity for a

Q1D device and the LSSL device can be calculated by means of yet another simple convolution.

As the first example of using the convolution method, we consider a 1D conductor in which the electrons are free to move in the y-direction. The energy-momentum dispersion relationship is

$$E_{1D}^y = \frac{\hbar^2 k_y^2}{2m}.$$

(21)

The group velocity is then

$$v^y(E) = \frac{1}{\hbar} \frac{\partial}{\partial k_y} E_{1D}^y = \sqrt{\frac{2E}{m}}.$$

(22)

The density of states for the electrons (including both spin states) is the usual 1D result (Kittel, 1986; Ashcroft and Mermin, 1976),

$$N_{1D}^y(E) = \sqrt{\frac{2m}{\pi^2 \hbar^2 E}},$$

(23)

which integrates to give the density

$$n_{1D} = \sqrt{\frac{8mE}{\pi^2 \hbar^2}}.$$

(24)

The conductivity is non-zero only in the y-direction so that

$$\sigma_{1D}^{yy} = e^2 v_y^2(E) \tau(E) N_{1D}^y(E).$$

(25)

These zero temperature results are displayed in the first column of Figure 5 where the scattering time has been assumed to be a constant for all energies. The density of states $N_{1D}^y(E)$ displays the inverse square root singularity in E, while the density $n_{1D}(E)$ has the resulting $\sqrt{E}$ dependence. The conductivity $\sigma_{1D}^{yy}(E)$ also has a $\sqrt{E}$ dependence because the energy dependence of the density of states cancels the energy dependence of one factor of velocity in Equation 25. Therefore, the conductivity for a constant scattering time has the same energy dependence as the velocity. Conductivity versus density is obtained by parametrically combining the conductivity-vs-energy and the density-vs-energy plots. This combination results in a conductivity versus density is that is linear, as shown in the bottom plot in the first column. This reflects the fact that $\sigma = en\mu$ and the mobility is a constant. This result also holds at non-zero temperatures.

The importance of taking τ to be inversely proportional to the density of states as in Equation 20 is illustrated in Figure 6. For zero temperature and for no impurity broadening, the density of states and the density are the same as in Figure 5. The linear increase of σ_{1D}^{yy} with energy is a direct result of τ being inversely proportional to $N_{1D}^y(E)$ so that these two factors cancel in Equation 25. The conductivity then has the energy dependence of v_y^2 which is proportional to E. Therefore, since n_{1D} depends on the square root of energy, the conductivity depends on the square of the density as shown in the bottom figure in the first column. The effect of temperature broadening is included by simply convolving the zero temperature results for $N_{1D}^y(E)$ and $\sigma_{1D}^{yy}(E)$ with $f'(E,T)$. These results are shown as the dashed lines in Figure 6. Although we have only shown the results for temperature broadening, the

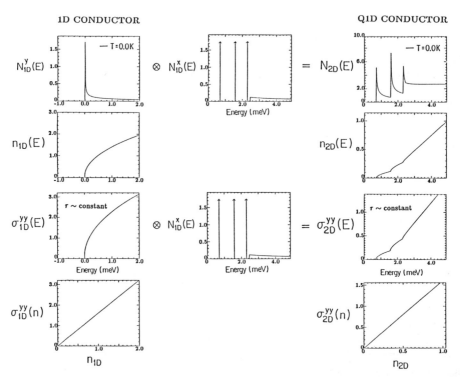

Figure 5. Calculated diffusive conductance for 1D and Q1D wires. The first column shows the density of states, density, and conductivity as a function of energy and the conductivity versus density for a 1D conductor where the electrons are free to move in the y-direction and τ is taken to be a constant. The second column is the density of states in the x-direction for a Q1D conductor. The third column shows the full results for the Q1D conductor. The symbol $\otimes$ denotes a convolution in energy. Units are arbitrary.

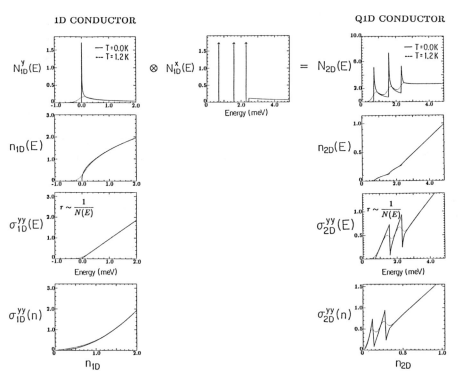

Figure 6. Calculated diffusive conductance of 1D and Q1D wires as Figure 5, but with the scattering time τ *taken to be inversely proportional to* $N_{2D}(E)$. The solid curves are for $T = 0$, and the dashed curves are for $T = 1.2\,\mathrm{K}$. Note that $\sigma_{2D}^{yy}(E) \neq \sigma_{1D}^{yy}(E) \otimes \frac{1}{2} N_{1D}^{x}(E)$ since the scattering time is now a function of energy.

results for impurity broadening are similar because the resulting convolutions will give qualitatively similar plots.

To find the conductivity for the Q1D and grating-gate LSSL devices (but not the grid-gate), the two dimensional problem must be solved. In both of these devices the total potential is of the separable form

$$V(x, y) = V_x(x) + V_y(y).$$ (26)

With this potential the total energy E of the 2D system is

$$E = E_{1D}^x + E_{1D}^y$$ (27)

where E_{1D}^x and E_{1D}^y are the eigenenergies for the corresponding 1D problems with potential $V_x(x)$ and $V_y(y)$ respectively. Let $N_{1D}^x(E^x)$ and $N_{1D}^y(E^y)$ be the corresponding density of states for the 1D problem. Then the total density of states $N_{2D}(E)$ for E consistent with Equation 27 is simply

$$N_{2D}(E) = \frac{1}{2} \int_{-\infty}^{\infty} N_{1D}^x(E_x) N_{1D}^y(E - E_x) \, dE_x.$$ (28)

The factor of $1/2$ compensates for the overcounting due to the spin states in the 1D densities of states. Equation 28 is a convolution so that it can be written as

$$N_{2D}(E) = N_{1D}^y(E) \otimes \frac{1}{2} N_{1D}^x(E).$$ (29)

Figure 7a shows the diagram for this convolution.

The convolutions for the conductivity are not as straightforward as for the density of states. This is because the scattering time depends on the total density of states. Nevertheless, the components of the conductivity tensor in 2D for a separable potential can be written as (Bagwell et al., 1989; Bagwell and Orlando, 1989b)

$$\frac{\sigma_{2D}^{yy}(E)}{\tau_{2D}(E)} = \frac{\sigma_{1D}^{yy}(E)}{\tau_{1D}(E)} \otimes \frac{1}{2} N_{1D}^x(E)$$ (30)

and

$$\frac{\sigma_{2D}^{xx}(E)}{\tau_{2D}(E)} = \frac{\sigma_{1D}^{xx}(E)}{\tau_{1D}(E)} \otimes \frac{1}{2} N_{1D}^y(E)$$ (31)

and

$$\sigma_{2D}^{xy}(E) = \sigma_{2D}^{yx}(E) = 0.$$ (32)

A convenient algorithm for doing the convolutions for the conductivity is to first do the problem for the case when the scattering times are constant and equal. Then Equations 30 and 31 are simply convolutions with the 1D conductivities. That result can be multiplied by the energy dependent scattering time (which is proportional to the inverse of the density of states) to give the conductivity. Figure 7b shows the diagram for the convolution in Equation 30 and Equation 31.

To see how the convolutions for going from 1D to 2D devices work, consider the case of the Q1D conductors shown in Figure 1. Each one of the 1D parallel conductors can be considered to have a number of subbands at discrete energy levels. For simplicity we take the number of subbands to be three. The corresponding density of states $N_{1D}^x(E)$ describing the subbands in the x-direction is shown in the second

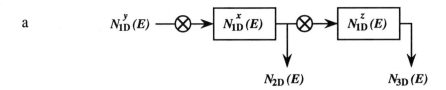

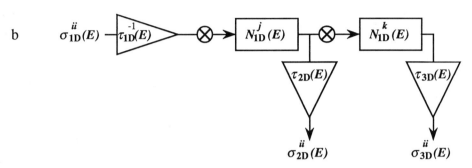

Figure 7. The diagrams for calculating (a) the density of states and (b) the conductivity when the potential is separable. These diagrams assume zero temperature and no impurity broadening. The symbol $\otimes$ denotes a convolution in energy.

column of Figure 5. The three delta functions correspond to the three energy levels of the subbands for a model potential. We have also included a portion of the 1D density of states going as $E^{-1/2}$ to describe the density of states when the electron's energy exceeds the confining potential, since at those high energies the electron is virtually free. Of course, this is only a model density of states. Calculations based on solving Schrödinger's equation for the confining potential give similar results (Bagwell and Orlando, 1989b). The density of states N_{2D} is calculated by convolving the two 1D density of states as is shown in the top row of Figure 5. The density $n_{2D}(E)$ is just the integral of the density of states. The conductivity $\sigma_{2D}^{yy}(E)$ is a convolution of the 1D conductivity with the density of states if the scattering time is taken to be a constant. The result of that convolution is shown in the third row of Figure 5. Note as a function of density that $\sigma_{2D}^{yy}(n)$ is linear even though the density of states and the conductivity both have structure in energy. This must be so for a constant scattering time, since the conductivity is given by $\sigma = en\mu$ and μ is a constant for a constant τ.

Taking $\tau(E)$ to be inversely proportional to $N_{2D}(E)$ gives $\sigma_{2D}^{yy}(E)$ as shown in Figure 6. $\sigma_{2D}^{yy}(E)$ in Figure 6 follows from multiplying $\sigma_{2D}^{yy}(E)$ in Figure 5 by the ratio $\tau(E)/\tau_o$ from Equation 20, where τ_o is the constant scattering time in Figure 5. Now when we plot $\sigma_{2D}^{yy}(E)$ versus density in Figure 6 the structure remains. The conductivity for a Q1D device is a minimum each time a new subband begins to be filled, because the scattering time (and mobility) is a minimum there due to the sudden increase in the density of states. Hence, we see that it is the modulation of the scattering time (mobility) that gives rise to the structure in the Q1D device.

By including temperature broadening and impurity broadening, we have been able to model the conductivity of the Q1D device shown in Figure 1. The model calculation semi-quantitatively reproduces the features in the measured conductance.

The convolution method can also be applied to find σ_{2D}^{xx}. If we consider a periodic array of wires, then this configuration is equivalent to finding the conductivity for the grating-gate LSSL. The model for this device has been discussed in (Bagwell and Orlando, 1989b).

3.2 Ballistic transport

In the ballistic limit of transport where $L_\phi \gg \ell \gg L$, the conductivity can be calculated using Landauer's formula. Landauer's formula connects the current in the device to the quantum mechanical transmission coefficient $T(E, V)$ through the device (Landauer, 1957; Landauer, 1970). In 1D the current flowing in the y-direction from a left contact, which is at a voltage V greater than the right contact, is given by (Wolf, 1985)

$$I_{1D}(E, V, T) = e \int v_{1D}^+(E') T_{1D}(E', V) N_{1D}^+(E')$$
$$[f(E' - E, T) - f(E' - (E - eV), T)] \, dE'. \tag{33}$$

Here $v_{1D}^+(E)$ is the group velocity for electrons moving in the positive y-direction, $N_{1D}^+(E)$ the density of states for electrons of both spins moving in the positive y-direction in a one dimensional free electron gas, $f(E, T)$ is the Fermi-Dirac distribution function, T the temperature of both contacts, and $T_{1D}(E, V)$ the transmission coefficient. Here the energy E is measured from the bottom of the band of electrons in the left contact.

The product of the group velocity and the electron density of states in one dimension is a constant given by

$$v_{1D}^+(E) N_{1D}^+(E) = \frac{1}{\pi \hbar}. \tag{34}$$

Note that this is a half of what Equations 22 and 23 imply since we are only interested in the electrons which are traveling in the positive direction, which is half the total density of states at a given energy.

The Fermi-Dirac function can also be expressed as the convolution

$$f(E, T) = [1 - \theta(E)] \otimes f'(E, T) \tag{35}$$

where θ is the unit step function. Equation 33 can therefore be rewritten as (Bagwell and Orlando, 1989a)

$$I_{1D}(E, V, T) = \frac{e}{\pi \hbar} T_{1D}(E, V) \otimes W(E, V) \otimes f'(E, T). \tag{36}$$

Here

$$W(E, V) = [\theta(E) - \theta(E - eV)] \tag{37}$$

is the voltage broadening function.[‡] Figure 8 depicts the convolution method for the current in 1D as given by Equation 36. Figure 8 is valid in the limit of ballistic transport.

[‡]If the applied voltage is greater than the Fermi energy then $W(E, V)$ must be cut off at the Fermi energy, namely

$$W(E, V) = \begin{cases} [\theta(E) - \theta(E - eV)] & eV < \mu \\ [\theta(E) - \theta(E - \mu)] & eV \geq \mu \end{cases}. \tag{38}$$

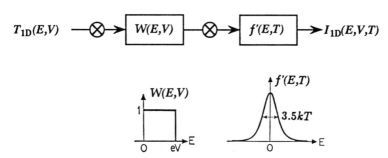

Figure 8. The convolution method for the current in 1D ballistic transport. The thermal broadening function $f'(E,T)$ and the voltage broadening function $W(E,V)$ are shown graphically along with their widths.

As an example of the convolution method, consider a ballistic Q1D conductor of uniform width. If E_1 is the energy of the first subband, then, neglecting all the higher subbands, the Q1D conductor has unity transmission coefficient for all energies greater than E_1 such that

$$T_{1D}(E,V) = \theta(E - E_1).$$
(39)

For small voltages the zero temperature current becomes

$$I_{1D}(E,V) = \frac{e^2}{\pi\hbar}V\theta(E - E_1),$$
(40)

where structures of order eV have been neglected. Hence, we see that a Q1D conductor with perfect transmission is characterized by a resistance $\pi\hbar/e^2$ for all energies greater than E_1. This is the quantum contact resistance (van Wees et al., 1988; Wharam et al., 1988). If there are two 1D subbands at E_1 and E_2 in a Q1D device, then for perfect transmission the contact conductance would be $e^2/\pi\hbar$ for each subband. The convolution broadening of Equation 40 with temperature and voltage, similar to Figure 9 below, is shown in Bagwell and Orlando (1989a).

Next, consider the case where the 1D subbands are formed in a constriction between two wider regions. This geometry describes the experiments of van Wees et al. (1988) and Wharam et al. (1988). For this case, there is an electrostatic potential difference between the wider regions and the constriction (Payne, 1989). As the applied voltage increases, the 1D subbands formed in the constriction shift to a lower energy with respect to the incident electron distribution in the wide regions. The transmission coefficient becomes approximately

$$T_{1D}(E,V) = \theta(E - E_1 + meV),$$
(41)

where $m = 1/2$ for the case of a symmetrical constriction. In general m is a phenomenological parameter where $0 \leq m \leq 1$ (Kouwenhoven et al., 1989). For two Q1D subbands at E_1 and E_2, the conductance when V is small is given by the solid line in Figure 9. We consider a symmetrical constriction so that, when the applied voltage is not negligible, the broadening of the conductance $G = I/V$ due to finite voltage is shown by the dashed line in Figure 9. The effect of temperature broadening

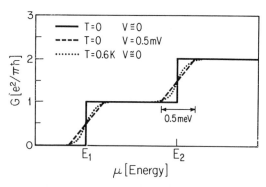

Figure 9. The ballistic conductance versus emitter Fermi energy for two 1D subbands at E_1 and E_2 formed in a constriction between two wider regions. The two subbands have perfect transmission as shown by the solid line. The energy of the left contact is assumed to be at a voltage V above the right contact. The dashed curved shows the broadening due to voltage and the dotted curve the broadening due to temperature following Bagwell and Orlando (1989a).

is also shown in the figure. Note that the shapes of the two broadenings are qualitatively different. Results similar to Figure 9 have also been obtained by Glazman and Khaetskii (1988, 1989). The nonlinear conductance of these ballistic devices has been measured and analyzed (Kouwenhoven et al., 1989). In addition, Landauer (1987, 1989) has addressed many fundamental issues relating to the conductance of these constrictions.

Many types of ballistic devices can be made in two and three dimensions. Again, if the transmission coefficient is known for the device, then the Landauer formula can be used to calculate the conductance. Unfortunately, the Landauer formula for higher dimensions, known as the multi-channel Landauer formula (Büttiker et al., 1985; Fischer and Lee, 1981), is in general quite complicated to evaluate. However, for separable potentials of the form given by Equation 26, the multichannel formula is equivalent to adding in parallel all the possible channels opened up by the second dimension. This parallel combination is equivalent to a convolution with the free electron density of states and hence, is similar to the diffusive result of Equations 29 and 30.

If the 1D transmission coefficient is known, then the convolution method will give the current for a 1D conductor. For a 2D conductor described by a separable potential as in Equation 26, the current density can be written as (Bagwell and Orlando, 1989a)

$$J_{2D}(E, V, T) = I_{1D}(E, V, T) \otimes \frac{1}{2} N_{1D}^z(E), \qquad (42)$$

where $I_{1D}(E, V, T)$ is known from Equation 36. Likewise, the current density in 3D can be written for a separable potential in 3D as (Bagwell and Orlando, 1989a)

$$J_{3D}(E, V, T) = I_{1D}(E, V, T) \otimes \frac{1}{2} N_{1D}^z(E) \otimes \frac{1}{2} N_{1D}^z(E). \tag{43}$$

These two equations have been used to correctly give the *I-V* characteristics for 3D resonant tunneling devices as well as for 2D resonant tunneling devices.

We have restricted our calculations to only the diffusive and ballistic limits. In the region between these two limits the effects of inelastic and elastic scattering must be treated at the same time. At present this has been done for a few isolated examples because of the complexity of the general problem. However, a rough estimate of how the transport is affected by energy level broadening in between these two limits can be given as follows: Divide the sample into blocks which have a linear dimension of the average L_ϕ. The conductivity can be calculated in each of these blocks by considering the transport in each block to be ballistic. The conductances of each block can then be added up as a network of resistors (Büttiker, 1988). The broadening due to temperature would be the same for each block. Hence, if there was structure on the order of ΔE in each block, then the temperature of about $\Delta E = 3.5kT$ would wash the structure away. However, the voltage broadening would only be due to that part of the voltage that was dropped across each block. Therefore, the voltage broadening would be much less if there were more inelastic lengths in the sample. For example, a 1D conductor which has $L = mL_\phi$ has a voltage broadening of width eV/m where V is the total voltage drop across the sample. Therefore, a voltage such that $eV = m\Delta E$ is necessary to wash away structure of the order ΔE. This dependence of voltage broadening on the number of inelastic lengths has been observed in grid gate LSSL devices and has been used to find L_ϕ (Ismail, 1989) for devices where m is small. As the number of inelastic blocks gets large, we expect the result from adding up the ballistic resistances of the blocks to merge into the result calculated from the diffusive limit when $L \gg L_\phi \gg \ell$.

4 Summary

In this paper we have shown how the convolution method provides a convenient framework for understanding and calculating the conductivity in the diffusive and ballistic regimes of transport. In the diffusive regime the effects of temperature and impurity broadening can be written as convolutions. In the ballistic regime the effects of temperature and finite voltage can also be written as convolutions. We have also seen that for separable potentials, which describe many but not all devices, that the density of states and conductivity for 2D and 3D structures can be written as simple convolutions with the results for the 1D structure. Hence, we have used the convolution method in two ways: (1) to include physical broadening effects, and (2) to include dimensional effects.

Acknowledgements

We thank D. A. Antoniadis and Henry I. Smith for useful discussions. We gratefully acknowledge financial support from an IBM fellowship, an Office of Naval Research Fellowship, an MIT-EECS Departmental special projects grant, and the support of U. S. Air Force contract AFOSR-88-0304. We also acknowledge partial support from Joint Services Electronics Program contract DAALO3-89-C-0001.

References

Abrikosov, A. A., Gorkov, L. P., and Dzyaloshinski, I. E., 1963, "Methods of Quantum Field Theory in Statistical Physics," Prentice-Hall, Englewood Cliffs, N.J.

Al'tshuler, B. L., 1985, Fluctuations in the Extrinsic Conductivity of Disordered Conductors, *Sov. Phys. JETP Lett.*, 41:648.

Antoniadis, D. A., Warren, A. C., and Smith, H. I., 1985, Quantum Mechanical Effects in Very Short and Very Narrow Channel MOSFETs, *IEDM Tech. Dig.*, 562.

Ashcroft, N. W. and Mermin, N. D., 1976, "Solid State Physics," Holt, Rinehart, and Winston, New York.

Bagwell, P. F., 1988, "Quantum Mechanical Transport Phenomenon in Nano-structured Inversion Layers," S. M. Thesis, MIT.

Bagwell, P. F., Antoniadis, D. A., and Orlando, T. P., 1989, Quantum Mechanical and Non-Stationary Transport Phenomenon in Nanostructured Silicon Inversion Layers, *in*: "Advanced MOS Device Physics," N. Einspruch and G. Gildenblat, ed., Academic Press, San Diego.

Bagwell, Phillip F., and Orlando, Terry P., 1989a, Landauer's Conductance Formula and its Generalization to Finite Voltages, *Phys. Rev. B*, 40:1456.

Bagwell, P. F., and Orlando, T. P., 1989b, Broadened Conductivity Tensor and Density of States for a Superlattice Potential in One, Two, and Three Dimensions, *Phys. Rev. B*, 40:3735.

Bernstein, G., and Ferry, D. K., 1987, Negative Differential Conductivity in Lateral Surface Superlattices, *J. Vac. Sci. Technol. B*, 5:964.

Büttiker, M., Imry, Y., Landauer, R., and Pinhas, S., 1985, Generalized Many-Channel Conductance Formula with Application to Small Rings, *Phys. Rev. B*, 31:6207.

Büttiker, M., 1986, Role of Quantum Coherence in Resistors, *Phys. Rev. B*, 33:3020. See also Büttiker, M., 1988, Coherent and Sequential Tuneling in Series Barriers, *IBM J. Res. Dev.*, 32:63.

Fischer, Daniel S., and Lee, Patrick A., 1981, Relation Between Conductivity and Transmission Matrix, *Phys. Rev. B*, 23:6851.

Glazman, L. I., and Khaetskii, A. V., 1988, Nonlinear Quantum Conductance of a Point Contact, *JETP Lett.*, 48:591.

Glazman, L. I., and Khaetskii, A. V., 1989, Nonlinear Quantum Conductance of a Lateral Microconstraint in a Heterostructure, *Europhys. Lett.*, 9:263.

Ismail, K., Chu, W., Antoniadis, D. A., and Smith, Henry I., 1988, Surface-Super-lattice Effects in a Grating-Gate GaAs/GaAlAs Modulation-Doped Field-Effect-Transistor, *Appl. Phys. Lett.*, 52:1071.

Ismail, K., Chu, W., Yen, A., Antoniadis, D. A., and Smith, Henry I., 1989a, Negative Transconductance and Negative Differential Resistance in a Grid-Gate Modulation Doped Field-Effect Transistor, *Appl. Phys. Lett.*, 54:460.

Ismail, K., Chu, W., Antoniadis, D. A., and Smith, Henry I., 1989b, One Dimensional Subbands and Mobility Modulation in GaAs/AlGaAs Quantum Wires, *Appl. Phys. Lett.*, 54:1130.

Ismail, Khalid, 1989, "The Study of Electron Transport in Field-Induced Quantum Wells on GaAs/AlAs", Ph. D. Thesis, MIT.

Kittel, C., 1986, "Introduction to Solid State Physics", Wiley, New York.

Kouwenhoven, L. P., van Wees, B. J., Harmans, C. J. P. M., Williamson, J. G., van Houten, H., Beenakker, C. W. J., Foxon, C. T., and Harris, J. J., 1989, Nonlinear Conductance of Quantum Point Contacts, *Phys. Rev. B.*, 39:8040.

Landauer, R., 1957, Spatial Variation of Currents and Fields Due to Localized Scatterers in Metallic Conduction, *IBM J. Res. Dev.*, 1:223.

Landauer, Rolf, 1970, Electrical Resistance of Disordered One Dimensional Lattices, *Phil. Mag.*, 21:683.

Landauer, Rolf, 1987, Electrical Transport in Open and Closed Systems, *Z. Phys. B.*, 68:217.

Landauer, Rolf, 1989, Conductance Determined by Transmission: Probes and Quantized Constriction Resistance, to appear in *J. Phys. Cond. Matt.*.

Lee, P. A., and Stone, A. D., 1985, Universal Conductance Fluctuations in Metals, *Phys. Rev. Lett.*, 55:1622.

Madelung, O., 1978, "Introduction to Solid State Theory," Springer-Verlag, New York.

Payne, M. C., 1989, Electrostatic and Electrochemical Potentials in Quantum Transport, *J. Phys. Cond. Matt.*, 1:4931.

Rickayzen, G., 1980, "Green's Functions and Condensed Matter," Academic Press, New York.

Scott-Thomas, J. H. F., Kastner, M. A., Antoniadis, D. A., Smith, Henry I., and Field, Stuart, 1988, Si MOSFETs with 70nm Slotted Gates for Study of Quasi One Dimensional Quantum Transport, *J. Vac. Sci. Technol. B*, 6:1841.

Skocpol, W. J., Mankiewich, P. M., Howard, R. E., Jackel, L. D., Tennant, D. M., and Stone, A. D., 1986, Universal Conductance Fluctuations in Silicon Inversion Layer Nanostructures, *Phys. Rev. Lett.*, 56:2865.

Tokura, Y., and Tsubaki, K., 1987, Conductivity Oscillation due to Quantum Interference in a Proposed Washboard Transistor, *Appl. Phys. Lett.*, 51:1807.

van Wees, B. J., van Houten, H., Beenakker, C. W. J., Williamson, J. G., Kouwenhoven, L. P., van der Marel, D., and Foxon, C. T., 1988, Quantized Conductance of Point Contacts in a Two Dimensional Electron Gas, *Phys. Rev. Lett.*, 60:848.

Warren, A. C., Antoniadis, D. A., Smith, H. I., and Melngailis, J., 1985, Surface Superlattice Formation in Silicon Inversion Layers Using 0.2-μm Period Grating-Gate Electrodes, *IEEE Electron Dev. Lett.*, EDL-6:294.

Warren, A. C., Antoniadis, D. A., and Smith, H. I., 1986, Quasi One-Dimensional Conduction in Multiple, Parallel Inversion Lines, *Phys. Rev. Lett.*, 56:1858.

Wharam, D. A., Thornton, T. J., Newbury, R., Pepper, M., Ahmed, H., Frost, J. E. F., Hasko, D. G., Peacock, D. C., Ritchie, D. A., and Jones, G. A. C., 1988, One Dimensional Transport and the Quantization of the Ballistic Resistance, *J. Phys. C: Solid State Phys.*, 21:L209.

Wolf, E. L., 1985, "Principles of Electron Tunnelling Spectroscopy", Oxford University Press, New York. See equations (2.5) and (2.6).

QUANTUM TRANSPORT AND DYNAMICS FOR BLOCH ELECTRONS IN ELECTRIC FIELDS

Gerald J. Iafrate

U.S. Army Electronics Technology and Devices Laboratory
Fort Monmouth, New Jersey 07703-5000

Joseph B. Krieger and Yan Li

City University of New York-Brooklyn College
Brooklyn, New York 11210

ABSTRACT

A novel formalism for treating Bloch electron dynamics and quantum transport in inhomogeneous electric fields of arbitrary strength and time dependence is reviewed. In this formalism, the electric field is described through the use of the vector potential. This choice of gauge leads to a natural set of basis functions for describing Bloch electron dynamics; in addition, a basis set of localized, electric field-dependent Wannier functions is established and utilized to derive a quantum "Boltzmann equation" which includes explicit band-mixing transients such as effective mass dressing and Zener tunneling. The application of this formalism to quantum transport and Bloch oscillations in homogeneous electric fields is emphasized; also, issues relevant to applications concerning spatially localized inhomogeneous electric fields such as occur in problems involving tunneling through "band-enginereed" tunneling barriers and impurity scattering is discussed.

INTRODUCTION

During the past decade, as microelectronics technology has continued to pursue the scaling down of IC device dimensions into the submicron and ultrasubmicron regions, many new and interesting questions have emerged[1] concerning the solid-state dynamics and quantum transport of carriers in semiconductors subjected to rapidly varying, spatially inhomogeneous electric fields and non-steady state temporal conditions.

Electronic Properties of Multilayers and Low-Dimensional Semiconductors Structures
Edited by J. M. Chamberlain *et al.*, Plenum Press, New York, 1990

In this paper, we report a novel formalism[2] for treating Bloch electron dynamics and quantum transport in inhomogeneous electric fields of arbitrary strength and time dependence. In this formalism, the electric field is described through the use of the vector potential. In this regard, this work is an expansion of methodology previously developed by the authors[3,4,5] to describe solid-state dynamics and quantum transport for Bloch electrons in a homogeneous electric field of arbitrary strength and time dependence, including weak scattering from randomly distributed impurities and phonons.

The Hamiltonian for a single electron in a periodic crystal potential subject to a general inhomogeneous electric field of arbitrary time dependence and strength is

$$H = \frac{1}{2m} \left(\vec{P} - \frac{e}{c} \vec{A}(\vec{x}, t) \right)^2 + V_c(\vec{x}) \tag{1}$$

Here $V_c(\vec{x})$ is the periodic crystal potential, and $\vec{A}(\vec{x}, t)$ is the vector potential for the inhomogeneous field $\vec{E}(\vec{x}, t)$ where

$$\vec{A}(\vec{x}, t) = -c \int_{t_o}^{t} \vec{E}(\vec{x}, t') dt'.$$

In developing a quantum transport formalism, we start with the Liouville equation for the density matrix

$$i\hbar \frac{\partial \hat{\rho}}{\partial t} = [H, \hat{\rho}] \tag{2}$$

from which we seek to derive an equation for the distribution function, i.e., the matrix elements of $\hat{\rho}, \rho_{m'm}$, in a convenient representation so that we can establish expectation values of any operator B through

$$\bar{B} = trace(\hat{\rho} \hat{B}) = \sum_{m'} \sum_{m} \rho_{m'm} B_{mm'} \tag{3}$$

Specifically, in this paper, we will choose $\vec{B} \equiv \vec{v} = \frac{1}{i\hbar} [\vec{x}, H]$, the velocity operator, which, for the Hamiltonian of Eq(1) is

$$\vec{v} = \frac{1}{m} \left(\vec{P} - \frac{e}{c} \vec{A} \right) \tag{4}$$

In previous work[3-5], when considering Bloch dynamics in a spatially homogeneous electric field, the authors exploited, as a basis set, the instantaneous eigenstates of the Hamiltonian of Eq(1) with

$$\vec{A} \equiv \vec{A}_o(t) = -c \int_{t_o}^{t} \vec{E}_o(t') dt' \quad .$$ Explicitly, these basis states are

$$\phi_{n\vec{K}}(\vec{r},t) = \frac{e^{i\vec{K}\cdot\vec{r}}}{\Omega^{1/2}} U_{n\vec{k}(t)}(\vec{r}). \tag{5}$$

Here, $U_{n\vec{k}(t)}$ is the periodic part of the usual Bloch function with band index n and wavevector $\vec{k}$; $\vec{k}(t)$ is determined by

$$\vec{k}(t) = \vec{K} - \frac{e}{\hbar c}\vec{A}_o(t) = \vec{K} + \frac{e}{\hbar} \int_{t_o}^{t} \vec{E}_o(t') dt', \tag{6}$$

where $\vec{K}$ is a constant determined by periodic boundary conditions in a box of volume Ω. (In subsequent equations, the vector nature of $\vec{K}$ and $\vec{k}$ will be understood but not always indicated.)

In this work, we note that a set of basis functions for Bloch electron dynamics in an inhomogeneous electric field can also be established based on the vector potential choice of gauge. Letting $\vec{A}(\vec{x},t) = \vec{A}_o(t) + \vec{A}_1(\vec{x},t)$ so that $\vec{E}(\vec{x},t) = -\frac{1}{c}\partial\vec{A}/\partial t = \vec{E}_o(t) + \vec{E}_1(\vec{x},t)$, a convenient basis for describing transport in this field having a spatially homogeneous and inhomogeneous part, is

$$\psi_{n\vec{K}} = e^{i\frac{e}{\hbar c}\int_0^{\vec{x}} \vec{A}_1 \cdot d\vec{l}} \phi_{n\vec{K}} \tag{7}$$

where $\phi_{n\vec{K}}$ is given by Eq(5) and $\vec{K}$ is defined previously. Here, the external magnetic field, defined through $\vec{B} = \vec{\nabla} \times \vec{A}$, is zero so that the line integral in Eq. (7) is independent of the path. This basis set, $\{\psi_{n\vec{K}}\}$ is complete and orthonormal since each element is obtained from the corresponding element of the complete orthonormal set $\{\phi_{n\vec{K}}\}$ by the multiplication of a common temporal and spatially dependent phase factor which is independent of "n" and "$\vec{K}$"

Taking matrix elements of Eq(2), the Liouville equation in this representation is found to be

$$i\hbar\frac{\partial \rho_{m'm}}{\partial t} = (\epsilon_{m'} - \epsilon_m)\rho_{m'm} + \sum_{m''} (S_{m''m}\rho_{m'm''} - S_{m'm''}\rho_{m''m}) \tag{8}$$

where $m \equiv (n\vec{K})$ also $\epsilon_m \equiv \epsilon_n\left(\vec{K} - \frac{e}{\hbar c}\vec{A}_o\right)$ is the energy band function of the nth band, and $S_{m'm}$ is given by

$$S_{m'm} \equiv S_{n'\vec{K}'n\vec{K}} = e\vec{E}_o(t) \cdot \vec{R}_{n'n}(\vec{k})\delta_{\vec{K}'\vec{K}} - eV_{n'\vec{K}'n\vec{K}}. \tag{9}$$

In Eq(9),

$$\vec{R}_{n'n}(\vec{k}) = \frac{i}{\Omega} \int d\vec{x} \, U^*_{n'\vec{k}} \vec{\nabla}_k U_{n\vec{k}}$$ (10)

and

$$V_{n'\vec{K}'n\vec{K}} = \int \psi^*_{n'\vec{K}'} V \psi_{n\vec{K}} d\vec{x} \equiv \int \phi^*_{n'\vec{K}'} V \phi_{n\vec{K}} d\vec{x}$$ (11)

where $V(\vec{x}, t)$ is the potential derived from the inhomogeneous electric field through the relation

$$\vec{E}_1(\vec{x}, t) = -\vec{\nabla} V(\vec{x}, t).$$ (12)

The Liouville equation given by Eq(8) includes an exact description of the time evolution of the density matrix elements for a Bloch electron in a spatially homogeneous electric field; included explicitly are the effects of Zener tunneling, effective mass dressing, and scattering from the inhomogeneous field. The authors have solved this equation analytically under the conditions of short times (after turning on the electric field) and weak inhomogeneity[4].

Although the basis states of Eq(7), along with the resultant Liouville equation, Eq(8), provide general applicability for problems involving Bloch dynamics in inhomogeneous electric fields, it is particularly difficult to use in situations where the inhomogeniety is localized and strong such as in the case of impurities, heterojunctions, or band-engineered quantum wells and barriers; this difficulty arises mainly due to the extended nature of the basis states. Therefore, we use the basis defined in Eq(7), along with the equivalent Wannier representation, to develop quantum transport equations for such local inhomogeneities[6,7].

DYNAMICAL WANNIER REPRESENTATION: QUANTUM TRANSPORT

In the Wannier representation[8-9], the Bloch functions are equivalently expressed as a Fourier decomposition of localized functions, the Wannier functions, defined on each periodic lattice site, $\vec{l}$, of the crystal. Wannier functions can also be defined for Bloch electrons in a spatially homogeneous electric field; as such, it can be shown [10] that

$$\phi_{n\vec{K}} = \frac{1}{\sqrt{N}} \sum_{\vec{l}} e^{i\vec{K}\cdot\vec{l}} A_n(\vec{x} - \vec{l}, t)$$ (13)

where the $\vec{K}$'s are defined in Eq(6) and N is the number of lattice sites. With the use of the completeness relations

$$\sum_{\vec{K}} e^{i\vec{K}\cdot(\vec{l}-\vec{l}')} = N \; \delta_{\vec{l},\vec{l}'} \tag{14a}$$

and

$$\sum_{\vec{l}} e^{-i\vec{l}\cdot(\vec{K}-\vec{K}')} = N \; \delta_{\vec{K},\vec{K}'} , \tag{14b}$$

Eq(13) can be inverted so that

$$A_n(\vec{x}-\vec{l},t) = \frac{1}{\sqrt{N}} \sum_{\vec{K}} e^{-i\vec{K}\cdot\vec{l}} \phi_{n\vec{K}} . \tag{15}$$

Since $(\phi_{n'\vec{K}'}, \phi_{n\vec{K}}) = \delta_{n'n}\delta_{\vec{K}'\vec{K}}$, it immediately follows from Eq(15) that

$$\int d\vec{x} \, A^*{}_{n'}(\vec{x}-\vec{l}',t) A_n(\vec{x}-\vec{l},t) = \delta_{n'n}\delta_{\vec{l}'\vec{l}} .$$

It is useful to note that

$$A_n(\vec{x}-\vec{l},t) = e^{i\frac{e}{\hbar c}\vec{A}_o\cdot(\vec{x}-\vec{l})} A_n^o(\vec{x}-\vec{l}),$$

where $A_n^o(\vec{x}-\vec{l})$ is the usual time-independent Wannier function.

The localized functions in Eq(15) are the instantaneous Wannier functions for Bloch electrons in a spatially homogeneous electric field. As such, it can be shown that

$$H_o A_n(\vec{x}-\vec{l},t) = \sum_{\vec{l}'} \epsilon_n(\vec{l}-\vec{l}',t) A_n(\vec{x}-\vec{l}',t) \tag{16}$$

where

$$\epsilon_n(\vec{l}-\vec{l}',t) = \frac{1}{N}\sum_{\vec{K}} e^{-i\vec{K}\cdot(\vec{l}-\vec{l}')} \epsilon_n\left(\vec{K}-\frac{e}{\hbar c}\vec{A}_o\right). \tag{17}$$

Here H_o is the Hamiltonian of Eq(1) with $\vec{A} = \vec{A}_o(t)$ and $\epsilon_n(\vec{K})$ is the nth Bloch energy band with crystal momentum $\vec{K}$. (In subsequent equations, the explicit time dependence will be assumed but not indicated).

A set of instantaneous Wannier functions for the inhomogeneous electric field can also be established. Using the basis set of Eq(7) for the inhomogeneous field and the Wannier expansion of Eq(13), we can identify these localized functions for the inhomogeneous electric field as

$$W_n(\vec{x},\vec{l},t) = \frac{1}{\sqrt{N}} \sum_{\vec{K}} e^{-i\vec{K}\cdot\vec{l}} \psi_{n\vec{K}} \equiv e^{i\frac{e}{\hbar c}\int_o^{\vec{x}} \vec{A}_1 \cdot d\vec{l}} A_n(\vec{x}-\vec{l},t) \qquad (18)$$

where

$$\int d\vec{x} W^*_{n'}(\vec{x},\vec{l}') W_n(\vec{x},\vec{l}) = \delta_{n'n}\delta_{\vec{l}',\vec{l}} \qquad (19)$$

Taking the matrix elements of Eq(2), the Liouville equation in the localized Wannier representation is

$$i\hbar \frac{\partial \rho_{n'n}(\vec{l}'l,t)}{\partial t} = \sum_{n''l''} \{[\epsilon_{n''}(\vec{l}'-\vec{l}'')\delta_{n'n''} - e\vec{E}_o\cdot\vec{\Delta}_{n'n''}(\vec{l}'-\vec{l}'') + eV_{n'n''}(\vec{l}',\vec{l}'')]\rho_{n''n}(\vec{l}'',l,t)$$

$$- [\epsilon_{n''}(\vec{l}''-l)\delta_{nn''} - e\vec{E}_o\cdot\vec{\Delta}_{n''n}(\vec{l}''-l) + eV_{n''n}(\vec{l}'',l)]\rho_{n'n''}(\vec{l}',\vec{l}'',t)\} \qquad (20)$$

where $\epsilon_n(\vec{l}-\vec{l}')$ is defined in Eq(17); also

$$\vec{\Delta}_{nn'}(\vec{l}-\vec{l}') = \frac{1}{N}\sum_{\vec{K}} e^{-i\vec{K}\cdot(\vec{l}-\vec{l}')} \vec{R}_{nn'}\left(\vec{K} - \frac{e}{\hbar c}\vec{A}_o\right) \qquad (21)$$

where $\vec{R}_{nn'}$ is defined in Eq(10), and

$$V_{n'n}(\vec{l}'l) = \int d\vec{x} W^*_{n'}(\vec{x},\vec{l}')V(\vec{x},t)W_n(\vec{x},\vec{l}) = \int d\vec{x} A^*_{n'}(\vec{x}-\vec{l}')V(\vec{x},t)A_n(\vec{x}-\vec{l}), \qquad (22)$$

the matrix elements of the inhomogeneous potential with respect to the localized basis.

It is interesting to note that the solution to Eq(20) is amenable to separation of variables[11] in the pure-state situation; seeking solutions to Eq(20) of the form,

$$\rho_{n'n}(\vec{l}',l,t) \equiv f^*_n(l,t)f_{n'}(\vec{l}',t) \qquad (23)$$

we find that $f_n(l,t)$ obeys the equation

$$i\hbar\frac{\partial f_n(l,t)}{\partial t} = \sum_{n''}\sum_{l''} [\epsilon_{n''}(\vec{l}''-l,t)\delta_{nn''} - e\vec{E}_o\cdot\vec{\Delta}_{nn''}(\vec{l}''-l,t) \qquad (24)$$

$$+ eV_{nn''}(l,l'',t)]f_{n''}(\vec{l}'',t).$$

216

Using a simple generalization of the well-known Wannier theorem[12],

$$\sum_{\vec{l}''} G_{n''}(\vec{l}''-\vec{l}) f_{n''}(\vec{l}'',t) = G_{n''}\left(-i\vec{\nabla}-\frac{e}{\hbar c}\vec{A}_0\right) f_{n''}(\vec{r})|_{\vec{r}=\vec{l}} \qquad (25)$$

where G is assumed to be an analytic function of $\vec{K}$, Eq(24) reduces to the differential equation

$$i\hbar\frac{\partial f_n(\vec{r},t)}{\partial t} = \epsilon_n\left(-i\vec{\nabla}-\frac{e}{\hbar c}\vec{A}_0\right) f_n(\vec{r},t) - e\vec{E}_0 \cdot \sum_{n''} \vec{R}_{nn''}\left(-i\vec{\nabla}-\frac{e}{\hbar c}\vec{A}_0\right) f_{n''}(\vec{r},t)$$

$$+ \sum_{n''} \sum_{\vec{l}''} eV_{nn''}(\vec{l}'',\vec{r}) f_{n''}(\vec{l}'',t). \qquad (26)$$

We point out that Eqs.(20,24,26) are identical to the result one would obtain by describing the <u>inhomogeneous</u> field by the scalar potential from the outset, and using, as a basis, the dynamical Wannier functions of Eq.(15) corresponding to the homogeneous field alone.

GENERAL DISCUSSION

For a general non-parabolic band structure and an inhomogeneous electric field, Eq(26) represents a set of coupled differential equations for the envelope functions $\{f_n\}$. We have studied the solutions to Eq(26) extensively and will report[10] detailed analysis elsewhere. However, to show the utility of the method in this concise report, we look at specific solutions to Eq(26) under assumption that interband mixing from both the homogeneous and inhomogeneous electric fields are negligible; then Eq(26) reduces to the single band equation

$$i\hbar\frac{\partial f_n(\vec{r},t)}{\partial t} = \epsilon_n\left(-i\vec{\nabla}-\frac{e}{\hbar c}\vec{A}_0\right) f_n(\vec{r},t) + \sum_{\vec{l}''} eV_{nn}(\vec{l}'',\vec{r}) f_n(\vec{l}'',t) \qquad (27)$$

Taking a trivial example to show the nature of the solution to Eq(27), consider $V_{nn}(\vec{l}'',\vec{r})=0$ for all $\vec{l}$; then

$$f_n(\vec{r},t) = \sum_{\vec{K}} e^{-\frac{i}{\hbar}\int_o^t \epsilon_n\left(\vec{K}-\frac{e}{\hbar c}\vec{A}_0\right)dt'} A_n(\vec{K},t) \qquad (28a)$$

where

$$A_n(\vec{K},t) = \frac{1}{\sqrt{N}} e^{i\vec{K}\cdot\vec{r}} \qquad (28b)$$

In a more significant application, if we consider the potential assumed by Koster and Slater[6] to describe a single impurity at the lattice site $\vec{l}_0$; namely

$$V_{n''n}(\vec{l}'',\vec{r}) = V_0 \delta_{i'',i_0} \delta_{\vec{r},i_0} \delta_{n''n} ,$$

then

$$f_n(\vec{r},t) = \sum_{\vec{K}} e^{-\frac{i}{\hbar}\int_0^t \epsilon_n\left(\vec{K}-\frac{e}{\hbar c}\vec{A}_0\right)dt'} \; e^{i\vec{K}\cdot(\vec{r}-\vec{l}_0)} \; A_n(\vec{K},t) \tag{29a}$$

where $A_n(\vec{K},t)$ satisfies

$$i\hbar \frac{\partial A_n(\vec{K},t)}{\partial t} = \frac{eV_0}{N} \sum_{\vec{K}'} A_n(\vec{K}',t) \; e^{\frac{i}{\hbar}\int_0^t \left[\epsilon_n\left(\vec{K}-\frac{e}{\hbar c}\vec{A}_0\right)-\epsilon_n\left(\vec{K}'-\frac{e}{\hbar c}\vec{A}_0\right)\right]dt'} \tag{29b}$$

In situations where the inhomogeneity is a positive or negative constant value over several contiguous lattice sites, we have the possibility of tunneling or scattering, depending upon the magnitude of the inhomogeneity relative to the incident energy of the particle; in this case[10] $f_n(\vec{r},t)$ will consist of the appropriate linear combination of plane waves or exponential functions.

Once $f_n(\vec{r},t)$ has been determined, the density matrix elements are then calculated from Eq(23) and the current density can be calculated with the use of Eqs (3,4,23). Explicitly, using $W_n(\vec{x},\vec{l})$ from Eq(18), it can be shown that the matrix elements of the velocity defined in Eq(4) are

$$(\vec{v})_{m'm} = v_{n'n}(\vec{l}',\vec{l}) = \frac{1}{m_0}\left\{\int A^*_{n'}(\vec{x}-\vec{l}')\vec{P}A_n(\vec{x}-\vec{l})d\vec{x} - \frac{e}{c}\vec{A}_0 \delta_{n'n}\delta_{\vec{l}'\vec{l}}\right\}, \tag{30}$$

where $A_n(\vec{x}-\vec{l})$ is defined in Eq(15) [the instantaneous Wannier functions for the homogeneous field, $E_0(t)$], and

$$\int A^*_{n'}(\vec{x}-\vec{l})\frac{\left(\vec{P}-\frac{e}{c}\vec{A}_0\right)}{m_0}A_n(\vec{x}-\vec{l})dx = \frac{1}{N}\sum_{\vec{K}} e^{-i\vec{K}\cdot(\vec{l}-\vec{l}')}\{\vec{V}_n(\vec{K},t)\delta_{n'n} + \vec{g}_{n'n}(\vec{K},t)\}$$

with

$$\vec{V}_n(\vec{K}) = \frac{1}{\hbar}\vec{\nabla}_{\vec{K}}\epsilon_n\left(\vec{K}-\frac{e}{\hbar c}\vec{A}_0\right) \tag{31}$$

and

$$\vec{g}_{n'n}(\vec{K}) = \frac{i}{\hbar}\left[\epsilon_n\left(\vec{K}-\frac{e}{\hbar c}\vec{A}_o\right)-\epsilon_n\left(\vec{K}-\frac{e}{\hbar c}\vec{A}_o\right)\right]\vec{R}_{n'n}\left(\vec{K}-\frac{e}{\hbar c}\vec{A}_o\right); \qquad (32)$$

here, $\vec{R}_{n'n}$ is defined in Eq(10). It then follows from Eqs(3,23,30) that

$$\dot{\vec{v}} = \sum_{\vec{K}}\left[\sum_n |\phi_n(\vec{K})|^2 \dot{\vec{V}}_n(\vec{K}) + \sum_{n'}\sum_n \phi_{n'}^*(\vec{K})\phi_n(\vec{K})\vec{g}_{n'n}(\vec{K},t)\right] \qquad (33)$$

where

$$\phi_n(\vec{K}) = \frac{1}{\sqrt{N}}\sum_{\vec{l}} f_n(\vec{l})e^{-i\vec{K}\cdot\vec{l}} \qquad (34)$$

with $\dot{\vec{V}}_n$ and $\vec{g}_{n'n}$ defined in Eqs (31,32) respectively.

In this formulation of quantum transport based on dynamical Wannier functions, the terms $\epsilon_n(\vec{l}-\vec{l}',t)$ and $\vec{\Delta}_{nn'}(\vec{l}-\vec{l}',t)$ defined in Eqs. (17,21), respectively, appear in the fundamental density matrix equation of Eq. (20). Upon substitution of variables $\vec{\xi} = \vec{K} - \frac{e}{\hbar c}\vec{A}_o$, and subsequent sum over $\vec{\xi}$, it follows that these terms can be re-expressed as

$$\epsilon_n(\vec{l}-\vec{l}',t) = e^{i\frac{e\vec{F}_o}{\hbar}\cdot(\vec{l}-\vec{l}')t}\epsilon_n(\vec{l}-\vec{l}',o), \qquad (35)$$

with

$$\epsilon_n(\vec{l}-\vec{l}',o) = \frac{1}{N}\sum_{\vec{K}} e^{-i\vec{K}\cdot(\vec{l}-\vec{l}')}\epsilon_n(\vec{K}); \qquad (35a)$$

and

$$\vec{\Delta}_{nn'}(\vec{l}-\vec{l}',t) = e^{i\frac{e\vec{F}_o}{\hbar}\cdot(\vec{l}-\vec{l}')t}\vec{\Delta}_{nn'}(\vec{l}-\vec{l}',o), \qquad (36)$$

with

$$\vec{\Delta}_{nn'}(\vec{l}-\vec{l}',o) = \frac{1}{N}\sum_{\vec{K}} e^{-i\vec{K}\cdot(\vec{l}-\vec{l}')}\vec{R}_{nn'}(\vec{K}). \qquad (36a)$$

As seen from Eqs. (35,36), for strong electric fields, $\epsilon_n(\vec{l}-\vec{l}',t)$ and $\vec{\Delta}_{nn'}(\vec{l}-\vec{l}',t)$ oscillate rapidly with time for $\vec{l} \neq \vec{l}'$; thus, only terms with $\vec{l} = \vec{l}'$ contribute to the sum over $\vec{l}'$ in Eqs.(20,24) thereby leading to cell localization and hopping from cell to cell through Stark ladder transitions.

The manifestation of lattice site localization and Stark-ladder hopping is evident by considering the time evolution of an electron initially in the energy-band state "n_o" at lattice position "$\vec{l}_o$", and with no inhomogeneous field present; then the solution to Eq(24) [the appropriate equation for pure-state initial conditions], in the early-time approximation, is

$$f_n(\vec{l},t) = \delta_{n,n_o}\delta_{\vec{l},\vec{l}_o} + \frac{F_{nn_o}(\vec{l}_o - \vec{l})}{e\vec{E}_o \cdot (\vec{l}_o - \vec{l})}\left[1 - e^{i\frac{e}{\hbar}\vec{E}_o \cdot (\vec{l}_o - \vec{l})t}\right]$$

where

$$F_{nn_o}(\vec{l}_o - \vec{l}) = \epsilon_{n_o}(\vec{l}_o - \vec{l}, o)\delta_{nn_o} - e\vec{E}_o \cdot \vec{\Delta}_{nn_o}(\vec{l}_o - \vec{l}, o),$$

with $\epsilon_{n_o}(\vec{l}_o - \vec{l}, o)$ and $\vec{\Delta}_{nn_o}(\vec{l}_o - \vec{l}, o)$ given by Eqs(35a,36a) respectively.

In this case, the density matrix in Eq (23) can be derived; for the diagonal elements of the density matrix, we find that

$$\rho_{nn}(\vec{l},\vec{l}) = \delta_{n,n_o}\left[\delta_{\vec{l},\vec{l}_o} + |\epsilon_{n_o}(\vec{l}_o - \vec{l}, o)|^2 \alpha^2(t)\right] + e^2 E_o^2 |\Delta_{nn_o}(\vec{l}_o - \vec{l}, o)|^2 \alpha^2(t)$$

where

$$\alpha(t) = \frac{\mathrm{Sin}\{e\vec{E}_o \cdot (\vec{l}_o - \vec{l})t/2\hbar\}}{e\vec{E}_o \cdot (\vec{l}_o - \vec{l})/\hbar}$$

The first term describes hopping from the initial site "$\vec{l}_o$" to a subsequent site "$\vec{l}$" within the initial energy-band "n_o", whereas the second term describes hopping from site "$\vec{l}_o$" to site "$\vec{l}$" accompanied by Zener tunneling from the initial energy-band "n_o" to the energy-band "n". As well, for large electric fields, the oscillatory term, $\alpha(t)$, will lead to vanishing contributions to the density maxtrix elements unless $\vec{l} = \vec{l}_o$ which results in localization at the initial site.

A similar analysis can be performed with the inhomogeneous field included; a detailed discussion of this case will be considered elsewhere[10].

The time independent contributions to $\epsilon_n(\vec{l} - \vec{l}', t)$ and $\vec{\Delta}_{nn'}(\vec{l} - \vec{l}', t)$, as defined in Eqs. (35a, 36a), are band-structure dependent quantities dependent upon the specific energy bands of the periodic potential in question as well as the related interband coupling, respectively. For a given energy band $\epsilon_n(\vec{K}), \epsilon_n(\vec{l}-\vec{l}',0)$, defined in Eq (35a), is the Fourier transform of $\epsilon_n(\vec{K})$ with respect to the lattice sites of the crystal potential. As an example, using the familiar nearest neighbor tight-binding approximation

$$\epsilon_n(\vec{K}) = \epsilon_n(0) + W \operatorname{Sin}^2\frac{\vec{K}\cdot\vec{a}}{2}, \tag{37a}$$

with "$\vec{a}$" denoted as the basic lattice vector in a specific crystallographic direction and with $\epsilon_n(0)$ and W denoted as the energy-band minimum and width, respectively, $\epsilon_n(\vec{l}-\vec{l}',0)$ in Eq (35a) becomes

$$\epsilon_n(\vec{l}-\vec{l}',0) = \left[\epsilon_n(0) + \frac{1}{2}W\right]\delta_{\vec{l}\vec{l}'} - \frac{1}{4}W[\delta_{\vec{l}',\vec{l}+\vec{a}} + \delta_{\vec{l}',\vec{l}-\vec{a}}] \tag{37b}$$

which includes the salient features of the specific band description.

As well, an explicit form for $\vec{\Delta}_{nn'}(\vec{l}-\vec{l}',0)$ can be obtained in terms of band-structure parameters by employing the two-band model first developed by Kane[13], and later extended by Krieger[14]. Specifically, Kane shows that an explicit form for $R_{nn'}(\vec{K})$, defined in Eq.(10), can be written, for a specific direction, as

$$R_{nn'}(\vec{K}) = \frac{\frac{1}{2}q_{nn'}}{K^2 + q_{nn'}^2} \tag{38}$$

with $q_{nn'} = \frac{1}{4}\frac{m}{\hbar}E_g/P_{nn'}(0)$; here E_g is the energy gap, and $P_{nn'}(0)$ is the momentum matrix element, between the nth and n'^{th} bands at the assumed band extrema, $\vec{K} = 0$. Utilizing $\vec{R}_{nn'}(\vec{K})$ in Eq.(36a) results in

$$\Delta_{nn'}(\vec{l}-\vec{l}',0) = \frac{V}{8\Pi N}q_{nn'}\ e^{-q_{nn'}\cdot|\vec{l}-\vec{l}'|}/|\vec{l}-\vec{l}'|, |\vec{l}-\vec{l}'| \neq 0 \tag{39}$$

in the direction of the electric field; $\Delta_{nn'}(\vec{l}-\vec{l}',0)$ depends explicitly on the interband parameter, $q_{nn'}$, and decreases exponentially with energy gap as $\vec{l}-\vec{l}'$ increases, a result which leads to the one-band approximation in the large energy-gap limit.

SOLID STATE DYNAMICS AND BLOCH OSCILLATIONS

Since the inception of solid-state physics, the acceleration theorems[12,15,16] and effective mass theory [8,12,15,16] have had a major role in guiding our understanding of electron behavior in perturbed crystals. Throughout the years, these concepts have been employed to provide an understanding of a wide variety of physical phenomena, such as the theory of electron transport in solids and the energy quantization of bound electrons in solids. Most recently these concepts have been used in studies concerning electron states and transport in superlattices and submicron/ultrasubmicron structures.

With the emphasis here on small spatial (submicron/ultrasubmicron) and fast temporal (pico/femtosecond) scales, it seems appropriate to discuss the validity of familiar solid-state concepts in the aforementioned space time regime. A critical review of solid-state dynamics has been published by the authors[3-5]. In this paper, we summarize salient features of electron dynamical response (acceleration theorems) to an applied dc external field, and also include the ranges of validity, appropriate corrections, and mathematical and physical assumptions implied.

Many years ago, Kretschmann[17] and, more recently, Pfirsch and Spenke[18] using Ehrenfest's theorem, independently showed that, for sufficiently short times, the dynamics of the electron is determined by its free-electron dynamics rather than its effective-mass dynamics (assuming the d.c. field is turned on instantaneously). Of course, after a sufficiently long time, the effective-mass dynamics prevails. In order to determine the time that must elapse before effective-mass dynamics gives the appropriate description of the electron motion, we make use of the correct time-dependent wave function to the first order in the applied d.c. field, employing Houston functions[19] as a basis, to obtain the acceleration theorems

$$\vec{v} = \frac{1}{\hbar} \frac{\partial \epsilon_n}{\partial k} + E \frac{\hbar}{m_o} \sum_m \frac{2}{m_o} \frac{|P_{nm}|^2}{(\epsilon_m - \epsilon_n)} \frac{\sin\left\{\frac{1}{\hbar} \int_o^t (\epsilon_m - \epsilon_n) dt'\right\}}{(\epsilon_m - \epsilon_n)} \tag{40}$$

and

$$\hbar \frac{dk}{dt} = E \tag{41}$$

Here, $\vec{v}$ is the velocity of a particle in the nth energy band, ϵ_n and E is the applied constant force, m_o is the free electron mass, P_{nm} is the momentum matrix element between the mth and nth band, $\hbar k$ is the electron quasi-momentum. Note that the second term in $\vec{v}$ is a sum over all bands with band index m), which implicitly includes interband mixing to the first order in E.

To determine the time that must elapse before effective mass dynamics is valid, we examine the acceleration, the total derivative of $\vec{v}$ with time, to the first order in E, which is

$$\frac{d\vec{v}}{dt} = \frac{1}{\hbar^2}\frac{\partial^2 \epsilon_n}{\partial k^2}E + \frac{E}{m_o}\sum_m \frac{2|P_{nm}|^2}{(\epsilon_m - \epsilon_n)}\cos\frac{1}{\hbar}\int_0^t (\epsilon_m - \epsilon_n)dt' \qquad (42)$$

This explicitly shows the acceleration as given by the usual effective mass, plus a term that oscillates with periods

$$\Delta t_{mn} = \frac{2\pi}{\Delta \epsilon_{mn}}$$

In the limit that $t \to 0^+$, the cosine terms in Eq. 42 go to unity and

$$\sum_m \frac{2}{m_o}\frac{|P_{nm}|^2}{(\epsilon_m - \epsilon_n)} = 1 - \frac{m_o}{\hbar^2}\frac{\partial^2 \epsilon_n}{\partial k^2} \qquad (43)$$

(the "f-sum" rule) so that $d\vec{v}/dt = E/m_o$, showing that the particle accelerates with free-electron dynamic, but in the limit $t \to \infty$, the cosine terms oscillate out of phase with each other and sum to zero, so that the usual effective-mass dynamics is achieved in the long time limit. The time necessary for the most slowly varying term in Eq.42 to go through one cycle is

$$\Delta t = \frac{2\pi\hbar}{\Delta E_g} \qquad (44)$$

where ΔE_g is the energy gap to the nearest band. (A plot may be found in Fig. 1). In bulk gallium arsenide, $\Delta E_g \sim 1.4eV$, so that $\Delta t \sim 3.0 \times 10^{-15}$ sec, during which time a conduction-band electron, moving at velocities of about 10^7 cm/sec, travels a distance of approximately 0.3 nm, i.e. of the order of a lattice constant. On the other hand, in superlattices where $\Delta E_g \sim 10-100\text{meVs}, \Delta t \sim 0.4-4$ psec, during which time a particle moving with a velocity of 10^7 cm/sec travels a distance of about 4-40 nm. Based upon these estimates, we conclude that additional corrections to effective-mass dynamics for bulk GaAs are not important in submicron structures, but certainly, in superlattices, where gaps are in the meV range, inertial contribution due to effective-mass dynamics may be nontrivial.

Finally, we note that the corrections to the semiclassical transport equation implied by these corrections to effective-mass dynamics have not been previously obtained by Kohn and Luttinger [20], employing a fully quantum mechanical treatment based on the time-dependent density matrix. More recent experimental work by Chukraborty et al.[21] has shown that, when the mean-free path of an electron in a metal is comparable to the distance traveled during the short time in which free electron dynamics is applicable, significant corrections to the resistivity occur.

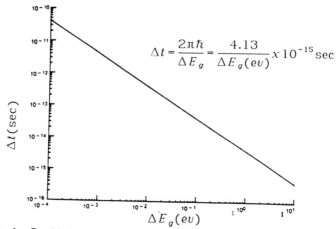

$$\Delta t = \frac{2\pi\hbar}{\Delta E_g} = \frac{4.13}{\Delta E_g(ev)} x\, 10^{-15}\,sec$$

Fig. 1. Graphical display of eq.(44) Δt is the time for the most slowly varying term in eq.(42) to go through one cycle; ΔE_g is the energy gap to the nearest band.

We now discuss the concept of Bloch oscillations, sometimes referred to as Zener oscillations, which arise from the solid-state theoretical hypothesis that electrons traversing a periodic lattice potential without scattering, while under the influence of a uniform dc electric field, will oscillate with a frequency that is linear in the electic field strength and the lattice constant.

The single-band equations of motion for an electron traversing a periodic potential, without scattering, under the influence of an external electric field are

$$\hbar\frac{dk}{dt} = eE$$

$$(45)$$

and

$$v = \frac{1}{\hbar}\nabla_k\epsilon(k)$$

$$(46)$$

where, $\hbar k$ is the electron quasi-momentum in the solid, E(t) is the external electric field, v is the electron velocity, $\epsilon(k)$ is the electron energy dispersion, which is periodic with reciprocal lattice vectors. (In the reduced zone scheme, k , in one dimension, varies between $-\pi/d$ to $+\pi/d$ with d being the lattice spacing.) It is clear from Eqs. 45 and 46 that, when a constant electric field is applied across the periodic potential of lattice spacing d (here, and through this discussion, a one-dimensional analysis is presented, and the extension of calculations to two-and three-dimensional analysis is straight-forward), the electron quasi-momentum ($\hbar k$) will increase linearly with time and the electron velocity will oscillate, since $\epsilon(k)$, hence $\nabla_k \epsilon(k)$, is periodic with the reciprocal lattice. The fundamental period of this Bloch oscillation is derived by integrating Eq. 45 with $E = E_o$ (constant electric field) over the entire first Brillouin zone to obtain

$$\tau = \frac{2\pi\hbar}{eE_o d} \tag{47}$$

The Bloch frequency is obtained from $\omega_\beta = 2\pi/\tau$ which yields

$$\hbar\omega_\beta = eE_o d \tag{48}$$

For lattice spacings d of 1nm, the Bloch frequency will range from about 10 to 100 GHz corresponding to a respective field variation of 1 to 10 kV/cm. Herein lies the potential impact of the Bloch oscillator: the possibility of a tunable millimeter-wave oscillator at modest electric fields!

Finally, we analyze Bloch electron motion in the dynamical Wannier representation; in this representation, the Bloch-oscillatory behavior is manifest as a space-time hopping in the periodic lattice rather than as k-space temporal oscillations, as is depicted in the Bloch representation As previously referred to in Eq. (28a), the time evolution of the single-band Wannier probability amplitude, normalized so that the electron is initially in the nth energy band, and localized at the lattice site "l_o", is

$$f_n(\vec{l},t) = \frac{1}{N} \sum_{\vec{K}} e^{-\frac{i}{\hbar}\int_o^t \epsilon_n\left(\vec{K} - \frac{e}{\hbar c}\vec{A}_o\right)dt'} e^{i\vec{K}\cdot(\vec{l}-\vec{l}_o)} \tag{49}$$

The general oscillatory nature of $f_n(\vec{l},t)$ in Eq. (49) can be shown explicitly by envoking the K-space periodicity of $\epsilon_n(\vec{k})$, namely, $\epsilon_n(\vec{K}) = \epsilon_n(\vec{K} + \vec{G}_j)$, where $|\vec{G}_j| = j2\pi/d$ are the appropriate reciprocal lattice vectors. In imposing this periodic condition in Eq. (49), it follows that $|f_n(\vec{l},t)| = |f_n(\vec{l},t+\tau)|$, where τ is the Bloch period given by Eq. (47); further imposition of this periodic condition shows that, after one Bloch period of time, $f_n(\vec{l},0)$ evolves to

$$f_n(\vec{l},\tau) = e^{-i\phi_n(\tau)//E_o}\, \delta_{\vec{l},\vec{l}_o} \tag{50}$$

where

$$\phi_n(\tau) = \int_o^{G_1} \epsilon_n(\vec{K})d\vec{K}, \tag{51}$$

225

independent of $\vec{K}$ and $\vec{l}$. It then follows from Eq.(50) that

$$|f_n(\vec{l},\tau)|^2 = \delta_{i,i_o};$$

(52)

more generally, it can be shown that

$$|f_n(\vec{l},s\tau)|^2 = \delta_{i,i_o}$$

(53)

where "s" refers to a positive integer, so that "$s\tau$" is an integral number of Bloch periods. Thus, it is evident that, in the single-band model with any $\epsilon_n(\vec{k})$, an electron in a Wannier state localized about a given site merely "breathes" about that site with the Bloch frequency, but does not propagate within the lattice.

Also, using the familiar symmetry condition that $\epsilon_n(\vec{K}) = \epsilon_n(-\vec{K})$ it is easy to show from Eq. 49 that for $\vec{l}_o = 0$ $f_n(\vec{l},t;-\vec{E}_o) = f_n(-\vec{l},t;\vec{E}_o)$ where $\vec{E}_o$ is a spatially homogeneous, and an arbitrarily time dependent, electric field.

Although we have indicated the qualitative behavior of $f_n(\vec{l},t)$ for a general $\epsilon_n(\vec{K})$ by studying its magnitude at times equal to integral multiples of Bloch periods, the explicit time dependence of $f_n(\vec{l},t)$ can be derived within the nearest-neighbor, tight binding approximation for $\epsilon_n(\vec{K})$. Noting that $\epsilon_n(\vec{K})$, in the nearest-neighbor, tight binding approximation, is given by Eq. 37 a, and utilizing this $\epsilon_n(\vec{K})$ in Eq. 49, we find that for $\vec{l}_o = 0$

$$|f_n(\vec{l},t)|^2 = J_l^2(\xi)$$

(54)

where J_l is a Bessel function of the first kind and

$$\xi = (W/\hbar\omega_\beta)\sin(\omega_\beta t/2)$$

(55)

with $\hbar\omega_\beta = eE_o d$

From the result in Eq. 54, we note that

$$\sum_l |f_n(\vec{l},t)|^2 = \sum_l |J_l(\xi)|^2 = 1$$

(56)

directly from the properties of Bessel functions; as well, ξ is a symmetric function of electric field, E_o, so that the direction of the uniform electric field does not affect the hopping probability; also, we note that

$$\frac{\partial}{\partial t}|f_n(\vec{l},t)|^2 = \frac{W}{2\hbar}\cos\frac{1}{2}\omega_\beta t \ \{J_l(\xi)J_{l-1}(\xi) - J_l(\xi)J_{l+1}(\xi)\}, \tag{57}$$

where ξ is defined in Eq. 55; Eq. 57 explicitly reveals the time dependence of the hopping rate.

There are many other interesting coherence characteristics that can be derived directly from $f_n(\vec{l},t)$ in Eq. 49, especially when the total electric field is a superimposed combination of a uniform and an oscillatory electric field; these characteristics will be discussed elsewhere[10].

Although we have reported a method for describing quantum transport and Bloch electron dynamics in the presence of a local, but strong inhomogeneous electric field, a more detailed illustration of the power of this method, which not only incorporates the ability to handle tunneling in the presence of a homogeneous electric field, but, along with tunneling, also allows for the simultaneous possibility of including weak scattering from a random distribution of defects or phonons during the tunneling process, will be published shortly.

ACKNOWLEDGEMENT

The authors acknowledge invaluable assistance from C.S. Kavina for the preparation of the manuscript.

REFERENCES

1. G. J. Iafrate, "The Physics of Submicron/Ultrasubmicron Dimensions," Gallium Arsenide Technology, edited by D.K. Ferry (H.W. Sams, Inc., Indianapolis, Indiana), 1985, Chapter 12.

2. G. J. Iafrate and J.B. Krieger, Bulletin of the American Physical Society, S13-3, P814, March 1988; Phys. Rev. B, 40, 6144 1989.

3. J. B. Krieger and G.J. Iafrate, Phys. Rev. B, 33, 5494 (1986).

4. J. B. Krieger and G.J. Iafrate, Phys. Rev. B, 35, 9644 (1987).

5. G. J. Iafrate and J.B. Krieger, Solid-State Electronics, 31, 517 (1988).

6. G. F. Koster and J.C. Slater, Phys. Rev. 95, 1167 (1954).

7. J. G. Gay and J.R. Smith, Phys. Rev. B11, 4906 (1975).

8. G. H. Wannier, Phys. Rev. 52, 191 (1937).

9. J. M. Ziman, Principles of the Theory of Solids (Cambridge Press Cambridge, England, 1964).

10. G. J. Iafrate and J.B. Krieger, to be published.

11. In the case where the initial state of the density matrix is not separable, as in mixed state analyses, such a separation may not be very useful.

12. P. T. Landsberg, <u>Solid State Theory-Methods and Applications</u> (Wiley-Interscience, New York, 1969).

13. E. O. Kane, J. Phys. Chem. Solids, <u>12</u>, 181 (1959).

14. J. B. Krieger, Phys. Rev., <u>156</u>, 776 (1967).

15. J. M. Ziman, Electrons and Phonons, Oxford Press.

16. J. M. Ziman, Principles of the Theory of Solids, 2d ed., p.172, Cambridge (1972).

17. E. Kretschmann, Z. Physik <u>7</u>, 518 (1934).

18. D. Pfirsch and E. Spenke, Z. Physik <u>137</u>, 309 (1954).

19. W. V. Houston, Phys. Rev. <u>57</u>, 184 (1940).

20. W. Kohn and J. N. Luttinger, Phys. Rev., <u>108</u>, 590 (1957).

21. B. Chakraborty and P.B. Allen, Phys. Rev. Lett. <u>42</u>, 736 (1979).

RECENT ADVANCES IN MICROFABRICATION

M. VAN ROSSUM, M. VAN HOVE, W. DE RAEDT, M. DE POTTER
AND P. JANSEN

IMEC vzw
Kapeldreef 75
3030 Leuven
Belgium

ABSTRACT. This paper discusses various new developments in III-V materials processing and their application to microdevice fabrication. In the field of patterning, the main tools are lift-off and dry etching. Both are scalable down to nanometer dimensions and have been used to produce quantum-size devices. Current control requires both Schottky and ohmic contacts on the device. Progress in processing technology now allows better control of the contact parameters, and the increased flexibility hereby obtained can be considered as a first step towards real "contact engineering". Finally, flexibility in the use of substrates has also been increased by the development of novel techniques for composite substrates fabrication and for device transplantation from one substrate to another.

1. Introduction

The field of microfabrication has witnessed a spectacular growth over the recent years. This expansion has been spurred by a combination of new developments in lithography, patterning and layer growth techniques. New technology tools have triggered the imagination of device physicists, who have come up with exciting ideas for novel devices based on the latest technological achievements. An innovation loop has hereby been set in motion, whose accelerating pace can be followed in the recent scientific literature.

Although progress towards miniaturization of Si processing has been impressive, it remains true that the concept of microfabrication is still mainly associated with compound (and more particularly III-V) semiconductors, because most device innovations have been proposed in this area. In order to keep our review reasonably self-contained, we will therefore mostly limit our discussion to the processing of III-V based structures and devices. In the next pages, we will pay attention to recent developments in patterning, contact and substrate technology and we will illustrate these with some practical device applications.

2. Patterning

2.1. LIFT-OFF

The lift-off technique has become very popular in III-V device processing for the definition of metallic patterns on semiconductor substrates [1]. It is a simple, flexible and reliable technique which has the important advantage of being applicable down to the nanometer scale. It can be performed with optical as well as e-beam resist (PMMA). Most important for the optimization of a lift-off process is the definition of an undercut resist profile to avoid metal edges on the resist walls. This can be achieved with a chlorobenzene soak or special resist bake-out procedures [2]. In more elaborate processes, multilayer resist schemes or dielectric assisted lift-off can also be useful [1]. In the case of e-beam lithography, an overhang profile is automatically obtained by the scattering of the electrons in the resist.

A major application of lift-off is the gate definition on FET devices. In the simplest case, the gate is defined by depositing a narrow metal strip between source and drain, lifting off the unwanted portions. State-of-the-art MESFET's or HEMT's require more elaborate gate structures where the gate is positioned inside a recess step etched into the device channel [3,4]. In many cases, a single lithographic step is used for gate metal and gate recess definition. Combination of the metal lift-off together with the (wet) recess etch results in so-called self-aligned gate structures. This process ensures the proper positioning of the gate even on very small scale devices [5].

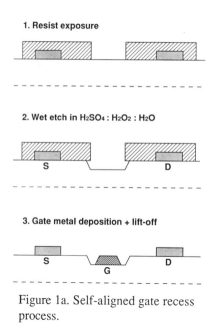

1. Resist exposure

2. Wet etch in $H_2SO_4 : H_2O_2 : H_2O$

3. Gate metal deposition + lift-off

Figure 1a. Self-aligned gate recess process.

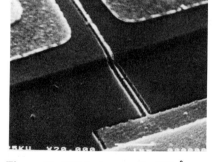

Figure 1b. SEM picture of a 600 Å gate deposited by lift-off in the self-aligned gate recess.

Another important development in submicrometer lithography is the so-called mushroom or T-shaped gate. This configuration is used to lower the gate access resistance, which increases with decreasing gatelength. Various fabrication schemes for T-shaped gates have been proposed, all based on extensions of the basic lift-off procedure. Many of these schemes imply the use of multilayer resists, or a combination of resist with metallic or dielectric layers [6]. Electron beam lithography is used in most cases, since mushroom gates are essentially relevant for submicron devices. However, processes based on focused ion beams [7] or a combination of ion and electron beams have recently been demonstrated [8].

2.2. SELECTIVE DEPOSITION

An alternative to the lift-off process is the selective deposition of metals on semiconductor substrates. In this case, the deposited metal originates from the decomposition of a precursor gas flowing over the substrate. This process is the result of a chemical reaction taking place at the semiconductor surface (CVD process). In order to form the desired patterns, this reaction must be locally activated by light or particle beams. Lasers, electron and focused ion beams have all succesfully been employed for writing metal line patterns on Si or GaAs. Laser deposition is the most mature technology, but is limited to rather broad lines (several microns) [9]. For this reason, its utility is more in the field of optical mask repair. Similarly, ion beams have been used to activate local deposition [10]. Nanometer line structures have been written with electron beams, using a SEM [11], or, as was recently proposed, a scanning tunneling microscope [12]. The major advantage of selective deposition is its flexibility, its major disadvantage is slowness. Moreover, most of these advanced techniques are still in an early development stage and at this time their usefulness in device fabrication has not yet been ascertained.

2.3. DRY PATTERNING

2.3.1. *Fundamentals.* Dry etching has become increasingly popular for lateral patterning of III-V semiconductor layers. One of the major advantages of this technique is the high resolution and the high degree of control that can be obtained over etch profiles and pattern dimensions. This allows the transfer of submicron or even quantum size patterns from the etch mask into the substrate. The diversity of the III-V device structures and the structural and chemical complexity of the epitaxial layers to be patterned presents a real challenge to the dry etching processes, which are expected to meet several requirements. For many applications (e.g. via holes), anisotropic etching with vertical sidewalls is highly desirable. Depending on the structure to be patterned, etch selectivity is an important requirement or must be avoided. The etched surfaces should be smooth and damage to the underlying substrate must be minimized.

Several technologies fulfilling these requirements are presently under development, most of which are plasma and/or ion assisted. The most mature and also the most versatile of these techniques is plasma-enhanced reactive ion etching [13]. It has been shown to be capable of high resolution patterning without affecting too much the integrity of the substrate. Although the details of the RIE mechanisms on III-V substrates are not yet completely understood, it is possible to distinguish some basic etching mechanisms. Plasma-enhanced RIE usually results from a combination between chemical and ion-assisted etching [14]. Chemical etching takes place when active species from the plasma react with the substrate surface and form a volatile reaction product. In the case of III-V compounds, the reactive species are usually free radicals such as F or Cl. The most popular etchants are therefore halogen-based gases, often mixed with rare gas diluents such as Ar or He. In RIE of III-V compounds by chlorine-source plasmas, both type-III ($GaCl_3$, $InCl_3$) and type-V ($AsCl_3$) reaction products are formed, the latter being the more volatile. During ion-assisted etching, substrate atoms are sputtered off by ions accelerated from the plasma. The etching mechanism is strongly influenced by the etching variables. Low pressures (below 0.1 torr) tend to favor ion bombardment and anisotropic etching. High temperatures increase the chemical reaction rates and therefore promote pure chemical etching over ion assisted reactions.

Ion bombardment can interfere with the chemical reaction in various ways. By damaging the surface, the impinging ions increase the reactivity of the substrate towards incident neutral radicals. The directionality of the ion beam therefore tends to wipe out the crystallographic orientation effects which are typical of pure chemical reactions. During some reactions, the substrate becomes covered with an inhibitor film consisting of involatile reaction products, which strongly affects the etch rate. In this case, the role of the ions is to clear the inhibitor from horizontal surfaces. Since the protective film is not removed from the vertical walls, a high degree of directionality can be achieved. BCl_3, CCl_4 and freon (CF_2Cl_2) are typical examples of inhibitor forming gases [15].

Another important aspect of RIE is etch selectivity. Selectivity is a consequence of the formation of involatile species at the substrate surface. As a typical example, CF_2Cl_2, which etches GaAs fast, will be nearly stopped by AlGaAs due to the formation of involatile AlF_3 [16]. Similar action is provided by a $SF_6/SiCl_4$ mixture, where the degree of selectivity can be adapted by changing the mixture ratio [17]. On the other hand, pure $SiCl_4$ is a good GaAs/AlGaAs etchant but the etching will be stopped by InGaAs layers, because of the low volatility of $InCl_3$ [17].

With shrinking device scale, radiation damage aspects of RIE tend to become increasingly important. Radiation damage effects in the substrate include higher resistivity, lower mobility and Schottky barrier degradation. Damage is produced by energetic ion bombardment and can therefore be minimized by reducing the ion penetration depth into the substrate. This can be achieved by using low ion energy, heavy ion species, or by forming a protective layer on the surface. Gases such as H_2, O_2 or Ar are known to produce a high amount of damage [18]. Halogen-based feeds are milder, although the resulting damage may strongly depend on the etch parameters. Very interesting results have recently been obtained with CH_4/H_2 mixtures (CH_4 concentration less than 25%) [19]. This gas seems to etch with the formation of metal-organic ($(CH_3)_3Ga$ and AsH_3 as the reaction products. It is believed that a thin polymeric film is formed at the sample surface, which acts as a protective layer against kinetically induced damage. However, penetration of H_2 into the GaAs produces dopant desactivation which must be countered by high-temperature annealing [20].

2.3.2. *Device applications*. The list of RIE applications for device fabrication is already very long. One of its first uses has been the processing of via holes for GaAs microwave circuits, where it replaces the less controllable wet etch process [21]. Another very important application is the fabrication of semiconductor laser facets, which requires altogether high anisotropy, smooth surface finish and low damage [22]. Attempts have been made to use RIE for gate recess on MESFET's or HEMT's, but the problem of Schottky barrier degradation has not yet completely been overcome [23]. Fabrication of thin membrane devices has been performed by the use of etch stops intercalated in the substrate layers [24].

Recently, RIE has become a very important component of nanofabrication technology, where it is used as the preferred method to transfer e-beam defined patterns into the substrate [25]. The most spectacular application of this technology is the processing of quantum wires and quantum dots. Since these structures are often characterized by very high aspect ratios, special attention should be given to the choice of the proper etch mask. PMMA resist has been used in some cases, but mask erosion may become a severe problem. Dielectric layers such as silicon oxide or silicon nitride are more stable, but they require a supplementary pattern transfer step, which may limit the resolution. The prefered solution is the definition of a metallic mask by metal deposition and lift-off. Moreover, by using metals suitable for ohmic contact formation, a self-aligned fabrication process becomes possible [26]. The high resolution of the lift-off step allows a very accurate transfer of the e-beam written patterns. In this way, multiple quantum well columns with 0.04 micron diameter and a 6:1 aspect ratio have been fabricated using a BCl_3/Ar etch mixture [27]. At this scale, radiation damage along the etched sidewalls becomes a limiting factor since the damage layer may easily reach a few tens of nm in depth.

3. Contact technology

3.1. OHMIC CONTACTS

Ohmic contacts on III-V materials have been a major cause of concern since the early days of compound semiconductor technology. Although the empirically developed eutectic AuGe-Ni alloy scheme has found widespread use in the fabrication of n-type ohmic contacts to III-V devices, it must be admitted that this type of contact remains unsatisfactory in many respects. The contact resistivity lies rather consistently in the 10^{-6} Ohm cm^2 range, which lags one order of magnitude behind the advanced contacts on Si. Because of the liquid phase regrowth during alloying, both the surface morphology and the lateral uniformity are very poor. The alloyed contact region consists of various intermixed crystalline phases, which even today have not been characterized completely. The presence of AuGa and NiAs(Ge) crystallites has been detected, the latter of which is thought to play an important role in the conduction mechanism [28]. Typical for these alloyed contacts is the occurence of Ge-rich spikes, which protrude deeply into the GaAs matrix [29]. The Braslau theory predicts that the contact resistance is dominated by the spreading resistance of these protrusions [30]. It therefore appears that the non-uniform morphology of the alloyed AuGe-Ni contact is strongly linked with its electrical properties.

With shrinking dimensions, these structural non-uniformities become increasingly detrimental to the device operation. As an example, submicron scaled FET's may suffer from the lateral diffusion of Ge dopants into the channel region, as well as from the vertical extension of the contact region well below the channel depth. On epitaxial structures, the existence of metallic spikes makes it essentially impossible to contact very thin epitaxial layers without penetrating into the underlying layer. One way to improve the uniformity would be to lower the temperature at which the contacts are formed. Successful formation of AuGe-Ni contacts at annealing temperature below 350°C has been achieved by optimizing the amount of Ni in the contact [31]. It appears that such low-temperature contacts also display a much smoother surface morphology. Good morphology has also been obtained by (cw or pulsed) laser alloying [32].

In recent years, more radical departures from the traditional contact systems have been proposed. One of the most interesting concepts is based on the mechanism of solid-state regrowth. A prime example of the latter is the Pd/Ge contact [33]. This contact scheme involves the deposition of a Pd layer onto which a layer of (amorphous) Ge is then evaporated without breaking the vacuum. Upon annealing (typical conditions are 325°C for 30 minutes), the Pd reacts with Ge at the top interface to form Pd$_2$Ge and then PdGe. At the same time, a limited amount of reaction is taking place at the Pd-GaAs interface, forming a Ga-rich Pd-Ga-As ternary compound. With sufficient Ge present, the Pd layer is completely consumed and the excess Ge is transported across the germanide towards the GaAs. When the Ge reaches the lower interface, the Pd-Ga-As compound is dissociated and transformed into

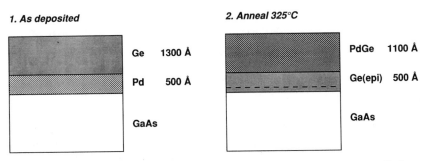

Figure 2. Schematic representation of the Pd-Ge solid-state reaction on GaAs.

PdGe and epitaxially regrown Ge on GaAs. During this process, some amount of Ge becomes incorporated into the regrowing GaAs and forms a thin heavily doped n⁺-layer, due to the presence of excess Ga vacancies resulting from the Ga-rich Pd-Ga-As phase. This type of contact displays both a low resistivity (comparable to or even better than AuGe-Ni) and a very uniform morphology, because of the solid-state regrowth and the epitaxial nature of the Ge-GaAs interface.

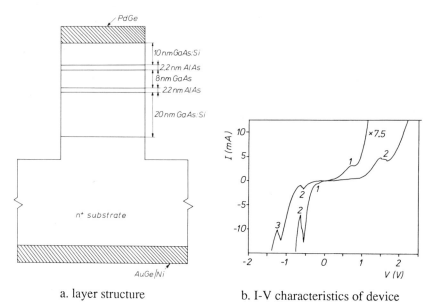

a. layer structure   b. I-V characteristics of device

Figure 3. Resonant tunneling diode with Pd-Ge contacts

Although the contact mechanism has not yet been completely elucidated, it is thought to be based essentially on tunneling conduction due to the presence of the Ge-doped n⁺-layer [34]. Recent investigations have shown that Pd/Ge belongs to a wider class of solid-state grown contacts, which also include Pd/Si, Ni/Si, Pd/In and others [35].

Several device applications of Pd/Ge contacts have been reported. These include MESFET's, HEMT's and MQW absorbers [36-38]. Recently, we have demonstrated the application of Pd/Ge as the top contact of a GaAs/AlAs resonant tunneling diode [39]. In this device, the Pd/Ge layers were deposited directly on top of the lowly doped spacer. In spite of the narrowness of this layer (100 Å), the device showed well-behaved resonances both at RT and at 77 K. This example proves the extreme sharpness and uniformity of the contact/semiconductor interface, which should make it ideally suited for contacting thin epitaxial layers.

Other approaches towards sharp and uniform ohmic contacts have been proposed, which involve the use of an epitaxial growth technique such as MBE. An early proposal was that of a graded bandgap contact to n-type GaAs [40]. This contact consists of a heavily doped epitaxial layer which is compositionally graded from GaAs to InAs. The compositional grading provides a smooth conduction band transition from GaAs to InAs, and since the surface Fermi level of InAs is pinned in the conduction band, it allows easy contact formation by simple metal deposition. Similar results can be obtained by growing InAs directly on GaAs, followed by a short-time anneal step to obtain interface mixing [41]. These contacts show typical examples of ohmic behavior due to the lowering of the effective electron barrier in the conduction band. Another type of ohmic contact produced by MBE involves the growth of a heavily doped n$^+$-layer right under the metal-semiconductor interface, thereby promoting a tunneling mechanism [42]. These n$^+$-layers can be either uniformly doped, or contain one or more layers of planar doping [43]. In both cases, physical deposition of a metal on top of the doped layer suffices to form an ohmic contact.

3.2. SCHOTTKY CONTACTS

Schottky contacts have been an important part of III-V microfabrication from the early days, through their use as gate structures in FET devices (essentially MESFET's and HEMT's). Recently, new applications of Schottky contacts have been found in microdevices based on electrostatic potential confinement or lateral surface superlattices. The structural simplicity of these contacts and the fact that they can be fabricated by a lift-off procedure makes them easy to scale down for nanostructure applications.

It is well known that the Schottky barrier height to GaAs and related ternaries is relatively insensitive to the metal used [44]. The invariance of the barrier height has been explained by a very large localized state density at the metal-semiconductor interface, which pins the semiconductor Fermi level at the interface near midgap. The origin of these localized states is still a subject of debate, and several models have been put forward, based e.g. on surface defects, adatom clusters, tunneling bandgap states and others [45]. Whatever the origin of the Fermi level pinning, it results in considerable loss of flexibility in the design and the operation of Schottky-based devices. To overcome this limitation, techniques for fabricating controlled Schottky barrier heights have recently been developed. It is well known theoretically (and it has in fact first been demonstrated by Shannon on Si diodes [46]) that the Schottky barrier height can be influenced by the presence of a thin, highly doped layer at the metal-semiconductor interface [47-49]. Opposite effects are expected for p$^+$ and n$^+$-doped layers. In the case of a metal-n$^+$-n structure, the width of the Schottky barrier is reduced and transport through the diode occurs mainly by quantummechanical tunneling. This effect, when analyzed by standard Schottky contact theory, corresponds to a lowering of the effective barrier height. In the limit of high doping, this structure behaves as an ohmic contact. In the metal-p$^+$-n structure, negative space charge resulting from ionized acceptors in the p$^+$-layer increase the effective barrier to thermionic emission. However, the p$^+$-layer should be thin enough to prevent the contact from operating as a p/n junction. For device applications, the p$^+$-layer is usually designed to be completely depleted. Although the theoretical mechanism behind Schottky barrier engineering is rather straightforward, its practical implementation requires excellent control over the width and the doping level of the thin interfacial layer. There have been attempts to achieve this by ion implantation, but with limited success [48]. More promising results have been achieved by MBE. Using this technique, Eglash et al. have produced engineered Schottky contacts on GaAs with barrier heights ranging from essentially zero to 1.4 eV (the GaAs bandgap) [49]. Avoiding MBE, Waldrop and Grant have fabricated 1 eV Schottky barriers on GaAs by evaporating a thin, p$^+$-doped Si layer between the metal and the semiconductor [50]. We have recently achieved similar results by rapid annealing of TiW gates on GaAs. The thermal step results in outdiffusion of the Ti into the GaAs, thus creating a p$^+$-region under the gate [51]. An alternative way towards control of Schottky barrier heigt has recently been proposed. This uses composite Schottky contacts with a very thin metal interlayer. By varying the thickness of this interlayer, the barrier height can be tuned continuously between its bulk values for the top and for the bottom metal. This effect has been demonstrated with Pt/Ti and Pd/Ti contacts [52]. These results show that a range of techniques are now available and that engineered Schottky contacts should become an important ingredient of future microdevices.

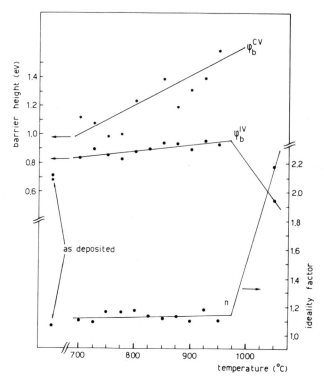

Figure 4. TiW Schottky contact on GaAs : barrier height and ideality factor as a function of annealing temperature.

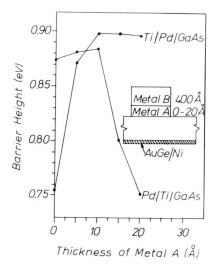

Figure 5. Composite Ti/Pd and Pd/Ti Schottky contacts on GaAs : barrier height dependence on metal thickness.

4. Substrates

The concept of "substrate engineering" is a relatively recent one. Its introduction is a result of the continuing advances of the epitaxial growth techniques, wich has produced a wide range of taylor-made semiconductor layers for various applications. Up to some years ago, epitaxial growth could only be envisaged in the case of lattice-matched crystal structures. Today, high-quality lattice-mismatched epitaxial films are becoming available, thereby triggering a variety of fundamental investigations and applications studies.

A major example of the hetero-epitaxial lattice-mismatched growth technology is GaAs on Si [53]. Two main application areas are being considered for this technology. The first one aims at taking advantage of the favorable properties of Si material (low cost, high mechanical strength, high thermal conductivity) by simply replacing the bulk GaAs wafers used for IC processing by their epitaxial GaAs on Si counterpart. This technology has already reached some level of maturity, up to the point that manufacturing issues have recently been addressed. The second approach is much more ambitious, as it intends to combine GaAs and Si on the same chip in order to join the best of both technologies for developing new IC configurations with improved performance and increased integration levels. To achieve this goal, composite substrates with GaAs islands surrounded by Si regions must be fabricated. The definition of these GaAs islands can be done in various ways. The most straightforward approach is to grow a uniform layer and then etch the GaAs away where it is not needed. However, this technique produces a non-planar substrate, which may complicate further processing. Another alternative is growth on patterned and etched Si substrates. Embedded or recessed growth has gained much interest as a coplanar integration technology [54].

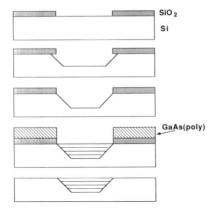

Figure 6. Embedded growth process for GaAs on Si.

In this approach wells are etched in the Si wafer prior to loading it in the deposition apparatus. This is done by using a SiO_2 etch mask to define the well patterns. The wells can be etched by wet or dry processing, but in both cases care must be taken to obtain a smooth and flat well bottom. Because this process tends to form an undercut profile, the SiO_2 corners need to be removed to avoid shadowing effects. Subsequently, GaAs is deposited on the whole wafer by MBE or MOCV. The GaAs grows epitaxial in the wells and polycrystalline over the SiO_2 mask. The latter one is removed by a lift-off step, thereby leaving nearly planar filled wells. This allows reliable interconnection between the GaAs devices processed on the islands and the surrounding Si circuitry [55].

A totally different approach towards composite substrates has recently been proposed by Yablonovitch et al [56]. It involves the removal of the epitaxial film from its growth substrate and the subsequent redeposition on a new host. This can be accomplished by incorporating a release layer during epitaxial growth of the structures. In the case of GaAs technology, a thin AlAs layer can be used for this purpose. After growth and subsequent device processing, the release layer is dissolved in a proper etchant (BHF), thereby lifting off the epitaxial film from the substrate. In order to perform this operation successfully, it has been shown to be essential to cover the film surface with a special wax, both for enhancing the lift-off procedure and for providing the free-floating film with a mechanical support. The film is finally redeposited on the substrate of choice. In this way, various devices, such as GaAs MESFET's [57], LED's [58] and even heterojunction lasers [59] have been redeposited on Si. This technique has also been proposed to fabricate thin GaAs solar cells on Si substrates e.g. for space applications.

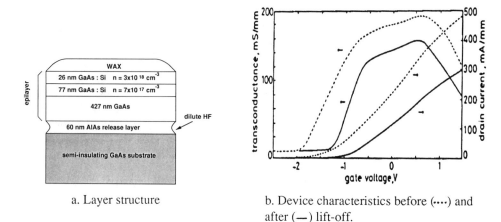

a. Layer structure

b. Device characteristics before (····) and after (—) lift-off.

Figure 7. Lift-off of epitaxially grown MESFET's

Conclusions

We have attempted to provide a short overview on some recent advances in III-V technology and their application to microdevices. New achievements have been simultaneously witnessed in several processing areas, thereby considerably extending the fabrication possibilities of submicron devices. Quite generally, development has been driven by the need for better dimensional control of the processed structures, both in the lateral and in the vertical direction. We have now come to the point where nanoscale patterns can be accurately and reproducibly fabricated. However, substantial qualitative improvements can still be expected, both by the reduction of processing damage and by the improvement of control over electrical parameters, especially on Schottky and ohmic contacts. Such a progress will require a better understanding (at the atomic scale) of the solid-state mechanisms which come into play during the fabrication and operation of interfacial structures.

References

1. See e.g. Williams, R.E. (1985) Gallium Arsenide Processing Techniques, Artech House, Dedham.

2. Hatzakis, M., Canavello, B.J. and Shaw, J.M. (1980) IBM J. Res. Develop. 24, 452.

3. Weitzel C.E. and Doane, D.A. (1986) 'A review of GaAs MESFET gate electrode fabrication technology', J. Electrochem. Soc. 133, 409C-415C.

4. See e.g. Thomas, H., Morgan, D.V., Thomas, B., Aubrey, J.E. and Morgan, G.B. (1986) Gallium Arsenide for Devices and Integrated Circuits, Proceedings of the 1986 UWIST GaAs School, Peter Peregrinus Ltd., London, U.K.

5. De Raedt, W, Jonckheere, R., Van Hove, M, Borghs, G., Van Rossum, M., Van den hove, L. and Born, R. (1989) 'Sub-100 nm gate MESFET's by electron beam mix&match', Microelectronic Engineering, 9341-9344.

6. Chao, P.C., Smith, P.M., Palmateer, S.C. and Hwang, J.C.M. (1985) 'Electron-beam fabrication of GaAs Low-noise MESFET's using a new trilayer resist technique', IEEE Trans. Electron Devices ED-32, 1042-1046.

7. Melngallis, J. (1987) 'Focused Ion Beam Technology and Applications', J. Vac. Sci. Technol. B5, 469-495.

8. Hosono, K., Fujino, T., Matsuda, S., Nagahama, K., Sasaki, Y., Morimoto, H. and Watakabe, Y. (1988) 'Hybrid Lithography of a focused ion beam and an electron beam for the fabrication of a GaAs field effect transistor with a mushroom gate', J. Vac. Sci. Technol. B6, 1828-1831.

9. See e.g. Bauerle, D. (1986) Chemical processing with lasers, Springer Series in Materials Science (1), Springer-Verlag, Berlin.

10. Kubena, R.L., Stratton, F.P. and Mayer, T.M. (1986) 'Selective area nucleation for metal chemical vapor deposition using focused ion beams', J. Vac. Sci. Technol. B6, 1865-1868.

11. Ichihashi T. and Matsui, S. (1988) 'In situ observation on electron beam induced chemical vapor deposition by transmission electron microscopy', J. Vac. Sci. Technol. B6, 1869-1872.

12. McCord, M.A., Kern, D.P. and Chang, T.H.P. (1988) 'Direct deposition of 10 nm-metallic features with the scanning tunneling microscope', J. Vac. Sci. Technol. B6, 1877-1880.

13. Flamm, D.L. and Manos, D.M. (1988) Plasma Etching: An introduction, Academic Press, Orlando.

14. Ibbotson, D.E. and Flamm, D.L. (1988) 'Plasma Etching for III-V Compound Devices : Part I', Solid State Technol., October, 77-79.

15. Ibbotson, D.E. and Flamm, D.L. (1988) 'Plasma Etching for III-V Compound Devices : Part II', Solid State Technol., November, 105-137.

16. Seaward, K.L., Moll, N.J., Coulman, D.J. and Stickle, W.F. (1987) 'An analytical study of etch and etch-stop reactions for GaAs and AlGaAs in CCl_2F_2 plasma' J. Appl. Phys. 61, 2358-2364.

17. Cooper III, C.B., Salimian, S. and MacMillan, H.F. (1987) 'Use of thin AlGaAs and InGaAs stop-etch layers for reactive ion etch processing of III-V compound semiconductors', Appl. Phys. Lett. 51, 2225-2226.

18. Pang, S.W., Geis, M.W., Efremow, N.N. and Lincoln, G.A. (1985) 'Effects of ion species and absorbed gas on dry etching induced damage in GaAs', J. Vac. Sci. Technol. B3, 398-401.

19. Cheung, R., Thoms, S., Beaumont, S.P., Doughty, G., Law, V. and Wilkinson, C.D.W. (1987) 'Reactive ion etching of GaAs using a mixture of methane and hydrogen', Electronics Lett. 23, 857-858.

20. Cheung, R., Thoms, S., McIntyre, I., Wilkinson, C.D.W. and Beaumont, S.P. (1988) 'Passivation of donors in electron beam lithographically defined nanostructures after methane/hydrogen reactive ion etching', J. Vac. Sci. Technol. B6, 1911-1915.

21. Cooper III, C.B., Salimian, S. and Day, M.E. (1989) 'Dry Etching for the Fabrication of Integrated Circuits in III-V Compound Semiconductors', Solid State Technol., January, 109-112.

22. Salzman, J., Venkatesan, T., Margulit, S. and Yariv, A. (1985) 'Double heterostructure lasers with facets formed by a hybrid wet and reactive-ion-etching technique', J. Appl. Phys. 57, 2948-2950.

23. Lecrosnier, D., Henry, L., Le Corre, A. and Vaudry, C. (1987) 'GaInAs FET fully dry etched by metal organic reactive ion etching techniques', Electronics Lett. 23, 1254-1255.

24. Ade, R.W., Fossum, E.R. and Tischler, M.A. (1988) 'Fabrication of epitaxial GaAs/AlGaAs diaphragms by selective dry etching', J. Vac. Sci. Technol. B6, 1592-1594.

25. Lee, K.Y., Smith III, T.P., Ford, J.B., Hansen, W., Knoedler, C.M., Hong, J.M. and Kern, D.P. (1989) 'Submicron trenching of semiconductor nanostructures', Appl. Phys. Lett. 55, 625-627.

26. Randall, J.N., Reed, M.A., Moore, T.M., Matyi, R.J. and Lee, J.W. (1988) 'Microstructure fabrication and transport through quantum dots', J. Vac. Sci. Technol. B6, 302-305.

27. Scherer, A. and Craighead, H.G. (1986) 'Fabrication of small laterally patterned multiple quantum wells', Appl. Phys. Lett. 49, 1284-1286.

28. Kuan, T.S., Batson, P.E., Jackson, T.N., Rupprecht, H. and Wilkie, E.L. (1983) 'Electron microscope studies of an alloyed Au/Ni/Au-Ge ohmic contact to GaAs', J. Appl. Phys. 54, 6952-6957.

29. Murakami, M., Childs, K.D., Baker, J.M. and Callegari, A. (1986) 'Microstructure studies of AuNiGe ohmic contacts to n-type GaAs', J. Vac. Sci. Technol. B4, 903-911.

30. Braslau, N. (1983) 'Ohmic contacts to GaAs', Thin Solid Films 104, 391-397.

31. Patrick, W., Mackie, W.S., Beaumont, S.P. and Wilkinson, C.D.W. (1986) 'Low-temperature annealed contacts to very thin GaAs epilayers', Appl. Phys. Lett. 48, 986-988.

32. Wuyts, K., Silverans, R.E., Van Hove, M. and Van Rossum, M. (1989) 'Characterization of pulsed laser beam mixed AuTe/GaAs ohmic contacts', to be published in Proceedings of the MRS 1989, Fall Meeting (Boston).

33. Marshall, E.D., Chen, W.X., Wu, C.S., Lau, S.S. and Kuech, T.F. (1985) 'Non-alloyed ohmic contacts to n-GaAs by solid phase epitaxy', Appl. Phys. Lett. 47, 298-300.

34. Yu, L.S., Wang, L.C., Marshall, E.D., Lau, S.S. and Kuech, T.F. (1989) 'The temperature dependence on contact resistivity of the Ge/Pd and the Si/Pd nonalloyed contact scheme on n-GaAs', J. Appl. Phys. 65, 1621-1625.

35. Sands, T., Marshall, E.D. and Wang, L.C. 'Solid-phase regrowth of compound semiconductors by reaction-driven decomposition of intermediate phases', J. Mater. Res. 3, 914-921.

36. Paccagnella, A., Canali, C., Donzelli, G., Zanoni, E., Wang, L.C. and Lau, S.S. (1988)'GaAs MESFET contacts : Technology and performances', Electronics Lett. 24, 708-709.

37. Wang, L.C., Lau, S.S., Hsieh, E.K. and Velebir, J.R. (1989) 'Low-resistance nonspiking ohmic contact for AlGaAs/GaAs high electron mobility transistors using the Ge/Pd scheme', Appl. Phys. Lett. 54, 2677-2679.

38. Van Eck, T.E., Chu, P., Chang, W.S.C. and Wieder, H.H., (1986) 'Electroabsorption in an InGaAs/GaAs strained-layer multiple quantum well', Appl. Phys. Lett. 49, 135-136.

39. Van Hoof, C., Van Hove, M., Jansen, P., Van Rossum, M. and Borghs, G. 'Non-alloyed Ge/Pd contacts for AlAs/GaAs resonant tunneling structures', to be published.

40. Woodall, J.M., Freeouf, J.L., Pettit, G.D., Jackson, T.and Kirchner, P. (1981) 'Ohmic contacts to n-GaAs using graded bandgap layers of $Ga_{1-x}In_xAs$ grown by molecular beam epitaxy' J. Vac. Sci. Technol. 19, 626-627.

41. Wright, S.L., Marks, R.F., Tiwari, S., Jackson, T.N. and Baratte, H. (1986) 'In situ contacts to GaAs based on InAs', Appl. Phys. Lett. 49, 1545-1547.

42. Kirchner, P.D., Jackson, T.N., Pettit , G.D. and Woodall, J.M. (1985) 'Low-resistance nonalloyed ohmic contacts to Si-doped molecular beam epitaxial GaAs', Appl. Phys. Lett. 47, 26-28.

43. Schubert, E.F., Cunningham, J.E., Tsang, W.T. and Chiu, T.H. (1986) 'Delta-doped ohmic contacts to n-GaAs', Appl. Phys. Lett. 49, 292-294.

44. See e.g. Sze, S.M. (1981) Physics of Semiconductor Devices, Wiley, New York.

45. Rhoderick, E.H. and Williams, R.H. (1988) Metal-semiconductor contacts, 2nd edition, Oxford University Press, Oxford.

46. Shannon, J.M. (1974) 'Increasing the effective height of a Schottky barrier using low-energy ion implantation' Appl. Phys. Lett. 25, 75-77.

47. Schwartz, G.P. and Gualtieri, G.J. (1986) 'Schottky barrier enhancement on M-P⁺-N structures including free carriers', J. Electrochem. Soc. 133, 1266-1268.

48. Stanchina, W.E., Clark, M.D., Vaidyanathan, K.V., Jullens, R.A. and Crowell, R.A. (1987) 'Effects and characterization of ion implantation enhanced GaAs Schottky barriers', J. Electrochem. Soc. 134, 967-971.

49. Eglash, S.J., Newman, N., Pan, S., Mo, D., Shenai, K., Spicer, W.E., Ponce, F.A. and Collins, D.M. (1987) 'Engineered Schottky barriers diodes for the modification and control of Schottky barrier heights', J. Appl. Phys. 61, 5159-5169.

50. Waldrop, J.R. and Grant, R.W. (1988) 'Metal Contacts to GaAs with 1 eV Schottky barrier height', Appl. Phys. Lett. 52, 1794-1796.

51. de Potter, M., De Raedt, W., Van Hove, M., Zou, G., Bender, H., Meuris, M. and Van Rossum, M. 'Characterization of the TiW-GaAs interface after rapid thermal annealing', to be published in J. Appl. Phys.

52. Wu, X., Schmidt, M.T. and Yang, E.S. (1989) 'Control of the Schottky barrier using an ultra thin interface metal layer', Appl. Phys. Lett. 54, 268-269.

53. Turner, G.W., Choi, H.K., Mattia, J.P., Chen, C.L., Eglash, S.J. and Tsaur, B.-Y. (1988) 'Monolithic GaAs/Si integration', Mat. Res. Soc. Symp. Proc. Vol. 116, 179-192.

54. Liang, J.B., De Boeck, J., Deneffe, K., Arent, D.J., Van Hoof, C., Vanhellemont, J. and Borghs, G. (1989) 'Embedded growth of gallium arsenide in silicon recesses for a coplanar GaAs on Si technology', J. Vac. Sci. Technol. B7, 116-119.

55. Sichijo, H., Matyi, R.J. and Taddiken, A.H. (1988) 'Co-integration of GaAs MESFET and Si CMOS circuits', IEEE Electron Device Lett., EDL-9(9), 444-446.

56. Yablonovitch, E., Gmitter, T., Harbison, J.P. and Bhat, R. (1987) 'Extreme selectivity in the lift-off of epitaxial GaAs films', Appl. Phys. Lett. 51, 2222-2224.

57. Van Hoof, C., De Raedt, W., Van Rossum, M. and Borghs, G. (1989) 'MESFET lift-off from GaAs substrate to glass host', Electronics Lett. 25, 136-137.

58. Pollentier, I., De Dobbelaere, P., De Pestel, F., Van Daele, P. and Demeester, P. (1989) 'Integration of GaAs LED's on Si by epi-lift-off', in H. Heuberger, H. Ryssel and P. Lange (eds.), ESSDERC '89, 19th European solid state device research conference, Springer Verlag, Berlin, pp. 401-404.

59. Yablonovitch, E., Kapon, E., Gmitter, T.J., Yun, C.P. and Bhat, R. (1989) 'Double heterostructure GaAs/AlGaAs thin film diode lasers on glass substrates', IEEE photonics techn. lett. 1, 41-42.

INTRODUCTION TO RESONANT TUNNELLING IN SEMICONDUCTOR HETEROSTRUCTURES

L. Eaves

Department of Physics
University of Nottingham
Nottingham NG7 2RD, UK

ABSTRACT

The effective mass approximation for conduction electrons tunnelling through a semiconductor heterostructure barrier is described. The current-voltage characteristics of a variety of double barrier resonant tunnelling devices with wide quantum wells (60-120 nm) and based on the (AlGa)As/GaAs system are presented. In these structures a large number of electron standing wave resonances are observed in I(V). The effect of a transverse magnetic field ($\underline{B}$ parallel to the plane of the barriers) on the resonances in I(V) is examined. Resonant tunnelling into hybrid magneto-electric states of the quantum well is observed and is interpreted using the effective mass approximation.

INTRODUCTION

One of the major themes of this Advanced Study Institute is concerned with electron tunnelling through potential barriers in semiconductor heterostructures. The most widely studied system is based on the alloy $Al_xGa_{1-x}As$, lattice matched to GaAs. Such a heterostructure is shown in a simplified and schematic way in the upper part of Figure 1. A thin layer of (AlGa)As, typically a few nm thick, is sandwiched between two thicker GaAs layers which are doped n-type. The lower part of the diagram considers the motion of a conduction electron moving directly towards the (AlGa)As layer which acts as a potential barrier. If the kinetic energy ε of the electron is less than the barrier height, V, it must traverse the barrier by quantum mechanical tunnelling. Remarkably, the tunnelling process can be treated reasonably well by means of a simple model, the effective mass approximation. In this model, the electron is assumed to tunnel through the (AlGa)As barrier as if it has an effective mass $m_\Gamma^* \simeq$ 0.1 m_e, corresponding to the Γ conduction band minimum of the barrier material. The height of the potential barrier $V = \Delta E_c$ is termed the conduction band discontinuity between the conduction band edge of GaAs and that of the Γ minimum of (AlGa)As. There is now a general consensus in the literature that $\Delta E_c(eV) = (0.9 \pm 0.1)x$ where x is the Al mole fraction. At values of $x > 0.4$, the (AlGa)As becomes indirect gap. In this case, tunnelling via the evanescent states associated with the indirect (X and L) minima needs to be taken into account[1]. However, for many practical cases and for the structures examined in this chapter, the transmission probability T is given by

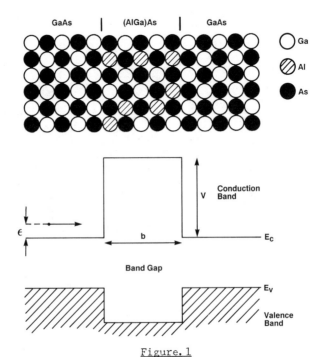

Figure. 1

Schematic diagram of a single tunnel barrier of (AlGa)As sandwiched
between two layers of GaAs. The lower part of the diagram illustrates the
conduction and valence band discontinuities.

$$\log_e T = -2 \int K(y)dy \tag{1}$$

where the integral is taken over the thickness of the barrier b, and where

$$K(y) = [2(U(y) - \varepsilon)m_\Gamma^*]^{\frac{1}{2}}/\hbar . \tag{2}$$

Here the potential height $U(y)$ includes not only ΔE_c but also the possibility of a varying Al composition in the barrier and of varying electrostatic potential across the barrier.

Since the earliest realisation of resonant tunnelling devices[2], double barrier structures have been regarded as the electronic analogues of the optical Fabry-Perot interferometer in which the two tunnel barriers play the role of partially reflecting mirrors for conduction electrons. The device is brought into resonance by means of an applied voltage which matches the energy of conduction electrons in the negatively biased contact with one of the standing wave resonances of the quantum well cavity. This is shown schematically in Figure 2. Electrons tunnel from the negatively biased emitter layer (doped n-type), through the emitter barrier and into the quantum well, eventually reaching the collector contact by tunnelling through the second barrier. The figure considers the case of a relatively wide well in which an electron is injected at an energy for which several de Broglie wavelengths fit into the well width. Note that de Broglie wavelength in the well becomes shorter from left to right due to the gain of kinetic energy arising from the electric field in the well region. In the Fabry-Perot or coherent tunnelling model, the peaks in the tunnel current as the voltage is varied simply reflect the transmission resonances. However, in certain cases, the build-up of electronic charge in the quantum well at resonance modifies the current-voltage characteristics, $I(V)$, and gives rise to an intrinsic bi-stability[3,4,5] effect, that is, a hysteresis loop in $I(V)$. Further details of the intrinsic bistability effect are given in the chapter by G. A. Toombs. The application of a magnetic field, $\underline{B}$, can also greatly modify the resonances observed in $I(V)$ in the configuration in which $\underline{B}$ is parallel to the plane of the barriers (i.e. $\underline{B} \perp \underline{J}$). In this case, hybrid magneto-electric states form in the quantum well and give rise to clear resonances in the tunnel current[6-10]. The basic properties of these magneto-electric resonances can be understood qualitatively by means of a semi-classical model. In this article, we shall examine (i) the Fabry-Perot resonances in devices with a range of different well thicknesses, (ii) hybrid magneto-electric resonances in the $\underline{B} \perp \underline{J}$ configuration.

FABRY-PEROT ELECTRON RESONANCES IN WIDE QUANTUM WELLS

The semi-classical criterion for observing a well-defined Fabry-Perot (standing wave) resonance in a double barrier structure is that the electron makes at least two traversals of the well before the coherence of its de Broglie wave is destroyed by a scattering process, e.g. emission of a phonon. Here we examine the current-voltage characteristics of a series of double barrier resonant tunnelling devices based on n-type GaAs/(AlGa)As and grown by molecular beam epitaxy (MBE) with well thicknesses between 60 and 180 nm. The layers were grown at a temperature of 630°C on silicon-doped GaAs (100)-substrates ($n = 2 \times 10^{18}$ cm^{-3}). The aluminium mole fraction in the (AlGa)As barrier layers was $x = 0.4$. Undoped GaAs spacer layers separate barrier and well regions from the doped contact layers. The spacer layers appear to improve the resonant tunnelling characteristics. A possible explanation is that the ionised impurity scattering in the resonant tunnelling region is thereby reduced. Structures A, B and C had barriers of thickness $b = 5.6$ nm whereas for structure D, $b = 8.5$ nm. The well width, w, for structures A, B, C and D

were 60, 120, 180 and 120 nm respectively. The composition of the devices
is given in Figure 3, below.

The devices were processed using optical lithography and chemical etching
to form mesas of diameter 100 and 200 μm.

A schematic diagram of the variation of the electron potential energy
across the structures is shown in Figure 2. Because of the light doping
in the regions close to the barriers, an applied voltage produces accumulation and depletion layers as shown. Electrons tunnel from the quasi-two
dimensional electron gas (2DEG) which forms in the accumulation layer of
the emitter contact. The low current density passed by the devices means

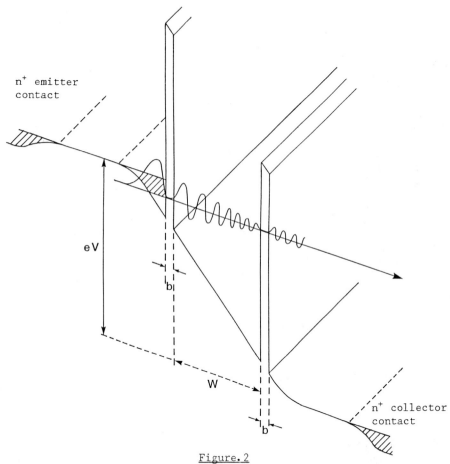

<u>Figure. 2</u>
The conduction band profile (electron potential energy) across a double
barrier resonant tunnelling structure. The left-hand contact acts as the
electron emitter. When there is an undoped interface between the heavily
doped contact and the barriers, an electron accumulation layer (shaded) is
formed adjacent to the emitter barrier. Beyond the collector barrier
(right hand side) a positively-charged depletion layer forms. The diagram
also shows the wavefunction of electrons bound in the accumulation layer
potential and that of a standing wave resonance in the quantum well.

The quasi-Fermi energies of electrons in the emitter and collector
contacts and in the accumulation layer correspond to the tops of the three
shaded regions.

that the average lifetime of an electron in the accumulation layer is relatively long (≈ 1 ns), so that electrons arriving from the heavily doped regions have time to thermalise before tunnelling. Resonant tunnelling occurs when the quasi-bound state energy of the 2DEG matches that of a standing wave state of the well. The well states are strongly affected by the large electric field and approximate to the eigenstates of a Stark ladder.

0.5 μm thick n$^+$GaAs top contact layer (2 x 10^{18} cm^{-3})
50 nm thick n-GaAs (2 x 10^{16} cm^{-3})
2.5 nm thick GaAs spacer layer (undoped)
(AlGa)As barrier of thickness b, [Al] = 0.4 (undoped)
GaAs well of thickness w (undoped)
(AlGa)As barrier of thickness b, [Al] = 0.4 (undoped)
2.5 nm thick GaAs spacer layer (undoped)
50 nm thick n-GaAs (2 x 10^{16} cm^{-3})
2 μm thick n$^+$GaAs buffer layer (2 x 10^{18} cm^{-3})
n$^+$GaAs substrate (2 x 10^{18} cm^{-3})

Figure. 3   Schematic diagram showing composition of resonant tunnelling devices A, B, C and D.

The current-voltage characteristics I(V) and derivatives (dI/dV or d^2I/dV^2) at 4 K are shown in Figure 4 for all four devices (boxes A to D). The substrate is biased positive. Devices A and B show regions of negative differential conductivity (NDC) in I(V). As the width of the well increases the number of resonances observed increases (up to 70 for device B) and their amplitude decreases. For structures C and D resonances are clearly seen only in the second derivative d^2I/dV^2. The existence of well-defined resonances implies coherent standing-wave states in the well. Therefore a significant fraction of the incident electrons must make at least two traversals of the well without scattering even when they have kinetic energies of several hundred meV. The widest well in which we have observed resonances is device C (180 nm), in which case some electrons would have to travel at least 360 nm ballistically. A devices with a 240 nm wide well was grown, but showed no quantum interference effects. For devices A and B resonance effects were still clearly identifiable at room temperature[11]. Note that for all four devices most resonances occur at voltages for which electrons reach the collector barrier with kinetic energies exceeding the height of the potential step. These "over the barrier" resonances arise from the partial reflection of de Broglie waves at the potential discontinuity[11,12].

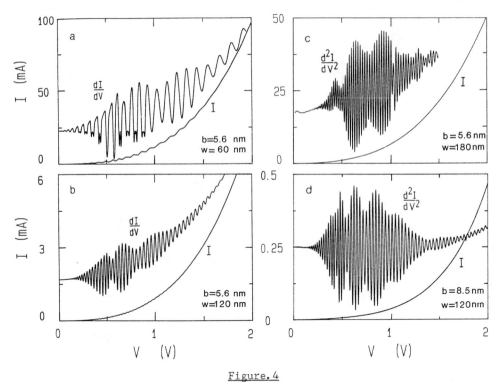

<u>Figure. 4</u>

Plots of the current-voltage characteristics I(V) and derivatives (dI/dV
or d^2I/dV^2) at 4 K for structures A to D. The diameters of the mesas are
100 μm.

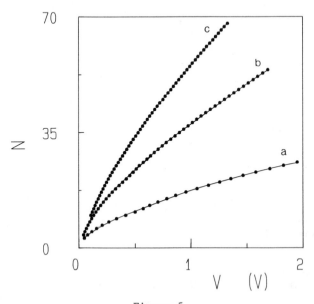

<u>Figure. 5</u>

Plots of peak number against peak voltage of the Fabry-Perot resonances in
I(V) for structures A (60 nm wide well), B (120 nm), C (180 nm).

The finite thickness of the collector barrier gives rise to an interesting quantum mechanical effect which produces the beating pattern in the amplitude of the resonances at high biases (see the dI/dV and d^2I/dV^2 plots in Figure 4). For a rectangular barrier, it is well known that the reflection coefficient falls to zero when an integral or half-integral number of de Broglie wavelengths fit within its width[13]. In our devices, the potential drop across the collector barrier means that the reflection coefficient is a minimum rather than zero when this condition is satisfied. For a varying potential the reflection coefficient is minimum when

$$\int_0^b k(x)\ dx = n\pi\ , \tag{3}$$

where the integral is taken over the collector barrier, n = 1,2,3.... and k(x), the electron wavenumber, is given by the kinetic energy and effective mass of the electron in the collector barrier region. Since the "over the barrier" resonant states of the well arise from a standing-wave interference between the waves incident on and reflected from the collector barrier, we expect the amplitude of the oscillatory structure in the conductance to be a minimum when equation 3 is satisfied.

In Figure 5 we have assigned an integer N to each resonant peak in the current. The figure plots N against peak voltage V for devices A, B and C. N essentially counts the number of de Broglie half-wavelengths of the standing resonance in the quantum well. As the well width increases the spacing between peaks decreases. This reflects the decrease in the energy spacing of the quasi-bound states of the well. At high biases the N(V) curves become approximately linear. A detailed analysis of these curves is complicated since it is necessary to model the non-parabolicity of the GaAs conduction band at electron energies up to and above 1 eV.

RESONANT TUNNELLING STUDIES OF MAGNETO-ELECTRIC QUANTISATION IN WIDE QUANTUM WELLS

In this section we use resonant tunnelling in double barrier hetero-structures with wide quantum wells to investigate the quantum mechanics of electron transport in crossed electric and magnetic fields. Resonances due to tunnelling into two distinct types of magneto-electric states are observed: 'traversing' orbits where the electron interacts with both potential barriers and 'skipping' states, in which the electrons interact with one barrier only. In addition, we observe magneto-oscillations in the current due to electrons tunnelling with different values of momentum in the plane of the barriers.

The voltage positions of the resonances observed in the I(V) characteristics are little changed by the presence of a magnetic field applied perpendicular to the plane of the barriers i.e. parallel to the direction of current flow ($\underline{B}||\underline{J}$). This result can be understood classically since there is no Lorentz force component associated with the motion of an electron in the direction of the applied electric field. The electron sheet density, n_s, in the 2DEG formed in the emitter accumulation layer can be measured at each voltage from the magneto-oscillations which are observed in the tunnel current I(B) when a magnetic field ($\underline{B}||\underline{J}$) is applied perpendicular to the plane of the barriers[11]. These oscillations arise from the passage of Landau levels through the quasi-Fermi level in the emitter 2DEG. However, when the magnetic field is parallel to the plane of the barriers, ($\underline{B}\perp\underline{J}$), the electronic motion in the quantum well is greatly modified. This can be seen clearly in Figure 6 which

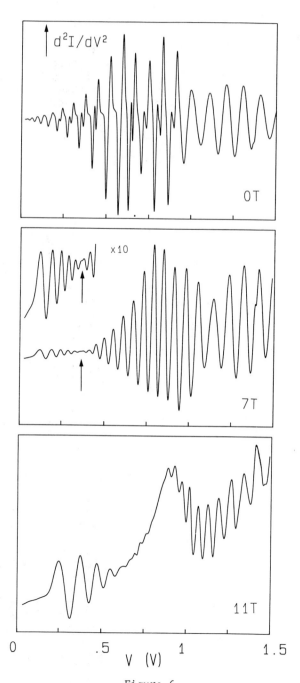

Figure. 6

Plots of d^2I/dV^2 versus V for Structure A (60 nm well) at 4 K for various magnetic fields ($\underline{B}\perp\underline{J}$). For B > 2 T, a series of oscillations due to tunnelling into magnetically quantised interface states can be observed. The crossover between tunnelling into skipping and traversing orbits is indicated by an arrow.

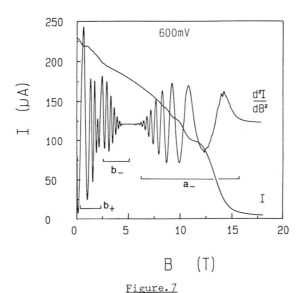

Figure. 7

Plots of I(B) and d^2I/dB^2 at V = 600 mV, T = 4 K for a 100 μm diameter
mesa of Structure B, showing oscillatory structure due to tunnelling into
magneto-electric states. The labels are explained in the text. Note the
quenching of the current at high magnetic fields.

shows the effect of increasing B on the plots of d^2I/dV^2 versus V for
structure A with a 60 nm wide quantum well. The data are presented as
second derivative plots since, in contrast to the case for $\underline{B}||\underline{J}$, the
transverse field ($\underline{B}\perp\underline{J}$) attenuates the resonances in I(V) relative to the
monotonically increasing background[14]. The effect of a magnetic field
on the tunnel current can also be seen by holding the applied voltage
constant and sweeping the magnetic field. Figure 7 shows a typical I(B)
plot for structure B (120 nm wide well) at a fixed bias voltage of 600 mV.
The tunnel current falls rapidly to zero with increasing transverse
magnetic field. Results with $\underline{B}||\underline{J}$ show that there is only a small
reduction in the tunnel current (< 10% at 18 T) in this configuration.
Oscillatory structure in I(B) is revealed more clearly in the derivative
d^2I/dB^2, also shown in Figure 7. There are three distinct series of
oscillations, labelled b_+, b_- and a_-. This labelling will be explained
below. None of the series is periodic in 1/B which would be the case for
tunnelling into bulk Landau levels. Figure 8 is a fan-chart plotting the
variation of the magnetic field positions of the resonances as a function
of applied voltage for structure A.

To understand the origin of the three series we use an approach
similar to that outlined in by Snell et al[8] and consider the effect of a
magnetic field on the energy levels in the quantum well, $\epsilon_n(k_y)$, and in
the accumulation layer $\epsilon_a(k_y,k_z)$ separately. We define a set of Cartesian
coordinates so that the x-axis is perpendicular to the plane of the
barriers and B is parallel to the z-axis i.e. $\underline{B}$ = (0,0,B), $\underline{E}$ = (-E,0,0).

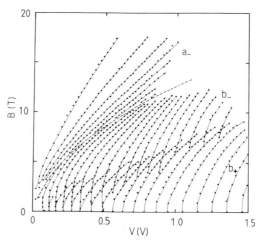

<u>Figure. 8</u>

Fan chart showing the positions of minima in d^2I/dB^2 and in d^2I/dV^2 in B-V space for Structure A. This illustrates the transition from electric to magnetic quantisation.

Using the magnetic vector potential in the Landau gauge ($\underline{A} = (0,Bx,0)$) and taking the origin of coordinates to be at the right hand interface of the emitter barrier, we can write the wavefunction in the well as $\psi(\underline{r}) = \exp[i(k_y y + k_z z)]\phi(x)$ where $\phi(x)$ is a solution of the one-dimensional Schrödinger equation,

$$\left\{ \frac{p_x^2}{2m^*} + \frac{\hbar^2 k_z^2}{2m^*} + m^*\omega_c^2(x - X)^2/2 - eEX + m^*v^2/2 + U(x) \right\}\phi(x)$$

$$= \left\{ \epsilon_n(k_y) + \frac{\hbar^2 k_z^2}{2m^*} \right\}\phi(x) \quad ,$$

E is the electric field, U(x) represents the discontinuities at the barriers, ω_c is the cyclotron frequency, $v = E/B$ is the velocity parallel to the interface and X, the origin of the simple harmonic oscillator potential, is given by $m^*E/eB^2 - \hbar k_y/eB$. The dependence of X on the transverse momentum $\hbar k_y$ is due to the action of the Lorentz force on the electron motion. We have calculated $\epsilon_n(k_y)$ using the WKB approximation in the simplified case of impenetrable barriers. The result of these calculations for structure A at a magnetic field of 10 T and an applied voltage of 1 V is shown in Figure 9. The orbit centre position X, which is related to k_y, is also plotted. As can be seen, there are four distinct groups of states. Those in the region of the ϵ-k_y diagram labelled d, with low energies and orbit centres near the middle of the well, corre- spond to bulk Landau levels of energy $(n + \frac{1}{2})\hbar\omega_c - eEX$. The electron orbit is unaffected by the presence of the barriers and, semiclassically, the electron executes cycloidal motion perpendicular to both the electric and magnetic fields with a radius of $2m^*E/eB^2$. In the region marked a, where the orbit centre is close to the emitter barrier, the electron interacts with the barrier and the energy level is increased. These states correspond to semiclassical skipping orbits which intersect with the interface as illustrated in the inset of Fig. 9. When the cyclotron orbit diameter exceeds the width of the well (i.e. when $2m^*E/eB^2 > w$) the electrons interact with both barriers forming traversing states (region b in the ϵ-k_y plane) in which the electron orbit extends across the well. At $B \to 0$ these evolve into the box quantised states of the well. Skipping orbits which intersect the collector barrier have energies in region c. Note that since the skipping orbits develop in a region of large electric field they are essentially different from those recently reported in single barrier heterostructures[8].

The energy of electrons in the emitter 2DEG is given by

$$\epsilon_a(k_y,k_z) = \epsilon_o + \frac{\hbar^2(k_y - k_o)^2}{2m^*} + \frac{\hbar^2k_z^2}{2m^*}$$

ϵ_o is the quasi-bound state energy, which has only a weak dependence on magnetic field. For most of the voltage and magnetic field range under consideration here the emitter state is strongly bound by the electrostatic potential and the magnetic field may be considered as a perturbation. This is in contrast to the experiments of Helm et al[10] where the emitter state was weakly bound. The shift in momentum of $\hbar k_o$ is caused by the action of the Lorentz force as the electron traverses the region of the barrier, k_o is given by $eB(b + a)/\hbar$ where b is the width of the barrier and a_o is the average distance of the 2DEG from the interface. The tunnelling process is governed by conservation of energy and of the transverse components of momentum, $\hbar k_z$ and $\hbar k_y = m^*v_y - eBx$. The condition for resonance given by the conservation rules is

$$\epsilon_o + \frac{\hbar^2(k_y - k_o)^2}{2m^*} = \epsilon_n(k_y).$$

Tunnelling can only occur from occupied emitter states which satisfy this condition. The occupancy of emitter states is given by the Fermi distribution function which, at low temperatures, is sharply cut off at $k_y = k_o \pm k_f$. Therefore, occupied emitter states can be represented by the parabola ϵ_a in Figure 9. The resonance condition may be interpreted graphically by looking for intersections in the ϵ-k_y plane of this parabola with the curves $\epsilon_n(k_y)$. This yields a discrete set of k_y values, each corresponding to a group of electrons in the emitter which are the only ones that contribute to the tunnel current. Sweeping either voltage or magnetic field causes the parabola to move relative to the dispersion curves and therefore the number of intersections (and hence the current) changes. This naturally leads to two sets of oscillations, one associated with each extremity of the parabola, i.e. at $k_y - k_o = +k_f$ and $k_y - k_o = -k_f$. When the parabola crosses into different regions of the ϵ-k_y plane the intersections will correspond to the different types of orbits described above and will have a distinct voltage and field dependence.

We can now interpret the I(B) curve of Figure 7. At low fields the energy levels are closely spaced and the parabola of emitter states lies

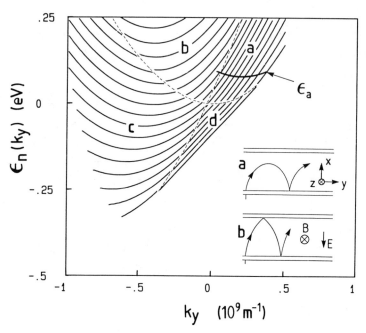

Figure 9

Plot of the energy eigenvalues $\epsilon_n(k_y)$ of the hybrid magneto-electric states in the 60 nm quantum well of Structure A for V = 1 V and B = 10 T. The energies were calculated using the WKB approximation. The parabola marked ϵ_a corresponds to the energies of the occupied states in the emitter accumulation layer. The inset shows the semiclassical orbits corresponding to (a) skipping and (b) traversing orbits.

in region b of the ϵ-k_y diagram so the electrons are tunnelling into traversing orbits. There are a large number of intersections and the current is high. As the magnetic field is increased the parabola ϵ_a shifts to higher k_y (i.e. to the right in Figure 9), due to the increase in k_o, and $\epsilon_n(k_y)$ shift to higher energies. This causes intersections to enter the right hand side of the parabola (at $k_y - k_o = +k_f$) giving rise to the series of oscillations between 0 and 2 T marked b_+ in Figure 7. The loss of intersections from the left hand side ($k_y - k_o = -k_f$) gives rise to the oscillations marked b_-. At higher fields the parabola is close to the right hand edge of the dispersion curves (in section a) and the oscillations are due to tunnelling into skipping states (labelled a_-). The observation of only one series at most values of B and V, as shown in Figure 9, is due to the different tunnelling probabilities for $k_y - k_o = \pm k_f$. In particular, the absence of a series corresponding to tunnelling into skipping orbits with $k_y - k_o = +k_f$ (i.e. a_+ type orbits) is due to the low matrix elements for these transitions[15]. As the number of intersections decreases, the current falls, until at B = 15 T (for this voltage) there are no intersections and the current is completely quenched. Note that in Figure 8 both the b_+ and b_- oscillations extrapolate back to the positions of the zero field 'box-quantised' resonances. The relatively low values of k_f in the 2DEG mean that tunnelling into bulk Landau levels cannot occur. In order to obtain more than qualitative agreement with the data it will be necessary to extend the model to include the effects of finite barrier height and the nonparabolicity and anisotropy of the conduction band at high energies.

The classical skipping orbit trajectory which just grazes the collector barrier has a path length of ~400 nm between intersections with the emitter barrier. The observation of magneto-oscillations due to electrons tunnelling into this state requires that a significant number of electrons have a ballistic path of at least this length. However, since electrons which tunnel into skipping states travel parallel to the interface, scattering is necessary for them to contribute to the measured current which flows perpendicular to the interface.

CONCLUSION

The previous section described oscillatory structure in the current due to electrons tunnelling from a 2DEG into a quantum well in the presence of crossed electric and magnetic fields. The resonances illustrate the development of hybrid magneto-electric quantisation. At high magnetic fields the electrons tunnel into interfacial Landau levels or skipping orbits in semiclassical terms. The formation of these states shows that conduction electrons accelerated to high kinetic energies (>1 eV) by large electric fields (~5 x 10^6 V m^{-1}) have exceptionally long ballistic path lengths (0.5 μm) in the pure GaAs layer which forms the quantum well of the resonant tunnelling device. Other chapters in these Proceedings describe even longer path levels in the high mobility two dimensional electron gas system based on single n-type modulation doped GaAs/(AlGa)As. In such structures, electrons at the Fermi energy can move in skipping orbits under the combined influence of a perpendicular magnetic field and a large potential formed near the edge of the metallic gate. Such orbits (edge states in quantum mechanical terms) appear to play a key role in the Quantum Hall Effect and in electron focussing between quantum point contacts.

ACKNOWLEDGEMENTS

The work on tunnelling into hybrid magneto-electric orbits was done in collaboration with M. L. Leadbeater, E. S. Alves, M. Henini, O. H. Hughes, A. Celeste and J. C. Portal and was supported by SERC (U.K.), CNRS (France) and the European Community.

REFERENCES

1. E. E. Mendez; "Physics and Applications of Quantum Wells and Superlattices", NATO ASI Series B 170:159 (1987).
2. L. L. Chang, L. Esaki and R. Tsu, Appl. Phys. Letts. 24(12) (1974).
3. E. S. Alves, L. Eaves, M. Henini, O. H. Hughes, M. L. Leadbeater, F. W. Sheard, G. A. Toombs, G. Hill and M. A. Pate, Electronics Letts. 24(18):1191 (1988).
4. M. L. Leadbeater, E. S. Alves, L. Eaves, M. Henini, O. H. Hughes, F. W. Sheard and G. A. Toombs, Semicond. Sci. Technol. 3:1060 (1988). See also M. L. Leadbeater, E. S. Alves, F. W. Sheard, L. Eaves, M. Henini, O. H. Hughes and G. A. Toombs, J. Phys.: Condens. Matter 1(51):10605 (1989).
5. A. Zaslavsky, V. J. Goldman and D. C. Tsui, Appl. Phys. Lett. 53:1408 (1988).
6. E. S. Alves, M. L. Leadbeater, L. Eaves, M. Henini, O. H. Hughes, A. Celeste, J. C. Portal, G. Hill and M. A. Pate, Superlattices and Microstructures 5(4):527 (1989).
7. L. Eaves, E. S. Alves, T. J. Foster, M. Henini, O. H. Hughes, M. L. Leadbeater, F. W. Sheard, G. A. Toombs, K. S. Chan, A. Celeste, J. C. Portal, G. Hill and M. A. Pate, Springer Series in Solid State Sciences 83: "Physics and Technology of Submicron Structures", ed. G. Bauer and F. Kuchar (Springer-Verlag, Berlin, 1988) p.74.
8. B. R. Snell, K. S. Chan, F. W. Sheard, L. Eaves, G. A. Toombs, D. K. Maude, J. C. Portal, S. J. Bass, P. A. Claxton, G. Hill and M. A. Pate, Phys. Rev. Lett. 59:2806 (1987).
9. M. L. Leadbeater, E. S. Alves, L. Eaves, M. Henini, O. H. Hughes, A. Celeste, J. C. Portal, G. Hill and M. A. Pate, J. Phys.: Condens. Matter 1:4865 (1989).
10. M. Helm, F. M. Peeters, P. England, J. R. Hayes and E. Colas, Phys. Rev. B 39:3427 (1989).
11. M. Henini, M. L. Leadbeater, E. S. Alves, L. Eaves and O. H. Hughes, J. Phys.: Condens. Matter 1:L3025 (1989).
12. S. Sen, F. Capasso, A. C. Gossard, R. A. Spah, A. L. Hutchinson and S. N. G. Chu, Appl. Phys. Lett. 51:18 (1987).
13. L. I. Schiff, Quantum Mechanics, 1st edition (1949), (New York, McGraw-Hill, chapter 5).
14. M. L. Leadbeater, L. Eaves, P. E. Simmonds, G. A. Toombs, F. W. Sheard, P. A. Claxton, G. Hill and M. A. Pate, Solid State Electron. 31:707 (1988).
15. F. W. Sheard, K. S. Chan, G. A. Toombs, L. Eaves and J. C. Portal, 14th Int. Symp. GaAs and Related Compounds, Inst. Phys. Conf. Series 91:387 (1988).

THE BACKGROUND TO RESONANT TUNNELLING THEORY

G. A. Toombs and F. W. Sheard

Department of Physics
University of Nottingham
Nottingham, NG7 2RD, England

INTRODUCTION

The concept of tunnelling is almost as old as quantum mechanics itself. Oppenheimer (1928) wrote a paper on a hydrogen atom in an electric field as early as 1927. He calculated the rate of dissociation of the hydrogen atom and developed a method for determining the transition probabilities between two almost orthogonal states of the same energy. This method was to be rediscovered years later by Bardeen (1961) and applied to many particle tunnelling between two metals separated by a thin oxide layer. It is normally referred to as the Bardeen transfer hamiltonian method and it is important in tunnelling theory. The most familiar example of tunnelling is undoubtedly the radioactive decay of a nucleus by α-particle emission. The enormous range of lifetimes observed experimentally for radioactive decay is a consequence of the sensitivity of the transmission coefficient to variations in the tunnel barrier. For the same reason, a wide range of lifetimes should also be expected for tunnelling in semiconductors even though the potential barriers are much smaller.

Tunnelling in solids has been reviewed by Duke (1969) and this review included an account of resonant tunnelling. However interest in resonant tunnelling has been greatly stimulated by recent advances in the techniques of growth of semiconductor heterostructures. Since the first observation by Chang et al (1974) of resonant tunnelling in a double-barrier semiconductor heterostructure there have been major improvements in sample quality which have extended the range for observation of negative differential resistance up to room temperature (Morkoc et al, 1986; Goodhue et al, 1986) and increased the observed peak-to-valley ratios in the current to 63:1 at a temperature of 77 K for a InGaAs/AlAs/InAs double-barrier structure (DBS) (Broekaert 1988). This increase in experimental activity has been accompanied by developments in the theory of resonant tunnelling. These developments are reviewed in this article.

The resonant tunnelling current is first calculated using the global transmission coefficient for a coherent wavefunction throughout the heterostructure. The transfer matrix method of calculating the transmission coefficient of a multi-barrier structure is introduced. The general properties of symmetric barrier structures are derived. The transfer matrix method is used to calculate the transmission coeffi-

Electronic Properties of Multilayers and Low-Dimensional Semiconductors Structures
Edited by J. M. Chamberlain *et al.,* Plenum Press, New York, 1990

257

cient of symmetric and asymmetric DBS. The current for an asymmetric
DBS is calculated. The WKB method for evaluating transmission coefficients for systems which do not have analytic solutions is outlined.
The tunnel current for a DBS is then calculated using sequential
tunnelling theory in which the transmission is regarded as two successive transitions, from the emitter into the quasi-bound state of the
well and then from the well into the collector contact. The Bardeen
transfer hamiltonian used in the sequential tunnelling theory is
derived. The results for the sequential and global approaches are
shown to be the same for an asymmetric DBS. The effects of charge
buildup in the quantum well on the current-voltage characteristics are
discussed using the sequential tunnelling theory. It is shown that
bistability in the current can occur and the conditions required to
observe it are considered. The use of high magnetic fields to investigate semiconductor heterostructures is described and some examples are
given. Finally, the factors limiting the peak-to-valley ratio in the
tunnel current are discussed.

THE TUNNELLING CURRENT

    We calculate first the current for electrons tunnelling from a
heavily doped emitter through a series of potential barriers formed by
a semiconductor heterostructure into a heavily doped collector. Such a
situation is illustrated in Fig. 1 by a DBS under a bias voltage V. The
electrons are assumed to be described be a one-electron hamiltonian.
The heavy doping ensures that the electrons tunnel from a Fermi sea of
three-dimensional states. The theory of tunnelling from a three-dimensional electron gas (3DEG) differs in detail from that for tunnelling
from a two-dimensional electron gas (2DEG). We consider here
tunnelling from a 3DEG and mention briefly the differences from 2DEG
tunnelling in the section on sequential tunnelling theory.

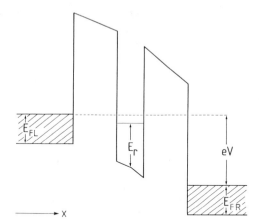

Fig. 1. Spatial variation of electron potential energy through a
        double-barrier structure, showing a bound state and the Fermi
        levels.

    Any accumulation or depletion of charge in a heterostructure can
be taken into account by introducing an electrostatic potential,
determined by Poisson's equation, into the hamiltonian. In Fig. 1, no
band-bending due to accumulation in the emitter or depletion in the
collector is shown since the heavy doping makes the screening lengths
short. The electric fields in the two barriers are shown to be different due to the buildup of charge in the undoped quantum well between

258

the barriers. The heterostructure is assumed to be grown ideally so that there are no defects or impurities and all the interfaces are perfectly planar. Under these conditions, the momentum components parallel to the interfaces are conserved. We discuss later the effects that imperfections have on the tunnel current.

The current density is given by the number of electrons per unit volume incident on the emitter barrier with velocity v_x perpendicular to the barrier and the probability of their transmission through the total system of barriers. Since $v_x = 1/\hbar(\partial E/\partial k_x)$, where E and $\mathbf{k}$ are respectively the electron energy and wavevector, the current density J_{LR} for tunnelling from left to right is given by

$$J_{LR} = \frac{e}{4\pi^3\hbar} \int dk_x \; dk^2_{||} \; f_L(E) \; [1 - f_R(E)] \; \frac{\partial E}{\partial k_x} \; T(E_x) \; , \tag{1}$$

where $E_x = \hbar^2 k_x^2/2m^*$, $T(E_x)$ is the global transmission coefficient for the set of barriers and f_L and f_R are respectively the Fermi-Dirac distributions for the left-hand emitter and right-hand collector. For a parabolic band, we can put

$$dk^2_{||} = \frac{2\pi m^*}{\hbar^2} \; dE_{||} \; , \tag{2}$$

where m^* is the effective mass of an electron. The current density J is then obtained by subtracting from J_{LR} a similar equation for tunnelling from right to left. We then find

$$J = \frac{em^*}{2\pi^2\hbar^3} \int dE_x \; dE_{||} \; [f_L(E) - f_R(E)] \; T(E_x). \tag{3}$$

Many experiments are carried out at low temperatures as the resonant effects are smeared out at high temperatures. We shall also take the temperature to be low since the zero-temperature limit simplifies the algebra. If the voltage is such that the Fermi energy in the emitter $E_{FL} \leq eV$, electrons in all the occupied states of the emitter may tunnel into unoccupied states of the collector. The current density (3) is then

$$J = \frac{em^*}{2\pi^2\hbar^3} \int_0^{E_{FL}} dE_x (E_{FL} - E_x) \; T(E_x) \; , \quad eV > E_{FL} \tag{4}$$

since the range of $E_{||}$ for the tunnel current is from 0 to $E_{FL} - E_x$. When $E_{FL} > eV$, the range of $E_{||}$ is restricted for tunnelling into unoccupied states. Therefore, in this case

$$J = \frac{em^*}{2\pi^2\hbar^3} \left[eV \int_0^{E_{FL}-eV} dE_x \; T(E_x) + \int_{E_{FL}-eV}^{E_{FL}} dE_x \; (E_{FL} - E_x)T(E_x) \right], \quad eV < E_{FL} \tag{5}$$

The tunnelling current can therefore be calculated provided the global transmission coefficient $T(E_x)$ is known. $T(E_x)$ for a series of potential barriers is readily determined by the transfer-matrix method.

259

THE TRANSFER-MATRIX METHOD

The transfer-matrix method is simply a systematic treatment of the textbook problem of tunnelling. If we assume that the electron potential $V(x)$ is only a function of the distance x then the time-independent Schrödinger equation

$$\frac{-\hbar^2}{2m^*} \nabla^2 \psi + V(x)\psi = E\psi \tag{6}$$

has eigenfunctions $\psi = \phi(x)\exp(ik_y y + ik_z z)$, where $\phi(x)$ is a solution of the equation

$$\frac{-\hbar^2}{2m^*} \frac{d^2\phi}{dx^2} + V(x)\phi = E_x \phi \tag{7}$$

and

$$E_x = E - \frac{\hbar^2 k_y^2}{2m^*} - \frac{\hbar^2 k_z^2}{2m^*}$$

The problem is then essentially a one-dimensional one.

The potential $V(x)$ will change discontinuously at each interface of the heterostructure. Therefore equation (7) is solved for each semiconductor region and the solutions are matched at the boundaries. The boundary conditions for a boundary at $x = x_0$ between two regions 1 and 2 with solutions $\phi_1(x)$ and $\phi_2(x)$ of equation (7) are

$$\phi_1(x_0) = \phi_2(x_0) \tag{8}$$

and

$$\frac{1}{m_1^*} \frac{d\phi_1}{dx}\bigg|_{x=x_0} = \frac{1}{m_2^*} \frac{d\phi_2}{dx}\bigg|_{x=x_0}. \tag{9}$$

Equations (8) and (9) guarantee conservation of flux and allow for different effective masses in the two regions. If the one-dimensional Schrödinger equation (7) cannot be solved analytically for each region then either an approximate solution such as obtained by the WKB method or a numerical solution has to be used for that region.

To illustrate the transfer-matrix method we consider first two regions of constant potential as shown in Fig. 2. The potential barrier formed at the interface between the regions is due to the different band gaps in the two semiconductors. The height of the barrier is determined by the band-gap offset (Kroemer, 1986; Van de Walle and Martin, 1987). When the potential is constant within a given region, the well-known general solution of equation (7) is

$$\phi(x) = A\exp(ik_x x) + B\exp(-ik_x x), \tag{10}$$

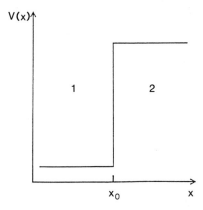

$$V(x)$$

1   2

x_0 x

<u>Fig. 2.</u>   Electron potential at the interface between two semiconductor
materials.

where

$$\frac{\hbar^2 k_x^2}{2m^*} = E_x - V_o \qquad (11)$$

and V_o is the constant potential. If $E_x > V_o$, then k_x is real and the
solution is made up of plane waves. If $E_x < V_o$, then k_x is imaginary
and the solution contains growing and decaying waves. The boundary
conditions (8) and (9) relate the values of the coefficients A and B on
the two sides of the boundary at x_o. This relationship can be written
as

$$\begin{pmatrix} A_1 \\ B_1 \end{pmatrix} = N \begin{pmatrix} A_2 \\ B_2 \end{pmatrix} \qquad (12)$$

where the matrix N is given by

$$N = \frac{1}{2} \begin{bmatrix} \left(1 + \dfrac{k_2 m_1^*}{k_1 m_2^*}\right) e^{i(k_2 - k_1)x_o}, & \left(1 - \dfrac{k_2 m_1^*}{k_1 m_2^*}\right) e^{-i(k_2 + k_1)x_o} \\[3mm] \left(1 - \dfrac{k_2 m_1^*}{k_1 m_2^*}\right) e^{i(k_2 + k_1)x_o}, & \left(1 + \dfrac{k_2 m_1^*}{k_1 m_2^*}\right) e^{i(k_1 - k_2)x_o} \end{bmatrix} \qquad (13)$$

and the subscript x has been omitted from the x-component of the
wavevector for simplicity. This form of matrix N has been used by Kane
(1969) to calculate the transmission coefficient for single- and
double-barrier structures. It has the disadvantage that the matrix N
depends on the absolute position x_o of the interface.

A better method is that used by Azbel (1983) and Ricco and Azbel
(1984). In this method, the matrix relates the waves evaluated at the
interface. If we put

$$u_n = A_n \exp(ik_n x) \qquad (14)$$

and

$$v_n = B_n \exp(-ik_n x) \tag{15}$$

then we have

$$\begin{pmatrix} u_1 \\ v_1 \end{pmatrix} = M \begin{pmatrix} u_2 \\ v_2 \end{pmatrix}, \tag{16}$$

where

$$M = \frac{1}{2} \begin{bmatrix} \left(1 + \dfrac{k_2 m_1^*}{k_1 m_2^*}\right) & \left(1 - \dfrac{k_2 m_1^*}{k_1 m_2^*}\right) \\[4mm] \left(1 - \dfrac{k_2 m_1^*}{k_1 m_2^*}\right) & \left(1 + \dfrac{k_2 m_1^*}{k_1 m_2^*}\right) \end{bmatrix}. \tag{17}$$

The matrix M does not depend on the position of the barrier. This information is now contained in the vectors (u_1, v_1) and (u_2, v_2) which are evaluated at $x = x_o$. To find out how the wavefunction changes across a sequence of barriers, equations similar to (16) are used in sequence. This requires matrices like (17) for each interface together with matrices for the regions between interfaces. If the potential be constant between interfaces at x_o and $x_o + w$, then

$$\begin{pmatrix} u_2 \\ v_2 \end{pmatrix} = \begin{pmatrix} \exp(-ikw) & 0 \\ 0 & \exp(ikw) \end{pmatrix} \begin{pmatrix} u_3 \\ v_3 \end{pmatrix}, \tag{18}$$

where u_2 and v_2 are evaluated at x_o and u_3 and v_3 at $x_o + w$. The matrix in (18) only depends on the relative position of the two interfaces. The absolute positions are again contained in the vectors (u_n, v_n).

For a sequence of potential steps, we have

$$\begin{pmatrix} u_1 \\ v_1 \end{pmatrix} = \prod_1 M_i \begin{pmatrix} u_n \\ v_n \end{pmatrix}, \tag{19}$$

where the matrices M_i are all similar to (17) or (18). The global transmission coefficient is simply given by setting $v_n = 0$ so that there is only a transmitted wave in the final region. Then

$$T(E_x) = \frac{k_n m_1^* |A_n|^2}{m_n^* k_1 |A_1|^2}$$

$$= \frac{k_n m_1^*}{m_n^* k_1} \frac{1}{|(M_G)_{11}|^2} \tag{20}$$

where

$$M_G = \prod_1 M_i. \tag{21}$$

The transfer-matrix method can thus be used for quite complicated problems if the potential profile is split up into a sufficient number of regions.

GENERAL PROPERTIES OF SYMMETRIC BARRIER STRUCTURES

We shall show that the transmission coefficient of a symmetric DBS can be exactly unity at certain energies so that a particle incident on the DBS has a 100% probability of transmission and zero probability of reflection. This is a remarkable result at first sight since the A_n and B_n of equation (10) are all complex functions of E_x and yet B_1 must go to zero for zero reflection. This requires the real part of B_1 which is a function of the energy E_x and the imaginary part which is a different function of E_x are zero for the same value of E_x. This is made possible as a result of the symmetry and therefore it is worthwhile to consider the general properties of barrier structures and symmetric ones in particular.

The solutions to the time-independent Schrödinger equation in the initial and final regions are related by

$$\begin{pmatrix} u_1 \\ v_1 \end{pmatrix} = \begin{pmatrix} M_{11} & M_{12} \\ M_{21} & M_{22} \end{pmatrix} \begin{pmatrix} u_n \\ v_n \end{pmatrix}, \tag{22}$$

where u_1 and v_1 are evaluated at the first interface and u_n and v_n at the final interface.

Since taking the complex conjugate of an eigenfunction is also a solution of the Schrödinger equation (time-reversal), $A_1^*\exp(-ik_x x) + B_1^*\exp(ik_x x)$ is a solution in the first region and $A_n^*\exp(-ik_x x) + B_n^*\exp(ik_x x)$ is a solution in the final region. B_1^* is now the complex amplitude for the incident wave and A_1^* for the reflected wave from the barrier system. B_n^* is the final transmitted amplitude.

Hence

$$\begin{pmatrix} v_1^* \\ u_1^* \end{pmatrix} = \begin{pmatrix} M_{11} & M_{12} \\ M_{21} & M_{22} \end{pmatrix} \begin{pmatrix} v_n^* \\ u_n^* \end{pmatrix}$$

which is equivalent to

$$\begin{pmatrix} u_1 \\ v_1 \end{pmatrix} = \begin{pmatrix} M_{22}^* & M_{21}^* \\ M_{12}^* & M_{11}^* \end{pmatrix} \begin{pmatrix} u_n \\ v_n \end{pmatrix} \tag{23}$$

Comparing equations (22) and (23), the matrix M for for any barrier structure must satisfy

$$M_{11}^* = M_{22}, \qquad M_{12}^* = M_{21}. \tag{24}$$

If we now turn our attention to a symmetric barrier structure and centre it on the origin $x = 0$, then another solution of the Schrödinger equation is obtained by replacing x by -x. Hence it is given by $A_n\exp(-ik_x x) + B_n\exp(ik_x x)$ in the initial region and $A_1\exp(-ik_x x) + B_1\exp(ik_x x)$ in the final region. We now have

$$\begin{pmatrix} v_n \\ u_n \end{pmatrix} = \begin{pmatrix} M_{11} & M_{12} \\ M_{21} & M_{22} \end{pmatrix} \begin{pmatrix} v_1 \\ u_1 \end{pmatrix}. \tag{25}$$

Therefore using equations (23) and (25)

$$\begin{pmatrix} v_1 \\ u_1 \end{pmatrix} = \begin{pmatrix} M_{11}^* & M_{12}^* \\ M_{12} & M_{11} \end{pmatrix} \begin{pmatrix} M_{11} & M_{12} \\ M_{12}^* & M_{11}^* \end{pmatrix} \begin{pmatrix} v_1 \\ u_1 \end{pmatrix} = \begin{pmatrix} 1 & 0 \\ 0 & 1 \end{pmatrix} \begin{pmatrix} v_1 \\ u_1 \end{pmatrix}. \tag{26}$$

263

Equation (26) shows that

$$|M_{11}|^2 + M_{12}^2 = 1 \tag{27}$$

and

$$M_{11}(M_{12} + M_{12}^*) = 0. \tag{28}$$

Another condition can be obtained by equating the initial and final fluxes for the symmetric barrier system:

$$|A_1|^2 - |B_1|^2 = |A_n|^2 - |B_n|^2$$

or

$$(u_1, v_1) \begin{pmatrix} u_1^* \\ -v_1^* \end{pmatrix} = (u_n, v_n) \begin{pmatrix} M_{11} & M_{12}^* \\ M_{12} & M_{11}^* \end{pmatrix} \begin{pmatrix} M_{11}^* & -M_{12}^* \\ -M_{12} & M_{11} \end{pmatrix} \begin{pmatrix} u_n^* \\ -v_n^* \end{pmatrix}$$

$$= (u_n, v_n) \begin{pmatrix} 1 & 0 \\ 0 & 1 \end{pmatrix} \begin{pmatrix} u_n^* \\ -v_n^* \end{pmatrix}. \tag{29}$$

Equation (29) shows that

$$|M_{11}|^2 - |M_{12}|^2 = 1. \tag{30}$$

Equations (24), (27), (28) and (30) can only be satisfied for the matrix M if M_{12} is purely imaginary for a symmetric system. Although this result has been derived for a potential centred on x = 0, it is always true since M only depends on the relative positions of interfaces.

We can define a transmission factor t and a reflection factor r by setting $B_n = 0$. We then have

$$t = \frac{u_n}{u_1} = \frac{1}{M_{11}} \tag{31}$$

and

$$r = \frac{v_1}{u_1} = \frac{M_{21}}{M_{11}}. \tag{32}$$

Therefore for any system of potential barriers we have

$$M = \begin{pmatrix} 1/t & r^*/t^* \\ r/t & 1/t^* \end{pmatrix}. \tag{33}$$

For a symmetric set of barriers we have the additional fact that r/t is purely imaginary. Therefore r contains a purely imaginary factor which is a function of energy. At certain energies E_x this single factor can go to zero and the reflection coefficient is zero. Therefore any symmetric barrier can have a transmission coefficient of unity. This is true even for one barrier but, as is well known, it occurs for values of E_x greater than the barrier height.

DOUBLE BARRIER STRUCTURES

We consider first the symmetric DBS shown in Fig. 3 and calculate the transmission coefficient using the transfer matrix method.

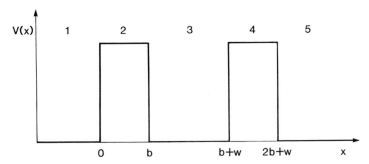

Fig. 3. The electron potential in a symmetric double-barrier structure. The regions of different potential are labelled 1 to 5.

The result is derived using transfer matrices M for the two identical single barriers. We then have

$$\begin{pmatrix} u_1 \\ v_1 \end{pmatrix} = \begin{pmatrix} M_{11}{}_* & M_{12}{}_* \\ M_{12}{}^* & M_{11}{}^* \end{pmatrix} \begin{pmatrix} \exp(-ikw) & 0 \\ 0 & \exp(ikw) \end{pmatrix} \begin{pmatrix} M_{11}{}_* & M_{12}{}_* \\ M_{12}{}^* & M_{11}{}^* \end{pmatrix} \begin{pmatrix} u_5 \\ v_5 \end{pmatrix}.$$

With $v_5 = 0$, we obtain

$$\frac{u_5}{u_1} = [M_{11}{}^2 \exp(-ikw) + |M_{12}|^2 \exp(ikw)]^{-1}$$

$$= \frac{t_B{}^2 \exp(ikw)}{1 - r_B{}^2 \exp(i2kw)} , \tag{34}$$

where the single-barrier transmission factor t_B and reflection factor r_B are defined by $M_{11} = t_B{}^{-1}$ and $M_{21} = r_B/t_B$. The transmission coefficient of the symmetric DBS is then

$$T_{2B} = \left[1 + \frac{4R_B}{T_B{}^2} \sin^2(kw - \theta) \right]^{-1} , \tag{35}$$

where $r_B{}^2 = R_B \exp(-i2\theta)$ and $|t_B|^2 = T_B$. R_B and T_B are thus the single-barrier reflection and transmission coefficient.

Fig. 4 shows T_{2B} for barriers of height 300 meV and width 4 nm. The well between the barriers is of width 10 nm and the effective mass $m^* = 0.07 m_e$. $T(E_x)$ has a series of sharp peaks as a function of E_x and the peak height is indeed 1. As can be seen from equation (35), this occurs whenever $kw - \theta = n\pi$ where n is an integer. This is almost exactly the same condition as that for the energy levels of a single isolated quantum well if the transmission coefficient of each barrier is small. Therefore the transmission coefficient peaks whenever the energy E_x is resonant with the quasi-bound states of the quantum well.

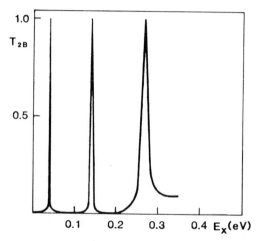

<u>Fig. 4.</u> The transmission coefficient T_{2B} of a symmetric double-barrier structure as a function of energy E_x. The potential barriers are of height 300 meV and width 4 nm. The well is of width 10 nm and $m^* = 0.07m_e$.

The transmission coefficient of a single barrier is approximately given by

$$T_B \simeq \frac{4E_x(V_o - E_x)}{V_o^2} \exp(-2\alpha b)$$

$$\approx \exp(-2\alpha b),\tag{36}$$

where

$$\alpha = [2m^*(V_o - E_x)]^{1/2}/\hbar\tag{37}$$

With the same values as before and $E_x = 50$ meV, $\alpha = 6.8\times10^8$ m^{-1} and $T_B = 4.4\times10^{-3}$. Therefore T_B is very small for values of the parameters which are typical for tunnelling through semiconductor single-barrier structures. Under these conditions R_B can be taken to be unity in equation (35).

The behaviour of T_{2B} close to resonance at energy $E_x = E_o$ can be found by expanding $\sin^2(kw - \theta)$ as a Taylor series in E_x since the other quantities in equation (35) vary relatively slowly with energy. E_o is the energy of the bound state relative to the emitter conduction band edge. Putting $\Phi = kw - \theta$, we have

$$T_{2B} = \left[1 + \frac{(E_x - E_o)^2}{\Delta E^2}\right]^{-1}\tag{38}$$

where the full width of the resonance at half-maximum is

$$2\Delta E = T_B(E_o)\left(\frac{\partial \Phi}{\partial E_o}\right)^{-1} = \frac{\hbar v_r\, T_B(E_o)}{(w + 2/\alpha)}\tag{39}$$

and

$$v_r = (2E_r/m^*)^{1/2}.\tag{40}$$

The width of the resonance and the lifetime τ of the quasi-bound state are related by $\tau = \hbar/2\Delta E$ (Price, 1988). τ is the combined lifetime for tunnelling out of the well through either barrier. Therefore the probability per second Γ of tunnelling through one barrier is $\Delta E/\hbar$ and, using equation (39),

$$\Gamma = \frac{v_r \, T_B}{2(w + 2/\alpha)} \, . \tag{41}$$

Γ is the product of T_B, the probability of tunnelling through a single barrier at one attempt, and the attempt rate $v_r/[2(w+2/\alpha)]$. The attempt rate is given by the electrons moving with velocity v_r backwards and forwards in not just the well but both the penetration distances α^{-1} for the decay of the wavefunction into the barriers. $v_r \simeq 5 \times 10^5 \, \mathrm{m \, s^{-1}}$ for $E_r = 50$ meV and using the same values of the parameters as before $\Gamma \simeq 10^{11} \, \mathrm{s^{-1}}$. Therefore the lifetime τ_1 for tunnelling from the quantum well through one barrier is $\tau_1 = \Gamma^{-1} \simeq 10$ ps. It should be remembered that τ_1 depends critically on the values which enter T_B, as can be seen from equations (36) and (37).

The transmission coefficient for an asymmetric DBS can also be derived using the transfer-matrix method. In this case the matrices for the left- and right-hand barriers are different. If the single-barrier transmission coefficients are small, then T_{2B} close to resonance is to a good approximation given by

$$T_{2B} = \frac{4 \, T_L \, T_R}{(T_L + T_R)^2} \left[1 + \frac{(E_x - E_o)^2}{\Delta E^2} \right]^{-1} , \tag{42}$$

where

$$\Delta E = \frac{\hbar v_r \, (T_L + T_R)}{4(w + 2/\alpha)} \tag{43}$$

and T_L and T_R are respectively the transmission coefficients of the left- and right-hand barriers. The lifetime $\hbar/2\Delta E$ is again the combined lifetime for tunnelling out of the well through either barrier. The transmission coefficient away from resonance is of order of the product of the probabilities of tunnelling through each barrier i.e. $T_{2B} \simeq T_L(E_x) T_R(E_x)$.

We can now calculate the tunnelling current for an asymmetric DBS using equation (42) and equation (4) which is valid since the tunnelling is into unoccupied states of the collector when the bound state is resonant. The integral over E_x is straightforward as it involves a Lorentzian of narrow width. We obtain

$$J = \frac{e m^* v_r}{2\pi \hbar^2 w} (E_{FL} - E_o) \frac{T_L \, T_R}{(T_L + T_R)} , \tag{44}$$

where we have dropped the penetration lengths from the attempt rate. Equation (44) is the resonant tunnelling current provided the bound state energy E_r lies in the energy range of the emitter Fermi sea. Therefore the current rises to a peak value J_p as the bound state energy passes the bottom of the Fermi sea;

$$J_p = \frac{em^* v_r E_{FL}}{2\pi\hbar^2 w} \frac{T_L T_R}{(T_L + T_R)}.$$

The current then drops rapidly with increasing voltage to the valley current which can be estimated by keeping the off-resonant transmission coefficient constant for the range of integration in equation (4). The valley current J_v is thus

$$J_v \simeq \frac{em^*}{4\pi^2\hbar^3} E_{FL}^2 T_L T_R$$

and the ideal low temperature peak-to-valley current ratio is typically $\geq 10^3$. This is more than a factor of 10 greater than the best values for structures grown to date.

THE WKB METHOD

The WKB approximation is a semiclassical approximation which is very useful when the Schrödinger equation cannot be solved analytically. It can be incorporated into the transfer-matrix method (Ricco and Azbel 1984). Details of the method can be found in many textbooks on quantum mechanics (e.g. Dicke and Wittke 1960) but we briefly discuss it here since it is so useful.

The solution to the one-dimensional, time-independent Schrödinger equation (7) is written as

$$\phi = \exp(\pm iz(x)), \tag{45}$$

where the phase $z(x)$ is assumed to vary slowly for a sufficiently slowly varying potential energy. Equation (45) is substituted into equation (7) and, if derivatives of z other than the first are ignored, we have

$$\frac{dz}{dx} = \frac{[2m^*\{E_x - V(x)\}]^{1/2}}{\hbar} = k(x).$$

Therefore

$$z = \int k \, dx.$$

This technique is now repeated by writing ϕ in the form

$$\phi = \exp\{\pm i \int k dx + \epsilon(x)\},$$

where $\epsilon(x)$ is now taken to vary slowly. We find

$$\frac{2d\epsilon}{dx} \simeq \frac{-1}{k}\frac{dk}{dx}$$

or

$$\epsilon = -\ln k^{1/2}$$

and

$$\phi = \frac{1}{k^{1/2}} \exp(\pm i \int k dx).$$ (46)

Equation (46) is the WKB approximation for the solutions of the Schrödinger equation. It assumes that $d^2k/dx^2 \ll k^3$ and $dk/dx \ll k^2$ as can be seen by direct substitution of equation (46) into the Schrödinger equation. The second condition requires that the change in potential energy in a distance of order a wavelength is very much less than the kinetic energy. The factor $k^{-1/2}$ ensures that the probability of finding the particle at a particular point in space is inversely proportional to the classical speed of the particle at that point.

Equation (46) is an approximate solution for $E_x >$ or $< V(x)$ but it fails when $E_x = V(x)$. This occurs for the positions shown in Fig. 5. They are the points where a classical particle moving in the potential would come to rest and turn around.

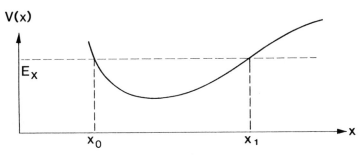

Fig. 5.   The classical turning points x_0 and x_1 for a particle of energy E_x moving in a potential $V(x)$.

Formulae can be derived connecting the solutions on either side of the classical turning points. The decaying solution to the left of x_0 is related to the solution

$$\frac{A_0}{k^{1/2}} \cos\left[\int_{x_0}^{x} k dx - \frac{\pi}{4} \right]$$ (47)

to the right of x_0. The decaying solution to the right of x_1 is connected to the solution

$$\frac{A_1}{k^{1/2}} \cos\left[\int_{x}^{x_1} k dx - \frac{\pi}{4} \right]$$ (48)

to the left of x_1. The solutions (47) and (48) must be identical between x_0 and x_1 and therefore

$$\int_{x_0}^{x_1} k\, dx = \left(n + \frac{1}{2} \right) \pi \quad , \quad n = 0, 1, 2 \ldots$$ (49)

Equation (49) determines the allowed energy levels for the situation shown in Fig. 5 where the turning points are "soft" boundaries. When there is a discontinuity of potential forming a hard boundary and one soft boundary the right-hand side of equation (49) becomes $(n + 3/4)\pi$. For two hard boundaries it is $(n + 1)\pi$.

BARDEEN TRANSFER HAMILTONIAN

In the sequential theory of resonant tunnelling from a DBS, the transmission is regarded as two separate processes. The first is a transition through the left-hand barrier into the well and the second is a transition out of the well through the right-hand barrier. Therefore we describe a method for calculating the required transition rates before discussing sequential tunnelling theory. It was proposed by Bardeen (1961) for problems in many-particle tunnelling but the essentials of the method were previously derived by Oppenheimer (1927).

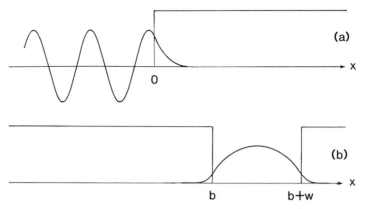

Fig. 6 (a) The initial state ψ_L and the potential for which it is a solution of the Schrödinger equation.

(b) The final state ψ_R to which the tunnelling particle makes a transition and its potential.

Since we wish to calculate the rate at which an electron tunnels through a potential barrier, such as the first barrier shown in Fig. 3, we prepare the electron in an initial state ψ_L which is localised to the left hand side of the barrier and calculate the rate at which the state evolves into a final state ψ_R localized to the right of the barrier. Fig. 6(a) shows the initial state of ψ_L. It is a correct solution of the hamiltonian with energy E_L for $x \leq b$ but it continues to decay for $x > b$. The potential for which ψ_L is an eigenstate is also shown in Fig. 6(a). Fig. 6(b) shows the final state ψ_R for tunnelling into a potential well. ψ_R is a correct solution of the hamiltonian for $0 \leq x \leq 2b + w$ but it is matched onto decaying solutions outside this range of x. It is therefore an eigenstate of the isolated potential well shown in Fig. 6(b).

The states ψ_L and ψ_R are not orthogonal and therefore an electron in the state ψ_L can make a transition to ψ_R. We now consider the evolution of the state

270

$$\psi = a(t)\psi_L \exp(-iE_L t/\hbar) + b(t)\psi_R \exp(-iE_R t/\hbar), \qquad (50)$$

where $a(0) = 1$ and $b(0) = 0$ since the electron is initially in ψ_L. The evolution of ψ is given by substituting expression (50) into the Schrödinger equation. We obtain

$$\left(i\hbar \frac{da}{dt} + E_L a\right) \psi_L \exp(-iE_L t/\hbar)$$

$$+ \quad \left(i\hbar \frac{db}{dt} + E_R b\right) \psi_R \exp(-iE_R t/\hbar) = H\psi \qquad (51)$$

where H is the hamiltonian of the actual tunnel-barrier structure. If the transition probability is small, $a(t) \simeq 1$ and $b(t) \simeq 0$. Therefore we put $a(t) = 1$, $b(t) = 0$ and $da/dt = 0$ which follows from the normalization of ψ since the overlap of ψ_L and ψ_R is taken to be small. Equation (51) becomes

$$i\hbar \frac{db}{dt} \psi_R = (H - E_L)\psi_L \exp\{i(E_R - E_L)t/\hbar\}$$

or

$$i\hbar \frac{db}{dt} = \int \psi_R^*(H-E_L)\psi_L \, dV \, \exp\{i(E_R-E_L)t/\hbar\}, \qquad (52)$$

if ψ_R is normalized.

Integration of equation (52) gives

$$b(t) = \int \psi_R^*(H-E_L)\psi_L \, dV \, \frac{(1 - \exp\{i(E_R - E_L)t/\hbar\})}{(E_R-E_L)}. \qquad (53)$$

The analysis now parallels that for the Fermi golden rule. The transition probability W per unit time is $|b(t)|^2/t$ and hence

$$W = \frac{2\pi}{\hbar} |M_{RL}|^2 \delta(E_R - E_L), \qquad (54)$$

where

$$M_{RL} = \int_{x_B}^{\infty} \psi_R^*(H - E_L)\psi_L \, dV \qquad (55)$$

and the lower limit of integration x_B can be set at any point in the barrier since $(H - E_L)\psi_L = 0$ for $x \leq b$. Equation (55) can be put in the more symmetric form

$$M_{RL} = \int_{x_B}^{\infty} [\psi_R^*(H - E_L)\psi_L - \psi_L(H - E_R)\psi_R^*] dV, \qquad (56)$$

as $(H - E_R)\psi_R^* = 0$ for $x \geq 0$. Putting $\psi_L = \phi_L(x)\exp\{i(k_y y + k_z z)\}/L^2$ and similarly for ψ_R, the integration of (55) over y and z gives con-

servation of the momentum components parallel to the interfaces as
before and the integration over x is carried out by parts. We find
that

$$
M_{RL} = \frac{-\hbar^2}{2m^*} \left[\phi_R^* \frac{d\phi_L}{dx} - \phi_L \frac{d\phi_R^*}{dx} \right]_{x = x_B} \delta_{q_{||}, k_{||}} \tag{57}
$$

where we have used $E_L = E_R$.

Evaluation of equation (57) for emitter and well states gives

$$
|M_{RL}|^2 = \frac{\hbar^2 v_x v_r T_L(E_x)}{4L(w + 2/\alpha)} \delta_{q_{||}, k_{||}} \tag{58}
$$

which involves the product of the two attempt rates from the left and
the right of the barrier and the transmission coefficient for the
barrier. Equations (54) and (57) enable us to calculate the transition
rates for tunnelling provided that the single-barrier transmission
coefficients are small.

SEQUENTIAL TUNNELLING THEORY

The sequential theory of resonant tunnelling has been developed
from a proposal by Luryi (1985) that the transmission be regarded as
two successive transitions. Payne (1986) and Weil and Vinter (1987)
used the Bardeen transfer hamiltonian to show that sequential
tunnelling theory gives the same current as the global transmission
coefficient for a symmetric DBS. We show here that the two approaches
give the same current for an asymmetric DBS. The sequential model
provides a natural framework for the calculation of the charge stored
in the quantum well and it has been used by Sheard and Toombs (1988) to
determine the effects of charge buildup on the resonant tunnelling
current.

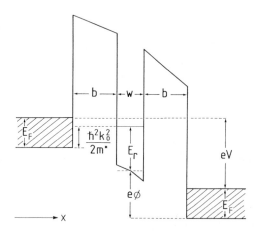

Fig. 7. Spatial variation of electron potential through the double-
barrier structure showing the bound state and longitudinal
kinetic energy of the resonant electrons in the emitter.

The model for a symmetric DBS under a bias voltage V is shown in
Fig. 7. The emitter and collector electrodes are heavily doped n^+
semiconductors with the same Fermi energy E_F. The quantum well is an

undoped layer of the same material and supports a quasibound state at an energy E_r which is measured from the bottom. For simplicity the screening charges in n^+ regions adjacent to the barriers and the stored charge in the well are represented as sheets of infinitesimal thickness. Electrostatically, the DBS is equivalent to two identical capacitors C connected in series, the charge -Q on the common central plate corresponding to the charge per unit area in the well. The potential drop across the right-hand capacitor is then

$$\varphi = \tfrac{1}{2}V + \tfrac{1}{2}(Q/C), \tag{59}$$

where $C = \epsilon_r \epsilon_o/(b + \tfrac{1}{2}w)$ and ϵ_r is the relative permittivity (assumed to be the same for barriers and well).

Taking the x axis perpendicular to the barrier interfaces, we label the electron states in the emitter by wave vectors $\mathbf{k} = (k_x, \mathbf{k}_{||})$ and in the collector by $\mathbf{p} = (p_x, \mathbf{p}_{||})$. In the well the longitudinal motion due to the bound state is the same for all states of transverse motion which are specified by two-dimensional wave vectors $\mathbf{q} = (0, \mathbf{q})$. For plane interfaces the transverse component of wave vector is again conserved in tunnelling so that $\mathbf{k}_{||} = \mathbf{q} = \mathbf{p}_{||}$. We describe the tunnelling in terms of transitions between the states $\mathbf{q}$ of the quantum well and the states $\mathbf{k}(\mathbf{p})$ of the emitter (collector). Introducing the corresponding transition rates $W_{\mathbf{kq}}$ ($= W_{\mathbf{qk}}$) and $W_{\mathbf{pq}}$ ($= W_{\mathbf{qp}}$), the rate equations for the occupancies $f_{\mathbf{k}}$ and $f_{\mathbf{q}}$ of the emitter and well states are

$$\dot{f}_{\mathbf{k}} = -\Sigma_{\mathbf{q}} (f_{\mathbf{k}} - f_{\mathbf{q}})W_{\mathbf{kq}},$$

$$\dot{f}_{\mathbf{q}} = -\Sigma_{\mathbf{k}} (f_{\mathbf{q}} - f_{\mathbf{k}})W_{\mathbf{qk}} - \Sigma_{\mathbf{p}} (f_{\mathbf{q}} - f_{\mathbf{p}})W_{\mathbf{qp}}.$$

The stored charge $Q = (e/A)\Sigma_{\mathbf{q}}f_{\mathbf{q}}$, where A is the interfacial area.

If the quasibound level is very narrow, conservation of energy and transverse momentum shows that an emitter electron can only tunnel into the well if its longitudinal kinetic energy $\hbar^2 k_x^2/2m^*$ is equal to

$$E_o = \hbar^2 k_o^2/2m^* = E_r + e\varphi - eV, \tag{60}$$

as shown in Fig. 7. We take $E_r > E_F$, so that at low temperatures when the Fermi surface is sharp, a finite bias is required before emitter states with $k_x^2 = k_o^2$ are occupied and resonant tunnelling into the well can ocur. Tunnelling out of the well then takes place into empty states in the collector contact. Hence we may put $f_{\mathbf{p}} = 0$. Furthermore, under DC conditions, $\dot{f}_{\mathbf{q}} = 0$. Solution of the rate equations then gives

$$f_{\mathbf{q}} = \Sigma_{\mathbf{k}} f_{\mathbf{k}}W_{\mathbf{qk}} (\Sigma_{\mathbf{k}} W_{\mathbf{qk}} + \Sigma_{\mathbf{p}} W_{\mathbf{qp}})^{-1}$$

and

$$Q = -\frac{e}{A} \Sigma_{\mathbf{k,q}} \frac{f_{\mathbf{k}}W_{\mathbf{kq}}}{\tau_L^{-1} + \tau_R^{-1}}, \tag{61}$$

where $\tau_L^{-1} = \sum_k W_{kq}$ and $\tau_R^{-1} = \sum_p W_{pq}$.

The current density $J = -(e/A)\sum_k \dot{f}_k$ and, under the same steady-state conditions, this gives

$$J = \frac{e}{A} \sum_{kpq} f_k W_{qk} W_{qp} \frac{\tau_L \tau_R}{\tau_L + \tau_R} . \tag{62}$$

The transition rates W_{kq} and W_{pq} are evaulated using the Bardeen transfer hamiltonian formalision of equations (54) and (58). We find that $J = Q/\tau_R$ where the rate of decay of stored charge into unoccupied collector states τ_R^{-1} is given by

$$1/\tau_R = (v_r/2w)T_R ,$$

where $v_r = (2E_r/m^*)^{1/2}$ and T_R is the transmission coefficient of the collector barrier and we have ommitted the penetration depth as before. The expression (61) may be similarly evaluated using a zero-temperature Fermi-Dirac distribution for f_k. This involves counting the number of occupied emitter states which have the same value of $k_x^2 = k_o^2$ and gives (including a factor 2 for spin degeneracy)

$$Q = Q_m \left(\frac{k_F^2 - k_o^2}{k_F^2} \right) \theta(k_o^2)\theta(k_F^2 - k_o^2), \tag{63}$$

$$Q_m = \frac{ek_F^2}{2\pi} \frac{T_L}{T_L + T_R} , \tag{64}$$

where T_L is the emitter-barrier transmission coefficient and $\theta(s) = 1(s > 0)$ or $0(s < 0)$ is the unit step function.

Evaluation of (62) gives for the resonant current

$$J = \frac{em^* v_r}{2\pi\hbar^2 w} (E_F - \frac{\hbar^2 k_o^2}{2m^*}) \frac{T_L T_R}{(T_L + T_R)} . \tag{65}$$

Equation (65) for the current density of an asymmetric DBS is the same as equation (44) which was derived using the global transmission coefficient.

The method used here for calculating the resonant tunnelling current for a 3DEG in the emitter can be used for tunnelling from a 2DEG provided the differences in the density of states and the transmission coefficients are taken into account. The range of voltage for which resonant tunnelling occurs is much smaller for a 2DEG if charge buildup is not included. We consider next the effects of charge buildup.

CHARGE BUILDUP AND CURRENT BISTABILITY

Equation (63) determines the charge $-Q$ in the well in terms of the tunnelling wave number k_o but k_o also depends on the stored charge Q through electrostatic feedback. Equations (59) and (60) give

$$Q = \frac{2C}{e} \left[\frac{\hbar^2 k_o{}^2}{2m^*} - E_r + \frac{1}{2} eV \right]. \tag{66}$$

The dependence of Q on bias voltage V is determined by the simultaneous solution of equations (63) and (66) in which $k_o{}^2$, which gives the position of the bound level relative to the bottom of the emitter conduction band, appears as a parameter. We illustrate this for the simplified case in which T_L and T_R are treated as constants independent of bias voltage. In Fig. 8 we plot Q vs $k_o{}^2$ for equation (63), which gives a triangular-shaped curve with maximum charge Q_m, and for equation (66), which gives a straight line whose position is voltage dependent. The intersection points of the two plots give the required solution.

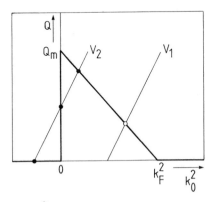

<u>Fig. 8.</u> Plots of Q vs $k_o{}^2$ for equation (63) (thick line) and equation (66) (thin lines) showing intersection points at two different voltages V_1 (open circle) and V_2 (solid circles).

The onset of resonant tunnelling corresponds to $k_o{}^2 = k_F{}^2$, $Q = 0$ and occurs at a threshold voltage $V_{th} = 2(E_r - E_F)/e$. For a higher bias, such as V_1 in Fig. 8, there is one intersection point and the charge, given by explicit solution of equations (63) and (66), is

$$Q = \frac{Q_m e}{2e_F} \frac{V - V_{th}}{1 + \alpha} , \tag{67}$$

where $\alpha = Q_m e / 2CE_F$ is a measure of the electrostatic feedback effect. When $V > 2E_r/e$, such as V_2 in Fig. 8, there is a range where three intersection points occur. The point with $0 < k_o{}^2 < k_F{}^2$ continues the above solution (67) while the point with $k_o{}^2 < 0$ gives $Q = 0$. The third solution is $k_o{}^2 = 0$ which indicates a pinning of the bound state at a level corresponding to the bottom of the emitter conduction band. This occurs for $2E_r < eV < 2(E_r + \alpha E_F)$. The pinning is sustained because a change in the applied voltage is electrostatically compensated by a change in the stored charge.

From these results we construct the explicit dependence of Q on V and hence, using $J = Q/\tau_C$, the DC current-voltage characteristic shown in Fig. 9. The dashed lines with arrows indicate the transitions between high-charge and zero-charge states which would occur on sweeping the voltage up and down through the regions of resonant tunnelling. In the sequential model the time required for such transitions is clearly $\sim \tau_R$. Hence we find a region of intrinsic bistability in the voltage range for which pinning of the bound state occurs. However, this pinning is not directly observable. We note that the voltage width $2\alpha E_F$ of the bistable region is just Q_m/C.

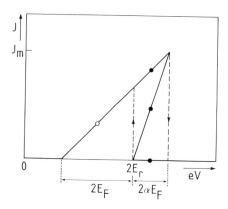

Fig. 9. Current-voltage characteristics J vs eV, showing region of bistability. The open and solid circles refer to the inter-section points of Fig. 8.

The intrinsic bistability range is controlled by the feedback parameter $\alpha = Q_m e/2CE_F$. Using equation (5) for Q_m, we can express

$$\alpha = \frac{2b + w + 2\lambda_s}{a_o} \frac{T_L}{T_L + T_R} , \tag{68}$$

where $a_o = 4\pi \epsilon_r \epsilon_o \hbar^2 /m^* e^2$ is the Bohr radius and we have used $C = \epsilon_r \epsilon_o /(b + \frac{1}{2}w + \lambda_s)$ to take approximate account of a finite screening length λ_s in the n^+ electrodes. For the GaAs/(AlGa)As DBS of Goldman et al (1987b) (w = 5.6 nm, b = 8.5 nm, $E_r \simeq 75$ meV, $E_F \simeq 20$meV) we have estimated T_L and T_R from the WKB approximation and find $\alpha \simeq 0.8$. This would give an intrinsic bistability range ~ 32 mV in this structure. However, this range could be reduced by inhomogeneous broadening of the bound-state level due to structural imperfections. We find that a broadening of a few meV is sufficient to wash out the bistability.

The feedback parameter α can be made large by making $T_L \gg T_R$. This asymmetry of the transmission through the barriers allows charge into the well from the emitter but inhibits its escape to the collector. There is therefore more charge in the well and the feedback is larger. α can also be increased by reducing the doping of the collector. This increases the screening length in the collector and magnifies the bistability by dropping more voltage in the collector. Intrinsic bistability has been observed by Alves et al (1988) and Leadbeater et al (1988 and 1989a) who decreased the transmission coefficient of the right-hand collector barrier by increasing the thickness of this barrier and by Zaslavsky et al (1988) who reduced T_R by increasing the height of the collector barrier.

When there is no intrinsic bistability, the current drops with increasing voltage from the resonant peak to a low value for off-resonant tunnelling. The differential resistance is therefore negative in this region. It was pointed out by Sollner (1987) that extrinsic bistability can occur in devices exhibiting negative differential resistance (NDR) due to current oscillations in the device and in the external circuit. This was confirmed by the computer simulations of Toombs et al (1988). The found that a circuit containing a NDR device in parallel with a capacitor and with resistance and inductive leads can break spontaneously into oscillation. Fig. 10 shows a typical simulation of the average current in the cicuit. There are in general, two regions of bistability for each region of NDR.

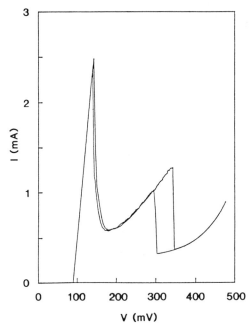

Fig. 10 The average current in a circuit as a function of applied voltage for a simulation of a NDR device in parallel with a capacitor and with resistive and inductive leads.

They arise from the difference between the turn-on and turn-off points for oscillatory behaviour when sweeping the applied DC voltage up or down through the NDR region. The voltage ranges over which the bi-stabilities occur are dependent on the dynamic resistances on either side of the NDR region and zero, one or two bistability regions may be observed. When the circuit oscillates the current is on average between the resonant peak currents and the low off-resonant currents. This results in a shoulder in the current in the NDR region as shown in Fig. 10. This shoulder is a sure sign that the circuit is in oscilla-tion and is a common feature of published resonant-tunnelling device characteristics (eg Eaves et al, 1988).

HIGH MAGNETIC FIELDS

High magnetic fields are very useful for investigating the properties of semiconductor heterostructures at low temperatures. As is well known, the energy levels of free electrons in a magnetic field are Landau levels. We give first a derivation of this result since the concept of the magnetic potential which occurs in the derivation is not so widely appreciated.

The vector potential A in the Landau gauge is $A = (0, -Bz, 0)$ for a magnetic field of magnitude B along the x-axis. If the electrons move in a potential $V(r)$, the hamiltonian H is

$$H = \frac{1}{2m^*} (p + eA)^2 + V(r)$$

$$= \frac{1}{2m^*} (p_x^2 + p_z^2) + \frac{1}{2m^*} (p_y^2 - 2eBzp_y + e^2B^2z^2) + V(r). \qquad (69)$$

The eigenstates of the hamiltonian (69) are given by

$\psi = v(x)\exp(ik_y y)u(z)$ if $V(r) = V(x)$ only.

$H\psi$ is therefore

$$H\psi = \left[\frac{p_z^2}{2m^*} + \frac{1}{2} m^* \omega_c^2 (z - Z)^2 \right] \psi + \left[\frac{p_x^2}{2m^*} + V(x) \right] \psi, \qquad (70)$$

where $\omega_c = eB/m^*$ and $Z(k_y) = \hbar k_y/m^*\omega_c = l_B^2 k_y$.

Examination of equation (70) shows the magnetic field gives rise to a magnetic potential which is a simple harmonic oscillator potential centred on the position $Z = l_B^2 k_y$. For bulk material, the energy levels for the z-motion are therefore Landau levels $(n + 1/2)\hbar\omega_c$, $n = 0, 1, 2...$, but if the wavevectors k_y of the occupied states and the magnetic length l_B are such that the centre of the magnetic potential lies close to a boundary the states become interfacial Landau levels. Interfacial Landau levels have been observed in resonant tunnelling through single-barrier structures (Snell et al, 1987; Chan et al, 1988; Sheard et al, 1988). The eigenstates for the x-motion can be plane waves or bound states according to the behaviour of $V(x)$.

The energy levels of an electron in a magnetic field with $V(r) = 0$ are $(n + 1/2)\hbar\omega_c + \hbar^2 k_x^2/2m^*$. As is well known there are peaks in the density of states due to the Landau levels. Therefore the density of states at the Fermi energy will oscillate as the magnetic field is increased and oscillations will be observed in the current, capacitance etc provided that the magnetic field is large enough for the Landau levels to be well defined.

Goldman et al (1987) have pointed out that evidence for space-charge buildup may be obtained from the magneto-oscillations in the current due to a magnetic field applied perpendicular to the barriers. The transverse kinetic energy of the electrons is then quantized into Landau levels. Since the resonant tunneling involves electrons with transverse kinetic energy up to $E_F - (\hbar^2 k_o^2/2m^*)$, the magneto-oscillations occur when

$$E_F - \hbar^2 k_o^2/2m^* = (n + 1/2)\hbar\omega. \qquad (71)$$

Using equations (66) and (67) this may be rewritten

$$V - V_{th} = [2\hbar(1 + \alpha)B/m^*](n + 1/2).\tag{72}$$

Thus the periodicity of the oscillations depends on the electrostatic feedback parameter α. The experimentally observed magneto-oscillations for the DBS of Goldman et al (1987a and b) give a value $\alpha \simeq 1$ which is broadly consistent with our theoretical estimate. The magnitude of α is rather sensitive to the material and structural parameters of the DBS. Earlier magneto-oscillation studies by Mendez et al (1986) on a similar DBS (w = 4.0 nm, b = 10 nm, $E_r \simeq 130$ meV) are consistent with a much smaller value $\alpha \simeq 0.1$. This is due to the asymmetry in the emitter and collector barrier heights relative to the higher tunneling energy which gives $T_R \gg T_L$ and, from equation (68), a small value of α in this case.

The magneto-oscillation studies of Payling et al (1988) show a series of oscillations at low bias due to off-resonant tunnelling from the three dimensional states of the emitter in addition to the oscillations due to resonant tunnelling. The periods of both sets of oscillations are $E_F m^*/\hbar e$ at the resonant peak in the current as expected from equation (71).

As pointed out by Mendez (1988) the predicted peak-to-valley ratios in the current are an order of magnitude higher than the best values realized to date even at low temperatures. At room temperature, thermal excitation to higher energy states will obviously increase the tunnelling current but at nitrogen temperatures and below this process can be neglected. The relatively low peak-to-valley ratios are therefore attributed to scattering processes. If an electron is scattered elastically or inelastically during the tunnelling process then conservation of momentum parallel to the interface is broken and electrons can be brought on to resonance even if the energy of the quasi-bound state is below the Fermi sea of the emitter. For inelastic scattering due to longitudinal-optic-phonon emission this results in a subsidiary peak beyond the resonant tunnelling peak (Goldman et al, 1987c; Bando et al, 1987; Leadbeater et al, 1989b) since an additional voltage equal to the longitudinal-optic (LO) phonon energy has to be dropped between the emitter contact and the quantum well before an LO phonon can be emitted. Leadbeater et al (1989b) have clearly revealed the LO-phonon-assisted peak and the contribution to the valley current of elastic scattering by applying a quantizing magnetic field parallel to the direction of the tunnel current. The emitter layer in their structure is lightly doped so that a 2DEG forms in the accumulation layer adjacent to the emitter barrier as shown in Fig. 11. The energy levels of the electrons in the accumulation layer of the emitter and in the quantum well are respectively given by $E_0 + (n + 1/2)\hbar\omega_c$ and $E_r + (n' + 1/2)\hbar\omega_c$ where n and n' are the Landau-level quantum numbers and E_0 and E_r are respectively the quasi-bound state energies of the emitter and well. Transitions which are only governed by conservation of energy occur when $E_0 = E_r + (n' - n)\hbar\omega_c + p\hbar\omega_{LO}$ where p = 0 for elastic scattering and p = 1 for the emission of an LO-phonon with energy $\hbar\omega_{LO}$. Magnetoquantum peaks are seen in the current-voltage characteristics which satisfy this condition. Since the quantizing magnetic field produces peaks in the density of state there are voltages for which the current is enhanced and others for which it is diminished.

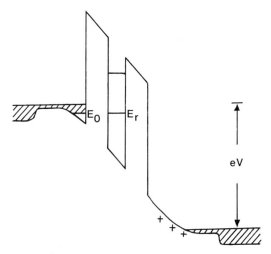

Fig. 11. The potential energy profile of a double-barrier device under bias showing a bound state in the emitter accumulation layer.

ACKNOWLEDGEMENT

We wish to thank our colleagues E S Alves, K S Chan, L Eaves, T J Foster, T M Fromhold, M Henini, O H Hughes, M L Leadbeater , D K Maude, C A Payling and B R Snell for their stimulation and collaboration.

REFERENCES

Alves, E.S., Eaves, L., Henini, M., Hughes, O.H., Leadbeater, M.L., Sheard, F.W., Toombs, G.A., Hill, G., and Pate, M.A., 1988, Observation of intrinsic bistability in resonant tunnelling devices, _Electronic Letters_, 24:1190.

Azbel, M. Ya., 1983, Eigenstates and properties of random systems in one dimension at zero temperature, _Phys. Rev. B_, 28:4106.

Bando, H., Nakagawa, T., Tokumoto, H., Ohta, K., Kajimura, K., 1987, Resonant magnetotunneling in AlGaAs/GaAs triple barrier diodes, _Japan J. Appl. Phys_. 26, Suppl. 26-3:765.

Bardeen, J., 1961, Tunnelling from a many-particle point of view, _Phys. Rev. Lett_., 6:57.

Broekaert, T.P.E., Lee, W., and Fonstad, C.G., 1988, Pseudomorphic $In_{0.53} Ga_{0.47}As/AlAs/InAs$ resonant tunneling diodes with peak-to-valley current ratios of 30 at room temperature, _Appl. Phys. Lett_., 53:1545.

Chan, K.S., Eaves, L., Maude, D.K., Sheard, F.W., Snell, B.R., Toombs, G.A., Alves, E.S., Portal, J.C., and Bass, S., 1988, Electron tunnelling into interfacial Landau states in single-barrier n-type (InGa)As/InP/(InGa)As heterostructures, _Solid-State Electronics,_ 31:711.

Chang, L.L., Esaki, L., and Tsu, R., 1974, Resonant tunneling in semiconductor double barriers, _Appl. Phys.Lett_., 24:593.

Dicke, R.H., and Wittke, J.P., 1960, "Introduction to Quantum mechanics", Addison-Wesley, Reading.

Duke, C.B., 1969, "Tunneling in Solids", Academic Press, New York.

Eaves, L., Toombs, G.A., Sheard, F.W., Payling, C.A., Leadbeater, M.L.,
Alves, E.S., Foster, T.J., Simmonds, P.E., Henini, M., Hughes, O.H.,
Portal, J.C., Hill, G., and Pate, M.A., 1988, Sequential tunneling due
to intersubband scattering in double-barrier resonant tunneling devices, Appl. Phys. Lett., 52:212.

Goldman, V.J., Tsui, D.C., and Cunningham, J.E., 1987a, Resonant
tunneling in magnetic fields: evidence for space-charge buildup, Phys.
Rev. B, 35:9387.

Goldman V.J., Tsui, D.C., and Cunningham, J.E., 1987b, Observation of
intrinsic bistability in resonant-tunneling structures, Phys. Rev.
Lett., 58:1256.

Goldman, V.J. Tsui, D.C., and Cunningham, J.E., 1987c, Evidence for LO-
phonon-assisted tunneling in double-barrier heterostructures, Phys.
Rev. B., 36:7635.

Goodhue, W.D., Sollner, T.C.L.G., Le, H.Q., Brown, E.R., and Vojak,
B.A., 1986, Large room-temperature effects from resonant tunneling
through AlAs barriers, Appl.Phys. Lett., 49:1086.

Kane, E.O., 1969, Basic concepts in tunneling, in: "Tunneling Phenomena
in Solids", E. Burstein and S. Lindqvist, ed., Plenum, New York.

Kroemer, K., 1986, Band offsets at heterointerfaces:theoretical basis
and review of recent experimental work, Surf. Sci, 174:299.

Leadbeater, M.L., Alves, E.S., Eaves, L., Henini, M., Hughes, O.H.,
Sheard, F.W., and Toombs, G.A., 1988, Charge build-up and intrinsic
bistability in an asymmetric resonant-tunneling structure, Semicond.
Sci. Technol., 3:1060.

Leadbeater, M.L., Alves, E.S., Eaves, L., Henini, M., Hughes, O.H.,
Sheard, F.W., and Toombs, G.A., Magnetic field and capacitance studies
of intrinsic bistability in double-barrier structures, 1989a,
Superlattices and Microstructures, 6:59.

Leadbeater, M.L., Alves, E.S., Eaves, L., Henini, M., Hughes, O.H.,
Celeste, A., Portal, J.C., Hill, G., and Pate, M.A., 1989b, Magnetic
field studies of elastic scattering and optic-phonon emission in resonant tunneling devices, Phys. Rev. B. 39:3438.

Luryi, S., 1985, Frequency limit of double-barrier resonant-tunneling
oscillators, Appl. Phys. Lett., 47:490.

Mendez, E.E., Esaki, L., and Wang, W.I., 1986, Resonant magnetotunneling in GaAlAs-GaAs-GaAlAs heterostructures, Phys. Rev. B,
33:2893.

Mendez, E.E., 1988, Physics of resonant tunneling in semiconductors,
in "Physics and Applications of Quantum Wells and Superlattices", E.E.
Mendez and K. von Klitizing, ed.,Plenum, New York.

Morkoc, H. Chen, J., Reddy, U.K., and Henderson, T., 1986, Observation of a negative differential resistance due to tunneling through a single barrier into a quantum well, Appl. Phys. Lett., 49:70.

Oppenheimer, J.R., 1928, Three notes on the quantum theory of aperiodic effects, Phys. Rev. 13:66.

Payling, C.A., Alves, E.S., Eaves, L., Foster, T.J., Henini, M., Hughes, O.H., Simmonds, P.E., Sheard, F.W., Toombs, G.A., Portal, J.C., 1988, Evidence for sequential tunnelling and charge build-up in double barrier resonant tunneling devices, Surface Science, 196:404.

Payne, M.C., 1986, Transfer hamiltonian description of resonant tunnelling, J. Phys. C: Solid State Phys., 19:1145.

Price, P.J., 1988, Theory of resonant tunneling in heterostructures, Phys. Rev. B, 38:1944.

Ricco, B., and Azbel, M. Ya., 1984, Physics of resonant tunneling:the one-dimensional double-barrier case, Phys. Rev. B, 29:1970.

Sheard, F.W., Chan, K.S., Toombs, G.A., Eaves, L., and Portal, J.C., 1988, Magnetotunnelling in single-barrier III-V semiconductor heterostructures, in:"Gallium Arsenide and Related Components 1987", A. Christou and H.S. Rupprecht, ed., Institute of Physics, Bristol.

Sheard, F.W., and Toombs G.A., 1988, Space charge buildup and bistability in resonant-tunneling double-barrier structures, Appl. Phys. Lett., 52, 1228.

Snell, B.R., Chan, K.S., Sheard, F.W., Eaves, L., Toombs, G.A., Maude, D.K., Portal, J.C., Bass, S.J., Claxton, P., Hill, G., and Pate, M.A., 1987, Observation of magnetoquantized interface states by electron tunnelling in single-barrier n^-(InGa)As-InP-n^+(InGa)As heterostructures, Phys. Rev. Lett., 59:2806.

Sollner, T.C.L.G., 1987, Comment on "Observation of intrinsic bistability in resonant-tunneling structures", Phys. Rev. Lett., 59:1622.

Toombs G.A., Alves, E.S, Eaves, L., Foster, T.J., Henini, M., Hughes, O.H., Leadbeater, M.L., Payling, C.A., Sheard, F.W., Claxton, P.A., Hill, G., Pate, M.A., and Portal, J.C., 1988, Magnetic field studies of resonant tunnelling double barrier structures, in: "Gallium Arsenide and Related Compounds 1987", A. Christou and H.S. Rupprecht, ed., Institute of Physics, Bristol.

Van de Walle, C.G., and Martin, R.M., 1987, Theoretical study of band offsets at semiconductor interfaces, Phys. Rev. B, 35:8154.

Weil, T., and Vinter, B., 1987, Equivalence between resonant tunneling and sequential tunneling in double-barrier diodes, Appl.Phys. Lett., 50:1281.

Zaslavsky, A., Goldman, V.J., and Tsui, D.C., 1988, Resonant tunneling and intrinsic bistability in asymmetric double-barrier heterostructures, Appl. Phys. Lett., 53:1408.

HIGH-FREQUENCY APPLICATIONS OF RESONANT-TUNNELING DEVICES

T. C. L. Gerhard Sollner, Elliott R. Brown, C.D. Parker, and W.D. Goodhue

Lincoln Laboratory, Massachusetts Institute of Technology
Lexington, Massachusetts 02173

1. INTRODUCTION

Since the prediction (Kazarinov and Suris, 1971; Tsu and Esaki, 1973) and first observation (Chang et al., 1974) of resonant tunneling through double-barrier heterostructures, there has been increasing interest in the field. This renewed interest arises in part because of the advances in the ease of fabrication by molecular beam epitaxy (MBE) of the atomically thin structures required, and in part because the charge-transport process has been shown to be extremely fast (Sollner et al., 1983). The quality of the heterostructure interfaces in the GaAs/AlAs system has been improved to the point that peak-to-valley (P/V) ratios of 3-4 at room temperature (Shewchuck et al., 1985; Goodhue et al., 1986) and nearly 10 at 77 K are easily achieved with current densities well above 10^4 A/cm². New material systems have been explored by several groups, the most notable being InGaAs lattice-matched to InP substrates with AlAs barriers. The Fujitsu group first showed the great promise of this material (Inata et al., 1987), and the best published results give a P/V ratio of about 30 at 300 K (Broekaert and Fonstad, 1988).

Figure 1 shows the structure of resonant-tunneling diodes (RTDs) used for high-frequency applications. The thin epitaxial layers are grown using MBE. A transmission electron micrograph of a GaAs double-barrier structure is shown in Fig. 2. In fabricating RTDs, the lateral device geometry, usually circular, is defined by an ohmic contact on the epitaxial material. Isolation is achieved by proton implantation, which renders the surrounding material semi-insulating, or by mesa etching. In both cases the ohmic contact is employed as a mask. The best room-temperature P/V ratios have been obtained with AlAs barriers, both in GaAs and in InGaAs. Peak current densities are in the range $5{\times}10^4$ to $2{\times}10^5$ A/cm², and the voltage excursion of the negative differential conductance (NDC) is typically a sizable fraction of a volt, so the device diameter must be kept below 10 μm to keep the negative conductance in a range that allows matching to normal circuit impedances.

Figure 3 shows the mechanisms of resonant tunneling responsible for the NDC region as well as the origins of some of the time-delay processes. With no voltage applied to the RTD, the Fermi level is at the same energy on both sides of the active region, as shown in the top part of the figure. The barriers act as partially reflecting mirrors for the electrons incident from these terminals. Electrons in the well between the

Electronic Properties of Multilayers and Low-Dimensional Semiconductors Structures
Edited by J. M. Chamberlain *et al.*, Plenum Press, New York, 1990

283

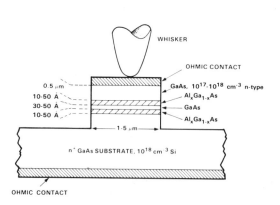

Figure 1. Cross-sectional view of a completed resonant-tunneling diode (RTD).

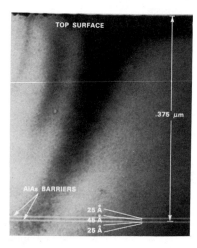

Figure 2. Transmission electron micrograph of an RTD wafer grown by molecular beam epitaxy.

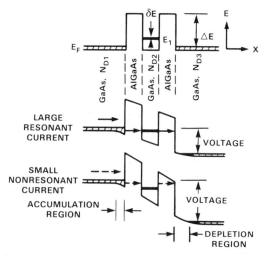

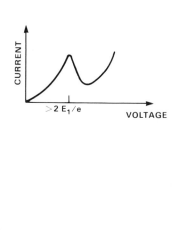

Figure 3. Electron energy as a function of position in a double-barrier resonant-tunneling structure and the resulting current-voltage (I-V) curve.

barriers are further confined in the direction perpendicular to the plane of the well, so their momentum in that direction is quantized. The (transverse) momentum in the plane of the well remains essentially continuous (quantized only by the size of the device), so the result is a sequence of subbands. The bottom of the first subband is shown as E_1. Electrons in this band have a finite lifetime τ in the well, and this leads to a finite width of the subband $\delta E = \hbar/\tau$. This energy width is the same as the width of the transmission as a function of incident energy for electrons impinging on the double-barrier structure (see, for example, Blatt and Weisskopf, 1979). When a voltage is applied between the two terminals so that the greatest number of incident electrons below the Fermi level of the cathode have the same energy as the subband minimum, a peak in the current results. At higher voltages the current decreases, creating the NDC region. Depletion and accumulation regions also occur, which will be discussed in more detail later. Both the finite electron lifetime in the well and the transit time across the depletion region contribute to phase delay between the current and the applied voltage.

Resonant-tunneling diodes possess several attractive features for high-speed devices. The charge-transport time can be of the order of 100 fs, while the intrinsic parasitics are low. A current density of about 2×10^5 A/cm^2 has been achieved (Broekaert and Fonstad, 1989; Brown et al., 1989a), and the specific capacitance of 0.1 μF/cm^2 is about an order of magnitude below that of pn (Esaki) tunnel diodes. The current-voltage (I-V) curve also has the useful properties of NDC and, for symmetric structures, antisymmetry about zero that will be useful for multipliers, as discussed below.

The NDC regions that exist in the (I-V) curve have been used for several proposed and fabricated devices. For example, the NDC region is capable of providing the gain necessary for high-frequency oscillations (Sollner et al., 1984). In our laboratory we have been working to increase the frequency and power of these oscillators, achieving a maximum oscillation frequency of 420 GHz to date (Brown et al., 1989a). A similar effort by Rydberg et al. (1988) at Chalmers Institute of Technology in Sweden has produced record power of 60 μW at 90 GHz. In the course of this work, we and others have considered the origins of delay processes that limit the maximum frequency (Luryi, 1985; Coon and Liu, 1986; Frensley, 1986; Sollner et al., 1987a), resulting in a more complete equivalent circuit for the diode (Brown et al., 1989b). The unique nonlinearity and symmetry of the RTD I-V curve have been used for resistive multiplication with output frequencies as high as 200 GHz (Sollner et al., 1987b; Batelaan and Frerking, 1987; Sollner et al., 1988). This method of power generation at submillimeter wavelengths may provide more power than fundamental oscillations using the same material, as will be discussed below. Three-terminal devices using resonant tunneling in various ways have also been proposed (for example, see Sollner et al., 1985; Capasso et al., 1986) and some very interesting resonant-tunneling transistors have been fabricated and tested (Yokoyama et al., 1985; Capasso et al., 1987; Reed et al., 1989). In this chapter we will describe recent results for oscillators and multipliers. The important questions of a complete equivalent circuit and the ultimate oscillation frequency will be addressed. We conclude with a look at new materials for RTDs.

2. RESONANT-TUNNELING OSCILLATORS

The waveguide oscillator structure used by Brown et al. (1987, 1988, 1989a) is shown in Fig. 4, along with the equivalent circuit. The diode is contacted by a 12-μm-diameter wire, which acts as part of the matching network to couple the fields from the waveguide to the diode. The backshort optimizes this coupling. Absorber is placed around the dc bias pin to prevent oscillations from occurring in the biasing circuit.

The equivalent circuit of the RTD shown in Fig. 4 has been modified by Brown et al. (1989b) from that of the usual pn tunnel diode to include the effects of electron storage in the well. The series resistance R_s includes any resistance encountered between the waveguide and the double barriers, e.g., ohmic contacts, finite resistivity of the semiconductor in the mesa, spreading resistance, and skin-effect resistance. The capacitance arises from the charge accumulated outside the barriers and the depletion region. (Accumulation of charge in the well is ignored here.) The "quantum-well inductance" is related to the lifetime of charge in the well, and is discussed in detail below.

We consider the conduction-current response $i_c(t)$ of a double-barrier structure to a voltage step of amplitude ΔV. Three essential assumptions are made: The sudden approximation is used, in which the Hamiltonian and its eigenvalues are assumed to change on a time scale short compared to that in which the wavefunctions evolve. It is further assumed that the current approaches its new equilibrium value exponentially with

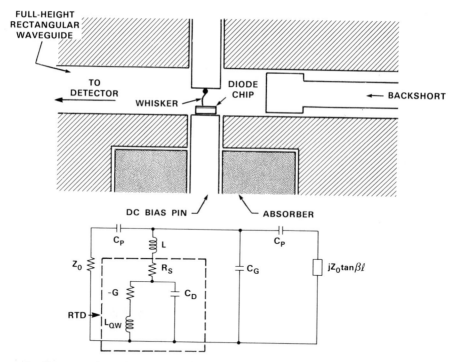

Figure 4. Cross section of a rectangular waveguide resonator used with RTDs to produce oscillators, along with the equivalent circuit.

a single time constant τ. (More complete calculations by Mains and Haddad (1988) show some small, rapidly damped ringing associated with the round-trip time of charge in the well superimposed on an exponential change, but we will neglect this ringing here.) The final assumption is that this time constant is equal to the lifetime of an electron in the well. This seems reasonable since it is the filling or emptying of charge in the well that will contribute to the current transient during the approach to equilibrium. Various aspects of this analysis are shown in Fig. 5. An expression for the current as a function of time, where I_1 and I_2 are the beginning and final equilibrium currents and $\theta(t)$ is the unit step function, is

$$i_c(t) = I_1\theta(-t) + [I_2 + (I_1 - I_2) \exp(-t/\tau)]\theta(t) \tag{1}$$

Equation (1) is the step-function response of the system. If ΔV is assumed small, linear response theory can be applied. The impulse response is

$$h(t) = (\Delta V)^{-1}\frac{di(t)}{dt} \tag{2}$$

and the admittance function is the Fourier transform of the impulse response, yielding

$$Y_c(\omega) = \frac{1}{R + i\omega L} \tag{3}$$

where $L = R\tau$, and R is the differential resistance from the I-V curve. The product $R\tau$ appears as an inductance in an equivalent circuit representing Eq. (3). However, note that it is a negative inductance in the NDC region where oscillations occur, so it cannot be used to form a resonant circuit with the capacitance under this condition. We call this inductance a "quantum-well inductance" to distinguish it from the usual positive inductance. The addition of this element to the equivalent circuit is shown in Fig. 5.

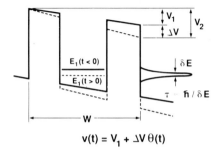
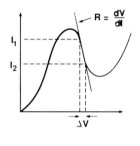

$$v(t) = V_1 + \Delta V\,\theta(t)$$

ASSUME : **(1) SUDDEN APPROXIMATION**

(2) $i_c(t) = I_1\theta(-t) + [I_2 + (I_1 - I_2)\exp(-t/\tau)]\theta(t)$

THEN, $\quad Y_c(\omega) = \dfrac{1}{\Delta V}\displaystyle\int_{-\infty}^{\infty}\left(\dfrac{di_c}{dt}\right)\exp(-i\omega t)\,dt = \dfrac{1}{R + i\omega L}$

Where $\quad L = R\,\tau$

NOTE: If $R < 0$, then $L < 0$

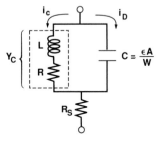

$$C = \frac{\epsilon A}{W}$$

Figure 5. Summary of the derivation of Eqs. (3) and (4) in the text, showing that the consequence of finite quasibound state lifetime is an additional "quantum well inductance" in the equivalent circuit.

The maximum oscillation frequency f_{RCL} is the highest frequency at which the real part of the differential impedance of the complete diode circuit is negative, allowing gain. For the equivalent circuit of Fig. 5 this is easily calculated to be

$$f_{RCL} = \frac{1}{2\pi}\left\{\left[\frac{1}{LC}(1 - C/2LG^2)\right]\left[1 - \left(1 - \frac{(GR_s + 1)/GR_s}{(C/2LG^2 - 1)^2}\right)^{1/2}\right]\right\}^{1/2} \quad (4)$$

where $G = 1/R$ and, as before, $L = R\tau$. In the limit in which the electron lifetime in the well is short compared to RC, this reduces to the more familiar expression

$$f_{RC} = \frac{1}{2\pi C}\left[\frac{-G}{R_s} - G^2\right]^{1/2} \quad (5)$$

A demonstration of the significance of the quantum-well inductance is shown in Fig. 6. Here the output power as a function of frequency is plotted for two different wafers. In the wafer with the highest values of f_{RC} and f_{RCL}, the barriers were comparatively thin so that τ was comparable to the other delay times of the RTD. Consequently, the cutoff frequency was shifted down very little by the inclusion of the quantum-well inductance (i.e., the difference between f_{RC} and f_{RCL} was small). However, the other wafer was grown with thicker barriers, so that τ was longer and constituted the dominant delay time. In that case the addition of the inductance to the equivalent circuit greatly reduces the cutoff frequency and provides much better agreement with experiment.

We note that Eqs. (4) and (5) increase without limit for decreasing R_s. The transit time for electrons across the depletion region will then begin to limit the speed of the RTD. It is possible to incorporate the effects of this transit time into the maximum

287

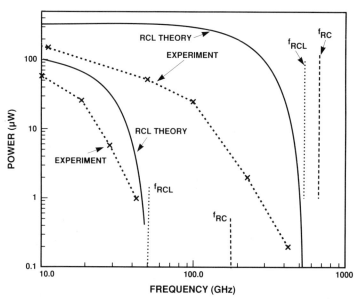

Figure 6. Output power as a function of frequency for two different double-barrier structures, with theoretical limits also marked. The quantum-well inductance is included in f_{RCL}, but not in f_{RC}.

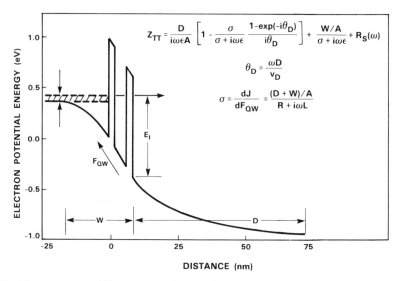

Figure 7. Electron potential energy at the conduction-band edge of an RTD near the bias voltage of the first current peak. The potential energy drop in the accumulation region is exaggerated for clarity.

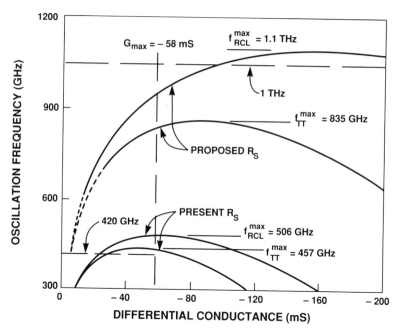

Figure 8. Maximum oscillation frequency as a function of differential conductance from the I-V curve. The lower two curves are for existing RTDs with series resistance R_s of 4 Ω at dc and 6 Ω at 600 GHz, while the upper two curves are for a proposed RTD with $R_s = 2\ \Omega$ at all frequencies. The maximum magnitude of the negative conductance is 58 mS in existing diodes. The phase delay associated with electron transit across the depletion region is included in f_T but not in f_{RCL}.

oscillation frequency (Brown et al., 1989a), although in this case it is not so easy to represent the resulting expression with an equivalent circuit. Figure 7 illustrates the approach for including the transit time across the depletion region. These techniques were first employed to analyze transit time devices, and a good description of the method appears in Sze (1981). The cathode and the double barriers are considered as an injector of electrons into the depletion region, with an injection conductance given by

$$\sigma = \frac{dJ}{dF_{QW}} = \frac{(D + W)/A}{R + i\omega L} \tag{6}$$

Here J is the current density, F_{QW} is the electric field across the double-barrier structure, and the other variables are defined in Fig. 7. This injection conductance contains the quantum-well inductance L discussed above. The use of Maxwell's equation relating the current density to the electric field leads to the following expression for the impedance of the total RTD structure:

$$Z_{TT} = \frac{D/A}{i\omega\varepsilon}\left[1 - \frac{\sigma}{\sigma + i\omega\varepsilon}\frac{1 - \exp(-i\theta_D)}{i\theta_D}\right] + \frac{W/A}{\sigma + i\omega\varepsilon} + R_s(\omega) \tag{7}$$

where $\theta_D = \omega D/v_D$, and ε is the material dielectric constant. A constant electron drift velocity v_D across the depletion region is assumed in this derivation, which will not be strictly true, but some average velocity should provide a good approximation.

The maximum oscillation frequency including all delay mechanisms, which we call f_T^{max}, is found by setting the real part of Eq. (7) equal to zero and solving for ω. In

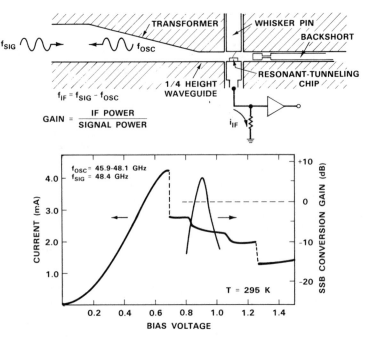

Figure 9. Waveguide cross section for an oscillator used as a self-oscillating mixer. The lower part of the figure shows a superposition of the measured single-sideband conversion gain as a function of voltage and the I-V curve of the RTD.

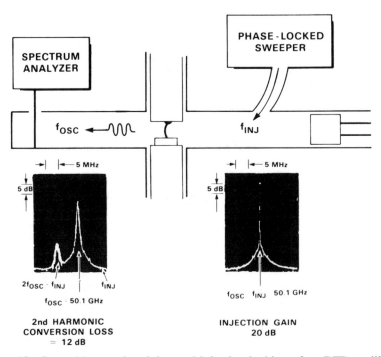

Figure 10. Second-harmonic mixing and injection locking of an RTD oscillator.

Fig. 8 we plot $f_T^{\max} = \omega/2\pi$ versus the differential conductance G that would be measured from the I-V curve. Also plotted are curves of f_{RCL} as an indication of the importance of including depletion-layer transit-time effects. The lower curves are for the parameters of our currently best diodes. They have a maximum magnitude of negative differential conductance of 58 mS, as marked on the plot, so the peak in $f_T^{\max}$ is attainable, yielding $f_T^{\max} = 457$ GHz. These diodes have a series resistance of 4 Ω at dc and 6 Ω at 600 GHz. If the series resistance could be reduced to 2 Ω at the operating frequency, the upper pair of curves in Fig. 8 would apply. These curves show the importance of transit delays across the depletion region as the other delay processes are reduced, as well as the large improvements that can be made by lowering the series resistance. It is expected that with additional increases in current density, the increase in the specific capacitance associated with thinner depletion regions can be tolerated, allowing maximum oscillation frequencies above 1 THz.

Since the resonant-tunneling I-V curve is nonlinear, especially near the negative differential resistance region, the same device can act as both a mixer and an oscillator (Sollner et al., 1990). Figure 9 shows the down-conversion efficiency for a self-oscillating mixer. The maximum conversion efficiency is +5 dB. Noise-temperature measurements have been made on this device, but it was not possible to obtain conversion gain when the low-noise IF amplifier was connected. The best noise figure obtained was a rather disappointing 19 dB. Shown in Fig. 10 is a result for second-harmonic conversion. In this case a signal at frequency f_{INJ} (50.11 GHz) is mixed with a local oscillator frequency f_{osc} (50.1 GHz) to produce a signal at a frequency $2f_{osc} - f_{INJ}$ (50.09 GHz). The conversion efficiency is about -12 dB, which is comparable to the best results for a pair of Schottky diodes in the antiparallel configuration (Carlson and Schneider, 1975).

Resonant-tunneling diode oscillators can also be injection locked. Figure 10 shows the case in which the injected frequency coincides with the oscillation frequency. The injection gain, defined as the locked output power divided by the injected power, is 20 dB at 50 GHz. One advantage of this scheme is that the bandwidth of this injection-locked oscillator is essentially the same as that of the injected signal.

3. RESISTIVE MULTIPLIERS

Another microwave application of RTDs, resistive multipliers (Sollner et al., 1988), takes advantage of the unique I-V curve of these devices. The undulations of the RTD I-V curve suggest that there should be large harmonic content to the current waveform, and the antisymmetry implies that only odd harmonics should appear. Figure 11 shows a

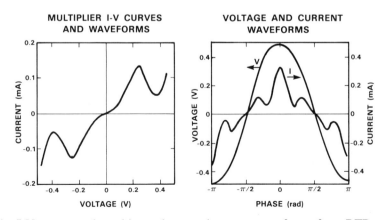

Figure 11. I-V curves and resulting voltage and current waveforms for a RTD quintupler.

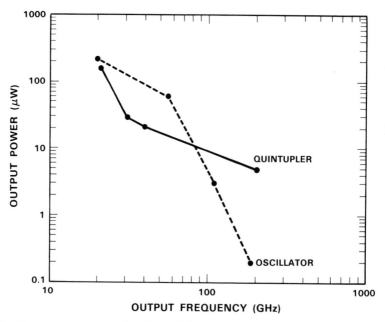

Figure 12. Output power as a function of frequency for both RTD oscillators and quintuplers using the same double-barrier structure.

dc I-V curve and the calculated voltage and current waveforms for this same device when driven by a 50-Ω source with an amplitude of about 0.5 V. With this pump amplitude the current contains a significant 5th-harmonic component. Figure 12 shows the measured output power of quintuplers and fundamental oscillators as a function of frequency. All these devices used the same RTD material. The multiplier power drops more slowly than that of the oscillator since multiplication still occurs when the conditions for oscillation, discussed above, are no longer met. Another advantage of the multiplier is that locking the pump at some lower frequency is usually easier than locking a fundamental oscillator at five times that frequency. These facts suggest that the multiplier may be a better source of power in the submillimeter spectrum than an RTD fundamental oscillator.

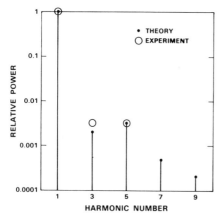

Figure 13. Theoretical and experimental spectrum of multiplier output. For the measurement the pump was at 4 GHz.

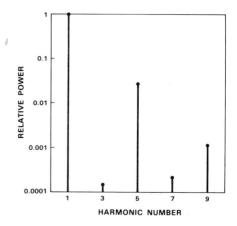

Figure 14. Theoretical output of a multiplier with an improved I-V curve.

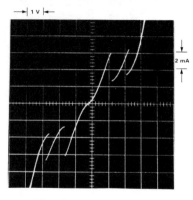

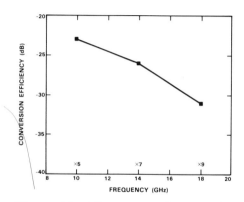

Figure 15. I-V curve of two double-barrier structures grown in series on one wafer.

Figure 16. The measured conversion efficiency of the series double-barrier structure of Fig. 15.

The output power spectrum of an RTD quintupler is shown in Fig. 13. There are several interesting features of this spectrum. One of the most difficult problems in harmonic-multiplier design is to reactively terminate all the harmonics at lower frequencies than the desired output frequency, since these harmonics are usually of larger amplitude than the higher-order ones. In the present case the even harmonics are entirely missing because of the symmetry of the I-V curve about zero, thus reducing by a factor of two the number of harmonics which need to be considered. Also, the 5th harmonic has the largest amplitude, so even if the power in the other harmonics were resistively terminated one would expect high conversion efficiency to the 5th harmonic. So far we have observed 1% conversion efficiency (output power at the desired harmonic divided by the input power) from 2 to 10 GHz. Batelaan and Frerking (1987) have produced 250 μW in a tripler to 191 GHz for 0.61% efficiency. Optimizing the multiplier structure to provide the proper impedances and engineering the shape of the I-V curve should raise the efficiency significantly. Figure 14 shows a calculated output spectrum in which the assumed I-V curve has been improved for multiplication.

It is, of course, possible to obtain several resonant current peaks in one RTD. When this is the case, even higher harmonics can be obtained. Figure 15 shows the I-V curve of two double-barrier structures grown in series on one wafer. One of these diodes was used in a coaxial circuit as a 5-times, 7-times, and 9-times multiplier. The resulting conversion efficiency is shown in Figure 16. For such a compact 9-times multiplier, -31 dB conversion efficiency may be quite acceptable for some applications.

3. OTHER MATERIALS

Significant advances in P/V ratio have been achieved for RTDs made with materials having $In_{0.53}Ga_{0.47}As$ for the contacts and wells and AlAs for the barriers (Inata et al., 1987; Broekaert et al., 1988). The InGaAs alloy composition is chosen to lattice-match the InP substrate on which it is grown by MBE. These RTDs have great promise for high-frequency applications because the InGaAs can be doped more heavily than GaAs, has a higher mobility, and provides a lower Schottky barrier height in contacts to metal. These all translate into much lower series resistance and hence higher f^{max}. A comparison of the I-V curves for the two materials is shown in Fig. 17. For the same peak current density InGaAs/AlAs RTDs have a much larger P/V ratio. This results in more power available for oscillations as well as possible logic applications where low power

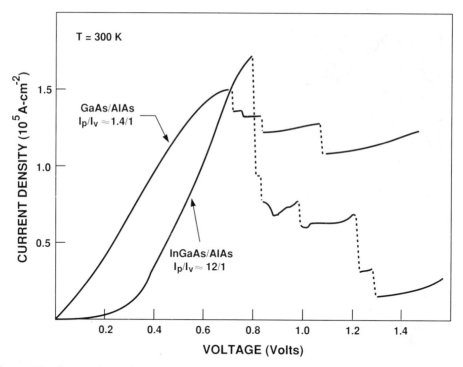

Figure 17. Comparison of the I-V curves of RTDs made from GaAs and InGaAs. Both diodes have AlAs barriers.

dissipation in the "off" state is desirable.

In addition to materials with lattice constants near 0.565 nm (GaAs, AlAs) and 0.587 (InP, InGaAs, AlGaAs), there is another group of semiconductors with with lattice constants near 0.610 nm: InAs, GaSb, and AlSb. These form Type II-staggered interfaces (the material that forms barriers for electrons forms wells for holes) and Type II-misaligned interfaces (the bandgaps do not overlap in energy). This system was studied as early as 1977 by Esaki's group at IBM (Sai-Halasz et al., 1977), and they investigated GaSb-AlSb superlattices in great detail. (See Voisin et al., 1985, for a review.) Very recently several novel tunneling structures have been made in this system (Soderström et al., 1989; Beresford et al., 1989), and several others are sure to follow as the growth parameters are better controlled.

4. CONCLUDING REMARKS

We have described a few interesting and useful high-speed devices based on resonant tunneling, and there are undoubtedly many more that await discovery. We expect the general area of resonant tunneling to provide a fertile ground for both solid state physics and active-device investigations for some years to come.

Acknowledgments

We wish to thank A.R. Calawa and M.J. Manfra for growing the InGaAs/AlAs RTDs as well as C. L. Chen, G.D. Johnson, K.M. Molvar, and N. Usiak for assistance in fabrication and packaging our devices. We are also indebted to C. L. Chen, P.J. Daniels, and P.E. Tannenwald for helpful discussions. This work was sponsored by the U.S.

Army Research Office, the Department of the Air Force (including specific support from the Air Force Office of Scientific Research), and NASA.

References

Batelaan, P.D., and Frerking, M.A., 1987, in *Conference Digest, Twelfth International Conference Infrared and Millimeter Waves*, edited by R.J. Temkin (IEEE, New York), p. 14.

Beresford, R., Luo, L.F., Longenbach, K., and Wang, W.I., 1989, *IEEE Int. Electron Devices Meeting Tech. Digest* (IEEE, New York), paper 21.2.

Blatt, J.,and Weisskopf, V.F., 1979, *Theoretical Nuclear Physics* (Springer, Berlin, Heidelberg).

Broekaert, T. and Fonstad, C., 1988, *Appl. Phys. Lett.* **53**, 1545.

Broekaert, T. and Fonstad, C., 1989, *IEEE Int. Electron Devices Meeting Tech. Digest* (IEEE, New York), paper 21.5.

Brown, E.R., Sollner, T.C.L.G., Goodhue, W.D., and Parker, C.D., 1987, *Appl. Phys. Lett.* **50**, 332.

Brown, E.R., Goodhue, W.D., and Sollner, T.C.L.G., 1988, *J. Appl. Phys.* **64**, 1519.

Brown, E.R., Sollner, T.C.L.G., Goodhue, W.D., Parker, C.D., and Chen, C.L., 1989a, *Appl. Phys. Lett.* **55**, 1777.

Brown, E.R., Parker, C.D., and Sollner, T.C.L.G., 1989b, *Appl. Phys. Lett.* **54**, 934.

Capasso, F., Sen, S., Gossard, A.C., Hutchinson, A.L., and English, J.E., 1986, *IEEE Electron Device Lett.* **EDL-7**, 573.

Capasso, F., Sen, S., and Cho, A.Y., 1987, *Appl. Phys. Lett.* **51**, 526.

Carlson, D., and Schneider, M.V., 1975, *IEEE Trans. Microwave Theory Tech.* **MTT-23**, 828.

Chang, L.L., Esaki, L., and Tsu, R., 1974, *Appl. Phys. Lett.* **24**, 593.

Coon, D.D., and Liu, H.C., 1986, *Appl. Phys. Lett.* **49**, 94.

Frensley, W., 1986, *IEEE Int. Electron Devices Meeting Tech. Digest* (IEEE, New York), paper 25.5.

Goodhue, W.D., Sollner, T.C.L.G., Lee, H.Q., Brown, E.R., and Vojak, B. A., 1986, *Appl. Phys. Lett.* **49**, 1086).

Inata, T., Muto, S., Nakata, Y., Sasa, S., Fujii, T., and Hiyamizu, S., 1987, *Jpn. J. Appl. Phys.* **26**, L1332.

Kazarinov, R.F., and Suris, R.A., 1971, *Sov. Phys. Semicond.* **5**, 707.

Luryi, S., 1985, *Appl. Phys. Lett.* **47**, 490.

Mains, R.K., and Haddad, G.I., 1988, *J. Appl. Phys.* **64**, 3564.

Reed, M.A., Frensley, W.R., Matyi, R.F., Randall, J.N., and Seabaugh, A.C., 1989, *Appl. Phys. Lett.* **54**, 1034.

Rydberg, A., Grönqvist, H., and Kollberg, E., 1988, *Microwave Opt. Techn. Lett.* **1**, 333.

Sai-Halasz, G.A., Tsu, R., and Esaki, L., 1977, *Appl. Phys. Lett.* **30**, 651.

Shewchuk, T.J., Chapin, P.C., Coleman, P.D., Kopp, W., Fischer, R., and Morkoç, H., 1985, *Appl. Phys. Lett.* **46**, 508.

Soderström, J.R., Chow, D.H., and McGill, T.C., 1989, *Appl. Phys. Lett.* **55**, 1094.

Sollner, T.C.L.G., Goodhue, W.D., Tannenwald, P.E., Parker, C.D., and Peck, D.D., 1983, *Appl. Phys. Lett.* **43**, 588.

Sollner, T.C.L.G., Tannenwald, P.E., Peck, D.D., and Goodhue, W.D., 1984, *Appl. Phys. Lett.* **45**, 1319.

Sollner, T.C.L.G., Le, H.Q., Correa, C.A., and Goodhue, W.D., 1985, *Proc. IEEE/Cornell Conf. Advanced Concepts in High Speed Semicond. Devices and Circuits* (IEEE, New York), p. 252.

Sollner, T.C.L.G., Brown, E.R., Goodhue, W.D., and Le, H.Q., 1987a, *Appl. Phys. Lett.* **50**, 332.

Sollner, T.C.L.G., Brown, E.R., and Goodhue, W.D., 1987b, in *Picosecond Electronics and Optoelectronics II*, vol. 24 in *Springer Series in Electronics and Photonics* (Springer, Berlin,), p. 102.

Sollner, T.C.L.G., Brown, E.R., Goodhue, W.D., and Correa, C.A., 1988, *J. Appl. Phys.* **64**, 4248.

Sollner, T.C.L.G., Brown, E.R., Goodhue, W.D., and Le, H.Q., 1990, in *Physics of Quantum Electron Devices, Springer Series in Electronics and Photonics, vol. 28*, p. 147 (Springer, Berlin,).

Sze, S.M., 1981, *Physics of Semiconductor Devices* (Wiley, New York), Ch. 10.

Tsu, R., and Esaki, L., 1973, *Appl. Phys. Lett.* **22**, 562.

Voisin, P., Delalande, C., Bastard, G., Voos, M., Chang, L.L., Segmuller, A., Chang, C.A., and Esaki, L., 1985, *Superlat. Microstr.* **1**, 155.

Yokoyama, N., Imamura, K., Muto, S., Hiyamizu, S., and Nishii, H., 1985, *Jpn. J. Appl. Phys.* **24**, L583.

TRAVERSAL, REFLECTION AND DWELL TIME FOR QUANTUM TUNNELING

M. Büttiker

IBM Research Division
Thomas J. Watson Research Center
Yorktown Heights, N.Y. 10598

I. INTRODUCTION

Many interesting questions in physics concern the calculation of time scales, i. e. elastic and inelastic scattering rates, and the comparison of such scales to find the most important process for the problem under consideration. Here we are specifically interested in characterizing the time scales for quantum tunneling. Most of the elementary textbooks on quantum mechanics discuss the time it takes a population of carriers to escape from a long lived state, such as a metastable state in a local potential valley or a resonant state in a double or multiple barrier structure. The textbooks, however, are silent when it comes to time scales which are not associated with a long lived state, such as tunneling through a single barrier or through a double barrier away from the resonant condition. A popular but unjustified method to investigate the time scales under the latter conditions, is the use of wave packets and the calculation and extrapolation of the peak motion. Büttiker and Landauer (1982) have criticized this technique, since the shape of a wave packet can differ appreciably from that of an incident packet. Moreover, this approach invariably invokes also an extrapolation procedure. The time of incidence of a packet is calculated by extrapolating the motion of the wave packet far from the barrier to the incident point. Such an extrapolation procedure is unwarranted as the motion of an incident wave packet near a barrier is strongly distorted by the interference with precursors of the packet which have already been reflected. Leavens and Aers (1989a) give a simple but striking example of this effect. Therefore, alternative methods which do not depend on wave packet motion, have been developed (Büttiker and Landauer, 1982; Büttiker, 1983). Below, a very brief review of these methods and their generalizations by Leavens and Aers (1987, 1988a) are presented. Some basic relations between different characteristic time scales are derived analyzing a simple model which describes absorption of particles in a tunneling barrier. A brief discussion of these time scales for multichannel scattering problems is given. This paper makes no attempt to present a complete review of the wide interest which these questions have found. For references to additional related work we refer the reader to Büttiker and Landauer (1986a, 1986b) and to a brief but instructive review paper by Leavens and Aers (1989b). In the last section of this paper we do, however, single out a number of theoretical papers which deserve critical comment. We also present a very brief description of the experimental efforts to measure characteristic tunneling times.

Electronic Properties of Multilayers and Low-Dimensional Semiconductors Structures
Edited by J. M. Chamberlain *et al.,* Plenum Press, New York, 1990

II. PHYSICAL DERIVATIONS OF CHARACTERISTIC TUNNELING TIMES

A. Time-Modulated Barrier

The basic idea of this approach is very simple. The static potential $V_0(x)$ is supplemented by a time dependent potential $V(x,t)$ which is zero except in a region of interest, where

$$V(x,t) = V_1 \cos(\omega t). \tag{2.1}$$

In Büttiker and Landauer (1982) this problem is solved for a rectangular barrier of height V_0 and extension d. The perturbation potential is taken to be non-zero over the whole width of the barrier. Suppose now that there is a characteristic time τ_T for barrier traversal. If the barrier modulation frequency is slow compared to this time, i. e. for $\omega \tau_T < 1$, carriers see a barrier of instantaneous height $V_0 + V_1 \cos(\omega t)$. The carriers traverse the barrier so fast that the potential does not change appreciably during tunneling. If the modulation frequency is fast compared to the traversal time, $\omega \tau_T > 1$, the barrier oscillates many times during barrier traversal. There is thus a crossover from a low-frequence behavior to a high frequency behavior. If this crossover can be found, we have a way to derive the traversal time.

To analyze this tunneling problem, we notice that carriers can acquire or lose a modulation quantum $\hbar\omega$ due to the oscillating barrier. In addition to carriers with energy E, the transmitted and reflected waves now also contain side bands at energies $E + \hbar\omega$ and $E - \hbar\omega$. For small modulation amplitudes it is only the first order side bands which are of interest. The transmitted wave function is time dependent and contains Fourier coefficients at the three energies E and $E \pm \hbar\omega$. For an opaque barrier, $\kappa d >> 1$, where $\kappa = (2m/\hbar^2)^{1/2}(V_0 - E)^{1/2}$ is the exponential decay factor of the wave function in the barrier region and d is the width of the barrier, a Fourier analysis gives for the intensities of the side bands

$$T_{\pm} = (V_1/2\hbar\omega)^2 (e^{\pm\omega\tau_T} - 1)^2 T. \tag{2.2}$$

Here, T is the transmission probability of the barrier at energy E. The characteristic time appearing in Eq. (2.2) is $\tau_T = md/\hbar\kappa$. This characteristic time is just the limiting result for a square barrier of the more general WKB expression,

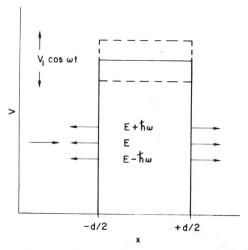

Figure 1.   Barrier with an oscillatory potential variation on top of a static potential. The potential oscillations give rise to side bands with energy $E + \hbar\omega$ and $E - \hbar\omega$ in the transmitted and reflected parts of the wave function.

$$\tau_T = \int_{x_1}^{x_2} dx (m/2(V_0(x) - E))^{1/2}. \tag{2.3}$$

To arrive at Eq. (2.2) it has been assumed that we deal with an opaque barrier which allows only for small transmission. In addition to $\kappa d >> 1$ we have assumed that $E + \hbar\omega < V_0$ and $E - \hbar\omega > 0$.

For small frequencies, $\omega\tau_T < 1$, Eq. (2.2) predicts that the upper and lower side band intensities are equal and given by

$$T_{\pm} = (V_1\tau_T/2\hbar)^2 T. \tag{2.4}$$

At high frequencies the upper side band is exponentially enhanced and the lower side band intensity is exponentially suppressed. Thus for an opaque barrier we do indeed have a rather well defined crossover between tunneling at high frequencies and tunneling at low frequencies.

While the crossover argument is a physically appealing procedure to arrive at a traversal time, we have not been able to extend it beyond the WKB regime of validity for which it was originally put forth. As the barrier becomes thinner, the traversal time becomes shorter and the upper side band allows for transmission of particles at an energy which exceeds the barrier height for the frequency at which one expects the crossover. Interpretation of the tunneling behavior becomes much more complex. Similarly, a crossover cannot be found to identify a reflection time. For the same reasons it is difficult to apply this procedure to identify a traversal time for all energies even for opaque barriers. Recent work by Stövneng and Hauge (1989) and Jauho and Jonson (1989) which investigates transmission through a time modulated barrier outside the limits stated above are briefly discussed in Section V. Below, we take a different approach and essentially use the low frequency behavior given by Eq. (2.4) to identify characteristic tunneling times valid at all energies and valid for barriers which are not opaque.

B. Larmor Precession

The use of the spin precession in a small magnetic field surrounding the target was proposed by Baz' (1966) to measure the duration of collision events (see also Baz' et al. 1969). The method was applied to one dimensional tunneling by Rybachenko (1967). His result does not agree with that obtained by Büttiker and Landauer (1982) described above. Büttiker (1983) reanalyzed the interesting proposal of Baz' and Rybachenko and below we give a brief exposition of Büttiker's interpretation of the Larmor clock. The "experiment" is sketched in Fig. 2. Carriers with spin 1/2 polarized in the x-direction are incident on a barrier. Motion of the carriers is one dimensional along the y-axis. A small magnetic field pointing in the z-direction is applied in the region of non-zero potential. Due to the magnetic field the Hamiltonian contains an additional energy

$$- (\hbar\omega_L/2)\sigma_z \tag{2.5}$$

which acts on the spin up and spin down components of the wave function. Here ω_L is the Larmor precession frequency and σ_z is a Pauli spin matrix. The basic idea now is that a carrier entering the barrier is also subject to the magnetic field and thus starts a spin precession which proceeds for the duration of the tunneling event. If the polarization of the transmitted and reflected waves are calculated and the angle of these polarizations is divided by the Larmor frequency we have a measure of the time spent in the barrier region. If the carriers move in a classically allowed region the Larmor precession consists of a rotation of the polarization in the x-y-plane (perpendicular to the magnetic field). This is what Rybachenko studied. However, as pointed out by Büttiker (1983), such a classical notion of a Larmor precession is not appropriate for our tunneling problem. The complete polarization of spin 1/2 particles in the x-direction can be represented by a spinor which yields a probability of 1/2 that carriers have a spin pointing in the direction of the magnetic field and a probability of 1/2 that carriers have a spin pointing in the direction opposite to the field. Particles with spin up see a barrier of effective height $V_0 - (\hbar\omega_L)/2$ whereas particles with spin down see a barrier of height $V_0 + (\hbar\omega_L)/2$. Spin up particles have a probability T_+ for transmission through the barrier which differs from the transmission probability T_- of particles with spin down. Consequently, the polarization of the transmitted particles also acquires a z-component, i.e. a component in the direction of the magnetic field. Thus in the presence of tunneling we do not have a conventional Larmor precession. In addition to a spin rotation in the x-y-plane there is also a "rotation" in the x-z-plane. For an opaque barrier the latter effect is dominant. Büttiker (1983) has, therefore, argued that in order to obtain a measure for the duration of the tunneling event, one should consider the total rotation angle rather than rely only on the precession angle in the x-y-plane.

To express the results of this calculation in an elegant way it is useful to parameterize the transmission and reflection amplitudes as follows,

$$t = T^{1/2}e^{i\Delta\phi} \tag{2.6a}$$

$$r = -iR^{1/2}e^{i\Delta\phi + \phi_a} \tag{2.6b}$$

$$r' = -iR^{1/2}e^{i\Delta\phi - \phi_a} \tag{2.6c}$$

Here, t and r are the transmission and reflection amplitudes for particles incident from the left, $t' = t$ and r' are the transmission and reflection amplitudes for particles incident from the right. T and R are the transmission and the reflection probability, $\Delta\phi$ is the phase accumulated by transmitted particles, and ϕ_a is an extra phase accumulated by reflected particles incident from the left and incident from the right. The polarization of the transmitted particles is found to be

$$< S_y > = -(\hbar/2)\omega_L \tau_{T,y} \tag{2.7a}$$

with a time

$$\tau_{T,y} = -\hbar\partial\Delta\phi/\partial V \tag{2.7b}$$

which characterizes rotation in the x-y-plane. The z-component is

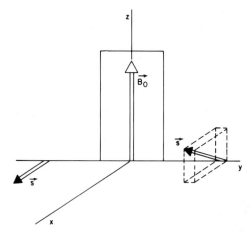

Figure 2.   Larmor clock. Particles propagating along the y-direction with spin polarized in the x-direction are incident on a barrier. A weak magnetic field pointing into the z-direction is non-zero only in a region of interest. The polarization of the transmitted (and reflected) carriers is determined by a conventional Larmor precession in the x-y-plane and, in addition, a rotation in the x-z-plane.

$$\langle S_z \rangle = (\hbar/2)\omega_L \tau_{T,z} \qquad (2.8a)$$

with a time which characterizes rotation in the x-z-plane,

$$\tau_{T,z} = - \hbar \partial \log T^{1/2}/\partial V. \qquad (2.8b)$$

The complete motion of the polarization during traversal is reflected in the x-component of the polarization,

$$\langle S_x \rangle = (\hbar/2)(1 - \omega_L^2 \tau_T^2) \qquad (2.9a)$$

which is characterized by the *traversal time*

$$\tau_T = (\tau_{T,y}^2 + \tau_{T,z}^2)^{1/2} = \hbar \left((\partial \log T^{1/2}/\partial V)^2 + (\partial \Delta \phi/\partial V)^2\right)^{1/2}. \qquad (2.9b)$$

The traversal time is determined by the variation of both the transmission probability and the phase with respect to a small change δV in the potential. The reflection of particles incident from the left is associated with the time scales

$$\tau_{R,y} = - \hbar \partial(\Delta \phi + \phi_a)/\partial V, \qquad (2.10)$$

$$\tau_{R,z} = - \hbar \partial \log R^{1/2}/\partial V. \qquad (2.11)$$

The complete motion of the polarization during reflection of particles is characterized by a *reflection time*,

$$\tau_R = (\tau_{R,y}^2 + \tau_{R,z}^2)^{1/2} = \hbar \left((\delta \log R^{1/2}/\delta V)^2 + (\delta(\Delta \phi + \phi_a)/\delta V)^2\right)^{1/2}. \qquad (2.12)$$

If particles are incident from the right, the polarization of the reflected particles is characterized by

$$\tau'_{R,y} = - \hbar \partial(\Delta \phi - \phi_a)/\partial V \qquad (2.13)$$

$$\tau'_{R,z} = - \hbar \partial \log R^{1/2}/\partial V. \qquad (2.14)$$

$$\tau'_R = (\tau'^2_{R,y} + \tau_{R,z}^2)^{1/2} = \hbar \left((\partial \log R^{1/2}/\partial V)^2 + (\partial(\Delta \phi - \phi_a)/\partial V)^2\right)^{1/2}. \qquad (2.15)$$

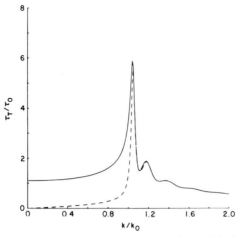

Figure 3. Traversal time (full line) and dwell time (dashed line) for a rectangular barrier of width d and height $V_0 \equiv \hbar^2 k_0^2/2m$ as a function of the wave vector of the incident carriers $k = (2mE/\hbar^2)$ for the case $k_0 d = 3\pi$. $\tau_0 \equiv md/\hbar k_0$.

For a barrier which is symmetric, $\phi_a = 0$, we have to distinguish only one reflection time. We emphasize that the traversal and reflection time are real and positive. The traversal time , as given in Eq. (2.9b) has the property that it gives for a barrier in the WKB limit for E < V the traversal time given by Eq. (2.3). In this limit the transmission amplitude is t≈ exp(− ∫dx | p |) with | p | = $(2m/\hbar^2)^{1/2}(V - E)^{1/2}$ with a phase $\Delta\phi$ which is independent of V. For E > V the transmission amplitude is t≈ exp(i$\Delta\phi$) with $\Delta\phi$ = ∫dxp and p = $(2m/\hbar^2)^{1/2}(E - V)^{1/2}$ ≡ mv and this gives τ_T = ∫dxv^{-1} , i. e. the traversal time which we expect on classical grounds. Thus Eq. (2.9b) gives the correct limiting behaviors both at low and at high energies. The exact analytical expressions for a rectangular symmetric barrier of height V_0 and width d are given in Büttiker (1983). We are not reproducing the complete analytical results here but only show a graphical representation in Figs. 3 and 4. Appendix A contains results for the rectangular barrier in the limit of very small incident energies and stresses the need to distinguish between derivatives of the transmission and reflection amplitudes with respect to V rather than to E. In Büttiker (1983) the results presented above and in Figs. 3 and 4 are expressed with the help of a derivative with respect to the exponential decay factor κ of the wave-function in the barrier. The more elegant notation used here is due to Leavens and Aers (1987, 1988a) who have studied the time scales discussed here for barriers of arbitrary shape. The broken line in Figure 3 and 4 is the dwell time which we will discuss later on. In Appendix B the traversal time and the reflection time, Eqs. (2.9b) and (2.12), are evaluated for a resonant double barrier.

C. Relation between the modulated barrier results and the Larmor precession results

In Section A we have studied a barrier subject to a modulated potential, V = V_0 + V_1 cos(ωt) and have identified the traversal time by searching for a crossover between the high frequency behavior and the low frequency behavior. We noted in Eq. (2.4) that the low frequency behavior yields a sideband intensity which is determined by the square of the traversal time. Now we extend this low frequency result beyond the WKB limit which concerned us in Section A. In the absence of a modulation the transmitted wave is

$$\psi(x, \tau) = t(E,V)e^{ikx-iE\tau/\hbar}. \tag{2.16}$$

To obtain the transmitted wave in the presence of a slowly modulated potential $\omega\tau_T$ < < 1, we can simply replace the static potential in the transmission amplitude by the time dependent potential. Therefore, the wave function to the right of the barrier is in the presence of a slowly varying potential,

$$\psi(x, \tau) = (t(E,V^0) + (\partial t/\partial V)V_1 \cos(\omega\tau))e^{ikx-iE\tau/\hbar}. \tag{2.17}$$

Now, using the parameterization for t as given in Eq. (2.6), and Eqs. (2.7b) and (2.8b) we find,

$$\partial t/\partial V = - (i/\hbar)t(\tau_{T,y} - i\tau_{T,z}). \tag{2.18}$$

To proceed, we Fourier analyze the transmitted wave and determine the amplitudes which multiply, E ± $\hbar\omega$. This immediately yields for the side band intensities $T_\pm$ the expression Eq. (2.4) with a traversal time τ_T = $(\tau_{T,y}^2 + \tau_{T,z}^2)^{1/2}$, as found for the Larmor clock, i.e. Eq. (2.9b). The low frequency behavior of a slowly modulated barrier is governed exactly by the traversal time found in the analysis of the Larmor clock. The equivalence of the low-frequency analysis and the Larmor clock analysis was pointed out in Büttiker and Landauer (1985). We can in a similar way analyze the reflected wave in the presence of a slowly modulated potential. The determining factor is $\partial r/\partial V$, for which we find,

$$\partial r/\partial V = -ir(\tau_{R,y} - i\tau_{R,z}). \tag{2.19}$$

Here, we have used Eq. (2.6b) and Eqs. (2.10) and (2.11). For the intensities of the reflected side bands this yields $R_\pm = (V_1\tau_R/2\hbar)^2$ with τ_R as given by Eq. (2.12). If the particles are incident from the right, the sidebands are determined by

$$\partial r'/\partial V = -ir(\tau'_{R,y} - i\tau_{R,z}), \qquad (2.20)$$

giving rise to a reflection time given by Eq. (2.15).

D. Absorption of Particles in the Barrier

In private communication, Pippard (1982) proposed that absorption of particles in the barrier might provide a measure for the duration of particle traversal. That is not the case. But the method proposed by Pippard is instead related to a time scale which we have called the dwell time (Büttiker, 1983). The dwell time is a measure of the time spent in the barrier region *independent* of whether the particles are in the end transmitted or reflected.

Let us assume that a small fraction of the particles is absorbed in the region of the barrier (by some traps or a weakly coupled additional terminal). The simplest way of describing an absorption of carriers is to include a small imaginary part in the potential. The Hamiltonian now contains an additional energy

$$-i\hbar\Gamma/2 \qquad (2.21)$$

where Γ is an absorption rate. First we calculate the absorbed particle flux. In the region of the barrier, we find from the Schrödinger equation

$$\partial \mid \hat{\psi} \mid^2/\partial t + dj/dy = -\Gamma \mid \hat{\psi} \mid^2 \qquad (2.22)$$

where $\hat{\psi}$ is the solution of the Schrödinger equation in the presence of absorption. We are considering a time independent problem. The first term in Eq. (2.6) vanishes. Integration over the extent of the barrier yields,

$$\Delta j \equiv j(-d/2) - j(d/2) = \Gamma \int_{-d/2}^{d/2} dy \mid \hat{\psi} \mid^2 \qquad (2.23)$$

Let us now introduce the dwell time defined as a ratio of the total integrated particle density N in the barrier region divided by the incident current j:

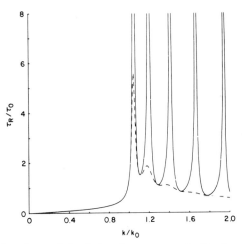

Figure 4.   Reflection time and dwell time for a rectangular barrier as a function of k for a barrier as specified in Fig. 3.

$$\tau_d = N/j = (1/v) \int_{-d/2}^{d/2} dy \mid \psi \mid^2 \qquad (2.24)$$

where $v = \hbar k/m$ is the velocity of the incident particles. (Smith (1960) introduced a "collision time" using a similar definition to characterize the duration of scattering events). If we replace in Eq. (2.23) the wave function in the presence of absorption with the wave function in the absence of absorption (which is permissible since we are dealing with a small absorption rate) we can express Eq. (2.23) with the help of the dwell time,

$$\Delta j = \Gamma \int_{-d/2}^{d/2} dy \mid \psi \mid^2 = (\hbar k/m)\Gamma\tau_d. \qquad (2.25)$$

Thus the total absorbed flux is determined by the dwell time. In the steady state the incoming flux of particles must equal the absorbed flux, the transmitted and the reflected flux,

$$j_{in} = \Delta j + j_T + j_R. \qquad (2.26)$$

We have already determined the absorbed flux. Let us now go on and calculate the reduction in the transmitted and the reflected flux. This is very simple. What we have to understand is again the variation of the transmitted wave and the reflected wave due to a small change in the potential $\delta V = i\hbar\Gamma/2$. The transmission amplitude in the presence of a small rate of absorption is

$$\hat{t} = t + (\partial t/\partial V)\delta V, \qquad (2.27)$$

which with Eq. (2.18) yields a transmission probability,

$$\hat{T} = T(1 - \Gamma\tau_{T,y}). \qquad (2.28)$$

Since the perturbation $\delta V = i\hbar\Gamma/2$ is purely imaginary, it is only one of the components (the rotation in the x-y-plane) which appears in Eq. (2.28). For the reflection probability in the presence of absorption we find similarly,

$$\hat{R} = R(1 - \Gamma\tau_{R,y}). \qquad (2.29)$$

If carriers are incident from the right, the transmission is as given by Eq. (2.28), but the reflected stream is now determined by,

$$\hat{R}' = R(1 - \Gamma\tau'_{R,y}). \qquad (2.30)$$

For a symmetric barrier $\tau'_{R,y} = \tau_{R,y}$ and hence $\hat{R}' = \hat{R}$. Furthermore, $\tau_{R,y} = \tau_{T,y}$ since both of these time scales depend only on the phase $\Delta\phi$. Hence, for a symmetric barrier, $\tau_d = \tau_{R,y} = \tau_{T,y}$, and both the reduction of the transmitted and reflected beam due to absorption are governed by the dwell time. Absorption of carriers in a barrier does not effectively distinguish particles which are finally transmitted or are finally reflected. Thus absorption of particles in the barrier cannot be used to arrive at a traversal and reflection time. While Pippard's suggestion is thus not a satisfactory procedure to define transmission and reflection times, it is nevertheless a nice and physical exercise to arrive at some important relations between the differing time scales, as we will demonstrate below. In Figs. 3 and 4 the dwell time τ_d for a rectangular barrier is compared to the traversal time and the reflection time.

The perturbation considered above, causing absorption in the barrier, was purely imaginary. For completeness, we mention briefly the results for a real perturbation which we take to be $\delta V = \hbar\gamma/2$. In this case current is of course conserved,

$$j_{in} = j_T + j_R. \tag{2.31}$$

The transmitted current is to first order in the perturbation δV proportional to

$$\hat{T} = T(1 - \gamma\tau_{T,z}), \tag{2.32}$$

and the reflected current is determined by

$$\hat{R} = R(1 - \gamma\tau_{R,z}). \tag{2.33}$$

We see that in linear response, it is either the spin rotation in the x-y-plane which appears or the spin rotation in the x-z-plane, depending on whether the perturbation is real or imaginary.

III. RELATIONS BETWEEN DIFFERING TIME SCALES AND GENERALIZATIONS

In Eq. (2.23) we expressed the dwell time as the total integrated density of carriers in the barrier divided by the incident flux. The traversal and reflection times, on the other hand, are expressed in terms of the transmission and reflection probabilities and phase factors characterizing transmission and reflection at the entire barrier. It would be very useful to have an expression for the dwell time in terms of the same parameters. Such a relationship can be found simply by using the results derived above. Consider Eq. (2.26) which states that the incident flux is equal to the absorbed flux and the transmitted and reflected fluxes. Using Eqs. (2.28) and (2.29) to express these fluxes in terms of transmission probabilities yields

$$1 = \Gamma\tau_d + (1 - \Gamma\tau_{T,y})T + (1 - \Gamma\tau_{R,y})R. \tag{3.1}$$

Here we have omitted a factor $\hbar k/m$ common to all the terms in Eq. (3.1). Therefore, Eq. (3.1) gives an expression for the dwell time in terms of the overall scattering amplitudes,

$$\tau_d = R\tau_{R,y} + T\tau_{T,y}. \tag{3.2}$$

Using this relation we can calculate the dwell time without knowing the wave function in the barrier. Similarly, if the carriers are incident from the right we find a dwell time

$$\tau'_d = R\tau'_{R,y} + T\tau_{T,y}. \tag{3.3}$$

Note that for a symmetric barrier the following simple relationship between the dwell time and the derivative of the phase with respect to a potential variation holds, $\tau_d = \tau'_d = -\hbar\partial\Delta\phi/\partial V$.

There is also a relation between the z components of the time scales introduced above. Using, current conservation, Eq. (2.31), i. e. R+T=1, and Eqs. (2.32, 2.33) we find

$$R\tau_{R,z} + T\tau_{T,z} = (1/2)\partial(R + T)/\partial V = 0. \tag{3.4}$$

We can combine the relations, Eq. (3.2) and Eq. (3.4), into one by writing,

$$\tau_d = R(\tau_{R,y} + i\tau_{R,z}) + T(\tau_{T,z} + i\tau_{T,z}). \tag{3.5}$$

Similarly, if carriers are incident from the right, we find

$$\tau'_d = R(\tau'_{R,y} + i\tau_{R,z}) + T(\tau_{T,y} + i\tau_{T,z}). \tag{3.6}$$

Eqs. (3.5) and (3.6) were found by Sokolovski and Baskin (1987) based on a path integral approach to tunneling time scales. Their results are entirely equivalent to those of Büttiker (1983) except that the times appear as in Eq. (3.5) as complex quantities. But the formal relationships Eqs. (3.5) and Eq. (3.6) made them conclude that tunneling times are complex quantities and not real. We disagree with this conclusion. Ultimately, it is always a real time which characterizes an experiment. The approach of Sokolovski and Baskin (1987) is, however, useful and has been invoked by Leavens (1988) to study distributions of characteristic tunneling times.

Until now we have always assumed that the perturbation δV is applied over the entire barrier. But this is not a basic requirement of our discussion. Leavens and Aers (1987; 1988a) have generalized the discussion of Büttiker (1983) to barriers of arbitrary shape and have considered perturbations δV which are applied only over a limited region. The magnetic field in Fig. 2 can be taken to be non-vanishing only in a region from y_1 to y_2 located inside or outside the barrier. Hence, Leavens and Aers obtain characteristic times which are associated with traversal of a small region. Instead of Eq. (3.5) and (3.6), the dwell time and the characteristic times are related by

$$\tau_d(y_1, y_2) = R(\tau_{R,y}(y_1, y_2) + i\tau_{R,z}(y_1, y_2)) + T(\tau_{T,z}(y_1, y_2) + i\tau_{T,z}(y_1, y_2)), \qquad (3.7)$$

and similarly for τ'_d. Most importantly, Leavens and Aers (1987; 1988a) find the law of addition for these characteristic times. Suppose we determine the characteristic times for two perturbations $\delta V(y_1, y_2)$ and $\delta V(y_2, y_3)$ which in the Larmor clock correspond to two regions of non-zero magnetic field extending from y_1 to y_2 and from y_2 to y_3. We can ask how these characteristic times are related to the time scales which we obtain if we apply a perturbation $\delta V(y_1, y_3)$ which extents over the entire interval from y_1 to y_3. Leavens and Aers (1987; 1988a) find

$$\tau_d(y_1, y_3) = \tau_d(y_1, y_2) + \tau_d(y_2, y_3) \qquad (3.8)$$

and

$$\tau_{R,y}(y_1, y_3) = \tau_{R,y}(y_1, y_2) + \tau_{R,y}(y_2, y_3) \qquad (3.9)$$

$$\tau_{T,y}(y_1, y_3) = \tau_{T,y}(y_1, y_2) + \tau_{T,y}(y_2, y_3). \qquad (3.10)$$

This in turn implies that the traversal time obeys the rule,

$$\tau_T(y_1, y_3) = (\tau_{T,y}^2(y_1, y_3) + \tau_{T,z}^2(y_1, y_3))^{1/2}$$

$$= ((\tau_{T,y}(y_1, y_2) + \tau_{T,y}(y_2, y_3))^2 + (\tau_{T,z}(y_1, y_2) + \tau_{T,z}(y_2, y_3))^2)^{1/2}. \qquad (3.11)$$

A similar rule applies for the reflection time. Thus the traversal and reflection times do, in general, not add in a direct way, but add like complex numbers with a real and an imaginary part. In classical mechanical transmission and reflection problems the dwell time, the traversal and the reflection time are directly related (Leavens and Aers, 1987; Landauer and Büttiker, 1987)

$$\tau_d = R\tau_R + T\tau_T. \qquad (3.12)$$

A carrier incident on the barrier is either transmitted with probability T or reflected with probability R. Quantum mechanically we do not find such a simple relation, but instead find Eq. (3.5) and (3.6). This is because we arrived at the transmission and reflection times not by studying a bare transmission and reflection problem, but by a somewhat more complicated process. In the Larmor clock the incident carriers have a probability of 1/2 for spin up and spin down and it is the transmission and reflection behavior of both spinor components which is investigated. Similarly, in the modulated barrier a carrier can be transmitted with either the original energy E, or with an energy which puts it in the upper side band or in the lower side band. An incident carrier has not just *two* alternatives but at least *four*. To characterize quantum mechanical tunneling we need to study the behavior with respect to a perturbation of the transmission and reflection process. This leads to an answer which is more intricate than we would expect from classical mechanical concepts.

IV. MANY CHANNEL SCATTERER

The time scales introduced above for the case of particle motion along a one-dimensional line can be generalized to treat transmission and reflection in more general situations. Smith (1960), Baz' et al. (1969) and more recently Tagliacozzo (1989) have considered this problem. The discussion given below provides such a generalization in direct analogy to the discussion of the single channel problem given above. For this purpose we want to assume, that as in problems of conduction, we have a scattering region, which is connected to perfect "waveguides" both to the right and to the left (See Büttiker, 1989). In the perfect waveguide regions the Hamiltonian is separable and the wave functions are of the form,

$$\psi_{n,k}(y) = e^{iky}f_n(x). \tag{4.1}$$

Here k is the wave vector along the prefect waveguide which extends along the y-axis, and x stands for all the transverse coordinates. The index n labels the quantum channel. The n-th mode becomes propagating above a threshold, $E_n(0)$. Above this threshold particles with wavevector k have an energy $E_n(k)$. Suppose that there are M_1 quantum channels in the asymptotic region to the left and M_2 quantum channels to the right with thresholds smaller than the energy E under consideration. Transmission and reflection at the scattering region is then characterized by a scattering matrix S, which in the absence of a magnetic field is symmetric, $S = S^T$. (Here T denotes the transposed matrix). We can now proceed as in the single channel case and associate with each matrix element of the scattering matrix the time-scales,

$$\delta s_{ij}/\delta V = - (i/\hbar)s_{ij}(\tau_{ij,y} + i\tau_{ij,z}) \tag{4.2}$$

where

$$\tau_{ij,y} = - \hbar\partial\phi_{ij}/\partial V, \ \tau_{ij,z} = - \hbar\partial \log | s_{ij} | /\partial V \tag{4.3}$$

and the phase ϕ_{ij} is defined by $s_{ij} = | s_{ij} | \exp(i\phi_{ij})$. In a Larmor precession approach, the first component in the parenthesis of Eq. (4.2) would be associated with precession in the x-y-plane and the second component with spin rotation in the x-z-plane. In Eq. (4.2) the scattering amplitude can describe either a transmission with an amplitude $t_{ij} = s_{ij}$ associated with a probability $T_{ij} = | t_{ij} |^2$ or a reflection amplitude $r_{ij} = s_{ij}$ associated with a probability $R_{ij} = | r_{ij} |^2$ depending on whether the indices refer to channels in differing asymptotic regions or in the same asymptotic region. Thus, we find traversal times and reflection times,

$$\tau_{ij,T} = (\tau_{ij,Ty}^2 + \tau_{ij,Tz}^2)^{1/2}, \ \tau_{ij,R} = (\tau_{ij,Ry}^2 + \tau_{ij,Rz}^2)^{1/2} \tag{4.4}$$

Suppose, a wave is incident from the left in channel j. In general, there are reflected waves and transmitted waves into all the channels. We can ask about the average time spent in the scattering region regardless of the channel into which carriers are eventually reflected or transmitted. This is the dwell time $\tau_{d,j}$ associated with particles incident in channel j. We find that the generalized Eq. (3.5) is

$$\tau_{d,j} = \sum_{i=1}^{i=M_2} | s_{ij} |^2(\tau_{ij,y} + i\tau_{ij,z}). \tag{4.5}$$

which in terms of the transmission and reflection probabilities for carriers incident in channel j becomes

$$\tau_{d,j} = \sum_{i=1}^{i=M_1} R_{ij}(\tau_{ij,R,y} + i\tau_{ij,R,z}) + \sum_{i=1}^{i=M_2} T_{ij}(\tau_{ij,T,y} + i\tau_{ij,T,z}). \tag{4.6}$$

To measure the traversal and reflection times introduced above, we must be able to make both a measurement which determines the velocity of particles (and hence the quantum channel into which they have been scattered) and a measurement which determines the spin polarization. In contrast, we can ask about the traversal and reflection times *if a measurement does not distinguish the channel to which the transmitted and reflected particles belong.* We define the total reflection probability R_j and the total transmission probability T_j if a unit particle flux is incident from the left in channel j,

$$R_j = \sum_{i=1}^{i=M_1} R_{ij}, \quad T_j = \sum_{i=1}^{i=M_2} T_{ij}. \tag{4.7}$$

We can introduce characteristic times averaged with respect to the quantum channels,

$$\tau_{j,Ry} = (1/R_j) \sum_{i=1}^{i=M_1} R_{ij}\tau_{ij,Ry}, \quad \tau_{j,Rz} = (1/R_j) \sum_{i=1}^{i=M_1} R_{ij}\tau_{ij,Rz}, \tag{4.8}$$

$$\tau_{j,Ty} = (1/T_j) \sum_{i=1}^{i=M_2} T_{ij}\tau_{ij,Ty}, \quad \tau_{j,Tz} = (1/T_j) \sum_{i=1}^{i=M_2} T_{ij}\tau_{ij,Tz}, \tag{4.9}$$

such that

$$\tau_{d,j} = R_j(\tau_{j,Ry} + i\tau_{j,Rz}) + T_j(\tau_{j,Ty} + +i\tau_{j,Tz}). \tag{4.10}$$

This in turn immediately yields the average traversal and reflection time. The imaginary part of this equation is zero on account of particle current conservation.

In solid state physics we often deal with situations where one, or more, or all the incoming quantum channels carry a unit current. Below we briefly discuss the case when all the incident channels carry a unit current. We consider only the case when particles are incident from the left. The dwell time is an average over all the dwell times associated with the individual quantum channels,

$$\tau_D = (1/M_1) \sum_{i=1}^{i=M_1} (1/v_i) \int dx \int_{-d/2}^{d/2} dy \mid \psi_i(x,y) \mid^2. \tag{4.11}$$

Note that the dwell time is again proportional to the total integrated carrier density in the scattering region. The characteristic times for this situation are

$$\tau_{T,y} = (1/T) \sum_{i=1,j=1}^{i=M_2,j=M_1} \tau_{ij,Ty}, \quad \tau_{R,y} = (1/R) \sum_{i=1,j=1}^{i=M_2,j=M_1} \tau_{ij,Ry} \tag{4.12}$$

and similar for $\tau_{R,z}$. The relation of the dwell time to these characteristic times is

$$\tau_d = (R/M_1)(\tau_{R,y} + i\tau_{R,z}) + (T/M_1)(\tau_{T,y} + i\tau_{T,z}). \tag{4.13}$$

The traversal and reflection time can be found from the square root of the sum of the characteristic times given above.

A further extension to scattering experiments in more complicated geometries such as structures with a number of terminals or ports at which particles are in going or out going is also of interest. In order to keep this paper within a reasonable length, we will not burden the reader with such a discussion. But the discussion given above makes it clear that the type of experiment performed is what determines the relevant time scales. If we want to compare the time scales for elastic transmission with an inelastic process for a multichannel scatterer, we must in generally compare the inelastic time with the time scales given by Eqs. (4.3-4.4) for the individual quantum channels.

V. DISCUSSION OF ADDITIONAL RELATED WORK AND OF EXPERIMENTS

Following are some remarks on papers which have in one way or another reached conclusions or advocated interpretations which disagree with the viewpoint advanced in this paper. The work of Sokolovski and Baskin (1987) has already been mentioned above. This paper agrees in the technical details completely with Büttiker (1983) and our difference with this paper concerns only the interpretation of these results. Some criticism (Collins et al. 1987) suffers from more elementary difficulties and has already been dealt with elsewhere (Büttiker and Landauer, 1988; Leavens and Aers, 1988a).

In addition to the two approaches discussed in Section II, Büttiker and Landauer (1985, 1986b) have considered a third approach in which two waves of slightly different energy are superimposed. It was shown by Leavens and Aers (1988b) that this approach leads to characteristic times as defined in Eq. (2.9b) and Eq. (2.12), but with the derivative with respect to the potential replaced by a derivative with respect to the total energy E. This is of no consequence as long as the transmission and reflection amplitudes depend on the energy and the potential only in the combination E-V (which is typical if the WKB approximation applies). But this is generally not the case and this third approach can give rise to results which differ strongly from those discussed in Section II (See Appendix A for an example). In contrast to the methods discussed in Section II this third approach is less suited to probe the interaction of the incident particles with the barrier. We emphasize that the arguments of Büttiker and Landauer (1985, 1986b) are very instructive and in a simple way demonstrate why it is both the phase and the amplitude of the transmission and reflection coefficients which must be taken into account.

Our criticism of approaches based on the analysis of wave packets (Büttiker and Landauer, 1982; Büttiker, 1983) has not prevented the further study along these very same lines (Hauge et al., 1987; Jaworski and Wardlaw, 1988). Hauge et al. 1987, instead of following the peak of wave packets, calculate the center of gravity of a wave packet. As in an analysis which follows the peak of a wave packet extrapolation methods are used which presume free wave packet motion even in a region where the leading part of the wave packet has already been reflected by the barrier. The pitfalls of such extrapolation methods are illustrated by Leavens and Aers (1989a) with a simple example. Based on their results Hauge et al. (1987) question the concept of the dwell time (rather than their phase-delay times). The abstract of their paper states: "This relation shows when the dwell time can and cannot be used". The answer to their criticism was also given in the paper by Leavens and Aers (1989a) and eventually caused Hauge and Stövneng (1989) to change their views. However, while Hauge and Stövneng (1989) came to accepted the dwell time it did not lead them to discard the phase-delay times, but at least caused an admission that the phase-delay times are not suitable to answer questions about transmission from A to B but must be understood in an "asymptotic sense" (Hauge and Stövneng, 1989).

In a paper by Falck and Hauge (1988) it is argued at length that the Larmor clock should be "properly set". They propose that in order to set the Larmor clock properly it is necessary to use wave packets. However, if one repeats the calculation with plane waves instead of wave packets, for the situation they consider and makes the same approximations that they make with plane waves (which consist of neglecting phase coherence at the entrance and exit point into the magnetic field region and at the entrance and exit of the barrier), one obtains precisely their results for spatially very extended wave packets (wave packets narrow in k-space). This was pointed out to us by Leavens (1989b). Thus there is no substance to the basic argument of this paper. Clearly, one can insist that the Larmor clock analysis just yields what we know from wave packet motion, if one is willing to completely disregard the non-classical spin rotation, and if one is willing to disregard differences between $\partial\Delta\phi/\partial V$ and $\partial\Delta\phi/\partial E$. All this, of course, has nothing to do with a Larmor clock that has to be "set correctly".

Huang et al. (1988) study tunneling in the presence of a small magnetic field which is not only applied to the barrier but throughout space. Their results for the characteristic tunneling times are the same

309

as those obtained by a wavepacket analysis. It is, therefore, instructive to compare this calculation with that of Büttiker (1983). As pointed out by Leavens and Aers (1989a), to arrive at the tunneling times, Huang et al. have to extrapolate "classical" spin precession into a region close to the barrier edge where the spin rotation is in reality affected by the interference of the incident and the reflected wave.

Stövneng and Hauge (1989) have recently applied the modulation technique of Büttiker and Landauer (1982) to a delta-function barrier. That seems particularly unwarranted since the traversal time for a delta-function barrier is just zero and hence the crossover energy associated with barrier traversal is infinite. In Büttiker (1983) it was pointed out that Eq. (2.9b) for a very thin barrier of height V_0 and an incident carrier energy so small that the thin barrier is opaque leads to $\tau_T = \hbar/V_0$ independent of the width of the (thin) barrier, as long as $\kappa d < < 1$. On the other hand, if the incident particle energy is sufficiently large such that the thin barrier is almost transparent, the traversal time is $\hbar \tau_T = md/\hbar k$. In either case in the limit of a delta function barrier the traversal time is zero. On the other hand the energy which governs the crossover from opaque transmission to transparent transmission is $E = (m/2\hbar^2)V_0^2 d^2$ and remains finite and non-zero in the limit of a delta function barrier. It is this energy scale which separates the opaque and the transparent barrier which governs the energy and frequency behavior of the side bands for a delta function barrier. Our papers (Büttiker and Landauer, 1982, 1985) state carefully the limits for which the crossover argument is applied. Jauho and Jonson (1989), in a paper in which they study the motion of wave packets through a modulated barrier, similarly criticize the modulated barrier discussion, based on the fact that it does not give a crossover for barriers which are not amenable to a WKB treatment. The fact that the crossover argument is not universally applicable is not, in our view, a reason to question its validity in the range where we invoked this procedure to find a traversal time.

Our discussion of a few dissenting papers should not give the impression that the work discussed above is not perceived as useful. We refer the reader once more to Büttiker and Landauer (1985, 1986b) and Leavens and Aers (1989a) for additional references and complement these by some more recent work. Kotler and Nitzan (1987) investigate a two channel scattering problem and discuss channel-mixing as a function of the traversal time. (Their problem is such that it can, by a suitable transformation, be mapped on the Larmor clock). Mugnai and Ranfagni (1988) discuss time scales in optically excited color centers. Mullen et al. (1989) investigate time scales in level crossing problems. Both Mugnai and Ranfagni (1988) and Schulman and Ziolkowski (1989) stress that Eq. (2.3) which appears often in the path integral formulation of tunneling problems as purely imaginary "bounce time" has in fact physical significance.

Fortunately, in the past few years, there has been growing experimental interest in the work discussed above. Before discussing this work it is worthwhile to mention that spin polarization experiments, as they are called for by the analysis of the Larmor clock, are entirely feasible although perhaps at the cutting edge of the present measurement technology. As shown by the work of Johnson and Silsbee (1988) ferromagnetic contacts can be used to inject spin polarized electrons into a paramagnetic conductor. Similarly, ferromagnetic contacts can be used to detect the spin polarization of carriers. Johnson and Silsbee have carried out such experiments for carriers in metallic diffusive samples. Introducing barriers into the sample would greatly reduce the measured signal. On the other hand, the use of high quality ballistic conductors, instead of metallic diffusive conductors, could perhaps partially compensate for the weaker signal in the presence of a barrier. These experiments are made in a four-terminal set-up and require an extension of the discussion given above to such a multi-terminal situation.

Let us now turn to experiments which are very closely related to the notions advanced in this paper. A brief review of the experimental work in this field can also be found in Landauer (1989). Eaves et al. (1986) and Gueret et al. (1987) have analyzed tunneling through a barrier in a weak magnetic field. The Lorentz force, which in the discussion given above has been entirely neglected, gives rise to a

310

tunneling path which is effectively longer than the path in the absence of the field. The ensuing exponential reduction of the transmission probability is determined by the square of the product of the cyclotron frequency and the traversal time. These experiments analyze a steady state situation similar to our discussion of the Larmor clock.

Cutler et al. 1987, have proposed using the measurement of the I-V-characteristic of a tunneling microscope under irradiation of a polarized beam of light. This proposal is very closely related to the approach of Büttiker and Landauer (1982) which investigates tunneling in the presence of a modulated barrier. If the frequency of radiation is small compared to the traversal time the radiation produces, due to rectification, a dc current. At high radiation frequencies the modulation is too fast for electrons to follow the field, and there should be no current flow between tip and sample. The crossover can give an indication of the traversal time. Experimental data are reported by Cutler et al. (1987). The data seem too preliminary to distinguish between differing theories.

A very interesting set of experiments by Gueret et al. 1988 compare a large number of tunneling barriers of differing widths. It is found that the transmission probability calculated without any image charge corrections (Jonson, 1980; Persson and Baratoff, 1988) agrees well with the experimental data as long as $\omega_p \tau_T \leq 3$, where ω_p is the plasma frequency and τ_T is the traversal time as given by Eq. (2.3), but shows marked deviations from calculations which neglect image forces for $\omega_p \tau_T$. It is only when the traversal time is long enough that the plasma waves are fast enough to establish an image force. These are very interesting experiments because they probe the concept of a traversal time in a dynamical fashion, and because the experiments reveal a crossover as proposed by Büttiker and Landauer (1982).

Martinis et al. (1988) have proposed a very intriguing experiment in which a Josephson junction is coupled to a transmission line which at some variable distance away from the junction is terminated. If the tunneling event is slow, waves generated by the tunneling process can travel up and down the transmission line and interfere with the tunneling process. If, on the other hand, the tunneling process is fast, waves sent into the transmission line will not return before the tunneling process is completed. Experiments by Esteve et al. (1989) do indeed exhibit a crossover. The crossover is proportional to the inverse of the small amplitude oscillation frequency near the bottom of the metastable state which is equal to the absolute value of the small amplitude frequency at the barrier top in this system. It seems, however, that the crossover time does not scale as predicted by Eq. (2.3). This requires further thought. Nevertheless, the fact that a crossover is indeed found is already a very interesting and very promising result.

We are hopeful that a continuation of the experimental effort will provide further support for the notions advanced in this paper.

APPENDIX

A. Opaque rectangular barrier

Here we give the characteristic times for an opaque rectangular barrier ($\kappa d >> 1$) of height V_0 and width d in the limit that the incident energy E approaches zero. In this limit, since $R \approx 1$, the dwell time and the reflection time are equal and given by

$$\tau_d = \tau_R \approx \frac{\hbar}{V_0} \sqrt{\frac{E}{V_0}}. \tag{A1}$$

Both the dwell time and the reflection time tend to zero proportional to $E^{1/2}$. The traversal time in this limit is

$$\tau_T = \frac{md}{\hbar\kappa} \tag{A2}$$

proportional to the width of the barrier. If instead of using the derivative of the phase $\Delta\phi$ with respect to a variation in the potential, an energy derivative is taken (as we would obtain by following the peak of a wave packet) a "time" $\tau_\phi = \hbar d\Delta/dE \approx \hbar/\sqrt{(EV_0)}$ results which is independent of the width of the barrier and diverges as the energy tends to zero. Thus in general, the distinction between a derivative with respect to a potential variation dV and an energy variation dE is crucial.

B. Time Scales for Resonant Tunneling

Consider for simplicity a symmetric double barrier with a well of width L. Let us also assume that we deal with a single channel transmission problem. Near resonance the transmission amplitude is $t = T^{1/2}e^{i\Delta\phi}e^{i\phi_0}$, with

$$T = \frac{\Gamma^2/4}{(E - E_r)^2 + \Gamma^2/4} \tag{B1}$$

where Γ is the decay width and E_r the energy of the resonance, and the accumulated phase is determined by

$$\tan \Delta\phi = 2(E - E_r)/\Gamma. \tag{B2}$$

ϕ_0 is an arbitrary phase which varies slowly with energy and near the resonance can be taken to be energy independent. The decay width is the sum of the rates for decay to the left and right through the barriers with transmission probability T_0 forming the double well. In WKB approximation

$$\Gamma = 2\hbar/\tau = 2\hbar\nu T_0 = 2\hbar(v/2L)T_0, \tag{B3}$$

where $\nu = 2L/v$ is the attempt frequency and v is the carrier velocity in the well at resonance. The formulae given above for the characteristic times require a derivative with respect to a potential variation δV. However, since near resonance the large contributions to these characteristic times comes from the carriers oscillating back and forth in the well and since the velocity in the well $v = (2/m)^{1/2}(E - V)^{1/2}$ depends only on V-E, we can obtain the important contributions to the characteristic times by taking the derivative with respect to E. This procedure neglects contributions to the characteristic times which are associated with the tunneling through the barriers forming the double well. Taking an energy derivative, we find for the dwell time

$$\tau_d = \hbar \frac{\Gamma/2}{(E - E_r)^2 + \Gamma^2/4} \tag{B4}$$

which at resonance is equal to $\tau_d/\hbar = 2/\Gamma$. Note that if we consider the decay of a carrier population initially in the well, the decay is governed by $e^{-\Gamma t/\hbar}$ with a decay time $\tau_d/2$. This exponential decay of the population assumes that there are two open channels: carriers can escape through the barriers on either side of the well. On the other hand, if we perform a scattering experiment with carriers incident from the left, we are at resonance left with only one open channel: transmission to the right. Hence the dwell time at resonance is twice as long as the life-time of a population in the well. For the traversal time, we find,

$$\tau_T = \frac{\hbar}{((E - E_r)^2 + \Gamma^2/4)^{1/2}}. \tag{B5}$$

At resonance the traversal time equals the dwell time. Taking into account that the reflection amplitude is also governed by the phase Eq. (B2), we find a reflection time

312

$$\tau_R = \frac{\hbar}{(E - E_r)} \frac{(\Gamma/2)}{((E - E_r)^2 + \Gamma^2/4)^{1/2}}. \tag{B6}$$

At resonance the reflection time diverges. As we move away from resonance both the reflection time and the traversal time decrease. For $E - E_r = \Gamma/2$, i.e. at an energy corresponding to half the peak height, the transmission probability is $1/2$. The dwell time is given by $\tau_d = 1/\Gamma$. The traversal and reflection time are equal and given by $\tau_T = \tau_R = \sqrt{2} \, (\hbar/\Gamma)$.

For a well formed by differing barriers the scattering matrix near resonance is in the Breit-Wigner limit (Büttiker, 1988) determined by

$$s_{mn} = (\delta_{mn} - i \frac{(\Gamma_n \Gamma_m)^{1/2}}{E - E_r + i\Gamma/2}) e^{i(\delta_m + \delta_n)}. \tag{B7}$$

For a single channel problem, $n = 1,2$ and $m = 1,2$ the diagonal elements of Eq. (B7) give the reflection amplitudes and the off-diagonal elements give the transmission amplitudes. The phases δ_1 and δ_2 are taken to be energy independent. Γ_1 is the partial width for decay to the left and Γ_2 is the partial width for decay to the right. $\Gamma = \Gamma_1 + \Gamma_2$ is the total decay width of the long lived state. Form Eq. (B7) we obtain the transmission probability, $T = | s_{12} |^2 = | s_{21} |^2$ the reflection probability, the phase which is accumulated during transmission, $\Delta\phi$ and the phases associated with reflection $\Delta\phi \pm \phi_a$ (see Eq. (2.6)). The phase which is accumulated during transmission, $\Delta\phi$ is determined by Eq. (B2) and the traversal time is given by Eq. (B5) with Γ equal to the total decay width. For the derivative of the phase associated with reflection of carriers incident from the left, we find from Eq. (B7)

$$d(\Delta\phi + \phi_a)/dE = \frac{\Gamma_1}{(E - E_r)^2 + \Gamma^2/4} \frac{(E - E_r)^2 + \Gamma(\Gamma_1 - \Gamma_2)/4}{(E - E_r)^2 + (\Gamma_1 - \Gamma_2)^2/4}. \tag{B8}$$

The derivative of the phase $\Delta\phi - \phi_a$ associated with reflection of carriers incident from the right is found from Eq. (B8) by interchanging Γ_1 and Γ_2. A wave packet analysis (Gracia-Calderon and Rubio, 1989) leads to reflection times, $\hbar d(\Delta\phi \pm \phi_a)/dE$, which is obviously not acceptable: Very close to resonance $E \approx E_r$ one of these derivatives is negative if $\Gamma_1 \neq \Gamma_2$. Using Eq. (B8) and the energy derivative of the reflection probability, we find with the help of Eq. (2.12) a reflection time,

$$\tau_R = \hbar \frac{\Gamma_1}{(E - E_r)^2 + \Gamma^2/4} \frac{((E - E_r)^4 + (\Gamma_1^2 + \Gamma_2^2)(E - E_r)^2/2 + \Gamma^2(\Gamma_2 - \Gamma_1)^2/16)^{1/2}}{(E - E_r)^2 + (\Gamma_1 - \Gamma_2)^2/4}. \tag{B9}$$

Note, that for $\Gamma_1 \neq \Gamma_2$ the reflection time is not singular at resonance but determined by $\tau_R = \hbar 4\Gamma_1/(\Gamma | \Gamma_1 - \Gamma_2 |)$. The dwell time, obtained from Eq. (3.2) with the help of Eqs. (B2) and (B8) is

$$\tau_d = \hbar \frac{\Gamma_1}{(E - E_r)^2 + \Gamma^2/4}. \tag{B10}$$

The reflection time τ'_R and the dwell time τ'_d are found by interchanging Γ_1 and Γ_2 in Eq. (B9) and (B10). Since we used energy derivatives a wave packet analysis (Garcia-Calderon and Rubio) yields the dwell times as obtained above. A much more detailed study of characteristic tunneling times in double well barriers is given by Leavens and Aers (1989c).

REFERENCES

Baz', A. I., 1967, Sov. Phys. Nucl. Phys. **4**, 182.

Baz', A. I., Zel'dovich, Ya. B., and Perelomov, A. M., 1969, *Scattering, Reactions and Decay in Nonrelativistic Quantum Mechanics*, Wiener Bindery Ltd., Jerusalem.

Büttiker, M., 1983, Phys. Rev. **B27**, 6178.

Büttiker, M., 1988, IBM J. Res. Develop. **32**, 63 (1988).

Büttiker, M., 1989 *Electronic Properties of Multilayers and Low Dimensional Semiconductor Structures*, edited by J. M. Chamberlain, L. Eaves and J. C. Portal (Plenum, New York).

Büttiker, M., and Landauer, R., 1982, Phys. Rev. Lett. **49**, 1739.

Büttiker, M., and Landauer, R., 1985, Phys. Scr. **32**, 429.

Büttiker, M., and Landauer, R., 1986a, *Festkörperprobleme (Advances in Solid State Physics)*, Vol. 25, ed. P. Grosse , (Vieweg,Braunschweig), 711.

Büttiker, M., and Landauer, R., 1986b, IBM J. Res. Develop. **30**, 451.

Büttiker, M., and Landauer, R., 1988, J. Phys. **C21**, 6207.

Collins, S., Lowe, D., and Barker, J. R., 1987, J. Phys. **C20**, 6213.

Cutler, P. H., Feuchtwang, T. E., Tsong, T. T., Nguyen, H., and Lucas, A. A., 1987, Phys. Rev. **B35**, 7774.

Cutler, P. H., Feuchtwang, T. E., Huang, Z., Tsong, T. T., Nguyen, H., Lucas, A. A., Sullivan, T. E., 1987, J. de Phys. Colloque C6 **48**, 101.

Eaves, L., Stevens, K. W. H., and Sheard, F. W., 1986, *The Physics and Fabrication of Microstructures and Microdevices*, ed. Kelly, M. J., and Weisbuch, C., (Springer, Berlin). 343.

Esteve, D., Martinis, J. M., Urbina, C., Turlot, E., Devoret, M. H., Grabert, H., and Linkwitz, S., 1989, Physica Scripta, (unpublished).

Falk, J. P., and Hauge, E. H., 1988, Phys. Rev. **B38**, 3287.

Garcia-Calderon, G., and Rubio, A., 1989, Solid State Commun. **71**, 237.

Gueret, P., Baratoff, A., and Marclay, E., 1987, Europhyis. Lett. **3**, 367.

Gueret, P., Marclay, E., and Meier, H., 1988 Solid State Communic. **68**, 977; Appl. Phys. Lett. **53**, 1617.

Hauge, E. H., Falck, J. P., and Fieldly, T. A., 1987, Phys. Rev. **B36**, 4203.

Hauge, E. H., and Stövneng, J. A., 1989, Rev. Mod. Phys. (unpublished).

Huang, Z., Cutler, P. H., Feuchtwang, T. E., Good, Jr. R. H., Kazes, E., Ngyuen H. Q., and Park, S. K., 1988, J. de Physique, Colloque **C6**, 17.

Jauho, A. P., and Jonson, M., 1989, (unpublished).

Johnson, M., and Silsbee, R. H., 1988, Phys. Rev. **B37**, 5326.

Jonson, M., 1980, Solid State Commun. **33**, 743.

Kotler, Z., and Nitzan, A., J. 1988, Chem. Phys. **88**, 3871.

Landauer, R., 1989, Nature (unpublished).

Landauer, R. and Büttiker, M., 1987, Phys. Rev. **B36**, 6255.

Leavens, C. R., 1988, Solid State Commun. **67**, 1135.

Leavens, C. R., 1989, private communication.

Leavens, C. R., and Aers, G. C., 1987, Solid State Commun. **63**, 1101.

Leavens, C. R., and Aers, G. C., 1988a, Solid State Commun. **68**, 13.

Leavens, C. R., and Aers, G. C., 1988b, J. Vac. Sci. Technol. **A6**, 305.

Leavens, C. R., and Aers, G. C., (1989a), in *Basic Concepts and Applications of Scanning Tunneling Microscopy and Related Techniques,* R. J. Behm, Kluwer Academic Publishers, Dordrecht.

Leavens, C. R., and Aers, G. C., 1989b, Phys. Rev. **B39**, 1202.

Leavens, C. R., and Ares, G. C., 1989c, Phys. Rev. **B40** , 5387.

Martinis, M., Devoret, M. H., Esteve, D., and Urbina, C., 1988, Physica **B152** 159.

Mugnai, D., and Ranfagni, A., 1988, Europhys. Lett. **6**, 1.

Mullen, K., Ben-Jacob, E. B., Gefen, Y., and Schuss, Z., 1989, Phys. Rev. Lett. **21**, 2543.

Persson, B. N. J. and Baratoff, A., 1988, Phys. Rev. **B38**, 9616.

Pippard, A. B., 1982, private communication.

Pollak, E., and Miller, W. H., 1984, Phys. Rev. Lett. **53**, 115.

Rybachenko, V. F., 1966, Sov. J. Nucl. Phys. **5**, 635.

Schulman, L. S., and Ziolkowski, R. W., 1989, in *3rd International Conference on Path Integrals from meV to MeV,* Bangkok, edited by V. Sa-yakarit et al., World Scientific, Singapore, (unpublished).

Smith, F. T., 1960, Phys. Rev. **118**, 349.

Sokolovski, D. and Baskin, L. M., 1987, Phys. Rev. **A36**, 4604.

Stövneng, J. A., and Hauge, E. H., 1989b, J. Stat. Phys. (unpublished).

Tagliacozzo, A., 1988, Il Nuovo Cimento, **D10**, 363.

DYNAMIC POLARIZATION EFFECTS IN TUNNELING

P. Guéret

IBM Research Division
Zurich Research Laboratory
8803 Rüschlikon, Switzerland

1. INTRODUCTION

Tunneling is a pure quantum-mechanical effect which allows particles to traverse potential barriers, in spite of not having sufficient energy to do so. It is fundamental to most transport and conduction processes. Particles (electrons) are seldom truly free. They are more often than not embedded within non-uniform potentials which constrain their motion. Even the so-called "free" electrons in the allowed bands of a solid are in fact (resonantly) tunneling through the periodic potential of the lattice ions. The simplicity of this phenomenon, as described in elementary textbooks on quantum mechanics, is deceiving. In reality, experimental situations involving tunneling are generally complicated by several (sometimes sizeable) side effects which may make a quantitative comparison with theoretical models difficult. Entire books (Duke, 1969; Wolf, 1985) have been devoted to tunneling-related issues, many of which are still unresolved or controversial, and still awaiting for unequivocal experimental confirmation.

The large and still growing body of literature on the subject is nevertheless a good measure of its significance in a variety of fields including condensed matter, surface and device physics. Photoelectric and field emission from solids are long known examples. More recently, quantum confinement in semiconductor heterostructures and barrier isolation in miniaturized field effect transistors have gained considerably in importance. Since the advent of scanning tunneling microscopy (STM) for the study of surfaces with atomic resolution, the requirement for a thorough understanding of the various aspects of tunneling has become even more crucial.

The main issue which we shall address in this article is that of the image potential correction on tunneling, and can be stated (in somewhat simplified terms) as follows. A test charge (electron) in front of a conducting half-space (metal or semiconductor) induces, by way of its own electric field, a positive charge on the surface of this body or, more precisely, a redistribution of the free charges within a few screening lengths of the solid surface. Since the conducting half-space is thus no longer perfectly neutral at every point, it then generates its own potential over the entire space which in turn modifies the original potential distribution which existed without the test charge. The total self-consistent potential which results from this situation is the image-corrected potential.

Electronic Properties of Multilayers and Low-Dimensional Semiconductors Structures
Edited by J. M. Chamberlain *et al.*, Plenum Press, New York, 1990

In the case of tunneling through a potential barrier between two conducting half-spaces, the situation is similar. Tunneling electrons, while in the barrier, induce charge redistributions in the electrodes which in turn modifies the size and shape of the original barrier. Since the barrier transmission coefficient may thus be changed substantially, it is expected that the image potential correction will produce measurable effects on the tunnel conductance. It is felt intuitively, moreover, that the dynamics of this polarization phenomenon will also play a role. The rearrangement of the mobile carriers in the electrodes is a dynamic process with a characteristic reaction time of the order of ω_p^{-1}, the inverse plasma frequency of the free carriers. One therefore expects that the image potential correction will approach the static one when the electron lifetime in the barrier is long compared to ω_p^{-1}, but that dynamic retardation will inhibit the full development of the electrode polarization when the electron traversal time through the barrier is short compared to ω_p^{-1}. The expected transition from a static to a dynamic regime thus stresses the significance of the "traversal time" for tunneling, a still rather controversial notion in spite of its long history (the first article on the subject dates back to 1931 (Condon, 1931)). According to experiments by the present author and coworkers (Guéret et al., 1988a, 1988b), it appears that the traversal time concept discussed recently by Büttiker and Landauer (1982) is the relevant one leading to a physically meaningful interpretation of experimental data.

There are many reasons why quantitative comparisons between experiment and theory in tunneling are difficult and not always conclusive. It is well known that tunnel conductances depend exponentially on barrier height and thickness. In a meaningful experiment, these parameters must therefore be known with sufficient accuracy, and in addition be fairly uniform over the entire barrier cross-sectional area. For similar reasons, the electrode-barrier interfaces should also be flat and defect-free. The above requirements point naturally to MBE (molecular beam epitaxy)-grown heterostructures (Parker, 1985) as a way to achieve the quality and uniformity necessary for systematic investigations. The lattice-matched GaAs/Al_xGa_{1-x}As system in particular satisfies most of the criteria: sharp interfaces, excellent control of barrier thicknesses, linear dependence of the GaAs/Al_xGa_{1-x}As conduction-band discontinuity on the Al content x. Systematic studies of distinct tunneling regimes covering wide ranges of barrier heights, thicknesses and shapes can be made with the system. By varying the doping, the Fermi energy in the (emitter) contacts can be tuned as well.

The present article is structured as follows. In Section II, the simple textbook treatment of tunneling is briefly reviewed. Section III describes the GaAs/Al_xGa_{1-x}As tunneling heterostructures we have used, together with the fabrication and measurement procedures. Section IV goes beyond the simple picture of Section II and discusses various corrections to this model such as band-bending and image potential. Section V introduces the notion of "traversal time" for tunneling and gives a brief summary of existing theoretical models for the image potential and its dynamic corrections. In Section VI, we discuss a systematic investigation of tunneling in GaAs/Al_xGa_{1-x}As heterostructures. Our data, presented in terms of the Büttiker-Landauer tunneling time, provide evidence for the occurrence of dynamic polarization effects.

II. TUNNELING THROUGH A POTENTIAL BARRIER

Figure 1 shows an electrostatic potential $V(x) = \phi(x)/q$ exhibiting a barrier of height $\phi_0 = qV_0$ and thickness ℓ, on which an electron of energy E and charge q impinges from the left. As shown in every elementary textbook on quantum mechanics, there is a finite probability for the electron to be transmitted to the right of

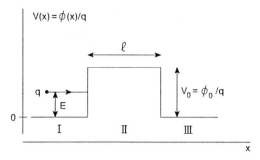

Fig. 1.   Tunnel barrier with electron of charge q and energy E impinging from the left.

the barrier, whatever its energy E. The transmission coefficient is obtained by solving Schrödinger's equation

$$\frac{\hbar^2}{2m}\frac{\partial^2 \psi}{\partial x^2} + (E - \phi(x))\psi = 0 \tag{1}$$

in the three regions I, II and III and matching the wave functions and their derivatives at the interfaces. The ratio of the probabilities for the outgoing wave in region III and the incoming wave in region I is the transmission coefficient and is given by

$$D(E) = \frac{1}{1 + \dfrac{\phi_0^2}{4E(\phi_0 - E)}\sinh^2\left(\sqrt{2m(\phi_0 - E)/\hbar^2}\;\ell\right)}. \tag{2}$$

$D(E)$ is exponentially small and of the order of

$$D(E) \simeq \frac{16E(\phi_0 - E)}{\phi_0^2}\; e^{-2\sqrt{2m(\phi_0 - E)/\hbar^2}\;\ell} \tag{3}$$

for $E \ll \phi_0$. It is close to 1 for $E > \phi_0$, reaching the classical limit $D(E) = 1$ for $E \gg \phi_0$.

III. TUNNEL BARRIERS USING GaAs/Al$_x$Ga$_{1-x}$As HETEROSTRUCTURES

For reasons given in the introduction, we have chosen MBE-grown GaAs/Al$_x$Ga$_{1-x}$As semiconductor heterostructures to fabricate tunnel barriers and investigate their properties systematically. The conduction-band diagram of such a tunnel barrier is shown schematically in Fig. 2. The n-GaAs contact electrodes are Si-doped to typically 10^{17} cm^{-3}, so that the Fermi level E_F is about 11.7 meV above the conduction-band edge in GaAs. The conduction-band discontinuity ΔE_c is a linear function of x (the Al content in the barrier layer) and is given by $\Delta E_c \simeq 0.67\Delta E_g$, where $\Delta E_g \simeq 1.247x$ is the band-gap discontinuity between GaAs and Al$_x$Ga$_{1-x}$As. The tunnel barrier height is then given by $\phi = (\Delta E_c - E_F)$.

The structures shown in Fig. 2 are grown by MBE on (100) n^+-GaAs substrates and include a highly-doped GaAs/AlGaAs superlattice buffer between the substrate and the tunnel structures. Growth rates are determined by measurement of intensity oscillations in a reflection high-energy electron diffraction pattern, so that tunnel

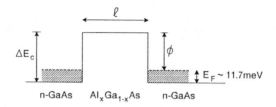

Fig. 2. Al$_x$Ga$_{1-x}$As semiconductor barrier with GaAs electrodes n-doped to 10^{17} cm^{-3}.

barrier thicknesses are believed to be well under control. There are many such structures on a chip, each isolated from the other by etching mesas, as shown in Fig. 3.
The ohmic contacts for transport measurements are made by alloying Ni/Au-Ge metallization on top of the mesas and on the back of the substrate. The contact and other
series resistances can be neglected since they are much smaller than the tunnel
resistances considered in our study. The typical mesa area is (125×125) μm^2. Our
experimental setup is a fairly standard one for measurements of I vs. V and dI/dV vs.
V characteristics from room temperature down to 1.5 K.

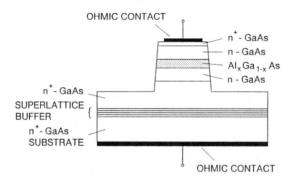

Fig. 3. Schematic view of the MBE-grown structures after mesa etching.

    The tunnel barrier heights ϕ (Fig. 2) are routinely obtained by measuring the
zero-bias conductance G_0 as a function of temperature. The data are plotted in an
Arrhenius diagram of log G_0 vs. 100/T (Fig. 4), which directly yields ϕ since
$G_0 \propto \exp(-\phi/kT)$ in the high-temperature regime. This provides an independent
check of the conduction-band discontinuity $\Delta E_c = (\phi + E_F)$ (Guéret et al., 1985). A
further check of the barrier parameters is also obtained from the transition temperature T_{tr}. As shown in Fig. 4, this temperature reflects the transition from the low-
temperature tunneling regime to the thermally assisted one and can be shown to be
related to the barrier parameters by the relation

$$kT_{tr} \sim \frac{\phi}{\gamma - \ln \gamma},\qquad(4)$$

where $\gamma = 2\sqrt{2m\phi/\hbar^2}\,\ell$ is the tunnel exponent.

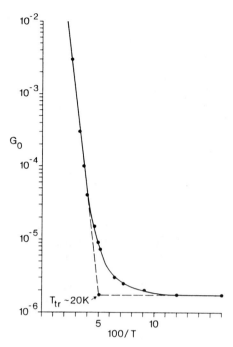

Fig. 4. Arrhenius plot of zero-bias conductance yields a barrier height $\phi \sim 38$ meV.

IV. BEYOND THE SIMPLE PICTURE

The picture of tunneling as given in Section II and also in Fig. 2 is perhaps the simplest case one can think of. But it is also not quite correct since various "side effects", to be described in this section, are not included.

If we first neglect quantum effects, and apply a bias voltage to the structure of Fig. 2, the resulting electric field changes the distribution of the mobile carriers in the n-GaAs electrodes. Carriers accumulate on the emitter side, whereas they are depleted on the collector side. As shown in Fig. 5, this gives rise to a bending of the conduction-band so that the voltage V_b which appears across the barrier is smaller than the total applied voltage V by an amount of the order of

$$\frac{V_b}{V} \sim \frac{\ell}{\ell + 2\lambda_s},\tag{5}$$

where $\lambda_s = v_F/\omega_p$ is the screening length in the n-GaAs electrodes and v_F is the Fermi velocity. On the other hand, it can also be seen in Fig. 5 that the bending of the band reduces the effective height ϕ of the tunnel barrier since the conduction-band discontinuity ΔE_c remains unchanged. Classical semiconductor model calculations of I vs. V characteristics show that both effects tend to cancel each other in the low voltage limit. At high biases, on the other hand, a proper model of band-bending effects is required.

The situation becomes somewhat more involved when quantum effects are taken into account. As was first noted by Baraff and Applebaum (1972), the presence of a potential step at the GaAs/AlGaAs interface considerably reduces the electron probability amplitude $|\psi|^2$ at this point ($|\psi|^2$ would actually be zero for an infinite potential

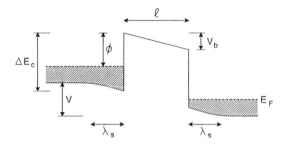

Fig. 5. Classical picture of band bending resulting from a bias voltage *V* applied to the tunnel structure.

step). In the vicinity of the interfaces, the *n*-GaAs is thus partially depleted of carriers which accumulate deeper within the bulk of the electrodes. The resulting band profile is shown in Fig. 6(a) and can be seen to yield a lower effective barrier height ϕ. This picture must be further modified, however, when the potential step is of finite height. In this case, the carriers in *n*-GaAs also have a finite probability of leaking (tunneling) into the barrier where they create an exponentially decaying space-charge distribution and thus raise the potential there. The band profile of Fig. 6(a), modified accordingly, is shown schematically in Fig. 6(b).

The various effects described above, which in some way tend to compensate each other, have been modelled in computer simulations (Zimmermann et al., 1988) using self-consistent semiclassical and quantum mechanical models. The charge rearrangements in the GaAs electrodes are indeed found to extend over distances of the order of the screening length $\lambda_s \simeq 100 \, \text{Å}$ for GaAs doped to $10^{17} \, \text{cm}^{-3}$. The results of these simulations have also been compared systematically with measurements (Guéret et al. 1985, 1988a, 1988b; Marclay, 1988) of *I* vs. *V* and d*I*/d*V* vs. *T* characteristics. Both the semiclassical model and the quantum mechanical model have turned out to yield roughly similar results in fairly good agreement with experimental data. It must be noted, however, that this good agreement was obtained only for tunnel barriers in which the image potential can be neglected. Barriers for which electrode polarization effects are expected to be significant could not be fitted (see Sections V

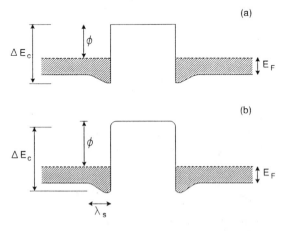

Fig. 6. Conduction-band profile (a) assuming that the electron probability amplitude at the *n*-GaAs/AlGaAs interface is zero; (b) taking into account space-charge leaking (tunneling) into the barrier.

and VI). As described in Section I, electrons tunneling through a barrier induce a charge redistribution in the n-GaAs electrodes which changes the barrier potential. In the limit of infinite electron lifetime in the barrier, the potential seen by an electron at each position is reduced by the static image potential $V_{im}(x)$ and given by (Simmons, 1963)

$$\phi(x) = \phi_0 + qV_{im}(x) = \phi_0 - A\frac{\ell}{x(\ell - x)}, \tag{6}$$

where $A = 1.15\,q^2 \ln 2/16\pi\varepsilon$, q is the electron charge, ε the dielectric constant of the medium assumed to be constant throughout, ℓ the barrier thickness and x the position of the electron in the barrier. This is shown in Fig. 7.

Now, since the tunneling conductance at zero bias voltage varies approximately like $\exp(-B\sqrt{\overline{\phi(x)}}\,\ell)$, where $\overline{\phi(x)} = \phi_0 + q\overline{V_{im}(x)}$ (the bars represent spatial averages), it can be immediately seen that for small $V_{im}(x)$, the dependence of the zero-bias tunnel conductance on the barrier parameters will be roughly given by

$$G(V \to 0) \propto \exp\left(-B\sqrt{\phi_0}\,\ell + C/\sqrt{\phi_0}\right), \tag{7}$$

where the last term in the parentheses represents the static image correction, and B and C are material constants. To first order, the correction is thus seen not to depend on barrier thickness, only on barrier height. This prompted us to perform an investigation of this effect by changing the barrier height (rather than the barrier thickness as in STM (Binnig et al., 1984)), and to use low barriers for which the correction is largest. This was also one of the reasons for us to choose the GaAs/AlGaAs system to perform the investigation on the image or polarization effect.

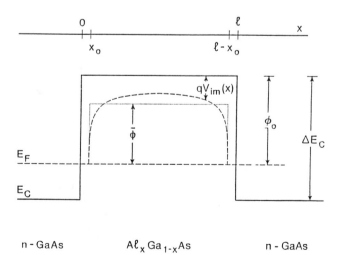

Fig. 7. Schematic of the conduction-band diagram of the GaAs/Al$_x$Ga$_{1-x}$As/GaAs heterostructure considered, with Fermi level E_F and static image correction $V_{im}(x)$. From Guéret et al. (1988a).

323

V. TUNNELING TIME AND DYNAMIC RETARDATION EFFECTS ON POLARIZATION

In the preceding section, we described the effect of a static electric charge, placed between conducting half-spaces (electrodes), on the effective potential distribution. Since this static "image" potential correction V_{im} results from the polarization of the electrodes by the test charge, its magnitude obviously depends on the location of the latter. This is expressed by Eq. (6) and is the local potential seen by an electron moving infinitely slowly through the barrier. This is clearly an approximation. In the tunnel structure shown in Fig. 7, the charge supply and the measured tunnel currents dQ/dt are both finite so that the charge transfer time τ_T through the barrier must be finite as well. If this is the case, then the static approximation Eq. (6) is no longer valid but must take into account the finite reaction (polarization) time of the free carriers in the electrodes, subject to the changing electric field of the electrons crossing the barrier.

It is generally believed that the polarization time is given by ω_p^{-1}, the inverse plasma frequency of the free carriers in the electrodes ($\omega_p^2 = q^2 N_D / \varepsilon m$, where N_D is the free carrier density). A short word of caution, however, is required at this point. Deviations from space-charge neutrality in a semiconductor give rise to damped oscillations at a complex frequency ω which is the solution of (Platzman and Wolff, 1973)

$$\omega(\omega + i v) = \omega_p^2 , \tag{8}$$

where $v = q/m\mu$ is the carrier collision frequency and μ their mobility. If $v \ll \omega_p$, Eq. (8) yields $\omega = \omega_p - iv/2$, i.e. a slightly damped plasma oscillation. If on the other hand $v \gg \omega_p$, Eq. (8) yields $\omega = -i\omega_p^2/v$, i.e. a fast exponential decay of the space-charge disturbance. In GaAs with a free carrier concentration of $N_D = 10^{17}$ cm^{-3}, a mobility of $\mu = 5000$ cm^2/Vs, and an effective mass of $m = 0.067 m_0$ (m_0 is the free electron mass), the collision frequency is $v \simeq 4.8 \times 10^{12}$ s^{-1} and the plasma frequency is $\omega_p \simeq 2 \times 10^{13}$ s^{-1}. The condition for the occurrence of slowly damped plasma oscillations is therefore met in this case.

Now, since the free charges in the electrodes have a finite reaction time of the order of ω_p^{-1} during the barrier traversal time τ_T, one expects intuitively that the importance of dynamic effects will simply be measured by the dimensionless parameter $\omega_p \tau_T$. For small values of this parameter, polarization will have little time to develop and the image potential correction to the tunneling probability can be neglected. For large values of this parameter, on the other hand, the correction is expected to take its full value.

What the charge traversal or tunneling time actually is, remains to this day a subject of much research and controversy (see for example, Büttiker and Landauer (1982) and references therein as well as the contribution of M. Büttiker to this volume). This is probably due to the fact that quantum mechanics deals with wave functions and probabilities and does not allow an electron and its trajectory to be localized within the tunnel barrier. There is also no time operator in quantum mechanics. One possible way around the conceptual difficulty is through physically meaningful *gedanken experiments* as done for example by Büttiker and Landauer (1982). According to these authors, the traversal time for tunneling is viewed as an interaction time with the barrier and is given by the simple expression (Fig. 7)

$$\tau_T = \int_{x_0}^{\ell - x_0} \frac{dx}{v(x)} , \tag{9}$$

where $v(x)$ is the local particle "velocity" in the inverted potential and given by $v(x) = \hbar\kappa/m = \sqrt{2\phi(x)/m}$. Whereas a particle of constant total energy *loses* kinetic energy while going over a potential barrier, in Eq. (9) it is as if the particle would *gain* kinetic energy while passing over the *inverted* potential. Although one can argue about the actual physical meaning of these concepts, the time scale τ_T defined in Eq. (9) appears nevertheless to be the relevant one in theories of dynamic polarization in tunneling. This has been found for example in recent work by Persson and Baratoff (1988), whose main conclusion is that "the contribution of the dynamic image potential to the tunneling exponent decreases monotonically for decreasing $\omega_p\tau_T$." This work is particularly interesting in that it not only confirms and extends earlier work by Sunjic et al. (1972), Heinrichs (1973) and Jonson (1980), to cite only a few, but also provides detailed calculations of specific cases, including rectangular and triangular (field-emission) barriers. It is impossible here to do justice to the large body of literature which has been published on the subject. The general consensus appears to be that coupling to plasmons leads to a renormalization of the barrier and enhanced transmission, and that the magnitude of the effect is related to the dimensionless parameter $\omega_p\tau_T$. The reader is referred to the cited literature and references therein for more details.

VI. SYSTEMATIC INVESTIGATIONS OF TUNNELING IN GaAs/AlGaAs HETERO-STRUCTURES. POSSIBLE EVIDENCE FOR DYNAMIC POLARIZATION EFFECTS

We have used the flexibility offered by the GaAs/Al$_x$Ga$_{1-x}$As system for a systematic study of tunneling and traversal time. Two distinct sets of experiments have been performed, using different clocks to measure the tunneling time.

In a first set of experiments (Guéret et al., 1987), a magnetic field B perpendicular to the current (Fig. 8) was used as a way to determine the tunneling time τ_T. The basic idea was to use the corresponding electron cyclotron frequency $\omega_c = qB/m$ as a clock to measure the traversal time. According to a first oversimplified but pictorial description, a tunneling electron is viewed as a ballistic object which is deflected sideways by the magnetic field, resulting in a longer effective tunneling path. An elementary calculation shows that the corresponding increase is of the order

$$\Delta\ell \sim \frac{\ell}{6}(\omega_c\tau_T/2)^2 . \tag{10}$$

This leads in turn to a change in the tunneling probability by a factor of $\exp(-2\sqrt{2m\phi}\,\Delta\ell/\hbar)$. A more formal analysis using a solution of Schrödinger's equation in a magnetic field yields the same result, with τ_T given by Eq. (9). According to this and Eq. (10), the cyclotron frequency is thus a clock to measure the traversal time. Since the tunneling time according to Eq. (9) can be quite large for thick and low barriers (as can be done with the GaAs/AlGaAs system), fairly low magnetic fields can already produce substantial effects on the current.

Figure 9 shows the logarithmic plots at 4.2 K of the calculated current I vs. B^2 for B between 0 and 4 Teslas and for different bias voltages. Experimental data are shown for comparison. The expected B^2 dependence of log I is well verified and is in accordance with Eq. (10). Barriers for which $\omega_c\tau_T$ is large (lines a in Fig. 9) exhibit a large dependence on the applied magnetic field, whereas those with small $\omega_c\tau_T$ (line b) are hardly affected. A value of $\tau_T \simeq 0.1$ ps is obtained for the traversal time from the slope of lines a in the figure.

In another set of experiments (Guéret et al., 1988a, 1988b), we have used the plasma frequency ω_p of the free carriers in the n-doped GaAs electrodes as a clock to

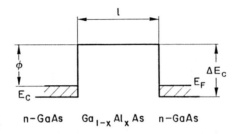

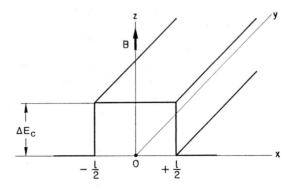

Fig. 8.   Conduction-band diagram for the heterostructures under investigation and the tunnel-barrier geometry with applied magnetic field. $E_F \sim 12$ meV. From Guéret et al. (1987).

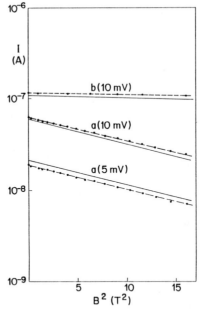

Fig. 9.   Theoretical (solid lines) and experimental (dots) field dependence of the tunnel current at 4.2 K for different applied biases and for two barriers with $\Delta E_c = 40$ meV, $\ell = 430$ Å (curves a) and $\Delta E_c = 83$ meV, $\ell = 250$ Å (curve b). Adapted from Guéret et al. (1987).

measure the tunneling time τ_T. We have already described in Sections I and V how tunneling electrons polarize the electrodes, and how the finite response time ω_p^{-1} of the latter leads to a retardation in the polarization. Since the polarization affects the potential barrier seen by an electron, these effects are expected to manifest themselves in tunnel conductance data.

For these reasons, we have made a systematic investigation of tunneling in GaAs/AlGaAs structures. Our investigation covers a broad range of values for the parameter $\omega_p\tau_T$, which is the relevant one in this case. The ranges of parameters covered in this study (Guéret et al., 1988b) (Fig. 7) are approximately

$0.02 < x < 0.12$ for the Al mole fraction in the $Al_xGa_{1-x}As$ barrier,
$8 < \phi_0 < 90$ (meV) for the barrier height,
$10 < \ell < 100$ (nm) for the barrier thickness, and
$0 < \omega_p\tau_T < 10$.

We thus span roughly seven orders of magnitude in the tunnel conductance, in good agreement with calculated values and confirming that we are indeed dealing with tunneling and not some spurious conduction mechanisms. The measured low-bias conductances are compared with those calculated from the current-voltage characteristic in the zero-bias limit, according to (Duke, 1969)

$$I(V) = A \frac{4\pi m^* qkT}{h^3} \int_{E_c} D(E)\, N(E,V)\, dE \,, \tag{11}$$

where A is the junction area, $D(E)$ is the transmission coefficient of the rectangular barrier, and $N(E,V)$ is the supply function given by

$$N(E,V) = \ln \left[\frac{1 + \exp(E_F - E)/kT}{1 + \exp(E_F - E - qV)/kT} \right]. \tag{12}$$

The totality of the data has then been used to infer the potential seen by an electron as it tunnels through the barrier. To do this, we start from the reasonable assumption that the tunnel conductance G varies as $G \propto \exp(-y)$. For a rectangular barrier of height ϕ_0 and length ℓ (Fig. 7), a good approximation of the tunneling exponent is given by $y_c = 0.085\sqrt{\phi_0(\text{meV})}\,\ell(\text{nm})$. Hence the ratio (G_{meas}/G_{calc}) of measured to calculated tunnel conductances defines a quantity $\Delta y \equiv (y_m - y_c) = \ln(G_{meas}/G_{calc})$ from which the effective tunneling exponent y_m can be obtained.

The dots in Fig. 10 form a plot of the thus derived ratio y_c/y_m vs. the dimensionless parameter $\omega_p\tau_T$. In spite of some scatter in the data, one observes that the tunneling exponent y_m derived from the experimental data is essentially equal to the calculated one y_c in the non-adiabatic limit, more precisely for values of $\omega_p\tau_T$ smaller than about 3. For larger values on the other hand, a clear dependence on this parameter is found.

The above suggests that the electron sees a dynamically varying (collapsing) potential barrier as it tunnels through. Working backwards from the data, we can then in principle try to infer the general features of this dynamic potential. If we assume with Büttiker and Landauer (1982) that a tunneling velocity $v(x)$ can be defined such that $dx = v(x)dt$ with $v(x) = \sqrt{2\phi(x)/m}$, the effective tunneling exponent y can be expressed as

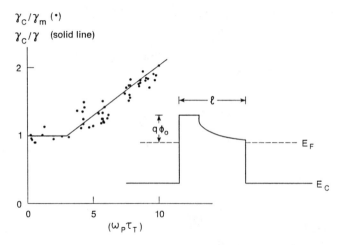

Fig. 10. The trend of the experimental data (•) for the tunneling exponent is well reproduced (solid line) by a dynamically variable potential barrier (inset). Reprinted with permission from Guéret et al. (1988b), © 1988 Pergamon Journals Ltd.

$$\gamma = \int_0^{\tau_T} 4\,\phi(x = vt)/\hbar\; dt, \tag{13}$$

where $\phi(x = vt)$ is the presumed dynamic potential. Based on the data in Fig. 10, we then take as a trial function

$$\phi(x = vt) = \phi_0 \qquad\qquad \text{for } \omega_p t = \omega_p x/v \lesssim 3$$
$$= \frac{\phi_0}{a^2 + b^2(\omega_p t)^2} \quad \text{for } \omega_p t > 3 \tag{14}$$

with $a^2 = 0.6$ and $b^2 = 0.1$. Using Eqs. (13) and (14), we find

$$\gamma_c/\gamma = 1 \qquad\qquad\qquad\qquad\qquad\qquad\qquad \text{for } \omega_p \tau_T \lesssim 3$$
$$= (\omega_p \tau_T)\left\{ 3 + \frac{1}{ab}\left(\tan^{-1}\omega_p \tau_T \frac{b}{a} - \tan^{-1} 3\frac{b}{a} \right)\right\}^{-1} \quad \text{for } \omega_p \tau_T > 3. \tag{15}$$

This ratio is plotted in Fig. 10 (solid line) for comparison, and appears to follow the trend indicated by the experimental data reasonably well.

   In summary, the above investigation provides evidence for substantial departures from the simple rectangular barrier model, which are interpreted as polarization corrections to the transmission probability. The broad range of barrier parameters covered has allowed us to make some inferences about the shape of the dynamical potential seen by a tunneling electron. The crossover is found to occur at a value of $\omega_p \tau_T \simeq 3$, below which the polarization correction becomes negligible.

   The above considerations would not be complete without citing an elegant alternative which has been used for the study of some dynamic aspects of tunneling and in which the variable parameter is simply a dynamic load attached to a single tunnel junction. This technique has been described in a beautiful experiment on macroscopic quantum tunneling (Estève et al., 1989) and has demonstrated good agreement with theoretical predictions.

VII. CONCLUSIONS

With this brief account, we hope to have convinced the newcomer to the field that the quantum phenomenon of tunneling is far more complicated than implied by the elementary textbook picture described in Section II. Band-bending, space-charge, screening and image potential must be taken into account. The situation is further complicated by the possible occurrence of dynamic retardation effects, as described in Sections V and VI. Which of these effects play a significant role must be analyzed for each situation, and depends much on electrode doping, barrier height and thickness. It is our observation that the issues raised in this article are most important in nonmetallic (lowly-doped), low-barrier systems. This, in turn, leads us to believe that they will acquire greater relevance as our knowledge of biological systems deepens.

ACKNOWLEDGMENT

The author is most greateful to his colleagues A. Baratoff, E. Marclay and H. P. Meier for their invaluable contributions to this work.

REFERENCES

Baraff, G. A., and Appelbaum, J. A., 1972, *Phys. Rev. B*, 5:475.
Binnig, G., Garcia, N., Rohrer, H., Soler, T. A., and Flores, F., 1984, *Phys. Rev. B*, 30:4816.
Büttiker, M., and Landauer, R., 1982, *Phys. Rev. Lett.*, 49:1739.
Büttiker, M., 1989, these proceedings.
Condon, E. U., 1931, *Rev. Modern Physics*, 3:43.
Duke, C. B., 1969, "Tunneling in Solids," Academic Press, New York.
Estève, D., Martinis, J. M., Urbina, C., Turlot, E., and Devoret, M. H., 1989, 9th General Conference of the Condensed Matter Division of the European Physical Society (Nice), to be published in *Physica Scripta*.
Guéret, P., Kaufmann, U., and Marclay, E., 1985, *Electron. Lett.*, 21:344.
Guéret, P., Baratoff, A., and Marclay, E., 1987, *Europhys. Lett.*, 3:367.
Guéret, P., Marclay, E., and Meier, H., 1988a, *Appl. Phys. Lett.*, 53:1617.
Guéret, P., Marclay, E., and Meier, H., 1988b, *Solid State Commun.*, 68:977.
Heinrichs, J., 1973, *Phys. Rev. B*, 8:1346.
Jonson, M., 1980, *Solid State Commun.*, 33:743.
Marclay, E., 1988, Ph.D. Thesis, unpublished.
Parker, E. H. C., 1985, "The Technology and Physics of Molecular Beam Epitaxy," Plenum Press, New York.
Persson, B. N. J., and Baratoff, A., 1988, *Phys. Rev. B* 38:9616.
Platzman, P. M., and Wolff, P. A., 1973, "Waves and Interactions in Solid-State Plasmas," Academic Press, New York.
Simmons, J. G., 1963, *J. Appl. Phys.*, 34:2581.
Sunjic, M., Toulouse, G., and Lucas, A. A., 1972, *Solid State Commun.*, 11:1629.
Wolf, E. L., 1985, "Principles of Electron Tunneling Spectroscopy", Oxford University Press, New York.
Zimmermann, B., Marclay, E., Ilegems, M., and Guéret, P., 1988, *J. Appl. Phys.*, 64:3581.

OPTICAL PROBES OF RESONANT TUNNELING STRUCTURES

Jeff F. Young, B.M. Wood and S. Charbonneau

Division of Physics
National Research Council
Ottawa, K1A 0R6, Canada

I INTRODUCTION

Semiconductor growth techniques have advanced to the point where layered structures can be accurately produced with sharp-edged potential energy steps that have widths comparable to the de Broglie wavelength of energetic electrons. Such structures are ideal for studying and exploiting the physics of electron tunneling in semiconductors. A wide range of potential energy profiles can be designed and implemented to facilitate the investigation of fundamental problems such as the time taken to tunnel, and the possible influences of scattering events on tunneling.

One class of tunneling structures, generically referred to as double barrier resonant tunneling (DBRT) diodes, has recently received a great deal of attention from both scientific and technological points of view (Sollner et al., 1983; Luryi, 1985; Goldman et al., 1987; Payling et al., 1988; Mendez et al., 1986; Weil and Vinter, 1987; Young et al., 1988b). A schematic diagram of the energy levels of a typical DBRT structure is shown in Fig. 1. The width of the central "well" region is typically on the order of 5 nm so that spatial confinement effects raise the energy of the lowest lying state in the well significantly above the Fermi energy, which is determined primarily by the doped contact regions. For the purpose of describing the basic principles of their operation, it is convenient to draw an analogy between DBRT structures and optical Fabry Perot etalons (Born and Wolf, 1959). The barriers form the "mirrors" of the cavity, and the confined electron energy levels correspond to its longitudinal modes. The emitter contact acts as a source of electrons incident on the cavity with an energy (relative to that of the cavity mode) that is tunable via the applied voltage across the structure. For applied potentials such that none of the emitter electrons are in resonance with the cavity mode, a very small current flows through the device due to "conventional", non-resonant tunneling. This occurs both for small voltages, at which the cavity mode energy is larger than the Fermi energy in the emitter, and for large voltages, at which the cavity mode energy is less than the conduction band energy in the emitter region. At intermediate voltages, where emitter electrons are resonant with the cavity mode, relatively large currents are passed in analogy with the high transmission coefficient of photons resonant with a Fabry Perot cavity mode. DBRT structures are therefore characterized by a negative differential resistance in their current-voltage behaviour, and it is this feature which forms the basis of their technological applications.

A substantial amount of electromagnetic energy can be stored in a high-Q Fabry Perot cavity when resonant photons are incident upon it. The characteristic response time of an optical cavity on-resonance is in fact determined by the time required to build up the stored energy. Similarly, the characteristic response time of a DBRT diode biased on-resonance should be determined by the time required to build up the electronic charge in the well appropriate for the Q of the given structure. Thus the

Electronic Properties of Multilayers and Low-Dimensional Semiconductors Structures
Edited by J. M. Chamberlain *et al.*, Plenum Press, New York, 1990

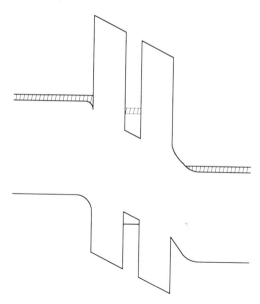

Fig. 1.   Schematic diagram of the energy level structure in a biased DBRT.

accumulation of electrons in the well region of a DBRT structure is intimately related to the response time of the device. An interesting and technologically important avenue of research has to do with understanding the influences that scattering events have on the accumulation of charge, and hence the characteristic tunneling time of DBRT structures (Buttiker, 1988; Price, 1987). This is one of the few areas in which existing knowledge about optical Fabry Perot cavities is of little use, since internal scattering is usually of negligible concern in etalons.

Charge accumulation is also important in determining the self-consistent potential energy distribution within the device. In contrast to an optical Fabry Perot wherein the cavity must be fabricated from a non-linear optic material in order to render its properties sensitive to the amount of stored energy, the accumulated charge in the well region *always* acts to modify the distribution of applied potential across DBRT structures. *Intrinsic* bistability in the current-voltage behaviour of DBRT structures has been shown to result from precisely this mechanism (Goldman et al., 1987b; Leadbeater et al., 1989).

Although the observation of intrinsic bistability can be used as an indicator of charge accumulation in special structures which exhibit the effect, it is not a generally applicable, quantitative tool for monitoring the buildup of charge in the well. The purpose of this paper is to describe how a variety of *optical* techniques have been systematically applied to study the accumulation of electrons in the well region of biased DBRT structures (Young et al., 1988b, 1989; Hayes et al., 1989). As will be demonstrated below, the spectroscopic nature of optical techniques allows for direct, unambiguous probing of the well region of operating devices.

Following a brief introduction to photoluminescence (PL) and PL excitation (PLE) spectroscopies as applied to semiconductor physics, in Section II, Section III addresses a number of issues specifically relevant to the use of these techniques to study DBRT structures. The application of steady state PL to monitor the accumulation of electrons in the well region of biased GaAs/AlGaAs DBRT structures is described in Section IV. The characteristic tunneling time of the structure is inferred from the degree of charge accumulation in analogy with the optical Fabry Perot. In Section V, time-resolved PL studies are described which allow the direct determination of the cavity lifetime, and, when combined with tunable wavelength excitation, can also be used to monitor reverse hole tunneling from the collector, through the DBRT structure.

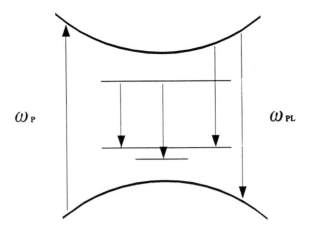

Fig. 2.   Schematic illustration of the excitation and luminescence
            transitions in typical direct gap semiconductors.

II GENERAL PRINCIPLES OF PL AND PLE SPECTROSCOPIES AS APPLIED TO SEMICONDUCTORS

PL and PLE spectroscopies provide relatively simple experimental means of obtaining information about the energy level structure of semiconductors. Both techniques involve exciting electronic transitions within the material of interest via dipole-allowed optical absorption, and monitoring the light emitted through radiative relaxation processes as the excited material returns to equilibrium. In the case of a direct gap bulk semiconductor illuminated with monochromatic radiation, the experimental conditions can be schematically represented as in Fig. 2.

PL is the simpler of the two experiments, but it generally does not yield as much information as PLE. All that is required in a PL experiment is a light source emitting photons with energies sufficient to generate electronic excitations in the material, ω_P, a spectrometer to disperse the luminescence signal, ω_{PL}, and a detector at the output of the spectrometer. PL usually only provides information about the lowest lying energy levels in the material since the photo-excited electrons and holes generally relax to their respective bandedges via non-radiative processes on picosecond or sub-picosecond timescales, whereas the radiative relaxation times of most states are on the order of a nanosecond or longer. Nevertheless, PL does provide very accurate (sub meV resolution) information about these bandedge states, and in the case of impurity transitions that can be saturated, a given spectrum can contain information about a number of different states (see Fig. 3).

Table 1. Physical Parameters of DBRT Structure

Layer	Composition	Doping (cm^{-3})	Thickness (nm)	Comments
1	$GaAs$	2×10^{17}		Substrate
2	$GaAs$	2×10^{18}	500	Epilayer
3	$GaAs$	none	7.5	Epilayer
4	$Al_{0.3}Ga_{0.7}As$	none	10	Epilayer
5	$GaAs$	none	5	Epilayer
6	$Al_{0.3}Ga_{0.7}As$	none	10	Epilayer
7	$GaAs$	2×10^{18}	1000	Epilayer

333

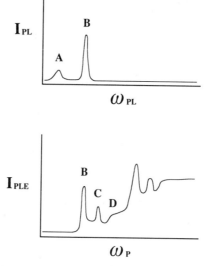

Fig. 3.   Illustration of the typical features observed in PL and PLE
spectra from quantum well structures. Feature A corresponds
bound excitons, feature B to free heavy hole excitons, feature C to
free light hole excitons and feature D to the continuum transitions,
all associated with the lowest subband.

In PLE experiments the spectrometer is set to pass a fixed band of the PL spectrum,
and the strength of the PL at that energy is monitored as the energy of the (monochromatic) excitation source is varied, keeping the incident intensity fixed. If every absorbed
photon resulted in an electronic excitation which relaxed to the state being monitored
by the detector, then PLE would provide a background-free method of obtaining the
electronic absorption spectrum of the material.   More typically, the PLE excitation
spectrum (PL signal as a function of excitation energy) represents the actual absorption spectrum, modulated by a function representing the probability that the as-excited
electron-hole pair eventually relax to the low-energy state being monitored. Nevertheless, PLE provides high-resolution (sub meV) information about high-lying electronic
states which usually cannot be seen in simple PL. The price for its relative simplicity
as compared to a true absorption measurement (no sample thinning, zero-background
signal) is the lack of an absolute calibration of the *magnitude* of the absorption.

III) GENERAL CONSIDERATIONS FOR PL AND PLE IN DBRT STRUCTURES

The majority of results discussed in this paper were obtained from the DBRT structure described in Table 1 (Young et al., 1988a, 1988b, 1989).   It consists of 10 nm
thick $Al_{0.3}Ga_{0.7}As$ barrier layers which define a 5 nm thick GaAs well region. A 7.5 nm
thick undoped GaAs buffer layer was grown on the substrate side of the double barrier
structure to prevent diffusion of the Si dopant from the 0.5 μm thick n+ contact region
beneath.   The top contact consists of 1 μm of n+ Si-doped GaAs, directly adjacent
to the double barrier structure. Annular-shaped contacts were fabricated on 325 μm
diameter mesas to allow optical access to the DBRT while a voltage was applied. The
experimental setup is schematically illustrated in Fig. 4. To ensure that the laser excitation was acting strictly as a probe, and not influencing the electrical properties of the
device in any way, a weak ($\sim$ 1 mW) HeNe beam was used for most of the PL work.
The resulting PL signal emanating from the well buried 1 μm beneath the top contact
was very weak, necessitating the use of sensitive detectors, including a Si charge coupled
device array, and an imaging photomultiplier tube used in photon counting mode.

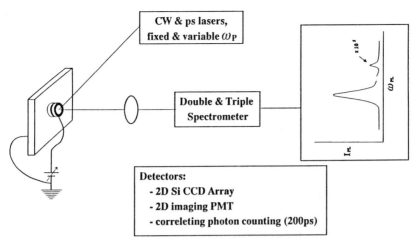

Fig. 4.   Experimental setup used to obtain spectra from the biased DBRT.

Fig. 5 illustrates the energy level structure of the DBRT under bias, with symbols representing the as-excited electrons and holes in the immediate vicinity of the well region. Also shown is a typical PL spectrum, which consists of a very broad, very strong signal at energies close to the bandgap of GaAs, and a single, very weak peak near 765 nm. This is characteristic of the PL spectra obtained at all applied voltages. The broadband signal comes from the thick contact region where most of the HeNe light is absorbed. From its spectral position, the peak near 765 nm is due to luminescence emitted from the well region, and thus represents a direct monitor of the local well environment, which can be studied as a function of the applied voltage.   The two greatest advantages of optical probe techniques are thus revealed; first, they can be used strictly as probes, not influencing the tunneling behaviour at all, and second, that by their spectroscopic nature, they allow direct monitoring of the *interior* of the DBRT.

In order for luminescence to be emitted from the well, there must exist populations of electrons in the lowest electronic subband, and holes in the lowest valence subband. These populations can in general derive either through direct optical absorption in the well followed by subsequent non-radiative relaxation to the respective bandedges (squares in Fig. 5), or through carriers which tunnel into the well from the contact regions. In the case of electrons, the density of photo-excited electrons in the emitter contact region (circles in Fig. 5) will be negligible compared to the doped-in electron population, hence any electrons in the well due to tunneling in from the contacts can be attributed to the basic behaviour of the DBRT, which is precisely what is under study. As for the holes, those photo-excited in the collector region (diamonds in Fig. 5) will in general be able to tunnel into the valence subbands under applied bias conditions, but only if the time required to do so is less than the recombination time of these holes with free electrons in the n+ contact region. Using time-resolved PL techniques as described in Section V below, it can be shown that this condition is only satisfied in the sample described in Table 1 when very large voltages, above those which bias the structure in the resonant tunneling regime, are applied. Under such conditions, a depletion region can be formed in the collector adjacent to the barrier, which significantly enhances the radiative lifetime of holes photo-excited nearby. At lower voltages, the population of holes directly excited in the well completely overwhelms that due to tunneling of holes into the well from the collector region. Therefore, for voltages which bias the structure in the resonant tunneling regime, the population of electrons in the well is in general due to both photo-excited electrons and those accumulated due to the resonant tunneling process, while the only source of holes is through direct photo-excitation in the well.

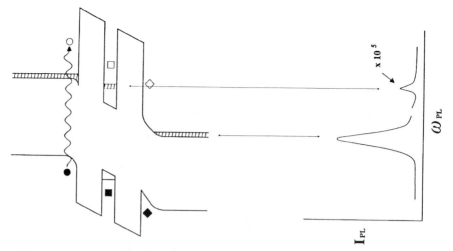

Fig. 5. The energy level structure of a biased DBRT showing the as-
excited electron and hole populations in the vicinity of the well.

It is worth noting here that the above conditions are particular to DBRT structures
with n+ doping very close to the barriers. In DBRT structures with much more lightly
doped contact regions adjacent to the barriers, the radiative lifetime of photo-excited
holes in the (much larger) depletion region of the collector need not be short compared
to the hole tunneling time, and holes tunneling in from the collector region can dominate
over those directly excited in the well even for relatively low applied voltages. This has
recently been experimentally verified by Hayes et al. (1989).

IV STEADY STATE PL FROM BIASED DBRT STRUCTURES

Using the experimental setup shown in Fig. 4 with $\sim$ 1 mW of HeNe laser power
focussed to a $\sim$ 325 μm diameter spot on the DBRT mesa, the PL spectra shown in
Fig. 6 were observed at 5K for applied voltages below, within and above the resonant
tunneling regime. Fig. 7 shows the PLE spectrum obtained from the same sample with
no applied voltage, using 1 mW of tunable dye laser power and with the spectrometer
set to monitor the low energy shoulder of the PL feature shown in Fig. 6. The strong,
low energy peak in the PLE spectrum identifies the heavy hole exciton state associated
with the lowest lying electron and hole subbands in the well (Weisbuch et al., 1981).
This peak in the PLE spectrum fits within the PL peak at zero bias, demonstrating
that the PL feature corresponds to radiative recombination of free heavy hole excitons
in the well.

There are three obvious effects on the HeNe PL spectra due to the application of
voltage on the DBRT structure. The peak position of the PL feature shifts to the red
by as much as 4 meV at the highest applied voltages, its linewidth increases by as much
as 50% in the resonant tunneling regime, but decreases almost to its zero bias value at
higher voltages, and the integrated area under the PL line increases by more than two
orders of magnitude. When a dye laser tuned to the high energy side of the PL feature is
used for excitation, similar shifts and broadening are observed when voltage is applied,
but the intensity of the PL peak changes by at most a factor of three. This dependence
on excitation wavelength implies that the shifts and broadening reflect effects of the
applied voltage on *intrinsic* properties of the exciton state, while the change in PL
intensity observed only with high energy excitation represents the influence of applied
voltage on the *kinetics* of exciton formation.

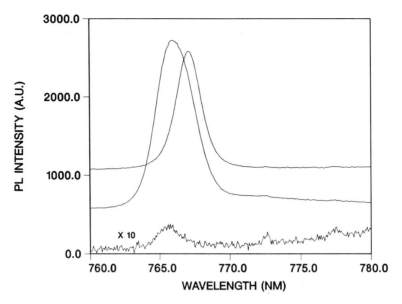

Fig. 6.   PL spectra obtained using HeNe excitation at voltages of 0V (bottom), 123 mV (middle) and 365 mV (top).

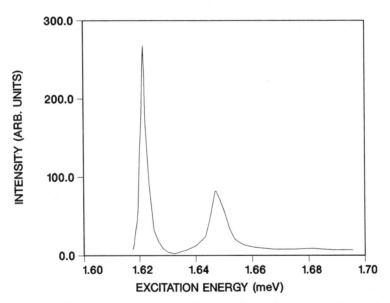

Fig. 7.   PLE spectra obtained with zero bias voltage. The low and high energy peaks correspond to heavy and light hole excitons.

Fig. 8 shows the voltage dependences of the integrated area under the PL feature (crosses), and the current flowing through the device (dots). At this stage the vertical scale for the integrated PL area is arbitrary; noteworthy is the fact that for both polarities, the PL area increases abruptly at the same point where the resonant current begins flowing (~ 80 mV), and it decreases abruptly beyond the peak of the resonant current. In this resonant tunneling regime therefore, the behaviour of the integrated PL area is strongly correlated with the resonant current flow. For voltages which bias the

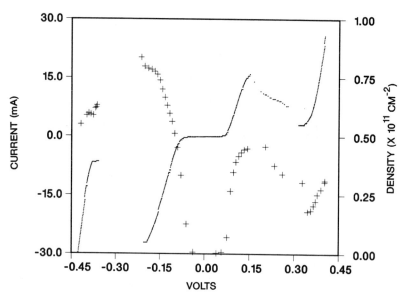

Fig. 8.   The current through the device (dots) and the integrated
            area under the PL peak (crosses) as a function of applied bias.
            The integrated PL area is scaled to give areal electron density in
            the well, as described in the text.

structure in the region of negative differential resistance (NDR), the circuit is known to
be oscillating (Young et al., 1988a), hence no significance can be attributed to the data
in this bias regime. For voltages beyond the NDR region, the integrated intensity of
the PL remains quite large, but is not well correlated with the current flowing through
the device.

Further insight into the nature of the PL excited by the HeNe laser is obtained from
the dependence of the PL intensity, I_{PL}, on the HeNe excitation intensity, I_L, as shown
in Fig. 9. With zero applied potential (bottom curve), I_{PL} increases quadratically over
two decades of I_L, at high I_L. This behaviour implies that over this range of excitation
intensities, the electron and hole populations optically excited in the quantum well
are determined by intensity independent decay processes that are unrelated to exciton
formation, i.e. $n_e \propto I_L \tau_e$ and $n_h \propto I_L \tau_h$ where τ_e and τ_h are decay times independent of I_L
and exciton formation processes. The steady state rate of exciton formation, which gives
rise to I_{PL}, is proportional to the product of free electron and hole populations; hence
$I_{PL} \propto I_L^2$. The change to a linear dependence at intermediate I_L signifies an intensity, or
equivalently, a density dependent change in the decay time associated with either the
electrons or holes. But the increasing lifetime at lower I_L must in fact be associated with
the electrons rather than the holes, since the introduction of a resonant current (top
curve in Fig. 9) which electrically injects additional electrons into the well, eradicates
the discontinuity in I_{PL} at these I_L. In fact, the linear behaviour of I_{PL} over five decades
of I_L when the structure is biased in the resonant tunneling regime indicates both that
the electron density accumulated due to the resonant tunneling process exceeds the
optically generated electron density for this range of I_L, and that the non-radiative
decay rate of optically excited holes is independent of excitation level over this same
range. This linear intensity dependence is observed for all voltages which bias the
structure in the resonant tunneling regime. At higher voltages which bias the structure
beyond the region of NDR, the intensity dependence of I_{PL} follows no simple functional
form, indicating a fundamental change in the nature of the exciton formation dynamics.
In the following Section, time-resolved PL is used to demonstrate that for $V > V_{NDR}$,
excitons can not only be formed from holes optically excited directly in the well, but
also from those created in the collector contact, which subsequently tunnel into the well.

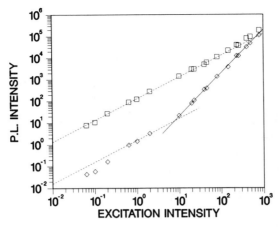

Fig. 9.   PL intensity from the DBRT as a function of the HeNe laser excitation intensity for 0 V (bottom) and 125 mV (top) applied voltages. Unity intensity corresponds to $\sim 1Wcm^{-2}$.

From Fig. 9 it is clear that there is a range of I_L over which the ratio of I_{PL} observed with and without an applied bias in the resonant tunneling regime is independent of I_L. This corresponds to a range of I_L over which the lifetime of the photo-excited holes at zero bias is increasing as I_L decreases, leaving the density of photo-excited electrons constant, independent of I_L. The increase in I_{PL} upon the application of the voltage therefore reflects the increased rate of exciton formation due to the accumulation of resonantly tunneling electrons. For $V < V_{NDR}$, $I_{PL} \propto (n_e^{op} + n_e^v)n_h^{op}$, where n_e^{op} is determined only by I_L and decay processes independent of exciton formation. The huge increase of I_{PL} in the resonant tunneling regime indicates that there $n_e^{op} \ll n_e^v$, so that $I_{PL}(V)/I_{PL}(0) = n_e^v/n_e^{op*}$ where n_e^{op*} is independent of I_L over the range of I_L where $I_{PL}(0)$ is linearly proportional to I_L. Hence for $V < V_{NDR}$, $I_{PL}(V)$ is proportional to the electron density accumulated in the well due to the resonant tunneling process. Knowledge of n_e^{op*} would enable an absolute calibration of the accumulated electron density in terms of $I_{PL}(V)$. No independent measurement of n_e^{op*} has been made, but one would expect a change in the photo-excited electron decay dynamics (Bimberg et al., 1985), when their density becomes comparable to the residual donor population in the well region, which are completely ionized in the dark at V=0. Estimates of the photo-excited electron density in the well at I_L corresponding to the linear part of Fig. 9 are in fact in good agreement with the known residual donor concentration in the well of $5 \times 10^{14} cm^{-3} \pm 50\%$.

The absolute density calibration of the integrated I_{PL} area curve in Fig. 8 was therefore obtained by scaling the ratio of $I_{PL}(V)/I_{PL}(0)$ by the residual donor concentration in the well. Thus to within a factor of two, the charge accumulated in the well is $\sim 5 \times 10^{10} cm^{-2}$. This corresponds to a screening potential of approximately 10% of the applied potential across the device; it is therefore significant, but the accumulated well charge in this particular device does not severely redistribute the applied potential across the barriers.

If one considers a DBRT structure operating in the resonant tunneling regime, then the response time of the device to a sudden change in applied potential is at least as long as the time required to reestablish the steady state value of the charge in the well at that new bias. This characteristic response time, τ, is simply related to the steady state charge density in the well and the corresponding current density through the structure as (Weil and Vinter, 1987),

$$\tau = Q/J. \tag{1}$$

The voltage dependence of τ obtained using the optically determined charge density is plotted in the top part of Fig. 10 as open squares. Note first that the value of τ decreases monotonically with increasing applied voltages of either polarity. This simply reflects the reduction of the effective collector barrier height as the applied voltage changes the barrier shape from square to trapezoidal. Under applied bias, the time τ basically corresponds to the cavity lifetime of an electron in the well; hence a reduced barrier height enables the electrons in the well to escape to the collector faster, reducing the characteristic tunneling time.

Mathematically, the transmission coefficient of the electron through the emitter and collector barriers depends on applied potential as (Goldman et al., 1987a),

$$T_{e/c} = 16(E_R/U)(1 - E_R/U)exp(-2d/d_{e/c}), \tag{2}$$

where,

$$d_{e/c} = \frac{3\hbar e V_{e/c}}{2(2m_b)^{1/2}}[\mp(U - E_R \pm eV_w/2)^{3/2} \pm (U - E_R \pm eV_w \pm eV_{e/c})^{3/2}]^{-1}. \tag{3}$$

Here E_R is the real part of the quasi-bound electron eigenenergy, $E_R = Re(E_0)$, U is the conduction bandedge discontinuity between the well and the barrier materials, d is the barrier thickness and $d_{e/c}$ are the effective depths to which the quasi-bound electron wavefunction penetrate into the emitter and collector barriers respectively. $V_{e/c}$ are the potentials dropped across the respective barriers, V_w is the potential dropped across the well, and m_b is the effective mass of the electron in the barriers.

The cavity lifetime of the well state is obtained by calculating the inverse of the imaginary part of the quasi-bound state's eigenenergy. In the limit that the resonance is well-defined, i.e. $Im(E_0) \ll Re(E_0)$, the result can be expressed analytically as (Price, 1986; Young et al., 1988b),

$$\tau = \frac{\hbar}{2Im(E_0)} \simeq \frac{2[(m_w/m_b)t + d_e + d_c]}{(2E_R/m_w)^{1/2}(T_e + T_c)}, \tag{4}$$

where t is the thickness of the well, and m_w is the effective mass of the electron in the well. The physical interpretation of Eqn. 4 is clear. Neglecting the (m_w/m_b) term in the numerator, which appears due to the effective mass boundary conditions, the escape rate is just the frequency at which the electron with effective velocity, $v = (2E_R/m_w)^{1/2}$, perpendicular to the barriers, make a round trip of distance $2(t + d_e + d_c)$, times the probability per round trip that the electron transmits out of the cavity. Using values of $m_w = 0.07m_0$, $m_b = 0.092m_0$, $U = 240$ meV, $E_R = 74$ meV and the dimensions from Table 1, the solid curve in the top of Fig. 10 was obtained from Eqn. 4. Thus the experimentally determined characteristic tunneling times not only qualitatively follow the calculated voltage dependence, they are also within approximately a factor of two of the absolute calculated times.

That the experimentally determined characteristic tunneling times are in agreement with these calculated times is significant, because it implies that scattering events do not significantly effect the characteristic tunneling time. More specifically, the scattering time of electrons in similar wells, as inferred from two dimensional Hall mobility measurements in modulation doped structures, is on the order of picoseconds. Scattering events which limit the two dimensional mobility of carriers in these wells therefore do not affect the characteristic cavity lifetime, and hence should not effect the response time of DBRT structures. This result is perhaps not surprising when one recognizes that with only one subband (mode) in the well (cavity), only intra-band transitions to other states in the same subband (longitudinal mode) are allowed. Since there is relatively little dependence of cavity lifetime on the lateral momentum of the state, the net escape rate of the carrier from the well is unaffected by these intra-band scattering events. The situation would be expected to change in thicker well structures where significant inter-subband scattering could occur between subbands with substantially different cavity lifetimes.

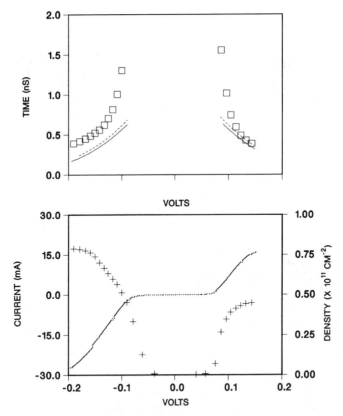

Fig. 10.   The bottom plot is identical to Fig. 8. The top plot shows
the experimentally determined value of τ obtained using Eqn. 1
and the data from the bottom plot (squares). The calculated times
obtained using Eqn.4, and from a solution of Schrodinger's equation
for Q/J, are shown as solid and dashed lines respectively.

Although the effect of biasing the structure is most dramatic in the case of the PL
intensity, valuable additional information can be obtained from the changes induced
in other spectral characteristics of the PL peak, specifically its peak energy and its
spectral width. Fig. 11 shows the measured position of the centroid of the PL line as a
function of applied voltage. For voltages beyond the NDR region, the shift of the line
centre can be fit to a parabola which also intersects points near zero bias. Identical
behaviour of the PLE spectra is observed for these elevated voltages. The curvature
associated with this quadratic, field-induced red shift is in good agreement with that
expected for the Stark effect in two dimensions, the so called quantum confined Stark
effect (QCSE) (Miller et al., 1985; Brum and Bastard, 1985). Thus, at least for voltages
beyond the NDR region, the spectral position of the PL can be used to measure the
internal field strength in the well. The fact that the parabola is offset by $\sim$ 26 meV
indicates that there is a built-in potential across the well, even at zero applied bias.
This is precisely what is expected in a structure with an undoped buffer region adjacent
to only one of the two barriers. In equilibrium, electrons spill into the undoped buffer
region from the contacts, raising the potential of that side of the structure with respect
to the unbuffered side. The measured built-in potential, 26 meV, agrees well with a
self-consistent calculation of the zero-bias potential energy distribution for the device
in Table 1, which yields a value of 30 meV.

341

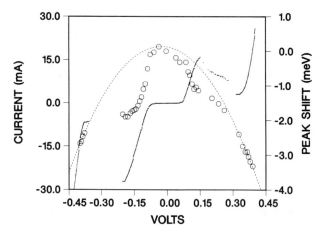

Fig. 11. The current through the device (dots) and the measured position of the centroid of the PL feature (circles) as a function of the applied voltage.

When the structure is biased within the resonant tunneling regime, the position of the line shifts much more to the red than expected due to the QCSE. This "anomalous" behaviour of the line shift is also apparent in plots of the linewidth of the PL feature as a function of applied voltage as shown in Fig. 12. Inspection of the actual PL spectra indicates that the anomalous shift in the resonant tunneling regime is entirely accounted for by the asymmetric broadening of the line to low energy. The dramatically different behaviours of these spectral characteristics of the PL feature for voltages above and below the resonant tunneling regime implies that the PL spectral parameters are very discriminating probes of resonant versus non-resonant current flow. Despite the fact that the absolute current flowing through the device is larger and spans a larger range in the non-resonant, high voltage regime, the PL linewidth remains only marginally larger than at zero bias. Within the resonant tunneling regime, its width increases by more than 50%.

Small, low energy shifts in PL spectra can often be associated with trapping of excitons on impurity sites (Yu et al., 1985) or, especially in the case of quantum wells, on local sites of low potential energy which arise due to structural imperfections in the material (Bastard et al., 1984). In good material these trap sites are usually at low concentrations, however they are thermodynamically favoured, hence they can still effect the PL spectra in principle. In order to rule out these extrinsic effects in the present case, PLE absorption spectra were obtained with the structure under external bias. Similar broadening was observed in the PLE spectra, again only when biased in the resonant tunneling regime. This result therefore implies that the broadening observed in the PL spectra must be an intrinsic effect associated with the many body nature of the luminescence from the two dimensional electron gas (Skolnick et al, 1987) accumulated in the well due to the resonant tunneling process.

The photoluminescent properties of two dimensional electron gases doped into quantum wells have been studied in detail by Skolnick et al. (1987). There is also a large body of literature on the general optical properties of two dimensional electron gases in quantum well systems (Chemla et al., 1988; Delalande et al., 1987; Egri, 1985;, Livescu et et al., 1988; Schmitt-Rink et al., 1985,1986; Yoshimura et al., 1988), however, most of the work has dealt with carrier densities in excess of $\sim 2 \times 10^{11} cm^{-2}$. At these densities and at low temperature, the electron gas is degenerate, and the photoluminescence

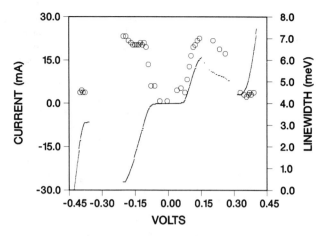

Fig. 12.   The current through the device (dots) and the measured
linewidth (FWHM) of the PL feature (circles) as a function of the
applied voltage.

usually consists of a low energy peak with a high energy shoulder which cuts off quite
abruptly. The detailed shape of the luminescence spectrum is complicated, reflecting
the many-body nature of the system. The density of carriers obtained by fitting the
spectra with a detailed many body calculation agrees quite well however, with the value
estimated by assuming that the low energy edge of the PL coincides with the (renormalized) bandedge, and the high energy cutoff occurs at the Fermi energy.

Recently Hayes et al. (1989) studied the PL from a DBRT sample which was designed with asymmetric barrier thicknesses to maximize the amount of charge accumulated in the well. Specifically, the collector barrier consisted of 11.1 nm of $Ga_{0.4}Al_{0.6}As$,
the well was 5.8 nm thick GaAs, and the emitter barrier was only 8.3 nm thick. The
lineshape of the PL observed from this sample, when biased resonant with the lowest
subband state, resembled that reported by Skolnick et al. (1987) in the modulation
doped sample. Using the same theory used by Skolnick to fit the modulation doped
lineshape, Hayes et al. estimated the peak charge density built up in the well to be
$\sim 2.2 \times 10^{11} cm^{-2}$ at the peak of the lowest resonance. This value was confirmed by observing the Landau level structure of the PL from the well as a function of the magnetic
field strength applied perpendicular to the plane of the layers. The disappearance of
PL from the second Landau level at an applied magnetic field of 4.6T implies that the
electron density is $\sim 2.3 \times 10^{11} cm^{-2}$, in good agreement with the value estimated from
the PL lineshape.

As shown in Fig. 6, there is absolutely no indication of a high energy tail with an
abrupt cutoff energy in the PL spectra obtained in the resonant tunneling regime from
the symmetric-barrier structure discussed above. This is consistent with the fact that
the estimated charge density in the symmetric-barrier structure is approximately four
times less than that reported for the asymmetric-barrier structure. Unfortunately there
is relatively little experimental or theoretical work related to the spectroscopic properties
of such low density two dimensional electron gas systems. In fact, at these low densities
it might be more appropriate to approach the problem from the point of view of how
a few free carriers can modify the linewidth of the "well-defined" exciton feature. The
influence of free carriers on the homogeneous linewidth of two dimensional excitons in
GaAs quantum wells has in fact been studied; theoretically by Feng and Spector (1987a
and 1987b), and experimentally, using picosecond self-diffraction techniques, by Honold

et al. (1988). The experimental results indicate that the homogeneous linewidth of the two dimensional exciton state increases linearly in the presence of an electron-hole plasma by 1.8 meV per $1 \times 10^{10} cm^{-2}$ electron-hole pairs at an ambient temperature of 2K. The theory also predicts a linear density dependence, with roughly equal contributions from electrons and heavy holes. If the incremental exciton PL linewidth observed in the resonant tunneling regime is attributed to this free carrier broadening mechanism, then the maximum implied carrier density in the well is $\sim 3.5 \times 10^{10} cm^{-2}$, within a factor of two of that estimated using the PL intensity data. The voltage dependence of the PL linewidth (Fig. 12) therefore provides independent confirmation of the manner in which charge accumulates in the well when the structure is biased in the resonant tunneling regime. In contrast to the PL intensity data which cannot be used to address the question of charge accumulation at higher voltages where the PL kinetics become complicated due to hole tunneling (see the following Section), the PL linewidth data clearly shows that the charge resonantly accumulated for $V < V_{NDR}$, almost entirely leaks out when biased in the non-resonant tunneling regime, for voltages beyond the region of NDR.

V TIME RESOLVED PL STUDIES OF DBRT STRUCTURES

So far it has been shown how the spectroscopic properties of steady state PL can be used to infer information about the resonant accumulation of electrons and the electric field strength internal to biased DBRT structures. From a knowledge of the steady state charge accumulation and the current density flowing through the device, the corresponding characteristic tunneling time can be deduced using the functional relationship in Eqn. 1. Drawing once again on the optical Fabry Perot analogy, this characteristic tunneling time can be associated with the effective cavity lifetime of electrons in the well, which is dictated by the leakage rate through the barriers via tunneling.

Picosecond laser pulses offer another means of directly measuring the leakage rate of electrons out of the well region of DBRT structures. By injecting electron-hole pairs using laser pulses of duration less than ~ 5 ps, separated by times long compared to that required for the excited system to relax back to equilibrium, the decay of the PL from the well can be time-resolved using a number of techniques. The lifetime of the PL excited in such a manner is in general determined by a combination of radiative decay, non-radiative trapping, and tunneling escape rates of the electrons and holes in the well. If the experimental conditions are such that the tunneling escape rate of at least one of the carrier species is greater than the radiative or non-radiative recombination rates, then the decay time of the PL signal yields a direct measure of the tunneling escape rate.

Tsuchiya et al. (1987) and Jackson et al. (1989) have independently studied the escape rate of electrons from AlAs/GaAs/AlAs DBRT structures using picosecond streak cameras and correlated excitation techniques respectively to time-resolve the PL signal following short-pulse excitation. Although there were some discrepancies in the quantitative results from these separate investigations, both groups generally found that the tunneling escape rates could be determined using these techniques, at least for samples held at 80K with barrier thicknesses such that the tunneling time was less than ~ 500 ps.

Fig. 13 shows the temporal dependence of the PL intensity obtained with the sample described in Table 1 held at 5K (solid) and 200K (dashed). The sample was excited with 4ps, 725 nm dye laser pulses separated by 450 ns, and the detection was via a time-correlated photon counting technique. The increase in the PL lifetime at higher temperatures is obvious. The decay curves exhibit single exponential behaviour at temperatures less than ~ 50K, and double exponential behaviour for higher temperatures. The faster of the decay times increases from 300 ps at 5K to 1.1 ns at 200K, and remains constant at 1.1 ns up to 250K. This limiting value of 1.1 ns is in very close agreement with the value of the electron escape time of 1.2 ns obtained using Eqn. 4 at zero applied bias.

PL decay curves similar to the solid line in Fig. 13 were obtained at 5K for applied voltages which biased the structure in the resonant tunneling regime. A slight (~ 50

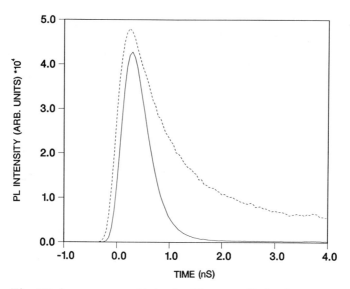

Fig. 13.   The PL decay curves obtained with no applied voltage at
5K (solid) and 200K (dashed). Zero time corresponds to the arrival
of the 3 ps laser pulse at the sample.

ps) decrease of the PL decay time was observed at the peak of the resonance, however,
this is not considered very significant since at 5K the lifetime of the PL is dominated
by decay processes unrelated to tunneling. Much more significant is the appearance of
a second, relatively long time constant in the PL decay curves as soon as the structure
is biased at voltages beyond the region of NDR. The solid decay curve shown in Fig.
14 was obtained using the same optical pulses used to measure the curves shown in
Fig. 13, at lower bias voltages. The nature of this slow decay component is revealed by
measuring the PL decay at the identical applied voltage, but using 784 nm laser pulses
(dashed curve in Fig. 14) which do not directly excite electron-hole pairs in the well
region of the device. The decay curve resulting from the subtraction of that obtained
using 784 nm pulses from that obtained at 725 nm is shown as a dotted line in Fig. 14.

The PL decay obtained using photons with energy greater than the subband gap in
the well region is thus the superposition of a fast component similar to that observed
at low bias voltages, and the identical slow component to that obtained using photons
with energy less than the subband gap in the well, but greater than the bulk GaAs
bandgap. The PL signal observed using the 784 nm excitation is spectrally identical
to that obtained at 725 nm; it is therefore due to exciton recombination in the well.
Since no electrons or holes can be directly excited in the well region, the excitons giving
rise to this PL signal must be formed from holes photo-excited in the collector contact
which subsequently tunnel into the well, and from electrons which tunnel into the well
from the emitter contact. The photo-excited carrier density in the contacts is much less
than the n+ doping level, so the photo-excited electrons in the emitter contact will not
alter the density of electrons normally resident in the well at $V > V_{NDR}$; those emitter
electrons which tunnel into the well off-resonance due to inelastic scattering processes.
The long PL decay shown as a dashed curve in Fig. 14 therefore directly reflects the
leakage of photo-excited holes from the collector contact region into the well via inelastic
tunneling.

Efficient tunneling of photo-excited holes from the collector into the well can only
occur if the characteristic time associated with this tunneling process is less than the
hole lifetime in the collector region due to radiative and/or non-radiative recombination
processes unrelated to tunneling. This explains why the slow time constant only occurs
for large applied voltages at which a true depletion region can form adjacent to the

345

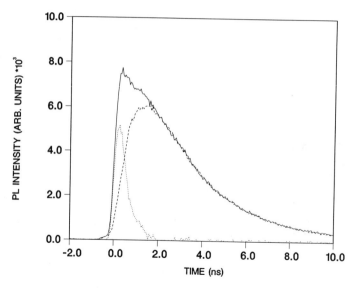

Fig. 14. The PL decay curves obtained with 500 mV applied voltage at 5K using 725 nm excitation (solid) and 784 nm excitation (dashed). The dotted line is the difference between the other two curves.

contact barrier. At voltages within the resonant tunneling regime, the radiative lifetime of photo-excited holes with electrons doped into the collector contact must be less than the hole tunneling time at those voltages. This argument is supported by the polarity dependence of the PL intensity and decay time obtained using 784 nm excitation. With the structure biased at 500 mV in such a way that the collector contact contains the 7.5 nm thick undoped buffer region (see Table 1), the PL intensity from the well is approximately an order of magnitude stronger than for the same bias of opposite polarity. The PL decay time constants for the polarities which yield high and low PL intensities are 2.7 ns and 1.4 ns respectively. Thus in the case where the undoped buffer region is in the collector contact, the degree of depletion is greater which increases the radiative lifetime of holes adjacent to the barrier, hence allowing more to tunnel into the well and contribute to the exciton luminescence.

This direct evidence of photo-excited hole tunneling from the collector into the well explains the anomalous behaviour of the PL intensity observed at voltages above the region of NDR, as discussed in Section IV. In particular, it explains why the intensity of the PL remains relatively high compared to that observed in the resonant tunneling regime (Fig. 8), even though the PL linewidth data (Fig. 12) and simple theories indicate that the charge accumulated in the well is much less when biased off as opposed to on resonance. The details of the PL kinetics, which determine the PL intensity, are completely different for $V < V_{NDR}$ than for $V > V_{NDR}$, as directly evidenced in the PL decay curves (Fig. 14 compared to Fig. 13).

These results also suggest a possible reason why the intensity dependence of the PL observed by Hayes et al. (1989) in their DBRT sample could not be correlated with the resonant tunneling properties of the device. As pointed out above in Section IV, the sample used by Hayes et al. (1989) had very lightly doped buffer regions in excess of 100 nm thick adjacent to the DBRT structure. The radiative lifetime of photo-excited holes in the collector depletion region of this sample will therefore be substantially longer than in the sample described herein. This, together with the fact that the depletion region itself will be much larger in the lightly doped sample (and also much thicker than the well region), means that leakage of photo-excited holes in the sample of Hayes et al. (1989) would be expected to be much more important than in the heavily doped sample, even at low applied voltages.

VI CONCLUSIONS

Some applications of PL and PLE optical spectroscopy techniques to study the interior state of biased double barrier resonant tunneling samples have been described. For structures with n+ doped contacts in close proximity to the resonant tunneling structure itself, the bias dependence of the steady state photoluminescence intensity, linewidth and peak energy from the confined well state can be used to quantitatively monitor the accumulation of electron charge when the structure is biased in the resonant tunneling regime. For structures with substantial buffer regions separating the resonant region from the n+ contacts, PL linewidth and its dependence on applied magnetic fields can be used to measure the accumulated electron density, but the photoluminescence intensity cannot. Simultaneous determination of the accumulated charge in the well and the current density flowing through the device allow the determination of the characteristic tunneling time of electrons in these devices. This characteristic tunneling time is found to exceed the scattering time associated with lateral motion of the electrons in the well by orders of magnitude, indicating that such intra-subband scattering events do not significantly effect this characteristic resonant tunneling time. Time-resolved PL techniques provide direct corroborative evidence that this characteristic tunneling time is in good agreement with calculated values of the time required for electrons to escape from the quasi-two dimensional well state. These time-resolved techniques also provide direct evidence that photo-excited holes in the collector contact region can tunnel into the well via inelastic tunneling processes, but only if the associated tunneling time is less than the recombination time in the contact.

VII REFERENCES

Bastard, G., Delalande, C., Meynadier, M.H., Frijlink, P.M., and Voos, M., 1984, Phys. Rev. B, 29: 7042.

Bimberg, D., Munzel, H., Steckenborn, A., and Christen, J. 1985, Kinetics of relaxation and recombination of non-equilibrium carriers in GaAs: Carrier capture by impurities, Phys. Rev. B, 31: 7788.

Born, M. and Wolf, E., 1959, "Principles of Optics", Pergamon, Oxford.

Brum, J.A., and Bastard, G., 1985, Electric-field-induced dissociation of excitons in semiconductor quantum wells, Phys. Rev. B, 31: 3893.

Buttiker, M., 1988, Coherent and sequential tunneling in series barriers, IBM J. of Res. and Devel., 32: 63.

Chemla, D.S., Bar-Joseph, I., Kuo, J.M., Chang, T.Y., Klingshirn, C., Livescu, G., and Miller, D.A.B., 1988, Modulation of absorption in field-effect quantum well structures, IEEE J. Quan. Elect., 24: 1664.

Delalande, C., Bastard, G., Orgonasi, J., Brum, J.A., Liu, H.W., and Voos, M., 1987, Many-body effects in a modulation-doped semiconductor quantum well, Phys. Rev. Lett., 59: 2690.

Egri, I., 1985, Burstein shift of the contact exciton, Phys. Rev. B, 32: 2631.

Feng, Y. and Spector, H.N., 1987a, Exciton linewidth due to scattering from free carriers in semiconducting quantum well structures, Superlattices and Microstructures, 3: 459.

Feng, Y. and Spector, H.N., 1987b, Scattering of screened excitons by free carriers in semiconducting quantum well structures, IEEE J. Quan. Elect., 24: 1659.

Goldman, V.J., Tsui, D.C., and Cunningham, J.E., 1987a, Resonant tunneling in magnetic fields: Evidence for space-charge buildup, Phys. Rev. B, 35: 9387.

Goldman, V.J., Tsui, D.C., and Cunningham, J.E., 1987b, Observation of intrinsic bistability in resonant tunneling structures, Phys. Rev. Lett., 58: 1256.

Hayes, D.G., Skolnick, M.S., Simmonds, P.E., Eaves, L., Halliday, D.P., Leadbeater, M.L., Henini, M., Hughes, O.H, Hill, G., and Pate, M.A., 1989, Optical investigation of charge accumulation and bistability in an asymmetric double barrier resonant tunneling heterostructure, to appear in the proceedings of the Modulated Structures and Superlattices Conference held in Ann Arbor, Michigan, July, 1989.

Honold, A., Schultheis, L., Kuhl, J. and Tu, C.W., 1988, Phase relaxation of two-dimensional excitons in a GaAs quantum well, in: "Springer Series in Chemical Physics, Vol. 48: Ultrafast Phenomena VI", Yajima, T., Yoshihara, K., Harris, C.B, and Shionoya, S. eds., Springer-Verlag, Berlin.

Jackson, M.K., Johnson, M.B., Chow, D.H., McGill, T.C., and Nieh, C.W., 1989, Electron tunneling time measured by photoluminescence excitation correlation spectroscopy, Appl. Phys. Lett., 54: 552.

Leadbeater, M.L., Alves, E.S., Eaves, L., Henini, M., Hughes, O.H., Sheard, F.W., and Toombs, G.A., 1989, Magnetic field and capacitance studies of intrinsic bistability in double-barrier structures, Superlattices and Microstructures, 6: 59.

Livescu, G., Miller, D.A.B, Chemla, D.S., Ramaswamy, M., Chang, T.Y., Sauer, N., Gossard, A.C., and English, J.H., 1988, Free carrier and many-body effects in absorption spectra of modulation-doped quantum wells, IEEE J. Quan. Elect., 24: 1677.

Luryi, S., 1985, Frequency limit of double-barrier resonant-tunneling oscillators, Appl. Phys. Lett., 47: 490.

Mendez, E.E., Calleja, E. and Wang, W.I., 1986, Tunneling through indirect-gap semiconductor barriers, Phys. Rev. B, 34: 6026.

Miller, D.A.B., Chemla, D.S., Damen, T.C., Gossard, A.C., Wiegmann, W., Wood, T.H., and Burrus, C.A., 1985, Electric field dependence of optical absorption near the band gap of quantum-well structures, Phys. Rev. B., 32: 1043.

Payling, C.A., Alves, E.S., Eaves, L., Foster, T.J., Henini, M, Hughes, O.H., Simmonds, P.E., Sheard, F.W., Toombs, G.A. and Portal, J.C., 1988, Evidence for sequential tunneling and charge build-up in double barrier resonant tunelling devices, Surf. Sci., 196: 404.

Price, P.J., 1986, Resonant tunneling properties of heterostructures, Superlattices and Microstructures, 2: 593.

Price, P.J., 1987, Coherence of resonant tunneling in heterostructures, Phys. Rev. B, 36: 1314.

Schmitt-Rink, S., Ell, C., and Haug, H., 1986, Many-body effects in the absorption, gain, and luminescence spectra of semiconductor quantum-well structures, Phys. Rev. B, 33: 1183.

Schmitt-Rink, S., and Ell, C., 1985, Excitons and electron-hole plasma in quasi-two-dimensional systems, J. Lumin., 30: 585.

Skolnick, M.S., Rorison, J.M, Nash, K.J., Mowbray, D.J., Tapster, P.R., Bass, S.J. and Pitt, A.D., 1987, Observation of a many-body edge singularity in quantum-well luminescence spectra, Phys. Rev. Lett., 58: 2130.

Sollner, T.C.L.G., Goodhue, W.D., Tannenwald, P.E., Parker, C.D., and Peck, D.D., 1983, Resonant tunneling through quantum wells at frequencies up to 2.5 THz, Appl. Phys. Lett., 43: 588. Appl. Phys. Lett., 47: 490.

Tsuchiya, M., Matsusue, T., and Sakaki, H., 1987, Tunneling escape rate of electrons from quantum well in double-barrier heterostructures, Phys. Rev. Lett., 59: 2356.

Weil, T. and Vinter, B., 1987, Equivalence between resonant tunneling and sequential tunneling in double-barrier diodes, Appl. Phys. Lett., 50: 1281.

Weisbuch, C., Miller, R.C., Dingle, R., Gossard, A.C. and Wiegmann, W., 1981, Intrinsic radiative recombination from quantum states in GaAs-AlGaAs multi-quantum well structures, Sol. State. Comm., 37: 219.

Yoshimura, H., Bauer, G.E.W., and Sakaki, H., 1988, Carrier-induced shift and broadening of optical spectra in an AlGaAs/GaAs quantum well with a gate electrode, Phys. Rev. B, 38: 10791.

Young, J.F., Wood, B.M., Liu, H.C., Buchanan, M., Landheer, D., SpringThorpe, A.J., and Mandeville, P., 1988a, Effect of circuit oscillations on the dc current-voltage characteristics of double barrier resonant tunneling structures, Appl. Phys. Lett., 52: 1398.

Young, J.F., Wood, B.M., Aers, G.C., Devine, R.L.S., Liu, H.C., Landheer, D., Buchanan, M., SpringThorpe, A.J., and Mandeville, P., 1988b, Determination of charge accumulation and its characteristic time in double-barrier resonant tunneling structures using steady-state photoluminescence, Phys. Rev. Lett., 60: 2085.

Young, J.F., Wood, B.M., Aers, G.C., Devine, R.L.S., Liu, H.C., Landheer, D., Buchanan, M., SpringThorpe, A.J., and Mandeville, P., 1989, Photoluminescence characterization of vertical transport in double barrier resonant tunneling structures, Superlattices and Microstructures, 5: 411.

Yu, P.W., Reynolds, D.C., Bajaj, K.K, Litton, C.W., Masselink, W.T., Fischer, R. and Morkoc, H., 1985, Effect of electric field on the sharp photoluminescence spectra of undoped GaAs-AlGaAs multiple quantum well strucutres, J. Vac. Sci. Technol. B, 3: 624.

ADVANCED MATERIALS TECHNOLOGY FOR THE NEXT DECADE'S OEICs

M. Razeghi and O. Acher

THOMSON-C.S.F - Laboratoire Central de Recherches
Domaine de Corbeville - B.P. 10
91401 ORSAY CEDEX (FRANCE)

ABSTRACT

This paper presents the recent advances on the in situ reflectance-difference spectroscopy on metalorganic chemical vapor deposition growth technique. The high sensitivity of this in situ diagnostic, allows the growth of high quality III-V semiconductors heterojunctions, quantum wells and superlattices for photonic and electronic devices on lattice matched or alternative substrates, such as Si for the next decade's OEICs.

1 - INTRODUCTION

The optoelectronic integrated circuit or OEIC is a circuit consisting of optical devices (such as laser diodes) and electronic devices (such as transistors) fabricated on a single chip of semiconducting crystal substrate.

Since the invention of the integrated circuit in 1958, the number of components on a state-of-the-art IC chip has grown exponentially, using Si substrate. ICs is an ensemble of active (ex: transistor) and passive devices (ex: resistor and capacitor) formed on and with a singlecrystal semiconductor substrate and interconnected by a metallization pattern.

Different categories of ICs are:
- SSI (Small Scale Integration), contains 100 components per chip,
- MSI (Medium-Scale-Integration), contains 1000 components per chip,
- LSI (Large-Scale-Integration), contains more than 100,000 components per chip.

OEICs are, precisely speaking, the combination of an optical element such as a laser, a photodetector and its driving elements all contained on the same chip.

One of the advantages of monolithic integration of multiple components is that parasitic capacitances associated with bondings and wirings are reduced considerably. This reduction results in the increase in the modulation frequencies through the decrease in the RC time delay of the circuits.

Electronic Properties of Multilayers and Low-Dimensional Semiconductors Structures
Edited by J. M. Chamberlain *et al.*, Plenum Press, New York, 1990

351

However, there are many technological problems to be solved in order to realize practical and highly integrated OEICs. The most important problem is concerned with materials, such as growing III-V materials on silicon substrates.

At present Si is used for VLSI, and the MOSFET is the dominant device used in VLSI circuits because it can be scaled to smaller dimensions than can other types of devices. The associated technologies are NMOS (n-channel MOSFET) and CMOS (complementary MOSFET), which provide n-channel and p-channel MOSFETs on the same chip.

There are different limitations in Si integrated circuits. Si has indirect energy gap and it is not possible to fabricate optical device such as laser. The mobility of electron in Si is low. Recent advances in GaAs, InP and related compounds processing techniques in conjunction with new fabrication and circuit approaches have made possible the development of Silicon-like GaAs IC technology.

Long-distance optical links use semiconductor lasers emitting at 1.3 µm and 1.55 µm fabricated using materials such as the GaInAsP-InP lattice-matched system on an InP substrate. Unfortunately, the technology of integrated circuits for signal treatment is more difficult on this material than on GaAs substrates. Since Silicon technology is more advanced than that for GaAs and InP based materials, the solution is to combine the advantages of InP and GaAs devices with the maturity of Si processing technology, using existing Si integrated circuit equipments. The most important advantages of III-V materials as compared to Si are:

- higher electron mobility, which results in lower series resistance for a given device geometry,
- higher drift velocity at a given electric field which improves device speed.

The disadvantages of III-V by comparison to Si are:

- very short minority carrier lifetime,
- lack of stable, passivating active oxide,
- crystal defects are many orders of magnitude higher than in Si.

In heterostructures, it is certainly desirable to select a pair of materials closely lattice-matched in order to minimize defect formation or stress. However, heterostructure lattice-mismatched to a limited extent can be grown with essentially no misfit dislocations, if the layers are sufficiently thin, because the mismatch is accommodated by a uniform lattice strain. Without the requirement of lattice matching, the number of available pairs for device applications and integrated circuits is greatly increased.

For future OEICs we need to control the doping level with high precision, to obtain high thickness and composition homogeneity, sharp interfaces, dislocation free materials, and to achieve localized epitaxy, and regrowth. Substantial improvements in material quality and processing technology are needed before OEICs can seriously challenge the performance of electronic devices.

Detailed investigations of the growth of materials, at the monolayer level, can need in situ measurements, in particular concerning the kinetic of the formation of an interface. The more conventional in situ electronic probes are only compatible with high vacuum processes. Then, they are incompatible with the gas pressures involved in the III-V compounds

epitaxy by the organometallic chemical vapor deposition method (MOCVD).
In the same way, they cannot be used in the reactive plasma environment
of the amorphous semiconductor thin films preparation. Thus, because of
its high precision, extreme sensitivity and the capability to record in
situ measurements, spectroellipsometry (SE) appears particularly well
adapted to address a number of problems in semiconductor material
processing[1-6]. In situ SE is generally used in the UV-visible range[1-6]. As
a consequence, this technique does not allow a direct measurement of the
vibrational properties of the materials. This can be a strong limitation
when dealing with complex compounds like amorphous semiconductor thin
films which include various chemical species. In particular, the electronic properties of plasma deposited amorphous silicon (a-Si:H) are controlled by the hydrogen incorporation into the silicon network. Thus, SE
was successfully extended towards infrared (IR)[7]. Chemical bonds have
been identified, using this last technique, on ultrathin amorphous semiconductor samples ($d_f < 100$ Å)[8,9]. An other powerful in situ optical
technique recently appeared: reflectance-difference spectroscopy (RDS) in
the UV-visible range[10,11,12]. RDS being sensitive to the crystal anisotropy, it only probes the III-V surface. Therefore, RDS returns information about the chemistry of the growth surface itself. The RDS technique
has recently been used successfully to study the MBE growth of GaAs and
AlAs. An extension of such in situ characterization to the MOCVD growth
of III-V compounds[13] would be very helpful.

In this paper, we report our preliminary results concerning RDS studies performed on a LP-MOCVD reactor under optimal growth conditions. It
is possible using the RDS technique to determine in situ the $\langle 011 \rangle$ and
$\langle 01\bar{1} \rangle$ directions of III-V compounds, in a non-destructive way, which is
very important for future OEICs. A study of InAs growth suggests that one
can have access to the III/V ratio on the surface using this technique.
The occurrence of very large surface anisotropies (10% or more) related
to the growth of lattice mismatched materials is described. The optical
anisotropy of buried interfaces is investigated.

2 - SPECTROSCOPIC ELLIPSOMETRY FROM UV TO IR

Ellipsometry measures the complex reflectance ratio:

$$\rho = r_p/r_s = \tan \Psi \exp(i\Delta) \tag{1}$$

where the subscripts p and s refer to the plane-wave electric field components, respectively, parallel and perpendicular to the plane of incidence. Ellipsometric measurements can be performed as a function of time
(kinetic ellipsometry) or as a function of the photon energy (spectroscopic ellipsometry). Let us recall that in the case of a homogeneous material with a sharp interface with ambient, ρ is directly related to its
complex dielectric function $\varepsilon = \varepsilon_1 + i\varepsilon_2$. Otherwise, a multiple-layer
calculation is necessary[14]. In the infrared, when dealing with ultrathin
films, the results are often presented in the form of the relative com-
plex optical density $D = \log(\bar{\rho}/\rho)$ where $\bar{\rho}$ refers to the substrate before
film deposition. It can be easily shown that D is simply related to the
film dielectric function ε_f by the relations[7]:

$$\text{Re } D \approx d_f (A \varepsilon_{2f} + B) \tag{2}$$

$$\text{Im } D \approx d_f (C \varepsilon_{1f} + D) \tag{3}$$

In situ SE measurements, from UV to IR, can be performed using two spectroscopic phase modulated ellipsometers (SPDME) coupled to growth chambers[8,15]. As compared to the modulation techniques based upon the mechanical rotation of a linear polarizer, SPME takes advantage of the fast modulation (35-50 kHz) provided by a photoelastic modulator. Thus SPME is particularly well adapted to real-time in situ applications. Moreover, a careful optical treatment of the photoelastic modulator allows high precision (Ψ, Δ) measurements[16]. The SPME technique can also be extended in the infrared up to 11 μm[8,17].

3 - REFLECTANCE-DIFFERENCE SPECTROSCOPY

Aspnes[11] underlined the interest of measuring the optical anisotropy of semiconductors at normal incidence. As bulk III-V compounds are optically isotropic, this technique is expected to be sensitive to the surface of the semiconductor, independently of the penetration depth of light in the material. Aspnes shows that Reflectance-Difference Spectroscopy (RDS), as it is called, could give indications on the MBE growth of GaAs and AlAs. He demonstrated the difference of optical anisotropy between As-rich and Ga-rich (100) surfaces of GaAs[10]. Results on MOCVD growth of GaAs at low temperature were also reported[12].

Metal Organic Chemical Vapour Deposition (MOCVD) is a powerful and widely used technique for growing III-V compounds[13]. However, very few characterization techniques are compatible with the MOCVD environment. Therefore, RDS would be a very attractive technique if it proves to give in situ information in realistic growth conditions. The aim of the present study is to survey the RDS behaviour related to the growth of some important materials and structures for future OEICs.

RDS is based on the analysis of the reflected beam after near-normal incidence of a polarization-modulated light on a sample. The study described here is performed on slightly misoriented (100) surfaces of III-V samples[18]. In this case, the optical eigenaxes are $\langle 011 \rangle$ and $\langle 01\bar{1} \rangle$ and the RDS experiment consists in the measurement of the relative complex difference $(r_{011} - r_{01\bar{1}})/r$ between the reflexion coefficients of light polarized along these directions. Eq. (1) shows that ellipsometry also measures the ratio of the reflection coefficients for two eigenaxes corresponding to the (p) and (s) directions. Therefore SE and RDS can be considered as similar techniques. Then the SPME formalism can be used for the same optical configuration in near-normal incidence. The RDS set-up, in the UV-visible range, is directly adapted to the MOCVD chamber without any modification of the quartz tube, as described elsewhere[18]. As previously mentioned, the real part of the anisotropy is not coupled at the first order to the quartz tube birefringence[19]. Then, variations of this quantity as low as 10^{-4} can be detected in real-time. Such high sensitivity measurements need the high frequency modulation provided by the photoelastic modulator.

Our RDS setup is derived from a previously described phase-modulated ellipsometer, with the different optical elements orientated in what is often called configuration III[15,16]. It was modified to work at near-normal incidence, and was adapted on a Low Pressure MOCVD reactor described elsewhere[13] (fig. 1).

The influence of the residual angle of incidence was minimized by reducing it to about 2°, and also by orienting the plane of incidence at $\pm$ 45° to the $\langle 011 \rangle$ axis.

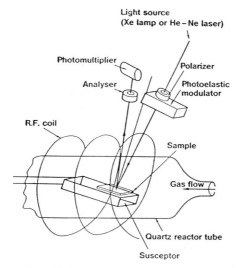

Figure 1.   Schematic view on the RDS setup
adapted to the MOCVD reactor tube.

The ability of RDS to work without any modification of the quartz
reactor tube is very important, since any change in the reaction chamber
design may have consequences on the flow pattern and on the epilayers
quality.

A wide variety of structures were investigated using RDS. The purpose was to determine whether the RDS signal connected to the growth of
different heterojunctions under optimal conditions was large enough to be
detected, and to what extent it could be related to growth conditions or
crystal quality.

4 - OPTIMIZATION OF III-V BINARIES BY RDS

The quality of III-V binary materials grown by metalorganic chemical
vapor epitaxy depends on the different growth parameters such as: growth
temperature, total flow rate, the ratio of III/V, the preparation of substrate and the purity of starting materials[13]. In the case of InAs layer,
the ratio of III/V is very an important parameter. We optimized the quality of InAs using in situ RDS.

RDS signals for InAs growth were observed for light wavelengths
between 4,500 and 6,000 Å, with a maximum amplitude in the range of 5,000
to 5,500 Å. Further studies were performed using He-Ne laser emitting at
5,435 Å. The laser power was 0.5 mW, and the spot size was roughly 2 mm
in diameter. Signals ranging from 3×10^{-4} to 2×10^{-3} were observed, with a
resolution of 10^{-4} and 0.1 seconds.

After growing an InAs buffer layer on an InAs substrate in the optimum conditions, the element III source (TMI) was switched off and on. The
element V flow (AsH$_3$) was never switched off, in order to avoid substrate
degradation. When the growth is resumed by turning the TMI on, the RDS
signal shows a sudden change and a few oscillations, then when switched
off, it resumes its former level. Figure 2 shows that the amplitude of
the step increases with the TMI flow for a given value of AsH$_3$ flow.

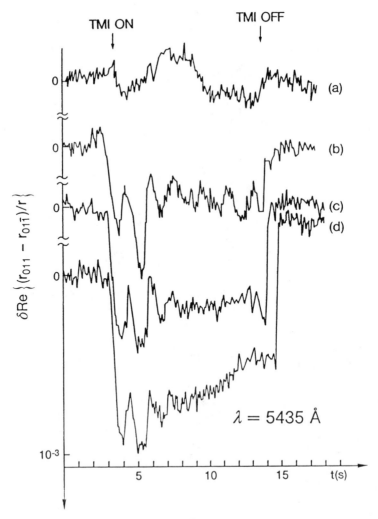

Figure 2. RDS record at λ = 5,435 Å of growth transients of InAs on InAs at T = 480°C, for a AsH_3 flow of 30 cm³/min, and different TMI flows. The AsH_3 is always supplied, and TMI is switched for 10 seconds to the reactor, and then switched off.

    a) H_2 through TMI bubbler flow = 50 cm³/min
    b) "   "   "   "   " = 100 cm³/min
    c) "   "   "   "   " = 150 cm³/min
    d) "   "   "   "   " = 200 cm³/min.

      Our preliminary results suggest that the amplitude of the step may be related to the III/V ratio at the surface of the sample. The nature of the oscillations have to be further investigated.

      The RDS can provide a valuable in situ control for InAs growth, since it gives indications on its surface. It can help in detecting drifts in growth conditions. It is important since InAs quality is very sensitive to the III/V ratio.

In LP-MOCVD growth of InAs, if we take the flow rate of AsH_3 and TMI constant, changing growth temperature can change the ratio of As/In in

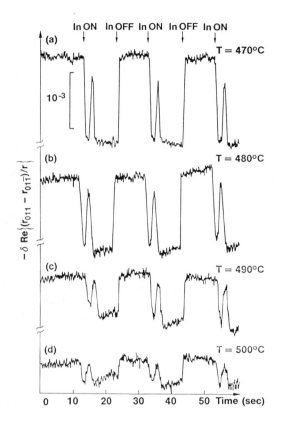

Figure 3. RDS record at λ = 5,435 Å of growth transients of InAs on InAs at different temperatures T. AsH_3 flow =10 cm^3/min, H_2 through TMI bubbler flow = 100 cm^3/min. The AsH_3 is always supplied, and TMI is switched off and on every 10 seconds.
  a) T = 470°C - b) T = 480°C - c) T = 490°C - d) T = 500°C.

the solid phase. The figure 3 shows the RDS signature of InAs layers grown by LP-MOCVD, at different growth temperatures, taking the flow of AsH_3 and TMI constant, and switching the TMI flow on and off energy 10 seconds. The intensity of RDS signals is changing as function of growth temperature. When the growth temperature is 480°C, the optical, electrical and structural properties of InAs epilayer are improved.

There seems to be a correlation between the surface anisotropy of InAs during its growth, and its quality.

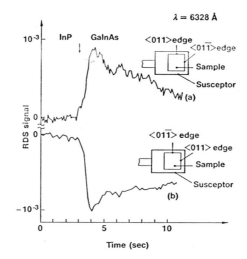

Figure 4. RDS signal at λ = 6,328 Å recorded during the growth of a GaInAs/InP heterojunction on two differently orientated substrates. RDS signal is $Re((r_{011}-r_{01\bar{1}})/r)$ for sample (a) and $Re((r_{01\bar{1}}-r_{011})/r)$ for sample (b).

5 - RDS INVESTIGATION OF III-V HETEROJUNCTIONS

The GaInAs-InP and GaInP-GaAs lattice-matched heterojunctions were investigated, mainly using He-Ne lasers emitting at 6,328 Å or 5,435 Å as a lightsource. Growth conditions are typically those described in refs.[13,20,21].

For all interfaces, the RDS signal changes $\delta Re(r_{011} - r_{01\bar{1}})/r)$ were in the range of 3×10^{-4} to 1.5×10^{-3}.

It is possible to use RDS to determine in situ substrate orientation. The RDS signal is either $Re((r_{011}-r_{01\bar{1}})/r)$, or the opposite, $Re((r_{01\bar{1}}-r_{011})/r)$, depending on substrate orientation (see Fig. 4). The sign of the RDS change corresponding to the growth of an heterojunction can indicate which of the cleaved facets on the sample is the $\langle011\rangle$ or $\langle01\bar{1}\rangle$ direction. This ability to determine crystal orientation in a non-destructive way is very important for technological applications. Besides, this orientation dependence of the RDS signal provides a simple way to check that the signal does not originate from parasitic contributions such as depositions on the quartz reactor tube. All RDS signatures presented here were checked in this way.

The growth conditions or the quality of the material can be related to the RDS signature. Figure 5 shows the different RDS signatures of GaInAs-on-InP heterojunctions grown at different temperatures. Increasing the temperature by 40°C from 5.a) to 5.c) leads to a significant increase in electronic mobility, revealing an improved interface quality.

A striking feature is that the RDS signature of the growth sequence material2->material1 is not mirrored by the signature of the reverse growth sequence (i.e. material1->material2) (see fig. 6). Another important aspect is that a given RDS signature can exhibit fast changes (less

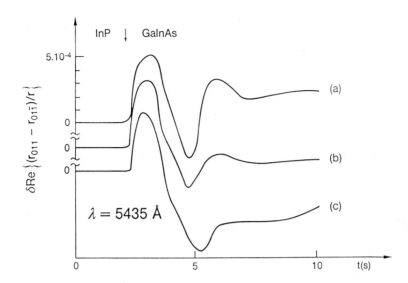

Figure 5.   RDS signature at $\lambda = 5{,}435$ Å of GaInAs/InP lattice matched heterostructures grown at 500°C (a), 520°C (b), 540°C (c). The corresponding interface quality was optimal in case c, and degraded at lower growth temperatures.

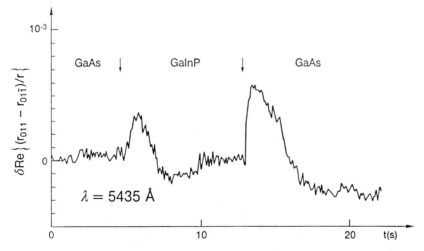

Figure 6.   RDS signature of a GaAs/GaInP/GaAs lattice-matched double heterostructure recorded at $\lambda = 5{,}435$ Å.

than 0.1 seconds, i.e. less than one monolayer), and slower ones (a few seconds to one minute). This is related to the diversity of contributions to the RDS signal. As pointed out by Aspnes[11], not only surface anisotropy, but also chemi- or physisorption, surface roughness, many-body screening and buried interface anisotropy may have a contribution. The time constant associated with these effects are different.

6 - RDS INVESTIGATION OF III/V ON Si

The growth of III/V compounds on lattice-mismatched substrates has emerged as an important subject. During the early stages of such depositions, three-dimensional growth may occur. In correct growth conditions, or after growing buffer structures to improve material quality[22,23], the surface becomes smooth, and the growth becomes two-dimensional.

Very large optical anisotropies were measured during the early stages of growth of lattice-mismatched epilayers on Si substrates. They are connected with the three-dimensional growth. As the surface improves, the signal returns to zero. This was observed during the growth of InP on GaAs on Si, (figure 7). Several GaInAs/InP superlattices were included in the structure, one of them is detailed in figure 7b. After growing a sufficiently thick InP buffer, the surface became completely smooth,

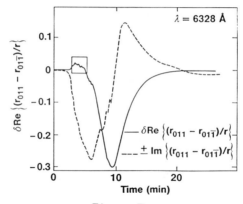

Figure 7.a

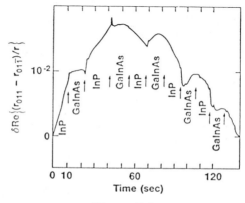

Figure 7.b

and another GaInAs/InP superlattice was made. Its RDS signature (shown in figure 7c) significantly differs from fig. 7b. This proves that the effect on the RDS signal of three-dimensional growth and of heterojunction do not merely superimpose.

The ability of RDS to give indications during such difficult growths may be valuable, both for fundamental understanding, and for in situ control.

7 - RDS INVESTIGATION OF BURIED INTERFACES

a) Experimental detail:

Several RDS studies have been performed using RDS on our MOCVD reactor[18,24], including the growth of lattice-matched heterostructures of InP-GaInAsP, InP-InAlAs or GaAs-GaInP under optimal conditions. As expected for a surface-sensitive technique, we were able to observe changes in optical anisotropy in very short time periods (less than 0.1 s to a few seconds) when growing heterostructures. It was possible in some cases to correlate RDS signature with interface quality, as mentioned before.

The point here is to focus on larger timescale variations of the RDS signal, and to show how it is possible to measure in situ the optical anisotropy of buried interfaces.

Figure 8 shows the RDS record of the growth of an InAlAs on InP structure during several minutes. Figure 8b gives the detail of the RDS signal during the first few seconds of InAlAs growth. This growth was performed under normal growth conditions, i.e. T = 540°C, the growth rate of InP being about 170 Å/min and that of InAlAs 340 Å/min. The InAlAs layer was checked to be lattice-matched to the substrate by X-ray diffraction.

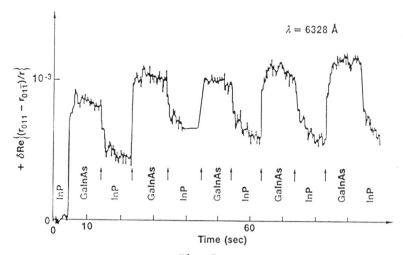

Fig. 7.c

Figure 7. a) RDS record at λ = 6,328 Å during the early stages of the growth of a InP on GaAs on Si structure. Some GaInAs/InP superlattices were included in the structure.
b) Detail of fig.7a corresponding to a GaInAs/InP superlattice.
c) RDS record of a superlattice grown on the same sample after the surface had smoothed.

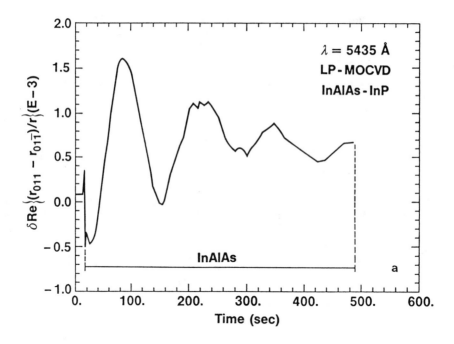

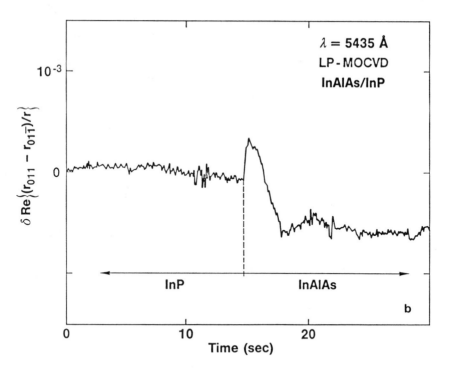

Figure 8. a) RDS record during the growth of a InAlAs on InP structure. Wavelength is 5,435 Å. The InP to InAlAs commutation is performed at the 15th second.
b) Detail of the signal corresponding to the switching from InP to InAlAs.

The RDS signal exhibits damped oscillations, with a pseudo-period of 131s (± 6s). This is due to interference phenomena combined with buried interface anisotropy effects. The growing layer acts as a Fabry-Perrot. The reflection coefficients associated with the bare surface and the InP/InAlAs interface are slightly anisotropic, and thus, the constructive interference conditions are slightly different for the $\langle 011 \rangle$ and $\langle 01\bar{1} \rangle$ light polarization. The pseudoperiod T of the oscillations can be related to the growth rate V by writing that the optical path increases by λ during a pseudoperiod:

$$\frac{V.T}{2.n} = \lambda \tag{4}$$

Assuming an optical index n of InAlAs equal to that of InP (i.e. n = 3.66 at 5,435 Å), one finds V = 340 Å/min (± 16 Å/min), which is in good agreement with determinations based on thickness measurements.

b) Comparison with model

The RDS oscillations produced by buried interface anisotropy and surface anisotropy can be calculated. A treatment based on the Abeles matrix formulation[25] is well suited for that purpose. Interfaces are treated as films with a very small thickness e compared to wavelength λ, and their optical properties are calculated to first order development in $\frac{e}{\lambda}$, as in ref.[26]. Working at normal incidence provides an appreciable simplification in the expressions. The bulk materials are assumed to be isotropic. Interface and surface are assumed to have the same optical eigenaxes, $\langle 011 \rangle$ and $\langle 01\bar{1} \rangle$.

$\delta\varepsilon_s$ is the difference between surface dielectric function associated to direction $\langle 011 \rangle$ and that associated to $\langle 01\bar{1} \rangle$. e_s is the thickness of this surface layer. The $e_s \delta\varepsilon_s$ product characterizes the surface anisotropy. $e_i \delta\varepsilon_i$ characterizes the interface anisotropy.

The calculation leads to:

$$\frac{\delta r}{r} = \frac{r_{011} - r_{01\bar{1}}}{r} = \frac{-j \pi t_{0I}^2}{\lambda r_{0I}} \left[\delta_1 + \delta_2 + \delta_3 \right] \tag{5}$$

with:

$$\delta_1 = \frac{t_{I\;II}^2 (1 - r_{0I})}{t_{0I} \left[1 + \dfrac{r_{I\;II}}{r_{0I}} X \right] (1 + r_{0I}\; r_{I\;II}\; X)} \; X \left(\frac{e_i \delta\varepsilon_i}{N_I} \right) \tag{6}$$

$$\delta_2 = - \frac{r_{I\;II} (1 - r_{0I})^2}{r_{0I} \left[1 + \dfrac{r_{I\;II}}{r_{0I}} X \right] (1 + r_{0I}\; r_{I\;II}\; X)} \; X \left(\frac{e_s \delta\varepsilon_s}{N_0} \right) \tag{7}$$

$$\delta_3 = \left(\frac{e_s \delta\varepsilon_s}{N_0} \right) \tag{8}$$

$$X = e^{-2 \cdot j \cdot \beta} \qquad \text{with } \beta = \frac{2 \cdot \pi \cdot d \cdot N_I}{\lambda} \tag{9}$$

N_0 is the index of ambient (=1), N_I is the index of the epilayer, N_{II} that of the bulk, d is the thickness, λ the wavelength. The reflexion coefficient of the surface is r_{0I}, that of the interface is $r_{I\ II}$, and the transmission coefficients of surface and interface are respectively t_{0I} and $t_{I\ II}$. They can be expressed as function of the index using Fresnel relations[14].

The real part of $\delta r/r$ gives the evolution of the RDS signal. The signal is the sum of three terms. The first one (eq. 6) arises from buried interface anisotropy, has a damped oscillating behavior and vanishes for layer thicknesses large compared to the light penetration depth. The others are related to surface anisotropy. The second one (eq. 7) also exhibits a damped oscillation behavior, corresponding to interference effects within the layer, and the third one (eq. 8) is simply the surface anisotropy of the epitaxial material without buried interface effects.

Using these results, the different features of the RDS signal corresponding to the growth of an InAlAs on InP layer can be accounted for. During the first seconds following the InP to InAlAs commutation, the RDS signal is governed by changes in surface chemistry and physical properties of surface and interface. After that, the surface is expected to reach its steady-state level of anisotropy, and the interface properties should stay the same. The RDS signal is then governed by the damped oscillation behavior described by Eq. 5. When the epilayer is sufficiently thick, the light does not penetrate enough to be sensitive to interface effects, and the RDS signal is related simply to surface anisotropy.

Using Eq. 5 to account for the RDS record shown in fig. 8, it is possible to get some information on surface anisotropy, and to calculate buried interface anisotropy.

The refractive index of InP at 5,435 Å was taken from ref.[16] to be 3.66-j0.40. Because of the absence of available data on InAlAs, its refractive index was taken as equal to that of InP. This approximation is expected to be fairly accurate since both materials have nearly the same band gap, and 5,435 Å corresponds to an energy below the critical point E_1. The variation of the optical index of III-V materials is roughly comparable in the E_0 (gap) to E_1 range. The fact that there is no important difference between optical index of InP and InAlAs is further confirmed by looking at reflectivity during the growth. No oscillation with an amplitude larger than 5% is detected, and a simple interference calculus shows that the difference in optical index between the two materials should be in that case less than 0.05.

Using Eq. 5, one can relate the amplitude and phase of the buried interface anisotropy to the amplitude and phase of the RDS oscillations. One finds that experimental measurements account for a buried interface anisotropy of:

$$e_i \delta\varepsilon_i = (-2 + j\ 7)\text{Å} \tag{10}$$

The difference between surface anisotropy of InAlAs and InP is:

$$\text{Re} \{\exp(j1.8) [(e_s\delta\varepsilon_s)_{InAlAs} - (e_s\delta\varepsilon_s)_{InP}]\} = 4 \text{ Å} \tag{11}$$

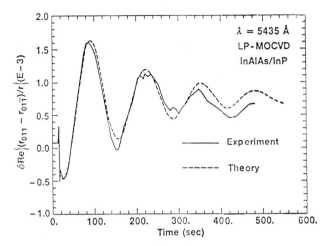

Figure 9. RDS record at λ = 5,435 Å during the growth of InAlAs on InP showing buried interface oscillations: comparison between the calculated curve, fitted with the appropriate value of the interface anisotropy, and the experimental curve. The InP to InAlAs commutation was performed at the 15th second.

The calculated curve using these values is also shown on Fig. 9, and fits well the measured curve. It is not possible to know the absolute value of surface anisotropy, since the zero of the RDS signal is not known.

Similar records were obtained at λ = 6,328 Å (Fig.10). With an optical index of 3.53-j0.3 for InP[16] and InAlAs, a growth rate of 330 Å/min, the buried interface anisotropy is found to be:

$$e_i \delta\varepsilon_i = (-5.6 + j\ 2.3)\ \text{Å} \tag{12}$$

The difference in surface anisotropy of InAlAs and InP is such as:

$$\text{Re}\ \{\exp(j1.75)\ [(e_s \delta\varepsilon_s)_{InAlAs} - (e_s \delta\varepsilon_s)_{InP}]\} = 10\ \text{Å} \tag{13}$$

Oscillations due to buried interfaces are also observed for GaInP/-GaAs (Fig. 11) and GaInAs/InP. However, there is one more difficulty in deducing the buried interface anisotropy of these samples. The reflectivity $r_{I\ II}$ at the interface is quite large, and unlike the case of InAlAs/InP, the oscillatory surface term given by Eq. 7 is not null or negligible. It means that the complex value of surface anisotropy has to be taken into account. This adds two more unknowns to be determined from experimental data, and would make the determination difficult - if not impossible -, and unreliable.

c) Reliability of measurements and model:

It is important to rule out some possible sources of error. Two parasitic contributions can lead to similar oscillations in the RDS spectrum. In the first place, parasitic coupling between the incoming flux and the signal would mix reflectance interference patterns with RDS oscillations. The detection system was checked not to produce such coupling, and the contribution of stray light to the incoming flux was measured and properly taken into account. Besides, for InAlAs on InP, the difference of refractive index is too small to create marked InAlAs interferences patterns on reflectivity, as already mentioned.

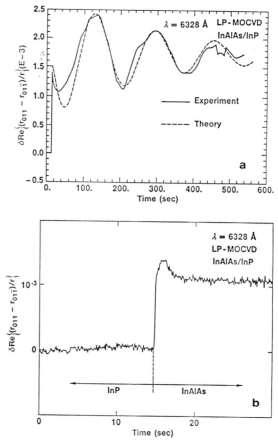

Figure 10 a) RDS record at λ = 6,328 Å during the growth of InAlAs on InP showing buried interface oscillations: comparison between the calculated curve, fitted with the appropriate value of the interface anisotropy, and the experimental curve. The InP to InAlAs commutation was performed at the 15th second.
b) Detail of the measured signal corresponding to the switching from InP to InAlAs.

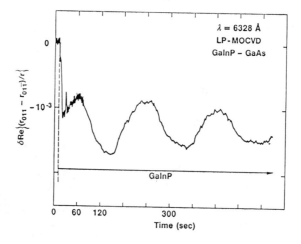

Figure 11. RDS record during the growth of GaInP on GaAs, at λ = 6,328 Å

In the second place, an ellipsometric contribution may arise from the non zero angle of incidence (about 2° on our setup), and would be expected to show this oscillatory behaviour. However, the ellipsometric contribution is small, proportional to the square of the residual angle of incidence, and can be cancelled by setting the plane of incidence at 45° from the ⟨011⟩ axe[24].

One may also be concerned with long term drift due to reactant deposition on the reactor tube.

All these parasitic contributions would affect the signal independently of sample orientation, whereas the sign of the RDS contribution changes when the substrate is rotated by 90°. In other words, $(r_{011} - r_{01\bar{1}})/r$ is measured for one orientation, changing the orientation by 90° corresponds to exchanging ⟨011⟩ and ⟨01$\bar{1}$⟩ direction, and the measure then yields $(r_{01\bar{1}} - r_{011})/r$. This suggests a very straightforward check: two similar growth are performed, with the substrate orientation changed by 90° between two runs. The two RDS signature should exhibit opposite variations (see fig. 4). This was checked to be the case for the materials presented here.

The model we used to take surface and interface effects into account, based on ref[26], is very simple. Del Sole[28] showed in the cases of surfaces that a more realistic and detailed treatment gave essentially the same results, provided that the quantity $(e_i \delta \varepsilon_i)$ is replaced by a more sophisticated expression of the average anisotropy of ε_i. For that reason, the present model is believed to be well adapted.

d) The origins of optical anisotropy, and its possible use to assess material quality:

There are several possible contributions to surface and interface anisotropy. Aspnes[30] mentioned the effect of many body screening, bulk spatial dispersion, physisorption, chemisorption, and microstructures on surface anisotropy. The doping level also has an influence[29]. For buried interface, the counterpart of physisorption is the nature of chemical bonding at interface. In the case of a perfectly abrupt InAlAs on InP interface, for example, the atomic layer sequence may be either ...In/P/(In,Al)/As..., or ...P/In/As/(In,Al)... . The nature and orientation of the chemical bonding differ between the two cases, and the associated RDS contribution is expected to be different. Thus, RDS signature of a heterostructure should be sensitive to interface chemistry and abruptness. It is also expected to be sensitive to micro-roughness, and in summary, it should be sensitive to interface quality. Further work will be carried to correlate RDS signatures with optical and electrical properties of interfaces. Quantitative evaluations of different contributions have been made[11,31,32,33] to account for ex situ measurements, but they are difficult, and require spectroscopic measurements for validation[34].

One important point highlighted by the present study is that the RDS signal corresponding to the growth of a heterostructure is related not only to the difference of optical anisotropy of material surfaces, but also to the effect of the buried interface. The signal observed during the first few seconds after the heterojunction is connected to changes in surface adsorbed layer and reconstruction, possibly in modification of electronic population at the interface, but also to buried interface effect. This effect is related not only to the real part of anisotropy, but also to its imaginary part. This allows the determination of both components of the buried interface anisotropy, whereas RDS is sensitive only to the real part of total anisotropy.

In contrast, there is no such effect for RDS studies of homoepitaxial growth.

In summary, oscillations in the RDS signal were found to be due to buried interface effects. Calculated curves agreed very well with the experimental data. From this information, it was possible to determine optical interface anisotropy of InAlAs on InP.

8 - STATE-OF-THE-ART OF LP-MOCVD GROWTH OF III-V SEMICONDUCTOR MATERIALS

Using this technique, Razeghi showed the feasibility of using various metalorganic sources of group III elements with hybride sources of group-V species for the following.

InP-InP System

1) The growth of high quality InP epilayer with carrier concentration as low as 3×10^{13} cm^{-3} and electron Hall mobility as high as 6,000 cm^2 V^{-1} s^{-1} at 300 K and 300,000 cm^2 V^{-1} s^{-1} at 50°K. Photoluminescence measurements at 2°K showed that this is the purest InP has yet been reported in the literature, with zero compensation ratio (Razeghi et al., 1987)[35].

These materials have been used for the fabrication of Gunn diode, and transferred already to the development and production department at THOMSON.

GaInAs-InP System

2) The first growth of high quality $Ga_{0.47}In_{0.53}As$-InP heterojunction, multiquantum wells (MQW) and superlattices (SL). The carrier concentration as low as 3×10^{14} cm^{-3} with electron Hall mobility as high as 13,000 cm^2 V^{-1} s^{-1} at 300 K and 700,000 cm^2 V^{-1} s^{-1} at 2°K, with photoluminescence linewidth at 2°K less than 2 meV have been measured (Razeghi et al., 1987)[36].

3) The first observation of two dimensional electron gas (2DEG) and Quantum Hall Effect with one, two, and three subband filled with 2DEG (Razeghi et al., 1982[37]; Razeghi et al., 1986[38]).

4) The first observation of two dimensional Hole gas with Hall mobility of 10,500 cm^2 V^{-1} s^{-1} at 2°K (Razeghi et al., 1986)[39].

5) The first observation of room temperature negative-resistance in double barrier resonant-tunnelling structure (Razeghi et al., 1987)[40].

6) The first observation of room temperature exciton in GaInAs-InP superlattices (Razeghi et al., 1986)[41].

7) The first growth of monoatomic layer (ALE) epitaxy of $(InAs)_n(GaAs)_n$-InP MQW (Razeghi et al., 1987)[42].

8) The first PIN photodetector with dark current density as low as 10^{-6} A cm^{-2}. The technology of PIN photodetector is already transferred to the production part (Razeghi et al., 1984[43]; Poulain et al., 1985[44]).

9) The first growth and fabrication of junction FET with excellent characteristic (Raulain et al., 1987)[45].

10) The first growth of very low loss optical waveguide (Delacourt et al., 1987)[46].

11) The first observation of Electron-spin resonance of the TDEG (M. Dobers et al., 1989)[47].

$Ga_x In_{1-x} As_y P_{1-y}$-InP System

12) The first growth of the entire compositional range of $Ga_x In_{1-x} As_y P_{1-y}$ lattice matched to InP with carrier concentration as low as 6×10^{14} cm^{-3} (for GaInAsP = λ = 1.3 μm), electron Hall mobility of 6,400 cm^2 V^{-1} s^{-1} at 300°K and 36,000 cm^2 V^{-1} s^{-1} at 300°K and 36,000 cm^2 V^{-1} s^{-1} at 77°K (Razeghi et al., 1984)[48].

13) The first observation of 2DEG in $Ga_{0.25} In_{0.75} As_{0.5} P_{0.5}$-InP heterostructures, MQW and SL (Razeghi et al., 1985)[49].

14) The first observation of QHE (Razeghi et al., 1987)[20], in a GaInAsP/InP heterojunction.

15) The first GaInAsP-InP, double heterostructure laser emitting at 1.3 μm with threshold current density as low as 430 A/cm^2 (Razeghi et al., 1983)[50].

16) First GaInAsP-InP laser emitting at 1.55 μm with threshold current density of 500 A/cm^2 (Razeghi et al., 1983)[51].

17) First BRS laser emitting at 1.3 μm and 1.5 μm with threshold current as low as 6 mA (Razeghi et al., 1985)[52].

18) First BRS-DFB laser emitting at 1.55 μm with threshold current of 10 mA (Razeghi et al., 1984[53]; Razeghi et al., 1987[54]).

19) First high power phase locked laser arrays emitting at 1.3 μm with output power of 300 mW without optical coating (Razeghi et al., 1986)[55].

20) First 1.5 μm GaInAsP-InP waveguide (Bourbin et al., 1987)[56].

21) First SCH GaInAsP-GaInAs-InP MQW laser (Razeghi, 1985[57]; Razeghi et al., 1986[58]).

Strained layer epitaxy

22) First growth of GaInAs-InP on GGG substrate (Razeghi et al., 1986)[59].

23) First growth of GaInAsP-InP on Si substrate (Razeghi et al., 1988)[60].

24) First growth of GaAs-GaInP on Si substrate (Razeghi et al., 1984[61], 1987[62]).

25) First growth of InP on GaAs, InAs substrates (Razeghi et al., 1987)[63].

26) First GaInAsP-InP laser emitting at 1.3 µm on GaAs substrate (Razeghi et al., 1984)[64].

27) First monolithic integration of a GaInAs/GaAs photoconductor with a GaAs F.E.T. for 1.3-1.55 µm wavelength applications (Razeghi et al., 1987)[65].

28) First monolithic integration of a Schottky photodiode and a F.E.T. using a $Ga_{0.49}In_{0.51}P/Ga_{0.47}In_{0.53}As$ strained material (Razeghi et al., 1987)[66].

29) First monolithic integration of planar monolithic integrated photoreceiver for 1.3-1.55 µm wavelength applications using GaInAs heteroepitaxies (Razeghi et al., 1986)[67].

30) First Room-temperature CW operation of a GaInAsP/InP (λ = 1.15 mm) light-emitting diode on silicon substrate (Razeghi et al., 1988)[68].

31) First GaInAsP-InP double-heterostructure laser emitting at 1.27 µm on a silicon substrate (Razeghi et al., 1988)[69].

32) First CW operation of a GaInAsP-InP laser on a silicon substrate, at room-temperature (Razeghi et al., 1988)[23].

33) High quantum efficiency GaInAs-InP photodetector on silicon substrate (Razeghi et al., 1988)[70].

$GaAs-Ga_{0.49}In_{0.51}P$ System

34) The growth of high quality GaAs with mobility of 335,000 cm^2 V^{-1} s^{-1} at 40°K which is the highest mobility have yet been reported independent of growth technique (Razeghi et al.)[71].

35) The growth of high quality $GaAs-Ga_{0.49}In_{0.51}P$, heterostructures, MQW and SL[62], with electron Hall mobility of 8,000 cm^2 V^{-1} s^{-1} at 300 K and 800,000 cm^2 V^{-1} s^{-1} at 2°K (Razeghi et al., 1987)[21].

36) The first observation of 2DEG in $GaAs-Ga_{0.49}In_{0.51}P$, heterostructures, MQW and SL, (Razeghi et al., 1986)[73].

All of these results are the first, most of them also not only the first, but also are the best independent of the growth technique.

9 - CONCLUSION

We developed a suitable material technology for the growth of high quality III-V semiconductors heterojunctions, quantum wells and superlattices for photonic and electronic devices of tomorrow.

RDS as an in situ technique, appears a very promising and useful for the improvement the quality of interfaces.

ACKNOWLEDGEMENTS

We would like to thank F. OMNES, S. KOCH, M. DEFOUR, P. MAUREL, P. BOVE, and J.C. GARCIA for their contributions to this work.

REFERENCES

1. D.E. Aspnes, Thin Solid Films 89, 249 (1982).
2. D.E. Aspnes, J. Phys. (Paris) 44, C10-3 (1983).
3. Y. Demay, P. Maurel and S. Gourier, ibid. 253.
4. B. Drevillon, Thin Solid Films 163, 157 (1988).
5. F. Hottier and R. Cadoret, J. Crystal Growth 56, 304 (1982).
6. R.W. Collins, Advance in disordered semiconductors, Vol. I, ed. H. Fritzsche (World Scientific, 1989), p. 1003.
7. B. Drevillon and R. Benferhat, J. Appl. Phys. 63, 5088 (1988).
8. R. Benferhat, B. Drevillon and P. Robin, Thin Solid Films 156, 295 (1988).
9. R. Benferhat and B. Drevillon, J. Non-Cryst. Solids 97&98, 835 (1987).
10. D.E. Aspnes, J.P. Harbison, A.A. Studna and L.T. Florez, Phys. Rev. Lett. 59, 1687 (1987).
    E. Colas, D.E. Aspnes, R. Bhat, A.A. Studna and V.G. Keramidas, J. Crystal Growth 94, 613 (1989).
11. D.E. Aspnes, J. Vac. Sci. Technol. B3, 1498 (1985).
12. D.E. Aspnes, E. Colas, A.A. Studna, R. Bhat, M.A. Koza, and V.G. Keramidas, Phys. Rev. Lett., 61, 2782 (1988).
13. M. Razeghi, the MOCVD challenge, Adam-Hilger (1989).
14. R.M.A. Azzam and N.M. Bashara, Ellipsometry and polarized light, North Holland ed. (Amsterdam 1977).
15. B. Drevillon, J. Perrin, R. Marbot, A. Violet and J.L. Dalby, Rev. Sci. Instrum. 53 (1982) 969.
16. O. Acher, E. Bigan and B. Drevillon, Rev. Sci. Instrum. 60 (1989) 65.
17. N. Blayo, B. Drevillon and J. Huc, Vacuum (in press).
18. O. Acher, F. Omnes, B. Drevillon and M. Razeghi, in E-MRS Strasbourg 89 Conference Proceeding, Symposium C (1989) (in press); and in Materials Science and Engineering, B (1989) (in press).
19. D.E. Aspnes, J.P. Harbison, A.A. Studna and L.T. Florez, Appl. Phys. Lett. 52, 957 (1988).
20. M. Razeghi, P. Maurel, M. Defour, F. Omnes, O. Acher, D. Tsui, H.P. Wei, Y. Guldner and J.P. Vieren, Appl. Phys. Lett. 51, 1821 (1987).
21. M. Razeghi, M. Defour, F. Omnes, M. Dobers, J.P. Vieren, and Y. Guldner, Appl. Phys. Lett. 55, 457 (1989).
22. M. Razeghi, Ph. Maurel, F. Omnes, and E. Thorngren, in Optical Properties of narrow gap low dimensional structures, Plenum Publishing Corporation (1987).
23. M. Razeghi, M. Defour, F. Omnes, Ph. Maurel, O. Acher, R. Blondeau, F. Brillouet, J.C.C. Fan, J. Salerno, Appl. Phys. Lett. 53 (1988), 2389.
24. O. Acher, F. Omnes, S. Koch and M. Razeghi, B. Drevillon, E. Bigan, to be published in "Revue Technique Thomson-CSF", Gauthier-Villars, Paris.
25. R.M.A. Azzam and N.M. Bashara, Ellipsometry and Polarized Light, North Holland, 1977, pp. 332-340.
26. J.D.E. McIntyre and D.E. Aspnes, Surf. Sci. 24, 417 (1971).
27. Handbook of Optical Constants of Solids, Palik E.D. Editor, Academic Press (1985).
28. R. Del Sole and Annabella Selloni, Solid State Comm. 9, 825 (1984).
29. S.E. Acosta-Ortiz and A. Lastras-Martinez, Solid State Comm. 64, 809 (1987).
30. D.E. Aspnes and A.A. Studna, Phys. Rev. Lett. 54, 1956 (1985).
31. A. Selloni, P. Marsella, and R. Del Sole, Phys. Rev., B33, 8885 (1986).

32. F. Manghi, E. Molinari, R. Del Sole and A. Selloni, Surf. Sci. 189/190, 1028 (1987).

33. F. Manghi, R. Del Sole, E. Molinari and A. Selloni, Surf. Sci. 211/212, 518 (1989).

34. V.L. Berkovits, I.V. Makarenko and V.I. Safarov, Solid State Comm. 56, 449 (1985).

35. M. Razeghi, Ph. Maurel, F. Omnes, M. Defour, G. Neu, and A. Kosacki, in Appl. Phys. Lett. 52 (1987), 117.

36. M. Razeghi, Ph. Maurel, F. Omnes. To be published in "Nato Advanced Research Workshop" on Properties of Impurity States in Semiconduction Superlattices, 1987.

37. Y. Guldner, J.P. Vicren, P. Voisin, M. Voos, M. Razeghi, Appl. Phys. Lett 40, 877 (1982).

38. M. Razeghi, J.P. Duchemin, J.P. Portal, L. Donovski, G. Remeni, R.J. Nicholas, A. Briggs, Appl. Phys. Lett. 48, 712 (1986).

39. M. Razeghi, Ph. Maurel, A. Tardella, L. Dinevski, D. Gauthier, J.C. Portal, J. Appl. Phys. 60, 2454 (1986).

40. M. Razeghi, A. Tardella, X.A. Davies, A.L. Long, M.J. Kelly, E. Britton, C. Boothroyd, W.A. Stobbes (1987). Electronic Letters, 23, 3, 1987.

41. M. Razeghi, J. Noyle, Ph. Maurel, F. Omnes, J.P. Pocholle, Appl. Phys. Lett. 49, 1110, 1986.

42. M. Razeghi, Ph. Maurel, F. Omnes, and J. Noyle, in Appl.Phys. Lett. 51, 2216 (1987).

43. M. Razeghi, Rev. Thomson-C.S.F., 16, 1 (1984).

44. P. Poulain, M. Razeghi, P. Hirtz, K. Kcymierski, B. Decremoux, 9th International Conf. on Semiconduction lasers, 7-11, August 1984, Rio, Brazil.

45. J.Y. Raulin, E. Vassilakis, M.A. Poisson, M. Razeghi, G. Colomer, Appl. Phys. Lett. 50, 535, 1987.

46. D. Delacourt, M. Papuchon, M.A. Poisson, M. Razeghi, G. Colomer, Electron. Lett., 23, 451, 1987.

47. M. Dobers, J.P. Vieren, Y. Guldner, P. Bove, F. Omnes, and M. Razeghi, Phys. Rev. B, 40 (1989), in press.

48. M. Razeghi, J.P. Duchemin, J. Cryst. Growth, 70, 145, 1984.

49. M. Razeghi, J.P. Duchemin, J.C. Portal, Appl. Phys. Lett., 46, 46, 1985.

50. M. Razeghi, P. Hirtz, R. Blondeau, J.P. Duchemin, Electron. Lett. 19, 481, 1983.

51. M. Razeghi, S. Hersee, R. Blondeau, P. Hirtz, J.P. Duchemin, Electron. Lett. 19, 336, 1983.

52. M. Razeghi, R. Blondeau, B. Decremoux, J.P. Duchemin, Appl. Phys. Lett. 46, 131, 1985.

53. M. Razeghi, R. Blondeau, J.C. Bouley, B. Decremoux, J.P. Duchemin, Proceeding of the 9th IEEE International laser conference, 1984.

54. M. Razeghi, F. Omnes, Ph. Maurel, R. Blondeau, M. Krakowski. SPIE Conference, Holland, 1987.

55. M. Razeghi, R. Blondeau, M. Krakowski, B. Decremoux, J.P. Duchemin, F. Lozes, M. Martinet, Bursoussan, Appl. Phys. Lett., 50, 230, 1987.

56. Y. Bourbin, A. Enard, R. Blondeau, M. Razeghi, M. Papuchon, B. Decremoux, to be published in Appl. Phys. Lett.

57. M. Razeghi, in Lightwave technology for communication, ed. W.T. Tsang (Academic, N.Y.C., 1985).

58. J. Nagle, S. Hersee, M. Razeghi, M. Krakowski, B. Decremoux, C. Weisbuch, Surf. Sci. 174, 148, 1986.

59. M. Razeghi, P.L. Meunier, Ph. Maurel, J. Appl. Phys. 59, 2261, 1986.

60. M. Razeghi, M. Defour, F. Omnes, Ph. Maurel, in Appl. Phys. Lett. 52, 209 (1988).

61. M. Razeghi, MRT Report, private communication (1984).

62. M. Razeghi, Ph. Maurel, F. Omnes, M. Defour, C. Boothroyd, W.M. Stobbs, and M. Kelly, in J. Appl. Phys. 63, 4511 (1988).

63. M. Razeghi, Ph. Maurel, F. Omnes, E. Vassilakis-Thorngren, from Optical Properties of Narrow Gap Low Dimensional Structures, ed. C.M. Sottomayor Plenum publishing Corporation, 1987.

64. M. Razeghi, R. Blondeau, J.P. Duchemin, Inst. Phys. Conf. Sci. n°74, chap. 9, p. 679.

65. M. Razeghi, J. Ramdami, P. Legry, J.P. Vilcot, D. Decoster, to be published in the Proceeding of the 1987 GaAs and related compounds conference, Creta.

66. M. Razeghi, A. Hosseini, J.P. Vilcot, D. Decoster, to be published in the Proceeding of the 1987 GaAs and related compounds Conference, Creta.

67. M. Razeghi, J. Ramdani, H. Verriele, D. Decoster, M. Constant, J. Vanbremeersh, Appl. Phys. Lett. 49, 215, 1986.

68. M. Razeghi, R. Blondeau, M. Defour, F. Omnes, P. Maurel, F. Brillouet, Appl. Phys. Lett. 53, 854 (1988).

69. M. Razeghi, M. Defour, F. Omnes, Ph. Maurel, J. Chazelas, and F. Brillouet, Appl. Phys. Lett. 53, 725 (1988).

70. M. Razeghi, F. Omnes, R. Blondeau, Ph. Maurel, M. Defour and O. Acher, J. Appl. Phys. 65, 4066 (1989).

71. M. Razeghi et al., to be published in Appl. Phys. Lett. (1989).

72. M. Razeghi, Ph. Maurel, F. Omnes, S. Ben Armor, L. Dnowski, J.C. Portal, Appl. Phys. Lett. Vol. 48, 19, 1267 (1986).

OPTICAL PROPERTIES OF Ge-Si SUPERLATTICES

T.P. PEARSALL
Dept. of Electrical Engineering
University of Washington
Seattle, Washington
98195, U.S.A.

ABSTRACT

The optical properties of strained Ge-Si heterostructures are reviewed with particular emphasis on both strain and ordering in determining the energy band structure. electro-reflectance and Raman Ge-Si superlattices with periodicities of a few atomic monolayers are now known to have properties quite distinct from those of Ge-Si alloys. Measurements of absorption, electro-reflectance and photoluminescence have produced complementary supporting evidence for the existence of a direct band gap in certain Ge-Si superlattices.

1.   Optical Properties of Strained Ge-Si Alloys

A.   Absorption and Photoconductivity

The strain tensor imposed by growth of Ge_xSi_{1-x} alloys on Si (001) substrates has both hydrostatic and uniaxial components. The hydrostatic stress is compressive and serves to increase the direct energy gaps. In addition, hydrostatic compressive strain will raise in energy all the levels in the alloy relative to the unstrained Si regions that surround the alloy, thus modifying the heterojunction band offset. The effect of hydrostatic pressure on the band structure of $Ge_x Si_{1-x}$ alloys, was first experimentally determined by Paul and Warschauer (1959).

Electronic Properties of Multilayers and Low-Dimensional Semiconductors Structures
Edited by J. M. Chamberlain *et al.*, Plenum Press, New York, 1990

The uniaxial strain component will lift the degeneracy of the valence band at $\mathbf{k}=0$. This splitting can be calculated quantitatively from standard perturbation theory using the strain deformation potentials for Si and Ge, appropriately interpolated for the alloy composition in question, (Pearsall, *et al.*, 1986a)

$$\delta\left[E_0(1)\right] = E_H + E_S \tag{2}$$

$$\delta\left[E_0(2)\right] = E_H + \frac{\Delta_0}{2} - \frac{E_0}{2} - \frac{1}{2}(\Delta_0^2 + 2\Delta_0 E_s + 9E_s^2)^{\frac{1}{2}} \tag{3}$$

$$\delta\left[E_0 + \Delta_0\right] = E_H + \frac{\Delta_0}{2} - \frac{E_s}{2} + \frac{1}{2}(\Delta_0^2 + 2\Delta_0 E_s + 9E_s^2)^{\frac{1}{2}} \tag{4}$$

where E_H = hydrostatic strain = $D_H(2 - K)\varepsilon$ \hfill (5)

E_S = uniaxial strain = $D_U(1 + K)\varepsilon$ \hfill (6)

D_H and D_U are the appropriate deformation potentials, and:

$$\varepsilon_{ij} = \begin{pmatrix} \varepsilon & 0 & 0 \\ 0 & \varepsilon & 0 \\ 0 & 0 & -K\varepsilon \end{pmatrix} \tag{7}$$

for growth on (001)-oriented Si substrates. An excellent and comprehensive review of strain effects in semiconductors has been prepared, (Pollak, 1990) and the reader is referred to this article for a more complete treatment including the most recent compilation of experimentally measured deformation potentials.

The splitting of the valence band at the zone center is the direct result of the strain induced lowering of the symmetry from cubic to tetragonal structure. The effect of the uniaxial strain on the conduction band minima however depends on the relative orientation of the uniaxial strain direction relative to that of the energy minima. In the case of Ge grown on (001) Si, the uniaxial strain has no net effect on the Ge <111> minima because the component of the [001] strain is the same along all four <111> axes. However, the effect on <100> conduction band minima is quite different. In this case the [001] direction will be under tensile strain while the [100] and [010] directions will be under compressive strain. Since the signs of the strains are opposite, these energy bands will move in opposite directions. Since some minima must move down in energy, (in this case, the [100] and [010] minima), the energy gap will be reduced. Note that if the sign of the strain is changed by growing on a substrate with a large lattice constant than that of the alloy, then the [001] minimum, now in compressive strain, will be lowered in energy. In either case, the effect of uniaxial strain oriented along a <100> direction is to lower the bandgap. This is a direct

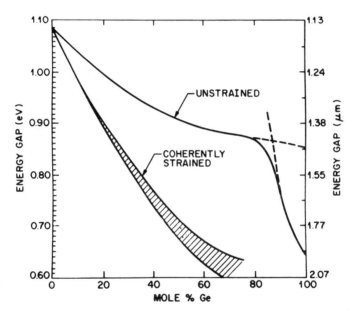

Fig. 1 Upper Curve: Indirect bandgap of Ge_xSi_{1-x} unstrained bulk alloys (RCA samples) at 295K. The near zero compositional variation of the <100> band edges in group IV (and group III-V) semiconductors is responsible for the small energy gap variation for Ge concentrations less that 0.8.

Fig. 1 Lower Curve: Indirect bandgap of Ge_xSi_{1-x} *strained* alloys grown on (001) Si. The uniaxial component of the strain tensor is responsible for the lowering of the bandgap of these structures relative to that of unstrained alloys. (reproduced from People (1985) with permission of the author.)

reflection of the indirect bandgap nature of Ge_xSi_{1-x} alloys.

The strain effect on the bandgap was calculated, (People, 1985) and confirmed experimentally in a series of measurements of photoconductive experiments, (Pearsall, *et al.*, 1986b; Temkin, *et al.*, 1986; Lang, *et al.*, 1985). The calculation of People is shown in Fig. 1, lower curve. Here it can be seen that the bandgap of $Ge_{0.5}Si_{0.5}$ alloy is lowered from about 1.0eV to about 0.8eV by the 2% strain imposed by growth on a Si(001) substrate. It can be seen in Fig. 1 that the bandgap variation with composition of unstrained Ge-Si alloys is rather uninteresting, whereas the compositional bandgap variation for strained layer alloys opens the possibility of photodetector applications in the technologically important 1.3μm to 1.5μm wavelength region.

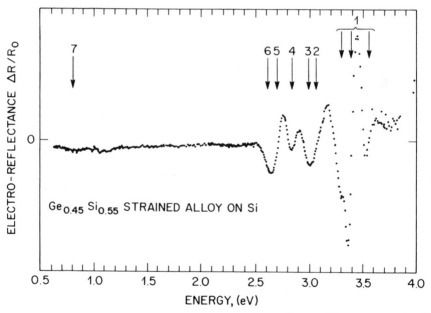

Fig. 2 Electro-reflectance measurement of a $Ge_{0.45}Si_{0.55}$ strained layer alloy. Oscillations in the spectrum result from direct optical transitions. The transitions from the Si buffer, labeled as 1, is the dominant feature in the spectrum. Since this is the lowest transition energy in Si, all the remaining transitions come from the alloy.

B.   Raman scattering.

The optical vibrations of Ge_xSi_{1-x} alloys consist of Si-Si, Ge-Ge, and Ge-Si modes. In the case of strained alloys, these three modes persist, but their characteristic energy depends on the level of strain as well, (Cerdeira, *et al.*, (1984). The additional Raman shift from strain is directly related to the change in bond length produced by the strain. Thus a 1% strain produces an additional 1% shift in Raman frequency: upwards for compressive strain, downwards for tensile strain. In principle the Raman effect could be used to measure strain. In order to determine the strain accurately in a $Ge_{0.5}Si_{0.5}$ alloy the uncertainty in the alloy composition must be less than 3 atomic percent. This level of accuracy may be difficult to achieve, particularly if compositional grading is present. As a result, the Raman effect is primarily useful as a rapid and non-destructive means of measuring alloy composition with an accuracy of about 10%. However, Abstreiter, *et al.*(1985, 1986) and Brugger, *et al.* (1986) have performed an elegant series of experiments using Raman scattering to reveal carrier confinement in heterostructures, interface quality, and folding of phonon modes by the superlattice periodicity.

Under controlled conditions, the Raman effect can be used to make comparative measurements of interfacial abruptness in Ge-Si superlattices, (Cerdeira, *et al.*,1989; Menendez, *et al.*,1988; Iyer, *et al.*, 1989). In a structure consisting of pure Ge and pure Si layers, Ge-Si vibrations occur only at the interfaces. The amplitude of Ge-Si vibrations depends on the scattering volume containing Ge-Si bonds. This is determined by the number of interfaces and by the interfacial roughness.

C.   Electro-reflectance Spectroscopy

Ge_xSi_{1-x} strained-layer structures with an alloy content of x = 0.5 have a critical layer thickness less than 200A when grown on Si or Ge substrates. Measurement of interband transition energies requires a technique of high sensitivity because the sample volume is necessarily small. Electro-reflectance is an ideal technique for characterization of Ge_xSi_{1-x} strained layer structures for two reasons. The first is sensitivity. Since it is a modulation technique in which only changes in the reflectance are measured, electro-reflectance has a sensitivity 4 to 5 orders of magnitude greater than the usual photoluminescence or absorption experiment. The second important feature of electro-reflectance is that it gives the entire spectrum of all band-to-band transitions that occur at

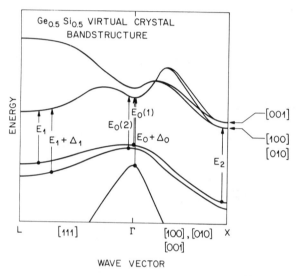

Fig. 3 Electronic bandstructure for $Ge_{0.5}Si_{0.5}$ alloy grown in the strained-layer mode on a Si(001) substrate. Note that the strain tensor splits the valence band at $\mathbf{k}=0$ and lifts the degeneracy of the <100> conduction band minima which form the band edge.

high symmetry points in the Brillouin zone. Photoluminescence and excitation luminescence spectroscopy give less information about transitions above the fundamental band edge.

The electro-reflectance spectrum of $Ge_{0.45}Si_{0.55}$ alloy grown on (001) is shown in Fig. 2. The spectrum is rich in structure below the 3.4eV E_1 transition in Si. Since the lowest energy transition in the Si substrate occurs at 3.4eV, the lower energy structure must

originate in the superlattice. Analysis of the spectrum proceeds by
calculation of strain induced perturbations of the conduction and
valence bands. Two-dimensional quantum confinement effects become
important only for strained-layer thicknesses less than 50Å, so that
for most alloy compositions (x<0.6) strain is the dominant
perturbation of the optical transition energies. Our results show
that linear elastic relations and the linear strain Hamiltonian
(Equations 1-6) account completely for observed spectra under
conditions quite close to the elastic strain limit. For comparison,

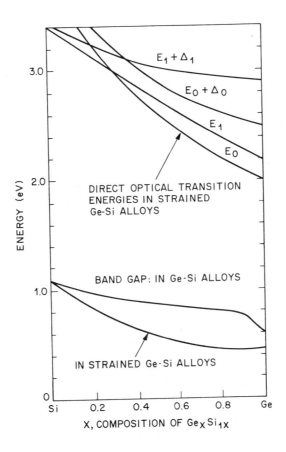

Fig. 4 Optical transition energies for Ge_xSi_{1-x} strained alloys grown on
(001) Si. Transition energies for the same alloy composition will vary if
the substrate orientation is changed because the strain tensor will no
longer by the same. These energies should be compared to those shown in
Fig. 2 for unstrained alloys. Unlike the behavior of the indirect gap,
the direct transition energies are all raised by the strain imposed by the
Si (001) substrate.

the electronic band structure of $Ge_{0.5}Si_{0.5}$ strained alloy in the virtual crystal approximation is shown in Fig. 3 with the transitions measured in electro-reflectance labeled to show their origin in the Brillouin zone.

The net hydrostatic strain imposed by growth on Si increases the energy of all direct transitions in the Brillouin zone. In Fig. 4 the variation of transition energies with composition is shown for the case of Ge_xSi_{1-x} alloys grown strained on (001) Si. Quantum confinement effects are not included in Fig.4 in order to isolate the effect of strain alone. These results show that the effect of strain is seen mainly on the E_0 and $E_0 + \Delta_0$ transitions, which are the lowest energy features over most of the compositional range. Note that the lowest energy transition that is seen in the electro-reflectance spectrum of a $Ge_{0.5}Si_{0.5}$ alloy occurs at about 2.6 eV. This is in contrast to the transitions measured in Ge-Si superlattices composed of alternating 4 monolayers of Ge and Si. Although the average composition is the same as that for the $Ge_{0.5}Si_{0.5}$ alloy, strong optical transitions are measured near 1.0 eV, at which Fig.4 indicates no transitions should occur. This is the result of ordering and will be discussed in the next section.

2.Ge-Si Strained-Layer Superlattices with Ordering on the Atomic Scale.

A.   Changing the Symmetry of the Unit Cell.

The effect of ordering on the electronic band structure had been treated by several groups, (Stroud and Ehrenreich, 1970; Gnutzmann and Clausecker, 1974; and Moriarty and Krishnamurthy, 1983) in advance of any experimental attempts to measure such effects. In 1986, Bevk, et al., published a seminal paper demonstrating that extended layers of Ge and Si, only a few atomic layers in thickness, could be grown epitaxially by M.B.E. With these structures it became possible to create Si and Ge superlattices with a tetragonal unit cell instead of the cubic diamond structure, (Bevk, et al., 1987a).

382

An important difference between these superlattices and previous work on Ge_xSi_{1-x}/Si strained layer superlattices is that the individual layers are composed of elemental, as opposed to alloy material. The inherent uncertainty in the position of the interface between an alloy and elemental material composed of one of the alloy constituents can be many atomic monolayers, depending on the alloy composition. For example, at the interface between Si and $Ge_{0.1}Si_{0.9}$, it is impossible to tell whether a particular Si atom is

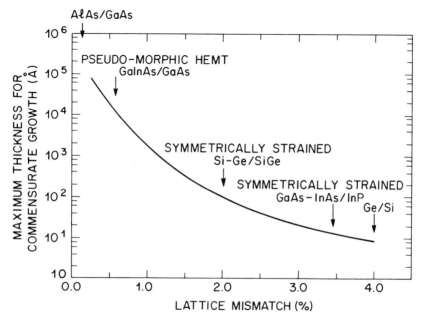

Fig. 5 Critical layer thickness for Ge-Si ordered superlattices as well as other strained layer systems. The critical layer thickness ordered superlattices grown on Si or Ge is small, about 10Å or 5 monolayers.

part of the alloy, or of the elemental region, plus or minus about 20 monolayers. As the concentration of Ge in the alloy is increased, the interface can be defined with greater precision. Finally for the direct growth of Ge on Si, the interface can be defined experimentally to about ± 1 monolayer. For this reason alone, studies of ordered strained layer superlattices are restricted to the growth of Ge on Si or vice-versa.

B.   Wavefunction Engineering

Strained epitaxial layers of Ge grown on Si are subject to a 4% strain and the critical layer thickness, as shown in Fig. 5, is about five monolayers. Extended alternating structures of Si and Ge of equal number of monolayers can be grown on a Si substrate to a strain-limited thickness of about 50Å. These superlattices have an average composition of 50% Si and 50% Ge and their electronic structure can be compared to that of a strained alloy whose electro-reflectance spectrum was shown in Fig.2.

Experimental measurements of a series of such Ge-Si strained layer ordered superlattices in fact showed clearly that these structures possess and electronic energy level spectrum quite distinct from that of Ge-Si alloys. In Fig. 6 we show a series of measurements,(Bevk, *et al.*, 1987; Pearsall, *et al.*, 1989)) that make this comparison clear. Three spectra are shown in this figure. The Ge-Si (1:1) structure is identical to the zinc-blende structure of GaAs. It possesses cubic symmetry and has an energy spectrum similar to, and probably indistinguishable from that of a random alloy, (Stroud and Ehrenreich, 1970). The (2:2) and (4:4) spectra are much more complex. The presence of additional transition in the 0.5eV – 1.0eV portion of the spectrum in the case of the (4:4) structure has no counterpart in the alloy spectra at all. These spectra show that atomic ordering creates new energy levels in superlattice structures rather than modifying existing levels. These experiments therefore differentiate the field of atomically ordered superlattices from the previous work on semiconductor superlattices where quantum confinement effects could be used to manipulate energy levels in the confined "well" region. The new, structurally-induced energy levels reflect the fact that the non-cubic symmetry of the unit cell selects a different basis set of electronic wave functions than those present in cubic materials. The growth and characterization of these samples introduced the field of *wavefunction engineering*, a domain where the crystal-grown can tailor to some extent the fundamental nature of the semiconductor electronic energy spectrum by appropriately choosing the symmetry of the unit cell.

C. Theoretical Studies

Following the experimental observation of structurally-induced states in Ge-Si superlattices, (Pearsall, *et al.*, 1987) numerous theoretical studies were made that have helped to build a framework for understanding of the relationship between unit cell symmetry and electronic band structure. Some of the important advances that have simplified this understanding are described below. People and Jackson (1987) suggested that the basic features of wavefunction

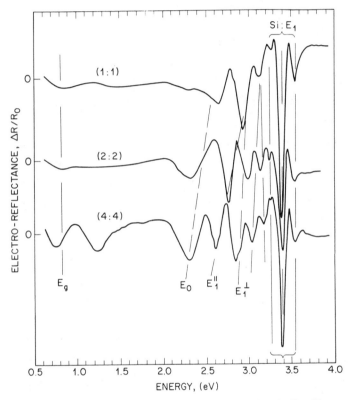

Fig. 6 Electro-reflectance spectra of three ordered Ge-Si superlattices (1:1), (2:2), and (4:4). All three superlattices have the same average composition (i.e. 50% of Si, 50% of Ge), yet their electro-reflectance spectra are quite distinct. The (1:1) "superlattice" is identical to the zinc-blende structure of GaAs. Since the compositional modulation takes place on a scale less than the Si or Ge unit cells the electronic structure is similar to that for the bulk random alloy shown in Fig. 7. The (2:2) and (4:4) structures are different, displaying new optical transitions below the 2.6eV lower limit for strained alloys (see Fig. 4).

engineering could be understood by folding of the Brillouin zone. Their contention was borne out by the full bandstructure calculation of Satpathy, Martin and van de Walle (1988). Froyen, Wood and Zunger, (1987, 1988) performed calculations for a series of superlattices showing the effects of layer thickness and unit cell symmetry on the energy levels. Their work pointed out the fundamental indirect bandgap character of Ge-Si superlattices grown on Si and suggested that superlattices grown on substrates with a larger lattice constant than that of Si could have a direct bandgap.

The possibility that wavefunction engineering could be used to create a direct bandgap semiconductor out of two indirect bandgap materials, Si and Ge is naturally a fascinating and controversial proposition. Conditions for achieving direct bandgap behavior have been studied by Satpathy et $al.$(1988), Friedel (1989), Hybertsen and Schlüter (1988), and by Gell (1988, 1989). This work has shown that the band alignment of Ge-Si superlattices grown on Ge or on Si substrates has a basic type-II nature. Together these works extend in a quantitative way the observations of Froyen, et $al.$ (1987, 1988) on observing direct bandgap behavior for superlattices grown on substrates with a lattice constant larger than that of Si. Gell, (1989) has recently shown, however, that it will be possible to create a direct bandgap Ge-Si superlattice of Ge and Si with a type-I bandgap alignment with the buffer provided that its lattice constant, intermediate between that of Si and Ge, is chosen carefully, in order to achieve the desired heterojunction band offset in both the conduction and valence bands. To understand this result, recall that the superlattice period is so short that carriers are delocalized in the superlattice region. A single quantum well is formed by the superlattice region and the buffer layer on either side of the superlattice. The band alignment between the superlattice and buffer depends on the buffer composition. A type-I band alignment is achieved by a balance between the amount of compressive uniaxial strain on the pure Si regions tending to lower the superlattice conduction band energy relative to that of the buffer, and the amount of compressive hydrostatic strain on the pure Ge regions, tending to raise the valence band edge relative to that of the buffer. This balance leading to type-I alignment occurs for alloy buffers of composition near $Ge_{0.8}Si_{0.2}$.

The magnitude of the optical matrix element between the valence and conduction bands in these structures remains a subject of contention. The transition probability is given by:

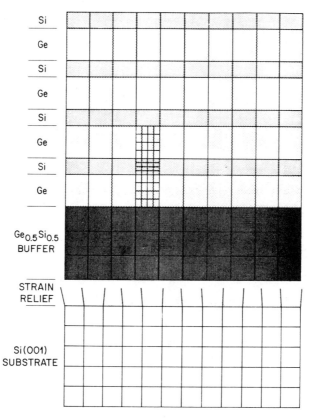

Si

Ge

Si

Ge

Si

Ge

Si

Ge

Ge$_{0.5}$Si$_{0.5}$
BUFFER

STRAIN
RELIEF

Si(001)
SUBSTRATE

Fig. 7 The principle of strain symmetrization is outlined in this figure. Growth of an alloy buffer layer beyond the critical thickness relieves strain through formation of dislocations. The subsequent growth of a Ge-Si superlattice results in Si regions being under expansion and Ge regions being under compression. The net strain per period can be kept near zero, permitting growth of thick superlattice regions (in principle without limit) without introducing additional dislocations.

$$T_{cv} = |M_{cv}|^2 = A \ [\langle \Psi_c|P|\Psi_v \rangle \ \delta(\mathbf{k}_c - \mathbf{k}_v) \ \delta(E - E_g) \ ]^2 \ , \quad (8)$$

where Ψ_c and Ψ_v are the conduction and valence band wavefunctions.

Changing the symmetry of the wavefunctions can be used to achieve conservation of momentum ($\mathbf{k}_c = \mathbf{k}_v = 0$) at the minimum of energy difference between the conduction band and valence band. While this will increase the transition probability, it does nothing in principle to change the interband matrix element which will depend on the atomic character of the two wavefunctions involved. Wong *et al.* (1988) have studied the effects of superlattice structure on this matrix element showing that strain-induced mixing of the conduction and valence band states with other electronic states at Γ can lead to greatly enhanced interband matrix elements for zone-folded states by introducing substantial amount of s-character in the conduction band state that would exhibit mostly p-character before folding. This work concludes that there is good reason to expect that particular Ge-Si superlattice structures, grown on substrates that strain Si in compression along the superlattice growth direction will have *bona fide* direct bandgap transitions with substantial transition probabilities.

D.   Direct Bandgap Behavior: Experimental Evidence

The average valence band edge energy of Ge is offset about 500 meV above that of Si, (van de Walle and Martin, 1986). The offset is so large that the Si-derived <100> conduction band minima are the lowest energy levels in the bandgap. To create the conditions for direct bandgap behavior, the [001] minimum which is folded by the (001) superlattice must lie lower in energy than the [100] and [010] energy band minima. Since these levels are degenerate in unstrained bulk materials, it is the presence of strain in Si along the [001] direction that will lift this degeneracy. In order for the [001] minimum to remain lowest in energy, the strain along this axis must be compressive relative to that along the [100] and [010] directions.

The necessary strain tensor is most easily achieved in strained-layer epitaxy by growing the Ge-Si superlattice on a (001)-oriented substrate of Ge, GaAs, or some substrate with a lattice constant larger than that of Si. Using the concept of strain symmetrization, (Kasper, *et al.*, 1988), the substrate can be a thick, strain-relieved buffer of Ge_xSi_{1-x} alloy grown on (001) Si. In this case, the critical thickness of the Ge-Si superlattice is in

principle unlimited because the strain tensor in Si is opposed by a
compensating strain tensor in Ge. The cumulative strain after each
period can be reduced to negligible levels, allowing the same
structure to be repeated. This effect is diagrammed in Fig. 7.

Having established the appropriate strain tensor for direct gap
behavior, it is next necessary to choose the superlattice period so
that the minimum conduction band energy will be folded back to the
center of the Brillouin zone. In bulk Si, this minimum occurs in a
region approximately 4/5 the distance from the zone center to the
<100> zone boundaries. Hence it is best to choose a period that will
subdivide the Brillouin zone by five. Since the basis of the Si unit
cell is two atoms, a superlattice periodicity of 10 atomic layers is
the shortest that will achieve the required folding. Abstreiter *et
al.,* (1989) have shown that such superlattices show strong

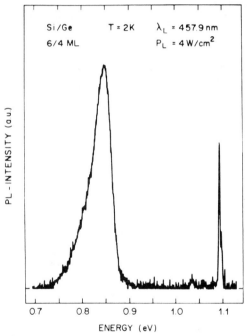

Fig. 8 Ge-Si superlattices with a 10 monolayer period along the (001)
direction will cause the lowest energy region of the conduction band to be
folded back to the center of the Brillouin zone. First principles
theoretical calculations predict these structures to have a direct bandgap
when they are grown strained on a substrate larger than that of Si. In
Abstreiter, *et al.*(1989) have shown that such materials exhibit strong
photoluminescence at the expected direct bandgap semiconductors.
(Reproduced from Abstreiter, *et al.*(1989) with permission of the authors.)

photoluminescence, (see Fig. 8). These samples consist of a Ge_4Si_6 superlattices grown on a Ge_xSi_{1-x} buffer. By using the principle of strain symmetrization, the superlattice thickness (~0.5μm) can be made comparable to the recombination length (~1μm), allowing photoluminescence to occur before the free carriers can diffuse out of the material. In the case of 10-atom period Ge-Si superlattices grown on pure Ge, the critical thickness is about 50Å. Because of the type-II band alignment of the superlattice with a pure Ge buffer, electrons and holes are separated spatially before they can recombine: with electrons confined to the superlattice and holes to the Ge buffer regions. In such samples we have measured no detectable photoluminescence. This result signals the importance of using either strain symmetrization or a type-I buffer composition in obtaining luminescence from Ge-Si superlattices. The recombination energy of 0.85eV measured by Abstreiter et al. (1989) is close to the bandgap expected for the Ge_4Si_6 superlattice structure. Since the substrate was Si (E_g = 1.2eV at 4.2K), the luminescence is not intrinsic to the substrate. It is unlikely that the luminescence would originate in the buffer because it is highly dislocated material. Furthermore the luminescence is strongly [001]-polarized supporting the idea that it originates from a superlattice-induced bandgap state that has $\mathbf{p}_z$-polarization symmetry.

The optical properties of Ge-Si superlattices grown on (001) Ge can be measured by electro-reflectance even though they are not well-suited to characterization by photoluminescence. In Fig. 9a we show these results for a Ge_6Si_4 superlattice at 40K grown on (001) Ge. For comparison, in Fig. 9b we show the electro-reflectance spectrum of the Ge substrate on which the superlattice was grown. Both spectra were taken under the same electric field condition: Vac = 200mV p-p, Vdc = -200mV. Each spectrum shows a sharp feature near 0.9 eV which appears to show splitting. This is the E_0 transition of Ge (0.89 eV) in both cases. The apparent splitting is an exciton effect of the Schottky barrier used to induce modulation, (Silberstein and Pollak, 1980). In Fig. 9a additional structure is apparent at 0.95eV which has its origin in the superlattice. At higher energies near 1.2eV there is an additional optical feature associated with the split off valence band. In Fig. 9a there are two features: one related to the substrate and one from the superlattice

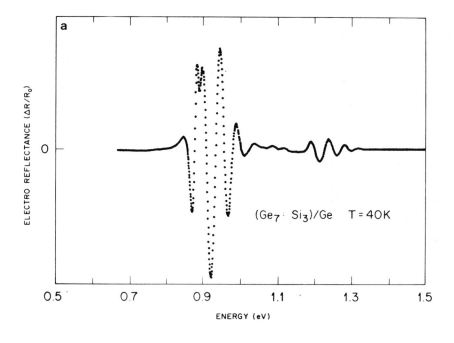

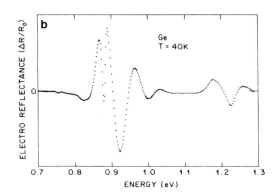

Fig. 9a. Electro-reflectance spectrum of a Ge-Si (7:3) superlattice grown on (001) Ge. Temperature of the measurement is 40k. Because the sample is grown on a Ge substrate, components of the electro-reflectance at 0.89 eV and 1.21 eV can be attributed to bulk Ge. The superlattice optical transitions appear at slightly higher energies: 0.95eV and 1.24eV. Because of the geometry of the experiments, only transitions with $p_{x,y}$ symmetry can be seen. The bandgap transition which has p_z symmetry cannot be excited in our experimental geometry.

Fig. 9b Electro-reflectance spectrum of the Ge substrate used in the growth of the superlattice shown in Fig. 15. Comparison of Fig. 15a and Fig. 15b makes it clear that additional optical transitions are present in the superlattice sample.

at 1.24eV. In Fig. 9b, only the substrate feature near 1.2eV is present. Because of the normal incidence geometry of the incident light used in electro-reflectance it is not possible to excite the transition at the superlattice bandgap at 0.78eV because it has p_z polarization symmetry.

E.   Modelling the Superlattice using the Kronig-Penney Method

Satpathy, et al. (1988), have shown that the Kronig-Penney model gives accurate band-to-band transition energies for these superlattices using the bulk properties of Si and Ge. To set up the calculation, we use the band offsets calculated by van de Walle and Martin (1986) for unstrained Si and Ge. Applying elasticity theory as outlined in Pearsall, et al., (1986a), the splitting of the valence band by the strain can be calculated. Following the work of People (1985), the splitting of the conduction band by the strain can be computed. These results are sufficient to model the conduction and valence bands of the Ge-Si superlattice as a series of square-well potentials. These results agree within a few meV with those obtained by Gell (1988). Using these values for the periodic potential and the effective mass values for bulk Si and Ge, the Kronig-Penney dispersion equation can be used to calculate the miniband E-$\mathbf{k}$ characteristic for each of the three valence bands and the conduction band.

$$\cos \mathbf{q}d = \cos(\mathbf{k_w}l_w) \cdot \cos h(\mathbf{k_b}l_b) + A \cdot \sin(\mathbf{k_w}l_w) \cdot \sin h(\mathbf{k_b}l_b), \quad (9)$$

$$\text{where } A = F(m_b{}^*\mathbf{k_w}, m_w{}^*\mathbf{k_b}), \quad \text{and} \quad (10)$$

$$l_w + l_b = d \text{ the superlattice period} \quad (11)$$

In equations (9-11), w refers to well and b refers to barrier regions. In a Ge-Si superlattice grown on (001) Ge, Si forms the well regions in the conduction band calculation. However, in the valence band, the wells are formed by the Ge -- a result of the type-II band alignment between Si and Ge.

The result of this calculation is diagrammed in Fig. 10. In this diagram the energy levels of Ge and strained Si are shown as solid lines. The dashed lines show the Kronig-Penney solutions at **q** = 0. The bold-face numbers give the calculated interband transition energies for the three allowed valence to conduction band optical transitions.

Ge-Si (7:3) ON (001)Ge

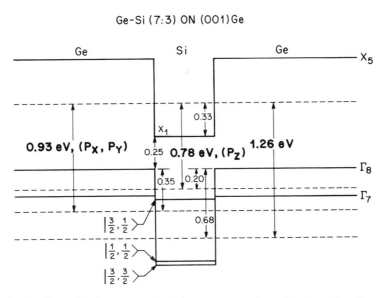

Fig. 10 Analysis of the superlattice energy levels by the Kronig-Penney model. In this diagram the strain-split energy levels of the Si regions, and the unstrained levels of Ge are shown. The solution to the Kronig-Penney equations for conduction and valence band are shown as dashed lines. The numbers in heavy-type are the transition energies in the superlattice between valence and conduction bands. The close correspondence between these calculated energy levels and those measured in experiment supports the notion that the Ge-Si (7:3) superlattice is a direct bandgap material.

The comparison between the Kronig Penney calculations and experiment is given in Table 1 for two Ge-Si superlattices grown on (001) Ge. In each case the agreement between theory and experiment is quite good, supporting the notion that the transitions measured in experiment originate from direct optical transitions across the superlattice bandgap. While the Kronig-Penney method is a simple and accurate scheme for calculating the transition energies, it gives no information about the matrix element for these transitions. If no

393

TABLE 1

Optical transition energy parameters deduced from 40K lineshape analysis

Transition Number	Energy (eV)	Amplitude ($\Delta R/R_0$)	Linewidth (eV)	Krönig-Penney Energy (eV)	Identification
A. Bulk Ge					
1.	0.88	7.1×10^{-4}	0.04		Ge: E_0
2.	1.19	1.3×10^{-4}	0.04		Ge: $E_0 + \Delta_0$
B. Ge_6Si_4					
1.	Not observed	--	--	0.745	S.L.: $\mathbf{p_z}$
2.	0.88	1.6×10^{-4}	0.017	--	Ge: E_0
3.	0.96	8×10^{-5}	0.015	0.94	S.L.: $\mathbf{p_x, p_y}$
4.	1.18	9.4×10^{-6}	0.024	--	Ge: $E_0 + \Delta_0$
5.	1.22	8×10^{-6}	0.020	1.24	S.L.: split-off
C. Ge_7Si_3/Ge (001)					
1.	Not observed	--	--	0.780	S.L.: $\mathbf{p_z}$
2.	0.89	6×10^{-6}	0.017	--	Ge: E_0
3.	0.93	6×10^{-6}	0.014	0.93	S.L.: $\mathbf{p_x, p_y}$
4.	1.20	5×10^{-7}	0.016	--	Ge: $E_0 + \Delta_0$
5.	1.25	3×10^{-7}	0.013	1.24	S.L.: split-off

mixing of the superlattice-induced conduction band state occurs, then the transition probability will not change from that for an indirect band gap transition. In the discussion of Eqn. 8, it was pointed out that such mixing does occur, introducing s-character into this state.

F.   Discussion

The experimental and theoretical evidence obtained on 10-monolayer Ge-Si superlattices and summarized above supports the idea that these structures have a direct bandgap at about 0.8eV. While this evidence may appear quite conclusive, it is not yet possible to rule out entirely interfacial or defect levels in these experiments. Further work is necessary to demonstrate all the features associated with direct bandgap materials. In this regard the measurement of the absorption coefficient and electroluminescence would bring additional credence to the claim for the creation of direct bandgap material out of Ge and Si. Even though these strained materials have larger concentrations of dislocations ($\sim 10^8$ cm^{-2}) than epitaxial layers of unstrained materials, the average distance between dislocations is about 1μm which is in most cases larger than the thickness of the superlattice region. This level of defects is too low to affect the electronic bandstructure.

In the studies reviewed here, there is conclusive evidence for the formation of a new conduction band state in the band gap with transition amplitudes comparable to those for direct transitions. This state shows the correct symmetry behavior for electronic wavefunctions imposed by the superlattice periodic structure. This initial demonstration of wavefunction engineering in Si and Ge superlattices may be extended to other superlattice systems using different materials.

References

Abstreiter, G., Brugger, H., Wolf, T., Jorke, H., and Herzog, H.-J., 1985, Phys. Rev. Lett., 54:2441.

Abstreiter, G., Brugger, H., Wolf, T., Jorke, H., and Herzog, H.-J., 1986, Surface Sci., 174:640.

Abstreiter, G., Eberl, K., Friess, E.,Wegscheider,W., and Zachai, R., 1989, J. Cryst. Growth, 95:431.

Bevk, J., Mannaerts, J.P., Feldman, L.C., Davidson, B.A., and Ourmazd, 1986, A., Appl. Phys. Lett., 49:1138.

Bevk, J., Davidson, B.A., Feldman, L.C., Gossmann, H.J., Mannaerts, J.P., Nakahara, N., and Ourmazd, A., 1987a, J. Vac. Sci Technol.B, B5:1147.

Bevk, J., Ourmazd, A., Feldman, L.C., Pearsall, T.P., Bonar, J.M., Davidson, B.A., and Mannaerts, J.P., 1987b, Appl. Phys. Lett., 50:760.

Braunstein, R., Moore, A.R., and Herman, F., 1958, Phys. Rev., 95:846.

Brugger, H., Abstreiter, G., Jorke, H., Herzog, H.-J., and Kasper, E., 1986, Phys. Rev. B, 33:5928.

Cerdeira, F., Pinczuk, A., Bean, J.C., Batlogg, B., and Wilson, B.A., 1984, Appl. Phys. Lett., 45:1138.

Cerdeira, F., Alonso, M.I., Niles, D., Garriga, M., and Cardona, M., 1989, Phys Rev. B, 40:xxx.

Friedel, P., Hybertsen, M.S., and Schlüter, M., 1989, Phys. Rev. B., to be published.

Froyen, S., Wood, S.M., and Zunger, A., 1987, Phys. Rev. B, 36:4574.

Froyen, S., Wood, S.M., and Zunger, A., 1988, Phys. Rev. B., 37:6893.

Gell, M.A., 1988, Phys. Rev. B., 38:7535.

Gell, M.A., 1989, Appl. Phys. Lett., to be published .

Gnutzmann, M., and Clausecker, K., 1974, Appl. Phys., 3:9.

Hybertsen, M.S., and Schlüter, M., 1987, Phys. Rev. B., 36:9683.

Hybertsen, M.S., and Schlüter, M., 1988, Proc. Mat. Sci. Res. Symp., 102:413.

Iyer, S.S., Tsang, J.C., Copel, M.W., Pukite, P.R., and Tromp, R.M., 1989, Appl. Phys. Lett., 54:220.

Kasper, E., Kibbel, H., Jorke, H., Brugger, H., Friess, E. and Abstreiter, G., 1988, Phys. Rev. B., 38:3599.

Lang, D.V., People, R., Bean, J.C., and Sergeant, A.M., 1985, Appl. Phys. Lett., 47:1333.

Menendez, J., Pinczuk, A., Bevk, J., and Mannaerts, J.P., 1988, J. Vac. Sci. Tech. B, B6:1306.

Moriarty, J.A., and Krishnamurthy, S., 1983, J. Appl. Phys., 54:1892.

Paul, W., and Warschauer, D.M., 1959, J. Phys. Chem. Solids, 6:6.

Pearsall, T.P., Pollak, F.H., Bean, J.C., and Hull, R.H., 1986a, Phys. Rev. B., 33:6821.

Pearsall, T.P., Temkin, H., Bean, J.C., and Luryi, S., 1986b, Electron Dev. Lett., EDL-7:330.

Pearsall, T.P., Bevk, J., Feldman, L.C., Bonar, J.M., Mannaerts, J.P., and Ourmazd, A., 1987, Phys. Rev. Lett., 58:729.

Pearsall, T.P.,Bevk, J., Bonar, J., and Mannaerts, J.P., 1989, Phys. Rev. B., 39:3741.

People, R., 1985, Phys. Rev. B., 32:1405.

People, R., and Jackson, S., 1987, Phys. Rev. B., 36:2114.

Pollak, F.H., 1990, "Strained-Layer Superlattices Vol. 1", in Semiconductors and Semimetals, ed. T.P. Pearsall, Academic Press, New York.

Renucci, J.B., Renucci, M.A., and Cardona, M., 1971, in Light Scattering in Solids, ed. M. Balkanski, Flammarion, Paris, p. 326.

Satpathy, S., Martin, R.M., and van de Walle, C.G., 1988, Phys. Rev. B., 38:13237.

Shen, D., Zhang, K., and Xie, X., 1988, Appl. Phys. Lett., 52:717.

Silberstein, R.P., and Pollak, F.H., 1980, Sol. State Comm., 33:1131.

Stroud, D., and Ehrenreich, H., 1970, Phys. Rev. B, 2:3197.

Temkin, H., Pearsall, T.P., Bean, J.C., and Luryi, S., 1985, Appl. Phys. Lett., 47:1333.

van de Walle, C.G., and Martin, R.M., 1986, Phys. Rev. B, 34:5621.

Wong, K.B., Jaros, M., Morrison, I., and Hagon, J.P., 1988, Phys. Rev. Lett., 60:2221.

ULTRA-FAST OPTICAL PROBES

IN QUANTUM WELLS AND SUPERLATTICES

Benoit Deveaud

Centre National d'Etudes des Télécommunications
22301 LANNION
FRANCE

Short optical pulses are the easiest way to generate and to study physical phenomena with very high temporal resolution. The shortest available pulses nowadays are, by far, provided by optical techniques. Indeed, these techniques have been able to generate pulses down to 6 fs only. This corresponds to only 3 cycles in the visible spectrum. If short electrical pulses can also be generated, the shortest ones (200 fs) use optical pulses to be generated and cannot be transferred over large distances. This ability, together with the eternal race towards higher and higher frequencies makes that short optical pulses must be used if the properties of materials are to be characterized in vew of high frequency applications. This is particularly true for the case of microstructures, where picosecond and femtosecond optical techniques have been greatly useful for their characterization as well as for the study of physical properties or device aspects.

This short account of a wide subject will be divided in three parts: in the first one, we shall briefly describe some of the possible ways to generate a short optical pulse. In the second part, we shall explain a few methods ways to measure the results of an experiment, if possible keeping the temporal resolution provided by the incoming pulse. In the last part, which is going to be the longest, we are going to describe a few recent experiments where short optical pulses have prooved being useful.

GENERATION TECHNIQUES

We shall briefly describe in this section the most widely used techniques to provide short optical pulses. We will not necessarily try to deal with world records, but rather put emphasis on techniques that are no too difficult to use on a day to day basis.

Electronic Properties of Multilayers and Low-Dimensional Semiconductors Structures
Edited by J. M. Chamberlain *et al.*, Plenum Press, New York, 1990

SYNCHRONOUS PUMPING

Generation of short pulses can be obtained by synchronous pumping of a dye laser by a modelocked laser [1,2,3]. By over simplifying, let's say that the cavity length of the dye is adjusted so as the travelling time of a pulse in the dye, fits the frequency of the incoming pulses. In such a system, one pulse generated in the dye cavity, returns back to the dye jet exactly when the next pulse from the modelocked laser comes to the jet. The dye jet gain needs being adjusted close to threshold in order to obtain the shortest pulses. Such a system easily gives pulses of about 5 ps, and possibly down to 1 ps. If a second jet of saturable absorber is included in the cavity [3], and possibly a set of 4 prisms in order to compensate for dispersion, further reduction of the pulsewidth can be obtained [4]. If the output power in the case of the single jet laser has to be low in order to provide short pulses, higher powers and still short pulses can be obtained with the dual jet configuration.

COLLIDING PASSIVE MODELOCKING

The most convenient way to produce very short pulses today is the use of passive mode locking in a ring dye laser. Such a system is called Colliding pulse Modelocking [5]. The principle of the setup has been described by a number of papers [6,7], and consists in a ring dye laser with two jets: one for the amplifying medium and one for a saturable absorber. The ring laser is pumped by a cw laser, and two counterpropagating pulses develop in the cavity. Shortening of the pulse can be obtained by compensating for dispersion effects inside the very cavity by the use of a set of 4 prisms. Such a system is able to deliver quite easily pulses down to 30 fs, at a frequency of about 100 MHz (the repetition rate of the pulses is given by the cavity length) and with an output power of 20 mW on each of the 2 output beams. This system, originally developped in the red with the couple rhodamine 6G, DODCI, has recently been extended to the near infrared region [8,9].

AMPLIFICATION

Once short pulses are generated, it is useful for some applications to amplify these pulses to obtain as high a power as possible. The principle of an amplification system is to pump a dye with a high power pulse having a duration long compared to the duration of the short pulse, and to pass the short pulse several times into this amplifying medium during its excitation by the high power pulse [10,11]. In general, 6 passes are used and allow to amplify the pulses up to an energy of 1 mJ/pulse [12]. Such an energy is very useful as it allows to generate white continuum [13] by focusing the laser into a water cell or a ethylene Glycol jet. One of the most remarkable properties of the continuum generation is that it allows the generation of a wide range of optical frequencies without loss of pulse width [14].

400

PULSE COMPRESSION

Pulse compression can be used at different stages in order to reduce the pulse length. The principle is to focus the pulse into an optical fibre [15,16]. In the fibre, non linear dispersion will produce a chirped pulse i.e. a pulse with a larger spectral width although keeping phase coherence. The blue part of the pulse will then be at the leading edge of the pulse and the red part at the trailing edge. Proper compensation for the dispersion, for example by passing the pulses on a pair of gratings, leads then to pulses having a width limited by the spectral width [17,18,11]. For example, a 100 ps pulse from a modelocked YAG laser can be compressed down to 1 ps with a fibre/grating compressor. The 30 fs pulses from a CPM laser have been compressed down to 6 fs by using a combination of fibre, prisms and gratings in order to compensate for dispersion up to the 3rd order [19].

DETECTION TECHNIQUES

Different techniques can be used in order to study the effect of a short pulse on a particular sample. As in cw experiments, absorption or luminescence spectra can be recorded, but useful information can also be obtained by other transient methods such as four wave mixing or electro-optic sampling. Time resolution of the detector is of course the main problem when comparing to steady-state measurements. In fact, most experiments use some kind of "trick" in order to obtain experimental results with a resolution limited mainly by the laser pulsewidth (typical examples of this are pump-probe or correlation techniques). We are first going to describe some of the techniques which are currently used, and then try to summarise their advantages and disadvantages.

STREAK CAMERA

Direct time resolved detection of fast optical transients is not an easy task. The most sensitive technique is certainly the streak camera [20,21]. Basically, a streak camera is a photomultiplier tube in which a voltage ramp is applied between two opposite electrodes. The photoelectrons will then be deflected by this field according to their generation delay. As a result, they will hit the back of the camera (which is a phosphorescent screen) at a position depending of the time at which they have been created. Such systems can be very sensitive if used in the synchroscan mode, and have a resolution of the order of 20 ps (see for example the review by Campillo and Shapiro [22] or by Schiller and Alfano [23]). Single shot streak camera are able of a much better time resolution (of the order of 1 ps) but have rather low detectivities.

PUMP-PROBE

One of the outstanding detection techniques is pump-probe experiments [24], the technique has been greatly improved by the use of white light continuum generation. A short pump pulse is sent first to excite the sample. A low intensity white pulse is then sent after a variable time delay on the same spot [25,24]. This provides the absorption spectrum of the sample at a certain time after the excitation pulse. No time resolution of the detector is needed, and photodiode arrays are generally used to record the whole spectrum in the same run. Time resolution is basically given by the pulse width.

KERR SHUTTER

The principle of the Kerr shutter has been known for a long time [26], if an intense optical pulse is focused on a medium such as CS_2, a change of the birefringence of this medium is observed with a time constant shorter than 1 ps. Thus, by properly polarising the signal to be analyzed, the Kerr cell exactly acts as a camera shutter [27,28]. The most well known example of the use of this technique was the photograph of a light pulse in flight [29]. The need to use quite high power pump pulses makes that device only operable with amplified pulses, thus at low repetition rates, but very good time resolution can be obtained [30].

UP-CONVERSION

Another technique to improve the time resolution of optical detection is to use upconversion techniques [31]. In such a system, the photons to be detected are mixed in a non linear crystal with a delayed beam from the laser [32,33]. Sum photons are then generated, depending on the phase-matching conditions, within a time window corresponding to the delayed pulse. The principle of this technique was proposed long ago. The practical implementation necessitates the use of a rather high power laser in order to increase the conversion efficiency of the non-linear crystal, and high repetition rate in order to acheive a good signal to noise ratio. Optical elements used to collect the luminescence and to focus it on the non-linear crystal must be non dispersive if high time resolution is seeked. Typical conversion efficiencies are of the order of 10^{-3} [33].

SIGNAL AMPLIFICATION

Such a technique derives from the same principle as just described for up-conversion; rather than mixing the signal with the laser pulse, the signal itself is amplified. This technique uses the newly developed non-linear organic crystals [34]. The drawback of the technique is the necessity to use high power pumping pulses (thus low rep rate), but this is compensated by the amplification factor which can be as high as 10^4.

CORRELATION TECHNIQUES

Another possible aspect of non-linear techniques, is the use of the sample itself as a non-linear medium. This technique uses the fact that recombination is a non linear process, and thus the luminescence emitted when the sample is excited by one laser pulse only is less than half the luminescence when the sample is excited by two pulses [35]. The technique only requires the separation of the laser pulses in two parts, with some delay between them, each beam being chopped at a different frequency. The advantage of the technique is that it does not require any time resolution of the detection system, the drawback is that it can only be used in samples where the non linearity of the luminescence has been carefully checked and is constant over the whole energy range and also, time resolved spectra cannot be obtained. Very useful informations can nevertheless be obtained in specially design samples, for example with a strong non radiative channel [36,37].

FOUR WAVE MIXING

Properties of materials can also been studied by four wave mixing or photon echoes techniques [38,39]. Such techniques present very interesting features. First, their resolution is directly given by the pulse-width. We will show in the next chapter that results have indeed been obtained down to less than 10 fs [40]. Second, these techniques provide direct information on mechanisms not easily accessible by other techniques such as dephasing times [41]. The principle of the experiment is to create an excitation grating in the material to be studied by interference between two excitation pulses, a third pulse, delayed with respect with the two others will probe the grating. Depending of the precise configuration, frequency and polarization relation between the pulses, different kinds of dephasing mechanisms can be studied.

ELECTRO-OPTIC SAMPLING

Electro-optic sampling is used to obtain the electrical prooperties of semiconductor devices in time or frequency domains where they are not accessible by usual electrical means. Two main techniques are used to probe the device characteristics. In the first configuration, a very short electrical pulse is created close to the sample by an electro-optic switch [42], carried to the device by a microstrip line. The response of the sample is then sent to another strip line, which is probed by the electrooptic effect induced for example in a piece of $LiTaO_3$ in contact with the second strip line [43]. In the second configuration, the device would be used in the usual way and the waveform is probed inside the very sample by using the electrooptic effect of the substrate itself[44].

ADVANTAGES-DISADVANTAGES

Time resolved absorption and luminescence give results which are easy to access due to the similarity with steady state spectra. These two techniques give complementary results as absorption signal depends on the value of $(1-f_e-f_h)$ whereas luminescence is given by the product $f_e x f_h$ (where f_e and f_h are the electron and hole distribution functions). Pump-probe technique easily allows for very short time resolution in absorption. On the contrary, luminescence signal at short times is small due to high carrier temperature effects, so that very high time resolution is difficult to achieve [45]. On the other hand, luminescence will be more suited to study fairly small amount of material: luminescence from a single quantum well can be resolved with a subpicosecond resolution at densities below $5x10^9$ cm^{-2} [46]. If time correlation techniques are very simple conceptually, it is impossible to get a time resolved spectrum, a information of great importance for the comprehension of the physics involved. As far as FWM techniques are concerned, informations of considerable interest can be extracted, such as different dephasing times. This technique has the advantage not to need any time resolved detection. Electro-optic sampling will give informations on the electrical propeties of systems or devices on a time scale not accessible by other techniques.

CHARACTERISTIC RESULTS

In this section we shall give characteristic examples of the results accessible to time resolved experiments. We do not intend to review all published results, but rather to give examples that evidence the advantages of the respective techniques for each case. Of course, we shall mainly give results for the case of superlattices or microstructures, however, in some cases of interest we will quote results which have been obtained on bulk materials.

RELAXATION IN QUANTUM WELLS

Relaxation in quantum wells is a subject of great current interest: most of the possible applications of quantum well systems indeed rely more or less on the possible relaxation mechanisms. Relaxation has then been studied by different means: Picosecond luminescence [47] and picosecond absorption [48] are only two possibilities amongst a number of other ones. Relaxation has been studied by different means, depending on the precise information that was seeked after. In this paragraph, we shall breefly describe hole burning, four wave mixing, photon echoes experiments, time resolved Raman scattering. The capture mechanisms in a QW will be dealt with in the next paragraph.

INTRASUBBAND RELAXATION

Hole burning experiments, especially in the case of GaAs/AlGaAs, are now quite well documented. We refer the reader to the existing literature [49,50], and we are only going to present a typical example showing the capabilities of the technique. Two possible cases can occur in a first rough classification depending if the excitation energy is more or less than one optical phonon above the bottom of the band. When the excitation energy is less than one LO phonon, the photoexcited carriers mainly interact by carrier carrier scattering (see Fig.1) [51]. A clear dependance on the excitation density is indeed observed in such experiments: thermalization occurs within 100 fs at a density of 2×10^{10} cm^{-2} and within 30 fs for a density of 5×10^{11} cm^{-2}. This small density dependance is a result of partial cancellation of: i) screening of the carrier-carrier interaction with density, and ii) increase of the mean energy exchange per collision [51].

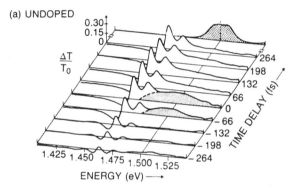

Fig.1. Hole burning experiments in an undoped quantum well (From Knox et al [51]).

If the excess energy is larger than an LO phonon, LO phonon emission can take place and the processes are much more complex. They have not yet been studied in detail yet in quantum wells, but interesting results have been obtained for example in bulk GaAs [40,49,52]. In particular, the times involved with the scattering from the   to the L or X valleys (or vice versa) are of great interest for many purposes. Such times have been approached by different cw techniques [53,54], but also more directly by time resolved luminescence [55] and pump-probe technique [52,49]. Some controversy still holds for the value of the deformation potential that ranges from 4×10^8 eV/cm to 9×10^8 eV/cm. Hot phonons effects have been evidenced that we cannot describe here, we refer the reader to the existing litterature [56].

INTERSUBBAND RELAXATION

Time resolved techniques have also been applied to measurements that we have not evoked up to now: Raman scattering. Two experiments exemplify quite well the usefulness of this kind of technique: hot phonon effects in GaAs [57] and time resolved electronic Raman scattering in quantum wells [58,59]. Electronic Raman scattering can probe the electron density of one of the levels of the quantum wells. If electrons are photoexcited in one of the upper levels of a quantum well, time resolved electronic Raman scattering will be able to determine the scattering time of an electron from the upper state to the ground state.

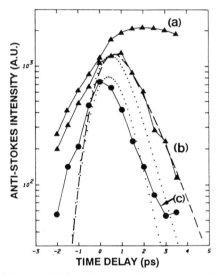

Fig.2. Time resolved population of the n=2 level of a 160Å quantum well as given by Raman scattering (from Tatham et al [59])

The determination of these scattering times has been performed in two steps, first in the case where the energy separation is smaller than an LO phonon energy [58], second for the case of a splitting larger than an LO phonon [59]. The results, of the order of 500 ps in the first case and less than 1 ps in the second one (see Fig.2) show the large efficiency of scattering via Frölich interaction when compared to interaction with acoustical phonons. The value of about 1 ps when LO phonons can be emitted compares well with the results of theoretical calculations [60]: scattering time from the n=2 to the n=1 level decreases as the well width increases, as a result of the smaller wavevector of the LO phonon involved in the scattering mechanism (1/q dependance of the Frölich matrix element). Calculated times range from 1.5 ps for well width of about 70 Å, to 0.3 ps for a well width of 200 Å (in the system GaAs/AlGaAs).

CAPTURE IN A QUANTUM WELL

Capture of carriers in a quantum well (QW) is a topic of great current interest because of the possible implications in terms of devices and of the basic physical mechanisms involved. Following the classical approach, the capture would depend on the well width and on the carrier thermal velocity: if a carrier can be scattered while it is crossing the QW, it will be captured. This process is highly unfavourable in very narrow QWs [61]. Following the quantum mechanical approach, the capture of an electron in a quantum well comes from the interaction between the "barrier" states and "well" states. The fastest process would be mediated by optical phonon scattering and is expected to give rise to resonances as a function of the well width [62,63]. These resonances occur when the lowest unconfined state is about to enter the well, thus presenting a larger probability density in the well region. These calculations also give a rather slow capture in very narrow wells.

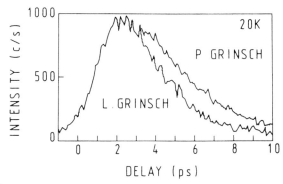

Fig.3. Decay time of the luminescence of the confinement layer of GRINSCH (linear and parabolic) structures. Time decays are 2.4 and 2.7 ps respectively (from Deveaud et al [67])

Neither of these two approaches has been checked by experiments. Quantum mechanical capture of carriers in a quantum well seems to be a very fast process as long as the barrier thickness is not too large. The upper limit for the capture time of holes has been estimated to be less than 300 fs, and is not slower in very narrow wells [64]. This was first observed for the case of InGaAs/InP quantum wells where no luminescence from the barriers is observed even with a time resolution of 300 fs. Equivalent results have been obtained recently for the case of InGaAs/GaAs quantum wells [65].

When the barrier thickness is larger, the capture of the carriers is not limited by the quantum mechanical process but rather by the diffusion of the carriers in the barriers, the observed time can be called "overall capture time". The transition between the two behaviours should occur for barrier thicknesses close to the mean free path of the carrier of interest. If the barrier luminescence is monitored, the larger the barrier, the longer the luminescence decay time.

It has been proposed by Tsang et al [66] that the operation of a QW laser would be improved by the use of a graded optical confinement region instead of the usual constant composition barrier. This kind of device has been named GRINSCH for Graded Index Separated Confinement Heterojunction. In such a device the capture of the carriers is supposed to be speeded up by the quasi-electric field in the graded region. We indeed found that, in such graded devices, the overall capture time is shorter than 3 ps at low temperatures. We find that a linear grading of the structure is better suited than a parabolic grading for fast capture [67]. We show on Fig.3 the time decay of the barrier luminescence obtained by luminescence up-conversion technique in GRINSCH structures.

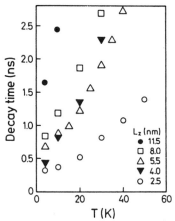

Fig.4. Exciton radiative lifetime as a function of temperature in GaAs quantum wells of different widths (from Feldmann et al [72])

EXCITONIC EFFECTS IN QUANTUM WELLS

It is now well known that, in quasi-two dimensional structures like quantum wells, excitons show different properties and especially an increase of their binding energy when compared with 3D systems [68]. The increase of the binding energy, together with the decrease of the interaction with optical phonons [69], allows the observation of excitonic resonances up to room temperature [70].

EXCITON SCATTERING AND RECOMBINATION

This increase of the binding energy of the exciton will result in a larger radiative rate an a shorter radiative lifetime. This has indeed been evidenced in time resolved luminescence experiments [71]: by using a streak camera, the time decay of the excitonic luminescence of a quantum well is monitored. It is observed that this time decreases with well width. As a function of temperature, an increase of the radiative lifetime is observed and explained by the increased interaction of excitons with phonons [72] (see Fig.4). Very good agreement is observed between the variation of the lifetime with temperature and the change in the exciton linewidth. The interpretation relies in the fact that the lifetime of the exciton is proportionnal to the homogeneous linewith, and inversely proportionnal to its binding energy as well as a temperature occupancy factor [72].

Direct insight into the behaviour of the exciton cannot be obtained by pump probe measurements which require a rather high carrier density. These high densities, as we will see in the next paragraph, give rise to a very rapid screening of the exciton. Luminescence techniques also are limited by the long binding time of a free electron-hole pair into an exciton [73,74]. Other techniques, using low power excitation levels, have then to be used if more specific properties of the exciton have to be obtain. A very useful category of techniques is four wave mixing: it has been quite succesfully applied to bulk GaAs [75,76], and also to the system of excitons in quantum wells [77,78]. Following the theory developped by Feldmann et al [72], the exciton lifetime should vary with the dephasing time T_2. A short radiative lifetime being an indication of the quality of the sample. This approach has been confirmed by DFWM [77] as a function of the density of photocreated electron-hole pairs.

EXCITON BLEACHING

Saturation of the excitonic absorption in MQWs is a phenomenon of great interest. However, as it has been dealt with quite extensively, we shall only briefly evoke it. Neither shall we deal with the now well documented optical Stark Effect, and will rather refer to the existing litterature [79]. Exciton bleaching has been studied using both cw and picosecond lasers [80,81]. More recently, the use of femtosecond experiment has allowed new insights into the physics of this system[82,83]. Results of these experiments show that the excitonic resonnances, which are still proeminent at room temperature, can be bleached by a short laser pulse.

This bleaching has a double origin: first the phase space filling of the excitons, and second the screening of these excitons by free carriers [84]. Another interesting phenomenon, linked to the 2D characteristics of the excitons, is the following: in 3D, excitonic resonances do not change position upon increasing excitation densities as a result of the almost perfect cancelation of the blue shift due to Pauli exclusion principle and the red shift due to screening. In 2D, the quenching of the screening should allow the observation of a blue shift of the exciton. This blue shift has indeed been observed experimentally in quite good agreement with theory [85]. Femtosecond experiments also showed that the excitons ionized within a time of 300 fs [68], a result already estimated from cw absorption measurements [86].

Direct measurement of the exciton diffusion is a difficult task. It has recently been studied by the dynamics of the luminescence as a function of the excitation spot diameter [87], as well as by FWM [77]. Another possibility is the use of the intrinsic properties of the sample itself: in high quality samples, a splitting of the excitonic lines is observed as a result of the quantification of the well thickness by integer number of monolayers [88]. The different lines correspond to exciton localized in regions of the well differing in width by one monolayer, such a localization may occur if the growth islands at each interface are large when compared to some electronic localization parameter [89].

Thermalization properties of the luminescence lines show that excitons are at least partly able to move from one region with a narrow well width to the next one where the well width is larger by one monolayer [90]. We have studied a sample, grown with interrupted techniques, where the size of the growth islands has been determined to be very large [91] (some are larger than 1 micron). In this sample, two of the wells (with width of 15 Å and 25 Å) evidence three luminescence lines as expected for this kind of splitting. We have studied these two wells by time resolved luminescence [46]. It is important in this kind of measurement to determine the surface ratio between the different regions of the well under study. This can be done by the study of the luminescence excitation spectra [92]. We have choosen to study the variation of the luminescence properties, under picosecond excitation conditions at saturation, i.e. when the Fermi level is higher than the upper energy level. The surface ratio is then simply given by the luminescence intensity at short times, and can be changed by moving the spot on the sample surface. Time dependance of the two

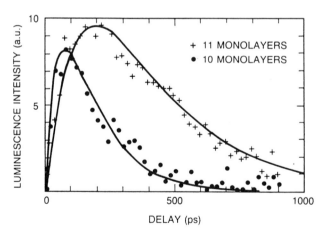

Fig. 5. Time dependance of the two luminescence lines of a 25 Å quantum well, respectively corresponding to well widths of 10 and 11 monolayers (from Deveaud et al [46]).

luminescence lines intensity is reported on Fig.5. This shows that the excitons are able to diffuse to islands larger by one monolayer in a characteristic time of about 200 ps, and indicates a rather high diffusion coefficient in agreement with the results of [87].

VERTICAL TRANSPORT IN SUPERLATTICES

Superlattices (SL) represent the particular case of quantum wells separated by barrier thin enough for the carriers in neighbouring wells to interact. In such a case, one of the first applications proposed by Esaki and Tsu [93], the carriers will be able to move in the growth direction of the SL (vertical transport). The realization of samples showing this type of behaviour has then be delayed by the high growth quality required for the observation of vertical transport [94]. As a matter of fact, any kind of disorder will lead to localization of the carriers and prevent the observation of the movement of the carriers in the growth direction of the SL [95]. One of the possible uses of time resolved luminescence is the determination of transport properties by techniques analogous to time of flight measurements [96,97].

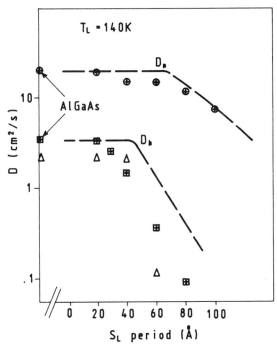

Fig.6. Electron and hole diffusion coefficients in GaAs/GaAlAs superlattices as a function of the SL period (from Lambert et al, [98]).

The principle of the experiment is to create carriers at the surface of the sample with a short laser pulse, the carriers then diffuse or drift to the interior of the sample.

411

Their arrival at a certain depth can be assessed by including a large well in the structure. The diffusion coefficient of the carriers is determined from the rise time of this large well luminescence [96]. Information on the hole transport is easily obtained as in the high photoexcitation density regime, the diffusion process is ambipolar and thus limited by the hole mobility. If the sample is doped p-type and if the excitation density is well below the doping density, the difusion process is governed by electron transport [97]. In this way it has been possible to estimate both electron and hole transport properties as a function of the superlattice period [98](see Fig.6). This figure shows that, both electron and hole transport properties behave as expected in superlattices, in particular, for periods of about 100 Å in GaAs/AlGaAs, electron mobility is still quite high, whereas the holes are completely localized: this is the concept of effective mass filtering introduced by Capasso et al [99].

Improvement on this technique can be obtained by grading (continuously or not) the composition of the sample [100,101]. In such a case, the luminescence energy of the photoexcited carriers depends on their position in the sample: analysis of the shape of the luminescence spectrum as a function of time delay then allows to get the displacement of the carriers as a function of time [101,102] (see Fig.7). Carriers do not move by diffusion alone, but also by drift if the grading is continuous. This is a typical example of the kind of techniques that can be used in order to obtain results not usually accessible with ordinary techniques, it partly relies on the ability to grow novel structures almost at will. This ability was named "band gap engineering by

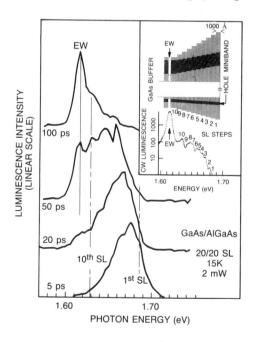

Fig.7. Luminescence spectra of a graded superlattice structure (see the inset) as a function of time delay. High excitation densities on undoped samples give hole diffusion coefficients plotted as triangles in Fig.6 (from Deveaud et al [101]).

412

Capasso [103]. By using this technique, we have been able to follow the movement of a carrier packet in a superlattice after excitation by a short pulse [101,102]. This has been performed in superlattices of different periods and provides information on the mobility of the holes in these superlattices. In order to obtain information on the electron motion, p-type doped samples have to be used [104].

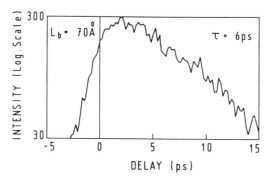

Fig.8. Time decay of the luminescence of a quantum well confined between two 70 Å barriers (from de Saint Pol et al [109]).

TUNNELING STUDIES:

DOUBLE BARRIER TUNNELING RESONANT DIODES

Double barrier resonant tunneling diodes (DBRTDs) are a topic of current interest due to the frequencies that might be attainable with such devices (devices oscillating up to frequencies above 400 GHz have been realized). The advantage of DBTRDs is that, due to the high quality of the growth techniques now available, very thin barriers can be made leading to very short tunneling times. The drwback comes from the fact that high transmission devices mean small peak to valley ratios.

The devices are so fast that it is even difficult to test them by conventional electrical techniques. The most remarkable result has been obtained by electro-optic sampling [105]. It this experiment, a double barrier tunnel junction is biased with a cw voltage just at the limit of the negative differential region. A short pulse generated by an electro-optic switch [106] is sent to the device. The response of the device is recorded by means of a strip-line on $LiNbO_3$. Switching of the device is observed to occur within less than 3 ps. Even such a very short switching time does not seem to be limited by the intrinsic tunneling time constant but rather by the RC time constant of the device. Other studies have been carried out on InGaAs/InAlAs structures by the same technique [107] and give results in qualitative agreement with those of Whitaker et al [105].

413

The same kind of structure can also be assessed by time resolved luminescence. Tunneling times in the case of GaAs/AlAs double barrier structures have been measured by using a streak camera [108]. Electron-heavy hole pairs were generated directly in the ground subband by properly tuning the excitation energy. Subsequent emission of the quantum well is monitored, and is easily separated from the luminescence of the GaAs layers on each side, occurring at lower energies. It is found that the decay times vary from 60 ps for a 28 Å thick barriers to 200ps for 40 Å thick barriers. Good agreement is found with theoretical predictions although it is difficult to estimate the relative importance of the gamma barrier and of the X barrier.

We have performed measurements on GaAs/AlGaAs double barrier structures where the X barrier does not influence the results, and where the barrier height is quite well known [109]. A characteristic result is presented on Fig.8, and shows that the tunnelling across a 70Å barrier occurs in 7 ps. In such a case where the barrier is much simpler to modelize, we find that the observed time is shorter than the calculated value. As the tunneling time exponentially depends on the barrier thickness, it is necessary to determine quite precisely the barrier thickness in order to be able to get useful information from such results.

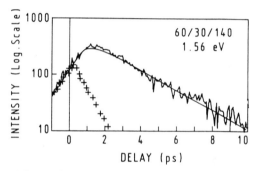

Fig.9. Time behaviour of the narrow well luminescence of a 60/30/140 Å CQW system. The solid line is a fit taking into account the pulse shape and the thermalization of carriers (from Deveaud et al [118])

TUNNELING BETWEEN TWO QUANTUM WELLS

Coupled quantum wells (CQWs) consist in two quantum wells coupled by a thin barrier (in most cases the two wells will have different thicknesses: a narrow well, NW and a wide one, WW). Such systems have attracted much interest in the past few years due to the possible applications [110,111] as well as due to more fundamental aspects. Of particular interest is the possible tunneling from one well to the other

(from the NW to the WW). Various possible mechanisms may lead
to the tunneling [112,113] but the resonance effect between
one level of the narrow well and one from the WW is of
particular interest. Time resolved luminescence is a good
tool to study such resonance effect as it allows to study the
decay time of the NW luminescence [37,114]. When an electric
field is applied to the sample, levels can be brought at will
on and off resonance [115,116]. In such an experiment, when
the electric field is tuned to bring the n=1 level of the NW
in resonance with the n=2 level from the WW, the NW
luminescence decay time is minimum. Oberli et al [106]
observed a change from about 200 ps off resonance down to a
value of 7 ps at the resonance (for a barrier width of 50 Å).

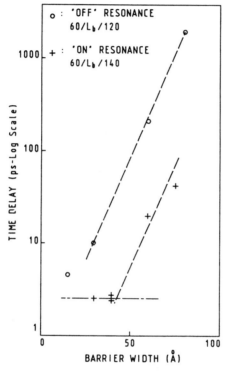

Fig.10. Decay time of the upper state
luminescence in a series of "on resonance "
CQWs for different barrier thiknesses (+).
For comparison, we show the behaviour of a
series of "off resonance samples (o) (from
Deveaud et al [118]).

Our approach has been different and we have choosed to
work on p-type QWs so as to obtain a very short
thermalization of the electrons, and also to be sure that we
are only looking at electron tunneling effects [118].
Resonance has then to be tuned by changing the width of one

of the wells. We show on Fig.9 the time decay behaviour of the "narrow well luminescence in a CQW system where a 60 Å QW coupled at resonance by a 30 Å barrier to a 140 Å QW. The time decay is 2.3 ps, and is not limited by the system resolution as shown by the shorter time decay recorded on an oxygen implanted sample (crosses). We show on Fig.10 the variation of the time decay at resonance as a function of the barrier thickness. Note that the time is limited to more than two ps which is not the instrumental resolution. This corresponds to the emission of one LO phonon from the upper coupled states two the ground state and agrees quite well with the value determined by Tatham et al [59] in isolated QWs (the factor 2 difference comes from the fact that in CQWs, half of te wavefunction is in the NW and will not give rise to an appreciable matrix element). The conclusion of our observation is that the lifetime of the upper states is not limited by the tunneling itself, but rather by the inelastic scattering down to the ground state [118].

ACKNOWLEDGEMENTS

I would like to acknowledge many useful discussions on short optical pulses with P. Becker, P. Georges, Ch. Hirlimann, W.H. Knox, A. Migus, F. Salin, J. Shah, and on the physics of microstructures with G. Bastard, D.S. Chemla, A. Chomette, F. Clérot, D. Hulin, E.O. Göbel, B. Lambert, J.C. Maan, A. Regreny, R. Romestain, B. Sermage, J. Shah, C. Tanguy.

REFERENCES

[1] A. Scavennec, *Opt. Comm.*, **17**, 14 (1976)

[2] J. Kluge, D. Wiechert, D. von der Linde, *Opt. Comm.*, **45**, 278 (1983)

[3] B. Couillaud, V. Fossati-Bellani, *Lasers and Appl.*, 79 and 91 (1985)

[4] W.H. Knox, *J. Opt. Soc. Am. B4*, 1771 (1987)

[5] R.L. Fork, B.I. Greene, C.V. Shank, *Appl. Phys. Lett.*, **38**, 671 (1981)

[6] E.P. Ippen, C.V. Shank, *Appl. Phys. Lett.*, **27**, 488 (1975)

[7] R.L. Fork, B.I. Greene, C.V. Shank, *Appl. Phys. Lett.*, **38**, 671 (1981)

[8] W.H. Knox, *CLEO*, Baltimore, 1989

[9] P. Georges, F. Salin, A. Brun, *Opt. Lett.*, (1989) to be published

[10] G. Mourou, T. Sizer, *Opt. Comm.*, **41**, 47 (1982)

[11] E.P. Ippen, C.V. Shank, in *Picosecond Phenomena*, Eds. C.V. Shank, E.P. Ippen, and S.L. Shapiro, Springer Verlag, New York, (1978)

[12] W.H. Knox, R.L. Fork, M.C. Downer, R.H. Stolen, C.V. Shank, J.A. Valdmanis, *Appl. Phys. Lett.*, **46**, 1120 (1985)

[13] R.R. Alfano, S.L. Shapiro, *Phys. Rev. Lett.*, **24**, 584 (1970)

[14] R.L. Fork, C.V. Shank, R. Yen, C.A. Hirlimann, *IEEE J. Quantum Electron.*, **QE-19**, 500 (1983)

[15] A. Laubereau, *Phys. Lett.*, **29A**, 539 (1969)
[16] W.J. Tomlinson, R.H. Stolen, C.V. Shank, *J. Opt. Soc. Am.*, **B1**, 139 (1984)
[17] C.V. Shank, R.L. Fork, J.P. Gordon, *Opt. Lett.*, **10**, 131 (1985)
[18] E.B. Treacy, *Phys. Lett.*, **28A**, 34 (1968)
[19] R.L. Fork, C.H. Brito Cruz, P.C. Becker, C.V. Shank, *Opt. Lett.*, **12**, 483 (1986)
[20] D.J. Bradley, G.H.C. New, *Proc. IEEE*, **62**, 313 (1974)
[21] V.V. Korobin, A.A. Malyutin, M.Ya. Schelev, *Sov. Phys. Tech. Phys.*, **16**, 165 (1971)
[22] A.J. Campillo, S.L. Shapiro, *IEEE J. Quantum Electron.*, **QE-19**, 585 (1983)
[23] N.H. Schiller, R.R. Alfano, *Laser Focus Mag.*, 43 (1982)
[24] E.P. Ippen, C.V. Shank, A. Bergman, *Chem. Phys. Lett.*, **38**, 611 (1976)
[25] J.W. Shelton, J.A. Armstrong, *IEEE J. Quantum Electron.*, **QE3**, 302 (1967)
[26] J. Kerr, *Phil. Mag. and J. Sci.*, **50**, 337 (1875)
[27] G. Mayer, F. Gires, *Compt. Rend. Acad. Sci. (Paris)*, **258**, 2039 (1964)
[28] M.A. Duguay, J.W. Hansen, *Appl. Phys. Lett.*, **15**, 192 (1969)
[29] M.A. Duguay, *Am. Scient.*, **59**, 550 (1971)
[30] D. Hulin, J. Etchepare, A. Antonetti, J.L. Chase, G. Grillon, A. Migus, A. Mysyrowicz, *Appl. Phys. Lett.*, **45**, 993 (1984)
[31] Midwinter, Warner, *J. Appl. Phys.*, **38**, 519 (1967)
[32] J. Shah, T.C. Damen, B. Deveaud, D. Block, *Appl. Phys. Lett.*, **50**, 1307 (1987)
[33] J. Shah, *IEEE J. Quantum Electron.*, **QE-24**, 276 (1988)
[34] D. Hulin, A. Migus, A. Antonetti, I. Ledoux, J. Badan, J.L. Oudar, J. Zyss, *Appl. Phys. Lett.*, **49**, 761 (1986)
[35] D. von der Linde, J. Kuhl, E. Rosengart, *J. Lum.*, **24/25**, 675 (1981)
[36] M.B. Johnson, T.C. McGill, A.T. Hunter, *J. Appl. Phys.*, **63**, 2077 (1988)
[37] N. Sawaki, R.A. Höpfel, E. Gornik, H. Kano, *6th Hot Carriers in Semiconductors Conf.*, Scottsdale, AZ, (1989)
[38] D.W. Phillion, D.K. Kuizenga, A.E. Siegmann, *Appl. Phys. Lett.*, **27**, 85 (1975)
[39] See for example: N.A. Kurnit, I.D. Abella, S.R. Hartmann, *Phys. Rev. Lett.*, **13**, 567 (1964), C.K.N. Patel, R.E. Slusher, *Phys. Rev. Lett.*, **20**, 1087 (1968), J. Hegarty, M. Boer, B. Golding, J.R. Simpson, J.B. Mac Chesney, *Phys. Rev. Lett.*, **51**, 2033 (1983)
[40] P. Becker, H.L. Fragnito, C.H. Brito Cruz, R.L. Fork, J.E. Cunningham, J.E. Henry, C.V. Shank, *Phys. Rev. Lett.*, **61**, 1647 (1988)
[41] B.S. Wherrett, A.L. Smirl, T.F. Boggess, *IEEE J. Quantum Electron.*, **QE-19**, 680 (1983)
[42] D.H. Auston, A.M. Johnson, P.R. Smith, J.C. Bean, *Appl. Phys. Lett.*, **37**, 371 (1980)
[43] A. Valdmanis, G.A. Mourou, C.W. Gabel, *IEEE J. Quantum Electron.*, **QE-19**, 664 (1983)
[44] B.H. Kolner, D.M. Bloom, *IEEE J. Quantum Electron.*, **QE-22**, 79 (1986)
[45] T.C. Damen, J. Shah, *Appl. Phys. Lett.*, **52**, 1291 (1988)
[46] B. Deveaud, T.C. Damen, J. Shah, C.W. Tu, *Appl. Phys. Lett.*, **51**, 828 (1987)

[47] see for example:
J. Shah, R.F. Leheny, in Semiconductors probed by ultrafast laser spectroscopy, Ed. R.R. Alfano, Academic press, 1984, p.45
S.A. Lyon, J. Lum., 35, 121 (1986)
J. Shah, IEEE J. Quantum Electron., **QE-22**, 1728 (1985)
J.F. Ryan, R.A. Taylor, A.J. Tuberfield, J.M. Worlock, *Surf. Sci.*, **170**, 511 (1986)

[48] C.V. Shank, R.L. Fork, R. Yen, J. Shah, B.I. Greene, A.C. Gossard, C. Weisbuch, *Solid State Commun.*, **47**, 981 (1983)

[49] J.L. Oudar, A. Migus, D. Hulin, G. Grillon, J. Etchepare, A. Antonetti, *Phys. Rev. Lett.*, **53**, 384, (1984)

[50] W.H. Knox, C. Hirlimann, D.A.B. Miller, J. Shah, D.S. Chemla, C.V. Shank, *Phys. Rev. Lett.*, **56**, 1191 (1986)

[51] W.H. Knox, D.S. Chemla, G. Livescu, J.E. Cunningham, J.E. Henry, *Phys. Rev. Lett.*, **61**, 1290 (1988)

[52] P.B. Becker, H.L. Fragnito, C.H. Brito Cruz, J. Shah, R.L. Fork, J.E. Cunningham, J.E. Henry, C.V. Shank, *Appl. Phys. Lett.*, **53**, 2089 (1988)

[53] D.N. Mirlin, I.J. Karlik, L.P. Nikitin, I.I. Reshina, V.F. Sapega, Solid State Commun., **37**, 757 (1981)

[54] R.G. Ulbrich, J.A. Kash, J.C. Tsang, Phys. Rev. Lett., **62**, 949 (1989)

[55] J. Shah, B. Deveaud, T.C. Damen, W.T. Tsang, A.C. Gossard, P. Lugli, *Phys. Rev. Lett.*, **59**, 2222 (1987)

[56] See for example the revue by J. Ryan, and references therein.

[57] J.A. Kash, J.C. Tsang, J.M. Hvam, *Phys. Rev. Lett.*, **54**, 2151 (1985)

[58] D.Y. Oberli, D.R. Wake, M.V. Klein, J. Klem, T. Henderson, H. Morkoç, *Phys. Rev. Lett.*, **59**, 696 (1987)

[59] N. Tatham, J.F. Ryan, C.T. Foxon, *4th Modulated Semiconductor Structures Conf.*, Ann Arbor, MI, (1989)

[60] R. Fereira, G. Bastard, *Phys. Rev.*, **B40**, 1074 (1989)

[61] Y. Tang, K. Hess, N. Holonyak Jr., P.D. Dakpus, *J. Appl. Phys.*, 53, 6043 (1982)

[62] A. Brum, G. Bastard, *Phys. Rev.*, **B33**, 1420, (1986)

[63] M. Babiker, B. Ridley, *Superlatt. Microstruct. and Microdev.*, 2, 287 (1986)

[64] B. Deveaud, T.C. Damen, J. Shah, W.T. Tsang, *Appl. Phys. Lett*, 52, 1886 (1988)

[65] D.Y. Oberli, J. Shah, J.L. Lewell, T.C. Damen, N. Chand, *Appl. Phys. Lett.*, 54, 1028 (1989)

[66] W.T. Tsang, *Appl. Phys. Lett.*, **39**, 134 (1981)

[67] B. Deveaud, F. Clérot, A. Regreny, K. Fujiwara, K. Mitsunaga, J. Otha, (to be published)

[68] see for example:
D.S. Chemla, *Helv. Phys. Acta*, **56**, 607 (1983), R.C. Miller, D.A. Kleinman, *J. Lumin.*, **30**, 520 (1985) and references therein.

[69] J.E. Zucker, A. Pinczuk, D.S. Chemla, A.C. Gossard, W. Wiegmann, *Phys. Rev. Lett.*, **53**, 1280 (1984)

[70] D.S. Chemla, D.A.B. Miller, *J. Opt. Soc. Am.*, **B2**, 1155 (1985)

[71] E.O. Göbel, H. Jung, J. Kuhl, K. Ploog, *Phys. Rev. Lett.*, **51**, 1588 (1983)

[72] J. Feldmann, G. Peter, E.O. Göbel, P. Dawson, K. Moore, C. Foxon, R.J. Eliott, *Phys. Rev. Lett.*, **59**, 2337 (1987)

[73] J.I. Kusano, Y. Segawa, Y. Aoyagi, S. Namba, H. Okamoto, *Phys. Rev.*, **B40**, 1685 (1989)

[74] T.C. Damen, J. Shah, D.Y. Oberli, D.S. Chemla, (private comm.)

[75] L. Schultheis, J. Kuhl, A. Honold, C.W. Tu, *Phys. Rev. Lett.*, **57**, 1797 (1986)

[76] L. Schultheis, J. Kuhl, A. Honold, C.W. Tu, *Phys. Rev. Lett.*, **57**, 1635 (1986)

[77] J. Hegarty, M.D. Sturge, A.C. Gossard, W. Wiegmann, Appl. *Phys. Lett.*, **40**, 132 (1982) and J. Hegarty, M.D. Sturge, C. Weisbuch, A.C. Gossard, W. Wiegmann, *Phys. Rev. Lett.*, **49**, 930 (1982)

[78] A. Honold, L. Schultheis, J. Kuhl, C.W. Tu, *IQEC Technical Digest*, Tokyo, (1988) and *Superlatt. and Microstruct.*, **2**, 441, (1986)

[79] See for example the NATO ASI series volume: "Switching in low dimensional systems", Ed. H. Haug and E.N. Banyay.

[80] D.S. Chemla, D.A.B. Miller, P.W. Smith, A.C. Gossard, W.Wiegmann, *IEEE J. Quantum Electron.* **QE-20**, 265 (1984)

[81] D.A.B Miller, D.S. Chemla, D.J. Eilenberger, P.W. Smith, A.C. Gossard, W. Wiegmann, *Appl. Phys. Lett.*, **42**, 925 (1983)

[82] W.H. Knox, R.L. Fork, M.C. Downer, D.A.B. Miller, D.S. Chemla, C.V. Shank, *Phys. Rev. Lett.*, **54**, 1306 (1985) and D.S. Chemla, D.A.B. Miller, *J. Opt. Soc. Am.*, **B2**, 1155 (1985)

[83] N. Peyghambarian, H.M. Gibbs, J.L. Lewell, A. Antonetti, A. Migus, D. Hulin, A. Mysyrowicz, *Phys. Rev. Lett.*, **53**, 2433 (1984)

[84] See for example: S. Schmitt-Rink, in *Interfaces, Quantum Wells and Superlattices*, Ed. C.R. Leavens and R. Taylor, Plenum Press, NATO ASI series Vol.179, NY and London, 1988, p. 211.

[85] D. Hulin, A. Mysyrowicz, A. Antonetti, A. Migus, W.T. Masselink, H. Morkoc, H.M. Gibbs, N. Peyghambarian, *Phys. Rev.*, **B33**, 4389 (1986)

[86] D.S. Chemla, D.A.B. Miller, P.W. Smith, A.C. Gossard, W. Wiegmann, *IEEE J. Quantum Electron.*, **QE-20**, 265 (1984)

[87] H. Hilmer, A. Forchel, S. Hansmann, M. Morohashi, H. Meier, K. Ploog, *Phys. Rev.*, **B39**, 10901 (1989)

[88] B. Deveaud, J-Y Emery, A. Chomette, B. Lambert, M. Baudet, *Appl. Phys. Lett.*, **45**,1078 (1984)

[89] D. Paquet, *Superlatt. and Microstruct.*,(1986)

[90] B. Deveaud, A. Regreny, A. Chomette, J-Y Emery, *J. Appl. Phys.*, **59**, 1633 (1986)

[91] P.M. Petroff, J. Cibert, A.C. Gossard, G.J. Dolan, C. W. Tu, *J. Vac. Sci. Technol.*, **B5**, 1191, (1987)

[92] M. Kohl, D. Heitmann, S. Tarucha, K. Leo, K. Ploog, *Phys. Rev.*, **B39**, 7736 (1989)

[93] L. Esaki, Tsu, *IBM Res. Dev.*, (1970)

[94] A. Chomette, B. Deveaud, B. Lambert, A. Regreny, *Solid State Commun.*, **54**, 75 (1985)

[95] A. Chomette, B. Deveaud, A. Regreny, G. Bastard, *Phys. Rev. Lett.*, **57**, 1464 (1986)

[96] B. Deveaud, A. Chomette, B. Lambert, A. Regreny, R. Romestain, P. Edel, *Solid State Commun.*, **57**, 885 (1986)

[97] B. Lambert, B. Deveaud, A. Chomette, A. Regreny, B. Sermage, *Semicond. Sci. Technol.*, **4**, 513 (1989)

[98] B. Lambert, F. Clérot, B. Deveaud, A. Chomette, G. Talalaeff, A. Regreny, (to be published in *J. Lum.*)

[99] F. Capasso, K. Mohamed, A.Y. Cho, R. Hull, A.L. Hutchinson, *Phys. Rev. Lett.*, **55**, 1152 (1985)

[100] B. Lambert, A. Chomette, B. Deveaud, A. Regreny, *Semicond. Sci. Technol.*, **2**, 705 (1987)

[101] B. Deveaud, T.C. Damen, J. Shah, B. Lambert, A. Regreny, *Phys. Rev. Lett.*, **58**,2582 (1987)

[102] B. Deveaud, J. Shah, T.C. Damen, B. Lambert, A. Chomette, A. Regreny, *IEEE J. Quantum Electron.*, **QE-24**, 1641 (1988)

[103] See for example, F. Capasso in *Physics and applications of Quantum Wells and Superlattices*, Ed. E.E. Mendez, K. von Klitzing, Plenum, NATO ASI Ser. Vol. 170, 1987, p. 377

[104] B. Deveaud, F. Clérot, B. Lambert, A. Regreny (to be published)

[105] J.F. Whitaker, G.A. Mourou J.F. Whitaker, T.C.L.G Sollner, W.D. Goodhue, *Appl. Phys. Lett.*, **53**,385 (1988)

[106] D.H. Auston, *Appl. Phys. Lett.*, **26**, 101 (1975)

[107] S Muto, A. Takeuchi, T. Inata, E. Miyachi, T. Fuji, *4th Modulated Semiconductor Structures Conf.*, Ann Arbor, (1989)

[108] M. Tsuchyia, T. Matsutsue, H. Sakaki, *Phys. Rev. Lett.*, **59**, 2356 (1987)

[109] L. de Saint Pol, D. Lippens, F. Clérot, B. Lambert, B. Deveaud, B. Sermage, GaAs and related compounds, 1989.

[110] B. Deveaud, A. Chomette, A. Regreny, J.L. Oudar, D. Hulin, A. Antonetti, in *High Speed Electronics*, Ed. B. Källback and H. Beneking, Springer Verlag, Berlin, 1986, p.101.

[111] A. Tackeuchi, S. Muto, T. Inata, T. Fujii, *Jap. J. Appl. Phys.*, **28**, (1989).

[112] T. Weil, B. Vinter, *J. Appl. Phys.*, **60**, 3227 (1986)

[113] R. Fereira, G. Bastard, *Phys. Rev.*, **B40**, 1074 (1989).

[114] M.G.W. Alexander, M. Nido, W.W. Ruhle, R. Sauer, K. Ploog, K. Köhler, W.T. Tsang, 6th Hot Carriers in Semiconductors Conf., Scottsdale (1989)

[115] D.Y. Oberli, J. Shah, T.C. Damen, D.A.B. Miller, C.W. Tu, Bull. Am. Phys. Soc., **34**, 501 (1989) and (to be published).

[116] T.B. Norris, N. Vodjani, B. Vinter, C. Weisbuch, G.A. Mourou, *Phys. Rev.*, **B40**, 1342 (1989).

BALLISTIC ELECTRON TRANSPORT IN THE PLANE

M. Heiblum

IBM Research Division, T. J. Watson Research Center
Yorktown Heights, New York 10598

A ballistic electron is that 'lucky electron' that succeeds traversing the solid without elastic or inelastic scattering events. The current interest in ballistic and quasi ballistic electron transport has emanated mostly due to the following characteristics of the transport: (a) Ballistic electrons move faster than diffusing ones, thus enabling the construction of high speed devices; (b) Ballistic electrons maintain their energy, direction, and phase, thus can interfere and be steered by electric or magnetic fields; (c) Ballistic electrons can be used as probes to study properties of semiconducting materials. These attractive features and the technology that has advanced rapidly have resulted in increased activities in the area of ballistic transport in the past few years. Significant recent discoveries in this arena were the direct demonstration of ballistic transport and the determination of the ballistic mean free path (mfp) via electron energy spectroscopy [1] in the 'vertical domain' (using hot electron transistors), and the observation of the quantized ballistic resistance [2] in the 'horizontal domain'. Here we summarize our recent results [3,4,5] of hot electron spectroscopy in the plane of the 2DEG and describe results of ballistic hot electron transport with an unexpectedly long mfp.

By injecting hot electrons one can maximize the average electron velocity and minimize the transit time. The hot electrons upon injection have also a relatively narrow, energy distribution, thus enabling energy spectroscopy to be done. This methodology was utilized in a GaAs-AlGaAs heterojunction structure supporting a high mobility 2DEG. Injection of hot electrons was done via a tunnel barrier induced in the 2DEG by a very narrow metal gate, continuous or with a small opening (point contact injector), deposited on the surface of the heterojunction. Applying a sufficiently negative gate voltage with respect to the 2DEG depletes the 2DEG underneath the gate, creating in effect a potential barrier for the Fermi electrons on both sides of the gate (like in a FET which is biased beyond pinchoff). If this potential barrier is narrow enough it can be used as a tunnel injector for hot electrons in the plane. A second metal gate, deposited some distance away from the injector, was used to form a spectrometer barrier. The region of the 2DEG between both barriers, where the hot electron transport took

Electronic Properties of Multilayers and Low-Dimensional Semiconductors Structures
Edited by J. M. Chamberlain *et al.*, Plenum Press, New York, 1990

421

place, is contacted also to form the common terminal. Thus, the potential diagram of the structure resembles that of a 'vertical' hot electron device: an injector (emitter), a transport region (base), and a spectrometer (collector).

In typical devices the metal gates, deposited on the surface of the heterojunction, were some 50 nm wide and were separated by 50-2000 nm. Energy spectroscopy of hot electrons has been performed on collected distributions of hot electrons. It was done by changing the spectrometer barrier height via the voltage applied to the spectrometer gate or the voltage applied between the two 2DEG domains on both sides of the spectrometer barrier. Electrons passing above this barrier relax in the collector layer that follows and the resultant current can be measured and differentiated with respect to the barrier height to give the energy distribution of the electron flux [1]. For injector barriers higher than the Fermi energy (which is about 10 meV) and injection energies around the potential height of the injector barrier, energy distributions about 3-5 meV wide, with median energy following the biasing voltage, were measured. This agrees well with the expected tunnelling distributions injected from relatively low (30 meV) but wide (50 nm) potential barrier injectors, suggesting transport with a minimal number of small angle elastic collisions, if any at all.

To estimate the ballistic mfp, geometrical effects had to be eliminated. In other words, all the electrons injected had to move forward toward the spectrometer, so that in the absence of scattering the collection efficiency would be unity. To accomplish that we had made the spectrometer gate longer than the injector gate and have estimated the ballistic currents via calculating the area under the ballistic peak in the collected distribution. The ballistic mfp's are found to be of the order of 0.5 μm for injection energies smaller than the LO phonon energy (36 meV). A similar experiment has been performed with the two gates being replaced with two point contact gates separated 2 μm apart. Sequential emission of three LO phonons was observed. For injection voltage slightly lower than 36 mV an inelastic mfp in the range of 2 μm was measured. However, when the voltage was increased slightly above this threshold, hot electrons did not arrive at the collector, suggesting many phonon emissions en route. For injection energy in the range $36 + \Phi$ to 72 meV above the Fermi energy (Φ is the spectrometer potential height), a relatively large fraction of the injected hot electrons that emitted one phonon were still collected, suggesting: (a) That scattering due to LO phonon emission is mostly a forward scattering event (collection necessitates a sufficiently large forward momentum), and (b) That the electrons, once relaxed below the phonon energy, have a long mfp. It is surprising that other scattering events such as electron-plasmon and electron-electron, expected theoretically, were not observed below 36 meV.

Steering and focusing experiments were done too. The focusing experiments were done by letting a diverging cold electron beam pass under a metal gate in a shape of a lens. Biasing the metal gate negatively depletes the 2DEG and reduces the electrons' velocity as they pass under the gate. It can be shown that electrons will interfere according to a modified "Snell's Law" for electrons and focus at the focal point of the lens. As the voltage applied to the gate (lens) was varied the voltage measured in a collector placed at the focal point of the lens reached a maximum and than decreased. Applying a magnetic field that curved the electrons' trajectories, or increasing the lattice temperature that reduced the ballistic mfp, had both destroyed the effect, suggesting indeed an electron focusing.

This work was done with the collaboration of A. Palevski, U. Sivan, C. P. Umbach, and H. Shtrikman.

REFERENCES

1. M. Heiblum and M. V. Fischetti, "Ballistic Electron Transport in Hot Electron Transistors", to appear in *Physics of Quantum Electron Devices*, Ed. F. Capasso, in **Topics in Current Physics**, Springer-Verlag, Berlin (1989).

2. D. A. Wharam, T. J. Thornton, R. Newbury, M. Pepper, H. Ahmed, J. E. F. Frost, D. J. Hasko, D. C. Peacock, D. A. Ritchie and G. A. Jones, J. Phys. C: Solid State Phys. **21**, L209 (1988); B. J. van Wees, H. van Houten, C. W. J. Beenakker, J. G. Williamson, L. P. Kouwenhoven, D. van der Marel, and C. T. Foxon, Phys. Rev. Lett. **60**, 848 (1988).

3. A. Palevski, M. Heiblum, C. P. Umbach, C. M. Knoedler, R. Koch, and A. Broers, Phys. Rev. Lett. **62**, 1776 (1989).

4. A. Palevski, C. P. Umbach, and M. Heiblum, Appl. Phys. Lett. **55**, 1421 (1989).

5. U. Sivan, M. Heiblum, and C. P. Umbach, Phys. Rev. Lett. **63** , 992 (1989).

STATIC AND DYNAMIC CONDUCTIVITY OF

INVERSION ELECTRONS IN LATERAL SUPERLATTICES

Jörg P. Kotthaus

Sektion Physik
Universität München
Geschwister-Scholl-Platz 1
8000 München 22, F.R. Germany

EXTENDED ABSTRACT

Two-dimensional (2D) electron inversion layers at semiconductor heterojunction interfaces can be transformed into lateral superlattices via field effect. The fabrication of laterally periodic field-effect devices with submicrometer periodicities on GaAs-AlGaAs heterojunctions containing a high-mobility 2D electron system as well as metal-oxide-semiconductor (MOS) structures on Si and InSb is discussed (Kotthaus, 1987). The effective lateral width of the confining potential in such devices can approach the Fermi wavelength of the inversion electrons and be much smaller than their elastic or inelastic mean free path. Hence they are well suited to investigate confinement to quantum wires (Hansen et al., 1987) and quantum dots (Sikorski and Merkt, 1989). On GaAs-AlGaAs heterojunctions the electronic mean free paths can also be much larger than the lateral periodicity such that lateral superlattice effects become observable. Here, recent low-temperature studies of the static and high frequency conductivity of laterally periodic field-effect devices are summarized that demonstrate the importance of quantum confinement and the large tunability of electronic properties in these devices (Kotthaus, 1989; Kotthaus and Merkt, 1989; Merkt et al., 1989).

Novel magnetoresistance oscillations observed in GaAs-AlGaAs heterojunctions with a one-dimensional (1D) modulation of the inversion electron density (Weiss et al., 1989) are shown to manifest 1D superlattice phenomena in a magnetic field applied perpendicularly to the sample plane (Winkler et al., 1989; Gerhardts et al., 1989; Beenakker, 1989). Magnetoresistance studies are also employed to analyse field-effect confinement to arrays of quantum wires via observation of magnetic depopulation of 1D subbands (Alsmeier et al., 1988; Brinkop et al., 1988).

The dynamic conductivity of laterally periodic structures is studied with far-infrared transmission experiments. In arrays of quantum wires strong infrared resonances are observed with resonance frequencies determined by the joint action of classical depolarization and quantum confinement. GaAs-AlGaAs heterojunctions are found to be particularly useful to investigate the transition from 2D to 1D confinement and to exhibit infrared excitations with dominantly collective character (Hansen, 1988). On InSb the infrared resonances are more single-particle-like and essentially reflect lateral quantization (Alsmeier et al., 1988). Using 2D periodic field-electrode arrays one can also electrostatically create and tune quantum dot arrays. The characteristic infrared excitations of such dot arrays realized in MOS-devices on InSb (Sikorski and Merkt, 1989) and Si (Alsmeier et al., 1989) are discussed in dependence on electron density, confinement size, and magnetic field strength.

Electronic Properties of Multilayers and Low-Dimensional Semiconductors Structures
Edited by J. M. Chamberlain *et al.*, Plenum Press, New York, 1990

425

REFERENCES

Alsmeier, J., Sikorski, Ch., and Merkt, U., 1988, Subband spacings of quasi-one-dimensional inversion channels on InSb, Phys. Rev., B37:4314

Alsmeier, J., Batke, E., and Kotthaus, J. P., 1989, Voltage tunable quantum dots on Silicon, Phys. Rev., B, in press

Beenakker, C. W. J., 1989, Guiding-center-drift resonances in a periodically modulated two-dimensional electron gas, Phys. Rev. Lett., 62:2020

Brinkop, F., Hansen, W., Kotthaus, J. P., and Ploog, K., 1988, One-dimensional subbands of narrow electron channels in gated AlGaAs/GaAs heterojunctions, Phys. Rev., B37:6547

Gerhardts, R. R., Weiss, D., and Klitzing, K. v., 1989, Novel magnetoresistance oscillations in a periodically modulated two-dimensional electron gas, Phys. Rev. Lett., 62:1173

Hansen, W., Horst, M., Kotthaus, J. P., Merkt, U., Sikorski, Ch. and Ploog, K, 1987, Intersubband resonance in quasi-one-dimensional inversion channels, Phys. Rev. Lett., 58:2586

Hansen, W., 1988, Quasi-one-dimensional electron systems on GaAs/AlGaAs heterojunctions, in: "Festkörperprobleme, Advances in Solid State Physics", Vol.28, U.Rössler, ed., Vieweg, Braunschweig

Kotthaus, J. P., 1987, Infrared spectroscopy of lower dimensional electron systems, Physica Scripta, T19:120

Kotthaus, J. P., 1989, Transport properties and infrared excitations of laterally periodic nanostrucutres, in: "Nanostructure Physics and Fabrication ", W. Kirk and M. Reed, eds., Academic Press, Boston

Kotthaus, J. P., and Merkt, U., 1989, Magnetotransport and infrared resonances in laterally periodic nanostructures, in: "Science and Engineering of 1- and 0-Dimensional Semiconductors", S. P. Beaumont and C.M. Sotomayor-Torres, eds., Plenum Press, New York, in press

Merkt, U., Sikorski, Ch., and Alsmeier, J., 1989, Electrons in lateral microstructures on indiumantimonide, in: "Spectroscopy of Semiconductor Microstructures", G. Fasol, A. Fasolino, and F. Luigi, eds., Plenum Press, New York, in press

Sikorski, Ch., and Merkt, U., 1989, Spectroscopy of electronic states in InSb quantum dots, Phys. Rev. Lett., 62:2164

Weiss, D., Klitzing, K. v., Ploog, K., and Weimann, G., 1989, Magnetoresistance oscillations in a two-dimensional electron gas induced by a submicrometer periodic potential, Europhys. Lett.,8:179

Winkler, R. W., Kotthaus, J. P., and Ploog, K., 1989, Landau band conductivity in a two-dimensional electron system modulated by an artificial one-dimensional superlattice potential, Phys. Rev. Lett., 62:1177

QUANTUM BALLISTIC ELECTRON TRANSPORT AND CONDUCTANCE

QUANTIZATION IN A CONSTRICTED TWO-DIMENSIONAL ELECTRON GAS

B.J. van Wees

Delft University of Technology
The Netherlands

The keywords of this paper are ballistic and adiabatic electron transport. Ballistic transport means that we consider the transport through devices with dimensions smaller than the mean free path between impurity scattering. This implies that the motion of the electrons is fully determined by the (smooth) electrostatic potential which confines the electrons in the conductor. Adiabatic tranport means that the electrons move through the conductor with conservation of their subband index. This occurs when the electrostatic potential of the conductor changes smoothly. As will be shown, the presence of a magnetic field plays an important role regarding the adiabaticity of the electron transport.

The quantum point contact (QPC) is the most elementary tool for the study of ballistic and adiabatic electron transport. The device is defined by fabricating a split-gate configuration[1,2] on top of a high-mobility $GaAs/AL_xGa_{1-x}$ As heterojunction. By applying a negative voltage to the gates, the two-dimensional electron gas underneath them is depleted and a narrow (250 nm) and short channel is defined in the 2DEG. By further reducing the gate voltage, the width as well as the electron density in the QPC can be reduced continuously.

The most dramatic phenomenon observed in these QPCs was the observation of quantized plateaux at multiples of $2e^2/h$ in the conductance of the QPCs, measured as function of the gate voltage[3,4]. This quantization, which was partially anticipated in the literature[5], can be explained by the formation of one-dimensional subbands, resulting from the lateral confinement of the electrons, each occupied subband contributing $2e^2/h$ to the conductance. The application of a perpendicular magnetic field to a QPC induces a continuous transition to the quantum Hall regime[6]. The fact that the quantized plateaux are preserved shows that the zero-field quantization and the quantum Hall effect are related.

An obvious geometry to study next is a series configuration of QPCs, connected by a cavity[7,8,9]. The important question is now whether or not adiabatic transport occurs. In the first case the electrons travel from one QPC to the other with conservation of their subband index. Whether or not electrons are transmitted through the device is determined by the QPC which transmits the least number of subbands. The resistance of the device is given by: R_{SER} = max (R_A,R_B). In the opposite limit the electrons which enter the cavity through a particular QPC are scattered into all available subbands. This implies Ohmic addition of the point contact resistances: $R_{SER} = R_A + R_B$. It should be mentioned here that deviations from the Ohmic addition rule can occur when part of the electrons are direcly transmitted from one QPC to the other. Electron beam collimation plays an important role here[10]. The experiments show that the zero-field transport is almost fully Ohmic. The adiabatic regime is reached when a sufficiently high magnetic field (>1T) is applied. In this regime the transport takes place in one-dimensional magnetic edge channels. The observation of adiabatic tranport in magnetic fields >1T means that the scattering between adjacent edge channels is suppressed in high magnetic fields.

Electronic Properties of Multilayers and Low-Dimensional Semiconductors Structures
Edited by J. M. Chamberlain *et al.*, Plenum Press, New York, 1990

A geometry with two adjacent QPCs, 1.5 μm apart, has been used to study electron focusing[11]. The observation of electron focusing in this device shows that ballistic transport occurs in the 2DEG region between the QPCs, and also that the reflection of the electrons at the 2DEG boundary is highly specular. Recently the quantum Hall effect was studied in this device. It was observed that QPCs can be used to selectively inject current into specific edge channels, or to selectively detect the occupation of specific edge channels. This, together with the absence of scattering between adjacent edge channels, has resulted in the observation of an anomalous integer quantum Hall effect, in which the quantization of the Hall resistance is not determined by the number of edge channels (or Landau levels) in the bulk 2DEG, but by the number of edge channels transmitted by the current and voltage probes instead[12]. This result shows that in micrometer scale devices the properties of the current and voltage probes are crucial for the quantization of the Hall resistance[13].

Recently, it was also shown[14] that the scattering between adjacent edge channels can also be weak on macroscopic length scales. In particular it was observed that scattering between the upper edge channel (the edge channel belonging to the Landau level with the highest quantum number) and the other edge channels can be almost absent on length scales of 100 μm or more.

The one-dimensional nature of the magnetic edge channels, combined with the absence of scattering between them, make it possible to study electron transport in a one-dimensional electron interferometer. Zero-dimensional states can be formed by confining edge channels in a cavity. These can be studied by attaching two QPCs to the cavity, which act as barriers with controllable transmission coefficients. Large oscillations were observed in the conductance of this device[15], which resulted from the resonant transmission when the Fermi energy coincides with the energy of a discrete, zero-dimensional electron state in the cavity.

The results described in this paper were obtained in a collaboration between the Delft University of Technology, The Netherlands, and the Philips Research Laboratories, Eindhoven, The Netherlands

REFERENCES

1. T.J. Thornton et al. Phys. Rev. Lett. 56, 1198 (1986)
2. H.Z. Zheng et al. , Phys. Rev. B34, 5635 (1986)
3. D.A. Wharam et al. , J. Phys. C21, L209 (1988)
4. B.J. van Wees et al. , Phys. Rev. Lett. 60, 848 (1988)
5. Y. Imry, in Directions in Condensed Matter Physics, ed. G. Grinstein and G. Mazenko, (World Scientific, Singapore, 1986), Vol. 1, p.102.
6. B.J. van Wees et al. , Phys. Rev. B38, 3625 (1988)
7. D.A. Wharam et al. , J. Phys. C21, L887 (1988)
8. P.H. Beton, et al. , J. Phys. Cond. Matt. 1, 7505 (1989)
9. L.P. Kouwenhoven et al. , Phys. Rev. B (oct. 1989).
10. C.W.J. Beenakker and H. van Houten, Phys. Rev. B39, 10445 (1989)
11. H. van Houten et al., Phys. Rev. B39, 8556 (1989)
12. B.J. van Wees et al., Phys. Rev. Lett. 62, 1181 (1989)
13. M. Büttiker, Phys. Rev. B38, 9375 (1988)
14. S. Komiyama et al., to be published, B.J. van Wees et al. , Phys. Rev. B39, 8066 (1989), B.W. Alphenaar et al, to be published.
15. B.J. van Wees et al. , Phys. Rev. Lett. 62, 2523 (1989)

ADIABATIC TRANSPORT IN THE FRACTIONAL QUANTUM HALL REGIME

Leo P. Kouwenhoven

Faculty of Applied Physics, Delft University of Technology
P.O.Box 5046, 2600 GA Delft, The Netherlands

A description of transport in the integer quantum Hall effect (IQHE) regime in terms of edge channels has recently gained much attention (Halperin, 1982; Streda et al.,1987; Büttiker, 1988). Experimentally, it has been shown that at a sufficiently high magnetic field and on an appropriate short lenght scale the transport in edge channels can take place adiabatically, i.e. without inter-channel scattering (van Wees et al., 1989). In this case edge channels can be treated as independent one-dimensional current carrying channels.

The single-particle description of the IQHE relates the bulk Landau levels in a one-to-one correspondence to the edge channels at the boundary of the two dimensional electron gas (2DEG). The many-body origin excludes a similar correspondence for the fractional quantum Hall effect (FQHE) (see for reviews: Prange and Girvin, 1987; Chakraborty and Pietiläinen, 1988). Therefore it is not obvious whether an adiabatic transport description can be applied to the FQHE.

To investigate the possibility of an edge channel description for the FQHE regime, we have used adjustable barriers as current and voltage probes, which are placed adjacently at 2μm distance (L.P. Kouwenhoven et al., 1989). The barriers in the probes are electrostatically induced in the 2DEG by means of metallic gates on top of the GaAs/AlGaAs heterostructure.

Fig.1 shows the Hall conductance as a function of (equal) gate voltage defining the barriers in the current and voltage probe. The magnetic field is kept fixed at 7.8T at which the filling factor of the 2DEG is approximately one. Although the filling factor of the bulk 2DEG is unchanged, the Hall conductance drops from the integer quantized value e^2/h to the fractional value $2/3*e^2/h$, when the probe characteristics are varied. To determine the characteristics of the probes we have measured their conductance as a function of the gate voltage, which will be used below to compare the Hall conductance quantitatively with the probe conductances.

To understand the deviation of the Hall conductance from the expected bulk 2DEG value, we propose the existence of fractional edge channels, which flow along equipotential lines at the 2DEG boundary. If a fractional edge channel is populated by the current probe or detected by the voltage probe it contributes $1/3*e^2/h$ to the Hall conductance (for simplicity we only consider the p/3-states, p=integer). However, if no coupling exists from the current probe nor from the voltage probe to a certain fractional edge channel it does not contribute to the Hall conductance. Note that for this selective population and detection of fractional edge channels also the absence of an indirect coupling via inter-channel scattering is required. The deviation of the Hall-conductance from the expected bulk value therefore demonstrates the occurrence of adiabatic transport over the distance between current and voltage probe.

We have calculated the Hall conductance from the measured probe conductance with the appropriate Büttiker-formulas for our geometry, which are derived by van Wees et al. (1989). The Büttiker formulas are generalized to the FQHE regime by taking $1/3*e^2/h$ as the conductance quantum and the number of occupied fractional edge channels as the number of contributing current channels. The

Electronic Properties of Multilayers and Low-Dimensional Semiconductors Structures
Edited by J. M. Chamberlain *et al.*, Plenum Press, New York, 1990

429

result is given in fig.1 (dashed line) and shows good agreement with the measured Hall conductance. The agreement demonstrates that the Hall conductance is completely determined by the probe caracteristics and that scattering between adjacent edge channels is absent over the 2μm distance between current and voltage probe.

In conclusion, it has been shown for the first time that adiabatic transport can occur through edge channels in the FQHE regime. The concept of fractional edge channels is supported by a recent generalization of the Landauer-Büttiker formalism to the FQHE regime by Beenakker (1989).

This work has been done in cooperation with C.E. Timmering of the Philips Research Laboratories, Eindhoven and C.J.P.M. Harmans, N.C. van der Vaart and B.J. van Wees of the Delft University of Technology.

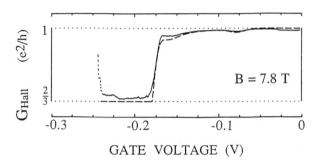

GATE VOLTAGE (V)

Figure 1. Hall conductance at a fixed magnetic field of 7.8T as a function of gate voltage defining the barrier heights of current and voltage probes. The dashed line is calculated from the probe conductances.

REFERENCES

Beenakker C.W.J., submitted to Phys. Rev. Lett.
Büttiker M., Phys. Rev. B 38, 9375 (1988).
Chakraborty T. and Pietiläinen P.,*The fractional quantum Hall effect* (Springer-Verlag, New York, 1988).
Halperin B.I., Phys. Rev. B 25, 2185 (1982).
Kouwenhoven L.P. et al., submitted to Phys. Rev. Lett.
Prange R.E. and Girvin S.M.,*The Quantum Hall Effect*, (Springer-Verlag, New York, 1987).
Streda P. et al., Phys. Rev. Lett. 59, 1973 (1987).
van Wees B.J. et al., Phys. Rev. Lett. 62,1181 (1989).

BALLISTIC ELECTRONIC TRANSPORT IN GaAs–AlGaAs HETEROJUNCTIONS

M. Pepper

Cavendish Laboratory
Madingley Road
Cambridge, CB3 0HE
U.K.

EXTENDED ABSTACT

The use of the high mobility electron gas in modulation doped GaAs–AlGaAs heterojunctions allows the fabrication of a range of structures whose size is less than the mean free path for electron scattering. This particular combination results in ballistic transport where the electron traverses the sample without scattering.

A particularly convenient way of converting two dimensional transport to one dimensional is by the method of split gates.[1] Here two Schottky gates are placed a distance apart which can be of order $\sim -10^{-4}$ cms. Application of a negative gate voltage results in the electron gas being "electrostatically squeezed" with the establishment of a series of quantised levels, (in exactly the same manner as the establishment of two dimensional levels by surface quantisation). This further quantisation can be seen by the application of a magnetic field which successively forces the levels through the Fermi energy. The resultant depopulation produces structure in the resistance[2] which can be distinguished from the 2D Shubikov-de Haas effect due to Landau levels.

The combination of one dimensional confinement, produced by quantisation in the other two dimensions, in the ballistic regime yields a range of quantum effects. The first is the quantisation of the ballistic resistance which we will now consider in more detail.

We consider the application of a voltage V across a one dimensional resistor, the electronic current diffusing down the resistor toward +V is given by

$$J_+ = \int_0^{E_F} \frac{eN(E)}{2} v \, dE \qquad (1)$$

and the current diffusing the opposite way is

Electronic Properties of Multilayers and Low-Dimensional Semiconductors Structures
Edited by J. M. Chamberlain *et al.*, Plenum Press, New York, 1990

431

$$J_- = \int_{E_F - eV}^{E_F} \frac{eN(E)}{2} v \, dE \qquad (2)$$

where N(E) is the 1D density of states at energy E and v is the group velocity at that energy, the factor 2 arises as only half the electrons diffuse in a particular direction. We note that the applied voltage V essentially provides a cut off energy (eV) below which electrons cannot diffuse against the applied voltage. As the 1D density of states including spin degeneracy is $\left(\dfrac{m}{2\pi^2\hbar^2 E}\right)^{1/2}$, which can be expressed as $\dfrac{2}{hv}$, the conductance becomes $\dfrac{2e}{h}$. If there are i 1D subbands then the conductance becomes $\dfrac{2e^2 i}{h}$. Each time a subband is forced through the Fermi energy, E_F, by changing the width, or E_F, directly so σ changes by $\dfrac{2e^2}{h}$. Experiments reporting the observation of the quantisation are described in references 3 and 4.

The factor of 2 in the conductance formula comes from the spin degeneracy and can be lifted by the application of a magnetic field. This is most clear when a parallel field is applied, which avoids the complication of the conversion of the ballistic quantisation to the Quantum Hall Effect. However, the transverse field can be used to investigate magnetic depopulation and the nature of the confining potential.

The above approach is simplistic in that it neglects the question of the transmission of electron waves through the sample. As pointed out by Landauer[5], Imry[6] and Buttiker[7] the correct expression is $\sigma = \dfrac{2e^2}{h} T$ where T is the transmission coefficient. The lifting of the spin degeneracy by a parallel magnetic field is discussed in reference 8.

Application of a high voltage across a sample such that $eV \approx E_F$ will, according to the derivation of the quantised conductance, result in a saturation of the current at a value of $\dfrac{2e^2}{h} E_F$. Calculation suggest that T will decrease with increasing V so producing a decreasing current, i.e. negative differential resistance. This is discussed in detail in references 9 and 10.

Modification of the transmission coefficient by reflection at entrance and exit regions of split gate devices has proved to be of importance in investigating "electrostatic squeezing". In the presence of a magnetic field, transport can arise from electrons confined to the edge regions so defining an effective ring geometry.[11, 12] The period of Aharonov–Bohm oscillations gives the area of the channel. In this way it was found that narrowing the conducting region by depletion also produced a lengthening, due to the depletion region extended out beyond the channel which was initially defined.

Quantum box structures in which a lateral box is interposed between two ballistic channels show striking evidence of resonance effects. Here, peaks in resistance rising above the quantised values are formed by a resonance in the quantum box. Application of a magnetic field results in Aharonov–Bohm oscillations modulated by a clear transmission resonance.[14,15] The well defined box potential allows both weak and strong coupling cases to be studied.[16]

Recently Smith et al[17] have constructed and measured the properties of Fabry Pérot interferometer which consists of a one dimensional resistor situated within a resonance cavity formed by reflector gates.

The large value of scattering mean free path in GaAs–AlGaAs heterostructures allows ballistic effects to be studied in integrated quantum structures. Early work showed that the resistance of two ballistic resistors in series was roughly equal to that of the narrowest, i.e. the resistance was determined by the smallest number of 1D subbands.[18] More recent work suggests that transmission effects may alter this result by $\approx 30\%$.[19] A surprising effect is displayed by two resistors in parallel.[20] When they are within a phase coherence length a coupling effect is found whereby they depopulate subbands together and show jumps in conductance of $\frac{4e^2}{h}$. This result is not fully understood but it may arise from Coulomb effects or the readiness of one dimensional systems to hybridise wavefunctions and so reduce the total energy.

Further details of the experiments described here are provided in the references listed.

ACKNOWLEDGEMENTS
This work was performed in collaboration with many colleagues including H Ahmed, R J Brown, C J B Ford, J E F Frost, D G Hasko, G A C Jones, M J Kelly, R Newbury, D C Peacock, D Ritchie, C G Smith, T J Thornton, D A Wharam.

Support was provided by the Science and Engineering Research Council of the United Kingdom, in part, by the European Research Office of the U.S. Army and, also in part, by the EEC.

REFERENCES

1. Thornton, T.J., Pepper, M., Ahmed, H., Andrews, D., Davies, G J., Phys. Rev. Lett., 56; 1198 (1986)

2. Berggren, K-F., Thornton, Pepper, M., Ahmed, H., Andrews, D., Davies, G. J., Phys. Rev. Lett. 57; 1769 (1986)

3. Wharam, D.A., Thornton, T.J., Newbury, R., Pepper, M., Ahmed, H., Frost, J.E.F., Hasko, D.G., Peacock, D.C., Ritchie, D.A., Jones, G.A.C., J. Phys. C Solid State Physics 21; L209 (1988)

4. van Wees, B.J., van Houton, H., Beenakker, C.W.J., Williamson, J.G., Kouwenhoven, L.P., van der Marel, D., Foxon, C.T., Phys. Rev. Lett. 60; 848 (1988)

5. Landauer, R., Phys. Rev. Lett., 85A; 91 (1981)

6. Imry, Y., in "Directions in Condensed Matter Physics", ed. G. Grinstein and G. Mazenko (World Scientific, Singapore, 1986) Vol. 1, 102

7. Buttiker, M., Phys. Rev.Lett. 57; 1761 (1986)

8. Glazman, L.J., Lesovik, G.B., Khmelnitskii, D.E., and Shekhter, R J. JETP Lett. 48; 238 (1988)

9. Kelly, M.J., Brown, R.J., Smith, C.G., Wharam, D.A., Pepper, M., Ahmed, H., Hasko, D.G., Peacock, D.C., Frost, J.E.F., Newbury, R., Ritchie, D.A., Jones, G.A.C., Electronics Letters, 25; 993 (1989)

10. Brown, R.J., Kelly, M.J., Pepper, M., Ahmed, H., Hasko, D.G., Peacock, D.C., Frost, J.E.F., Ritchie, D.A., Jones, G.A.C., J.Phys: Condens. Matter 1 (1989) 6285; Kelly, M.J., J. Phys; Condens. Matter 1; 7643 (1989)

11. Sivan, U., Imry, Y., Phys. Rev. Lett. 61; 1001 (1988)

12. Wharam, D.A., Pepper, M., Newbury, R., Ahmed, H., Hasko, D.G., Frost, J.E.F., Peacock, D.C., Ritchie, D.A., Jones, G.A.C., J.Phys; Condens. Matter 1; 3369 (1989)

13. Smith, C.G., Pepper, M., Ahmed, H., Frost, J.E.F., Hasko, D.G., Peacock, D.C., Ritchie, D.A., Jones, G.A.C., J.Phys. C21; L893 (1988)

14. van Wees, B.J., Kouwenhoven, L.P., Harmans, C.J.P.M., Williamson, J.G., Timmering, C.E., Broekaart, M.E.F, Foxon, C.T., Harris, J.J., Phys. Rev. Lett. 62; 2523 (1989)

15. Brown, R.J., Smith, C. G., Pepper, M., Kelly, M.J., Newbury, R., Ahmed, H., Hasko, D.G., Frost, J.E.F., Peacock, D.C., Ritchie, D.A., Jones, G.A.C., J.Phys: Condens. Matter 1; 6291 (1989)

16. Yosephin, M. and Kaveh, M., J. Phys; Condens. Matter, in the press

17. Smith, C.G., Pepper, M., Newbury, R., Ahmed, H., Hasko, D.G., Peacock, D.C., Frost, J.E.F., Ritchie, D.A., Jones, G.A.C., Hill, G., J. Phys: Condens. Matter 1; 6763 (1989)

18. Wharam, D.A., Pepper, M., Newbury, R., Ahmed, H., Hasko, D.G., Peacock, D.C., Ritchie, D.A., Frost, J.E.F., Jones, G.A.C., J. Phys: Condes. Matter, 21; L887 (1988)

19. Beton, P.H., Snell, B.R., Main, P.C., Neves, A., Owers-Bradley, J.R.. Eaves, L., Henini, M., Hughes, O.H., Beaumont, S.P., and Wilkinson, C.D.W., J. Phys: Condens. Matter 1; 7505 (1989)

20. Smith, C.G., Pepper, M., Newbury, R., Ahmed, H., Hasko, D.G., Peacock, D.C., Frost, J.E.F., Ritchie, D.A., Jones, G.A.C., Hill, G., J. Phys: Condens. Matter 1; 9035 (1989)

LECTURERS

Dr. S. J. Allen,
Bellcore,
331 Newman Springs Road,
REDBANK,
NJ 07701,
U.S.A.

Prof. A. Baldereschi,
IRRMA,
PH-Ecublens,
CH-1015 LAUSANNE,
Switzerland.

Dr. S. P. Beaumont,
Nanoelectronics Research Centre,
Department of Electronics &
Electrical Engineering,
University of Glasgow,
GLASGOW.
G12 8QQ, United Kingdom.

Dr. C. W. J. Beenakker,
Philips Research Laboratories,
P.O. Box 80.000,
5600 JA EINDHOVEN,
The Netherlands.

Dr. M. Büttiker,
IBM,
Thomas J. Watson Research Center,
P.O. Box 218,
YORKTOWN HEIGHTS,
NY 10598,
U.S.A.

Dr. F. Capasso,
AT&T Bell Laboratories,
600 Mountain Avenue,
MURRAY HILL,
NJ 07974-2070,
U.S.A.

Dr. B. Deveaud,
Centre National d'Études
   de Télécommunications,
Route de Trégastel,
BP 40,
22301 LANNION CEDEX,
France.

Prof. L. Eaves,
Department of Physics,
University of Nottingham,
NOTTINGHAM NG7 2RD,
United Kingdom.

Prof. L. Esaki,
IBM,
Thomas J. Watson Research Center,
P.O. Box 218,
YORKTOWN HEIGHTS,
NY 10598,
U.S.A.

Prof. D. K. Ferry,
Center for Solid State Electronics Research,
Arizona State University,
TEMPE,
Arizona 85287-6206,
U.S.A.

Dr. P. Guéret,
IBM Research Division,
Forschunglaboratorium Zürich,
Säumerstrasse 4,
CH-8830 RÜSCHLIKON,
Switzerland.

Dr. M. Heiblum,
IBM,
Thomas J. Watson Research Center,
P.O. Box 218,
YORKTOWN HEIGHTS,
NY 10598,
U.S.A.

Dr. D. Heitmann,
Max-Planck-Institut für Festkörperforschung,
Heisenbergstrasse 1,
7000 STUTTGART 80,
West Germany.

Dr. G. J. Iafrate,
Director, Electronic Devices Research,
U.S. Army Laboratory Command,
Electronics Technology & Devices Lab.,
FORT MONMOUTH,
NJ 07703-5302,
U.S.A.

Prof. Dr. K. von Klitzing,
Max-Planck-Institut für Festkörperforschung,
Heisenbergstrasse 1,
7000 STUTTGART 80,
West Germany.

Dr. J. P. Kotthaus,
Universität Hamburg,
Institut für Angewandte Physik,
Jungiusstraße 11,
D2000 HAMBURG 36,
West Germany.

Dr. L. P. Kouwenhouven,
Faculty of Applied Physics,
TU Delft,
P.O. Box 5046,
2600 GA DELFT,
The Netherlands.

Prof. Dr. G. Landwehr,
Physikalisches Institut der
  Universität Würzburg,
Lehrstuhl für Experimentelle Physik III,
D 8700 WÜRZBURG,
West Germany.

Dr. S. Luryi,
AT&T Bell Laboratories,
600 Mountain Avenue,
MURRAY HILL,
NJ 07974,
U.S.A.

Dr. P. C. Main,
Department of Physics,
University of Nottingham,
NOTTINGHAM NG7 2RD,
United Kingdom.

Dr. R. J. Malik,
AT&T Bell Laboratories,
600 Mountain Avenue,
MURRAY HILL,
NJ 07974-2070,
U.S.A.

Prof. T. P. Orlando,
Department of Electrical Engineering
  and Computer Science,
Massachusetts Institute of Technology,
CAMBRIDGE,
Massachusetts 02139,
U.S.A.

Dr. T. Pearsall,
AT&T Bell Laboratories,
600 Mountain Avenue,
MURRAY HILL,
NJ 07974,
U.S.A.

Prof. M. Pepper,
Department of Physics,
Cavendish Laboratory,
Madingley Road,
CAMBRIDGE.
CB3 0HE

Dr. M. Razeghi,
Laboratoire Central de Recherches,
Thomson-CSF,
Domaine de Corbeville, BP 10,
91401 ORSAY,
France.

Dr. M. van Rossum,
IMEC VZW,
Kapeldreef 75,
B-3030 LEUVEN,
Belgium.

Dr. M. Roukes,
Bellcore,
331 Newman Springs Road,
REDBANK,
NJ 07701,
U.S.A.

Prof. G. Sollner,
Massachusetts Institute of Technology,
Lincoln Laboratory,
LEXINGTON,
Mass. 02173-0073,
U.S.A.

Dr. G. A. Toombs,
Department of Physics,
University of Nottingham.
NOTTINGHAM NG7 2RD,
United Kingdom.

Dr. B. J. van Wees,
Faculty of Applied Physics,
Delft University of Technology,
P.O. Box 5046,
2600 GA DELFT,
The Netherlands.

Dr. D. Weiss,
Max-Planck-Institut für Festkörperforschung,
Heisenbergstraße 1,
7000 STUTTGART 80,
West Germany.

Dr. F. I. B. Williams,
Commissariat a l'Énergie Atomique
Institut de Recherche Fondamentale,
Service de Physique du Solide et de
Résonance Magnétique,
Orme des Merisiers,
91191 GIF-SUR-YVETTE CEDEX,
France.

Dr. Jeff Young,
Thermometry & Electrical Standards,
National Research Council,
Division of Physics,
Laboratory for Basic Standards,
OTTAWA,
Canada K1A 0R6

PARTICIPANTS

BELGIUM (B)

Mr. Dominique Neerinck,
Laboratory of Solid State Physics and Magnetism,
Physics Department,
Celestijnenlaan 200D,
3030 LEUVEN,
Belgium.

Mr. Kristiaan Temst,
Laboratory of Solid State Physics and Magnetism,
Physics Department,
Celestijnenlaan 200D,
3030 LEUVEN,
Belgium.

CANADA (CDN)

Dr. J. Beerens,
Department of Physics,
Faculty of Sciences,
University of Sherbrooke,
SHERBROOKE,
Quebec,
Canada J1K 2R1

Dr. Dave Inglis,
Division of Physics M36,
National Research Council of Canada,
OTTAWA,
Canada K1A 0R6

Dr. Barry M. Wood,
Section Head of Thermometry & Electrical Standards,
Division of Physics M36,
National Research Council of Canada,
OTTAWA,
Canada K1A 0R6

Mr. Michael G. W. Alexander,
Max-Planck-Institut für Festkörperforschung,
Heisenbergstrasse 1,
Postfach 80 06 65,
7000 STUTTGART 80,
West Germany.

Mr. Frank Brinkop,
Institut für Angewandte Physik,
Jungiusstraße 11,
D-2000 HAMBURG 36,
West Germany.

Mr. Valmir A. Chitta,
Max-Planck-Institut für Festkörperforschung,
Hochfeld-Magnetlabor,
CNRS,
Avenue des Martyrs,
166 X, F38042 GRENOBLE CEDEX,
France.

Mr. Claus Dahl,
Universität Hamburg,
Institut für Theoretische Physik,
Jungiusstraße 9,
D-2000 HAMBURG 36,
West Germany.

Mr. Thorsten Demel,
Max-Planck-Institut für Festkörperforschung,
Heisenbergstrasse 1,
Postfach 80 06 65,
7000 STUTTGART 80,
West Germany.

Mr. Klaus Greipel,
Universität Regensburg,
Institut für Physik I,
Theoretische Physik,
Universitätsstraße 31 - Postfach 397,
8400 REGENSBURG,
West Germany.

Dr. Rolf Haug,
Max-Planck-Institut für Festkörperforschung,
Heisenbergstrasse 1,
Postfach 80 06 65,
7000 STUTTGART 80,
West Germany.

Mr. Manfred Roßmanith,
Max-Planck-Institut für Festkörperforschung,
Heisenbergstrasse 1,
Postfach 80 06 65,
7000 STUTTGART 80,
West Germany.

Mr. Tobias Ruf,
Max-Planck-Institut für Festkörperforschung,
Heisenbergstrasse 1,
Postfach 80 06 65,
7000 STUTTGART 80,
West Germany.

Mr. Mathias Schlierkamp,
Universität Dortmund,
Fachbereich Physik,
Experimentelle Physik II,
Postfach 500500,
D-4600 DORTMUND,
West Germany.

Mr. Wolfgang Wilkening,
Fraunhofer-Institut für
  Angewandte Festkörperphysik,
Eckerstraße 4,
D-7800 FREIBURG,
West Germany.

DENMARK (DK)

Mr. Niels Chr. Alstrup,
Danish Institute of Fundamental Metrology,
Building 322,
Lundtoftevej 100,
DK-2800 LYNGBY,
Denmark.

Dr. P. E. Lindelof,
Physics Laboratory,
H. C. Ørsted Institute,
Universitetsparken 5,
DK-2100 COPENHAGEN Ø,
Denmark.

Mr. Rafael Taboryski,
Physics Laboratory,
H. C. Ørsted Institute,
Universitetsparken 5,
DK-2100 COPENHAGEN Ø,
Denmark.

SPAIN (E)

Dr. Ignacio Izpura,
Universidad Politecnica de Madrid,
E.T.S.I. Telecomunicación,
Ciudad Universitaria,
28040 MADRID,
Spain.

Dr. F. Messeguer,
Departemento de Optica,
Universidad Autónoma de Madrid,
28049 MADRID,
Spain

FRANCE (F)

Mr. Alain Celeste,
CNRS/INSA,
Département de Génie Physique des Matériaux,
156, Avenue de Rangueil,
31077 TOULOUSE CEDEX,
France.

Dr. J. Collet,
CNRS/INSA,
Département de Génie Physique des Matériaux,
156, Avenue de Rangueil,
31077 TOULOUSE CEDEX,
France.

Mr. Luiz Cury,
CNRS/INSA,
Département de Génie Physique des Matériaux,
156, Avenue de Rangueil,
31077 TOULOUSE CEDEX,
France.

Dr. Henri-Jean Drouhin,
DRET,
Laboratoire PMC École Polytechnique,
91128 PALAISEAU CEDEX,
France.

Mr. Lesek Dmowski,
CNRS/INSA,
Département de Génie Physique des Matériaux,
156, Avenue de Rangueil,
31077 TOULOUSE CEDEX,
France.

Mlle. Gohr Etemadi,
CNET,
196 Avenue Henri Ravera,
92220 BAGNEUX,
France.

Dr. Giancarlo Faini,
Centre National de la Recherche Scientifique,
Laboratoire de Microstructures et de Microelectronique,
196, Avenue Henri Ravera,
92220 BAGNEUX,
France.

Mr. Bernard Goutiers,
CNRS/INSA,
Département de Génie Physique des Matériaux,
156, Avenue de Rangueil,
31077 TOULOUSE CEDEX,
France.

Dr. Yves Lassailly,
École Polytechnique,
Laboratoire de la Matière Condensée,
91128 PALAISEAU CEDEX,
France.

Mr. Denis Lavielle,
CNRS/INSA,
Département de Génie Physique des Matériaux,
156, Avenue de Rangueil,
31077 TOULOUSE CEDEX,
France.

Mr. Emanuel Ranz,
CNRS/INSA,
Département de Génie Physique des Matériaux,
156, Avenue de Rangueil,
31077 TOULOUSE CEDEX,
France.

Professor J. B. Renucci,
Laboratoire de Physique des Solides/CNRS,
Université Paul Sabattier - INSA,
118 Route de Narbonne,
31062 TOULOUSE CEDEX,
France.

Mr. Michel Stohr,
CNRS/SNCI,
Avenue des Martyrs,
166 X, F38042 GRENOBLE CEDEX,
France.

Dr. Francis Therez,
I.U.T. Toulouse,
31077 TOULOUSE CEDEX,
France.

Mr. Mohamed Alickacem,
Department of Physics,
University of Nottingham,
NOTTINGHAM.
NG7 2RD

Ms. Debra Barnes,
Department of Physics,
University of Oxford,
Clarendon Laboratory,
Parks Road,
OXFORD.
OX1 3PU

Mr. Alexander C. Churchill,
Blackett Laboratory,
Imperial College,
Prince Consort Road,
LONDON.
SW7 2BZ

Mr. John Cooper,
Department of Physics,
University of of Nottingham,
NOTTINGHAM.
NG7 2RD

Mr. Igor Czajkowski,
Department of Physics,
University of Surrey,
GUILDFORD,
Surrey.
GU2 5XH

Dr. M. Elliott,
Department of Physics,
University College of Wales,
CARDIFF.
CF1 3TH

Mr. Trevor Galloway,
Department of Physics,
University of of Nottingham,
NOTTINGHAM.
NG7 2RD

Ms. Maureen Kinsler,
Department of Physics & Astronomy,
The University of Glasgow,
GLASGOW.
G12 8QQ

Dr. R. E. Miles,
Department of Electrical & Electronic Engineering,
University of Leeds,
LEEDS.
LS2 9JT

Mr. A. Neves,
Department of Physics,
University of Nottingham,
NOTTINGHAM.
NG7 2RD

Mr. David R. Richards,
Department of Physics,
Cavendish Laboratory,
Madingley Road,
CAMBRIDGE.
CB3 OHE

Mr. M. B. Stanaway,
Department of Physics,
University of Nottingham,
NOTTINGHAM.
NG7 2RD

Mr. Paul Steenson,
Department of Physics,
University of Nottingham,
NOTTINGHAM.
NG7 2RD

Dr. Robert I. Taylor,
Allen Clark Research Centre,
Plessey Research Caswell Ltd.,
Caswell,
TOWCESTER,
Northants.
NN12 8EQ

Mr. David Whittaker,
RSRE,
St. Andrews Road,
GREAT MALVERN,
Worcs.
WR14 3PS

Mr. C. R. H. White,
Department of Physics,
University of Nottingham,
NOTTINGHAM.
NG7 2RD

NORWAY (N)

Mr. Petter Helgesen,
Norwegian Defence Resesarch Establishment,
Division of Electronics,
P.O. Box 25,
N-2007 KJELLER,
Norway.

NETHERLANDS (NL)

Mr. R. de Bekker,
MPI/HML,
Avenue des Martyrs,
166X GRENOBLE CEDEX,
F38042,
France.

Mr. P. C. M. Christianen,
Physics Department,
Faculteit Natuurwetenschappen,
Katholieke Universiteit,
Toernooiveld,
6525 ED NIJMEGEN,
The Netherlands.

Mr. Leo P. Kouwenhoven,
Faculty of Applied Physics,
TU Delft,
P.O. Box 5046,
2600 GA DELFT,
The Netherlands.

Dr. L. W. Molenkamp,
Philips Research Laboratories,
P.O. Box 80.000,
5600 JA EINDHOVEN,
The Netherlands.

Mr. A. A. M. Staring,
Philips Research Labs.,
P.O. Box 80.000,
5600 JA EINDHOVEN,
The Netherlands.

Prof. Dr. J. H. Wolter,
Department of Physics,
TU Eindhoven,
Den Dolech 2,
P.O. Box 513,
5600 MB EINDHOVEN,
The Netherlands.

Ms. E. A. E. Zwaal,
Department of Physics,
TU Eindhoven,
Den Dolech 2,
P.O. Box 513,
5600 MB EINDHOVEN,
The Netherlands.

TURKEY (TK)

Ms. Habibe Mamikoglu,
Fizik Bölümü,
Middle East Technical University,
06531 ANKARA,
Turkey.

Mr. Erkan Tekman,
Department of Physics,
Bilkent University,
Bilkent,
06533 ANKARA,
Turkey.

UNITED STATES (US)

Ms. Julia J. Brown,
309 Vivian Hall,
Department of Electrophysics,
University of Southern California,
LOS ANGELES,
CA 90089-0241,
U.S.A.

Mr. Greg Brozak,
Physics Department,
Northeastern University,
BOSTON,
MA 02115,
U.S.A.

Dr. Paul M. Campbell,
Research Physicist,
Naval Research Laboratory,
Code 8832,
WASHINGTON DC, 20375-5000,
U.S.A.

Mr. Robert Grober,
Department of Physics,
University of Maryland,
College Park,
MD 20740,
U.S.A.

Mr. John Hinckley,
Department of Electrical Engineering,
University of Michigan,
ANN ARBOR,
MI 48109-2122,
U.S.A.

Mr. Daniel Howard,
School of Electrical Engineering,
Georgia Institute of Technology,
800 Atlantic Drive, N.W.,
ATLANTA,
Georgia 30322-0269,
U.S.A.

Dr. Alfred M. Kriman,
Center for Solid State Electronics Research,
Arizona State University,
TEMPE,
AZ 85287-6206,
U.S.A.

Mr. Kevin P. Martin,
Microelectronics Research Center,
Georgia Institute of Technology,
800 Atlantic Drive, N.W.,
ATLANTA,
Georgia 30322-0269,
U.S.A.

Mr. Jerome Rascol,
Microelectronics Research Center,
Georgia Institute of Technology,
800 Atlantic Drive, N.W.,
ATLANTA,
Georgia 30322-0269,
U.S.A.

Professor A. Torabi,
Physical Sciences Division,
Electromagnetics Laboratory,
Georgia Institute of Technology,
ATLANTA,
Georgia 30322,
U.S.A.

NON-NATO COUNTRIES

SWITZERLAND (CH)

Mr. Nicolas Blanc,
IBM Research Division,
Zurich Research Laboratory,
Saeumerst 4,
CH-8803 RUESCHLIKON,
Switzerland.

SWEDEN (SW)

Prof. Thorwald Andersson,
Department of Physics,
Chalmers University of Technology,
GOTHENBURG,
S-41296,
Sweden.

Mr. T. Swahn,
Institute of Theoretical Physics,
Chalmers University of Technology,
GOTHENBURG,
S-41296,
Sweden.

**ELECTRONIC PROPERTIES OF MULTILAYERS
AND
LOW DIMENSIONAL SEMICONDUCTOR STRUCTURES**

Château de Bonas, Castera-Verduzan, France
September 11th-22nd 1989

Photograph courtesy of "Photo Delinière", Vic-Fezensac, France

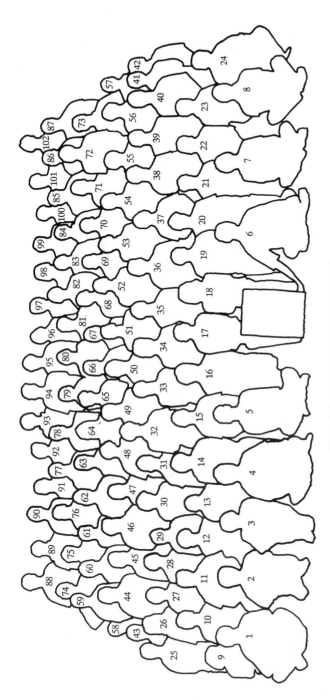

KEY TO GROUP PHOTOGRAPH

1. Weiss; 2. Schlierkamp; 3. Beenakker; 4. Hinckley; 5. Molenkamp; 6. Portal; 7. Helgesen; 8. Chitta; 9. Etemadi; 10. Orlando; 11. Iafrate; 12. Mrs. Iafrate; 13. Büttiker; 14. Landwehr; 15. Esaki; 16. Mrs. Esaki; 17. von Klitzing; 18. Razeghi; 19. Ferry; 20. Sollner; 21. Mrs. Allen; 22. Mrs. Landwehr; 23. Barnes; 24. Celeste; 25. Brown; 26. Mrs. Howard; 27. Kinsler; 28. Alstrup; 29. Mamikoglu; 30. Lassailly; 31. Drouhin; 32. Czjakowksi; 33. Faini; 34. Deveaud; 35. Taboryski; 36. Lindelof; 37. Chamberlain; 38. Cury; 39. Ruf; 40. Neves; 41. Stanaway; 42. Beerens; 43. Whittaker; 44. Howard; 45. Richards; 46. Guéret; 47. Young; 48. Baldereschi; 49. Wood; 50. Grober; 51. de Bekker; 52. Allen; 53. Eaves; 54. Kotthaus; 55. Heitmann; 56. van Rossum; 57. Stohr; 58. Andersson; 59. Temst; 60. Haug; 61. Campbell; 62. Blanc; 63. Staring; 64. Zwaal; 65. Cooper; 66. Capasso; 67. White; 68. van Wees; 69. Steenson; 70. Miles; 71. Christianen; 72. Brozak; 73. Ranz; 74. Toombs; 75. Kriman; 76. Izpura; 77. Alexander; 78. Greipel; 79. Luryi; 80. Elliott; 81. Dmowski; 82. Rascol; 83. Martin; 84. Torabi; 85. Wolter; 86. Swahn; 87. Galloway; 88. Brinkop; 89. Rossmanith; 90. Dahl; 91. Therez; 92. Taylor; 93. Demel; 94. Wilkening; 95. Collet; 96. Goutiers; 97. Lavielle; 98. Tekman; 99. Inglis; 100. Churchill; 101. Alikacem; 102. Kouwenhoven.

449

Epitaxial growth, 95 (see also
        CBE, MBE, MOCVD)
Equilibration length
  and contact spacing, 67
Equivalent circuit of RTD
        oscillator, 285-288
Esaki diodes, 285
Escape rate from well, 340
Etalons
  internal scattering in, 332
Etch selectivity, 232
"Etch tunnel", 156
Etching (see also Electron beam,
        Ion beam, Lithography,
        Mesas)
Etching
  rates, 158
  techniques, 152-158
Evanescent modes, 13
Evanescent states, 243
Excimer lasers, 153
Excitation grating, 403
Excitation transition
  in direct gap material, 333
Exciton
  binding energy, 171
  bleaching, 409
  blue shift, 409
  confined two dimensional, 170
  diffusion in quantum wells, 410
  effective masses of, 43
  formation kinetics, 336, 338
  heavy hole, 10, 11, 43
  in high magnetic fields, 41
  light hole, 10, 11, 43
  one dimensional, 171
  phonon interaction, 409
  photoluminescence, 344
  polaritons
    in quantum wells, 151, 170-171
  in quantum wells, 43, 408-409
  recombination, 408
  scattering, 408, 409
  shift, 161
  splitting of states, 43
  trapping on impurity sites, 342
Extended well luminescence, 412

Fabry-Perot, 363
  analogy, 344
  cavity, 331
  etalon, 331
  interferometer, 245, 433
  resonances, 245, 248
Far infrared (see also FIR)
  absorption, 127
    background, 122
    in lateral superlattices, 152,
        425
  investigations of impurities, 41

Far infrared (continued)
  ISR spectroscopy, 168-169
  magnetoabsorption in GaSb-InAs,
        15
  response
    in one dimensional electronic
        system, 151, 155
    in two dimensional electronic
        system, 158
    in zero dimensional electronic
        system, 151
  spectroscopy of quantum wires,
        168-169
  transmission of laterally
        periodic structure, 425
Faraday configuration, 48
Feedback, see Electrostatic
        feedback
Femtosecond
  experiment, 409
  techniques, 399
Fermi disc, 109
Fermi energy
  in degenerate semiconductor, 35
  pinning in 2DEGs, 26
  transmission probabilities at,
        106
Fermi function and resistivity
        oscillations, 36
Fermi gas
  finite compressibility, 161
Fermi Golden Rule, 271
Fermi level
  in localised states, 13
  pinning, 156
Fermi momentum, 100, 107
Fermi pressure, 161
Fermi sea in DBS emitter, 258, 267
Fermi velocity, 193, 195
  of injected electrons, 80
  at junction, 107
Fermi wavelength, 1
  and conductor width, 55
  and confining potential length,
        425
  and enhanced diffraction, 86
  in narrow structures, 86
Fermi wavenumber, 134
Fermi wavevector, 1
Ferromagnetic contacts, 310
Few-subband transport, 107-108
Fibre/grating compressor, 401
Field electrode arrays, 425
Field-emission
  barrier, 325
  from solids, 317
Filling factor, 134, 429
Fine structure constant, 13
FIR see Far infrared

Substrates
  lattice-matched, 351
  lattice-mismatched, 360
Sudden approximation, 285, 287
Superlattice, 118-120 (see also
      Lateral superlattice, SL,
      Strained layer
      superlattice)
  buffer
    GaAs/AlGaAs, 319
  compositional, 1, 2
  current-voltage characteristic
      of, 4, 7
  direction
    transport in, 118
  discovery of, 4
  doping, 2
  effective mass dynamics in, 223
  engineered, 1
  magnetotransport in, 118
  mini-band transport in, 117, 118
  periodicity and mode folding,
      379
  potential, 187
  type I, 6
    band diagram, 5
  type II
    staggered and misaligned, 5,
      6, 294
  type III, 6
    band diagram, 5
  ultrashort period, 17
  ultrathin, 1
Surface adsorbed layer, 367
Surface anisotropy, 363,367
Surface bandstructure, 161
Surface dielectric function, 363
Surface quantisation, 431
Surface reconstruction, 95,367
Surface roughness, 115
  atomic scale, 114
  and confinement, 95
Surface scattering
  quantum enhancement of, 114
Susceptor, 355
Symmetric barrier structure, 263-
      274
  resonant tunnelling in, 257
Symmetrically strained GaAs-
      InAs/InP, 383
Symmetrically strained Si-Ge/SiGe,
      383
Symmetry breaking and impurity
      configurations, 62
Symmetry relation for transmission
      probabilities, 79
Synchronous pumping, 400

T-shaped gate, 230

Temporal scale
  femtosecond, 222
  picosecond, 222
Tensile strain, 377
Tetragonal unit cell, 382
Thermally activated transport, 118
Thermionic emission, 235
Three terminal conductor, 58
Three terminal devices, 285
Three terminal double barrier, 59
Ti/Au Schottky barrier, 192
Ti outdiffusion, 235
Ti/Pd contacts, 236
Tight binding approximation, 226
Tight binding Hamiltonian, 178
Tight binding model, 119
Tilted angle evaporation, 152
Tilted high index surface growth,
      151
Time delay processes in RTDs, 283
Time modulated barrier, 298-299
Time of flight measurements, 411
Time reversal symmetry, 263
Time scales for quantum
      tunnelling, 297
Time resolved absorption, 404
Time resolved photoluminescence,
      332, 335, 338, 347, 404,
      405, 410
  and DBRT structures, 344
Time resolved Raman scattering,
      404, 406
TiW Schottky contact, 236
TMI, 355, 356, 357
Topologically inequivalent paths,
      175
Total internal reflection, 96
Trajectories
  electronic
    numerical simulation of, 76
    skipping, 96
  localisation of, 324
Transconductance of LSSL device,
      193
Transfer matrix method, 257, 259,
      260-262, 265, 268
Transfer resistance (R_T), 90, 105,
      106, 108, 109, 112
  ballistic, 105
    spatial decay of, 109-113
  fluctuations in, 92
  for hard-wall geometry, 88, 89
  of narrow wires with junctions,
      96
Transformer waveguide, 240
Transit time
  across depletion region in RTD,
      285, 287, 289, 291
  minimisation, 421